SCALAR LAPLACIAN

Rectangular $$\nabla^2 V = \frac{\partial^2 V}{\partial x^2} + \frac{\partial^2 V}{\partial y^2} + \frac{\partial^2 V}{\partial z^2}$$

Cylindrical $$\nabla^2 V = \frac{1}{r}\frac{\partial}{\partial r}\left(r\frac{\partial V}{\partial r}\right) + \frac{1}{r^2}\frac{\partial^2 V}{\partial \phi^2} + \frac{\partial^2 V}{\partial z^2}$$

Spherical $$\nabla^2 V = \frac{1}{r^2}\frac{\partial}{\partial r}\left(r^2\frac{\partial V}{\partial r}\right) + \frac{1}{r^2 \sin\theta}\frac{\partial}{\partial \theta}\left(\sin\theta\frac{\partial V}{\partial \theta}\right) + \frac{1}{r^2 \sin^2\theta}\frac{\partial^2 V}{\partial \phi^2}$$

VECTOR LAPLACIAN

Rectangular $$\nabla^2 \vec{E} = \left(\frac{\partial^2 E_x}{\partial x^2} + \frac{\partial^2 E_x}{\partial y^2} + \frac{\partial^2 E_x}{\partial z^2}\right)\vec{e}_x + \left(\frac{\partial^2 E_y}{\partial x^2} + \frac{\partial^2 E_y}{\partial y^2} + \frac{\partial^2 E_y}{\partial z^2}\right)\vec{e}_y$$
$$+ \left(\frac{\partial^2 E_z}{\partial x^2} + \frac{\partial^2 E_z}{\partial y^2} + \frac{\partial^2 E_z}{\partial z^2}\right)\vec{e}_z$$
$$= (\nabla^2 E_x)\,\vec{e}_x + (\nabla^2 E_y)\,\vec{e}_y + (\nabla^2 E_z)\,\vec{e}_z$$

Cylindrical $$\nabla^2 \vec{E} = \left(\nabla^2 E_r - \frac{2}{r^2}\frac{\partial E_\phi}{\partial \phi} - \frac{E_r}{r^2}\right)\vec{e}_r$$
$$+ \left(\nabla^2 E_\phi + \frac{2}{r^2}\frac{\partial E_r}{\partial \phi} - \frac{E_\phi}{r^2}\right)\vec{e}_\phi + (\nabla^2 E_z)\,\vec{e}_z$$

Spherical $$\nabla^2 \vec{E} = \left[\nabla^2 E_r - \frac{2}{r^2}\left(E_r + \cot\theta\, E_\theta + \csc\theta\frac{\partial E_\phi}{\partial \phi} + \frac{\partial E_\theta}{\partial \theta}\right)\right]\vec{e}_r$$
$$+ \left[\nabla^2 E_\theta - \frac{1}{r^2}\left(\csc^2\theta\, E_\theta - 2\frac{\partial E_r}{\partial \theta} + 2\cot\theta\csc\theta\frac{\partial E_\phi}{\partial \phi}\right)\right]\vec{e}_\theta$$
$$+ \left[\nabla^2 E_\phi - \frac{1}{r^2}\left(\csc^2\theta\, E_\phi - 2\csc\theta\frac{\partial E_r}{\partial \phi} - 2\cot\theta\csc\theta\frac{\partial E_\theta}{\partial \phi}\right)\right]\vec{e}_\phi$$

ELECTROMAGNETICS FOR ENGINEERS

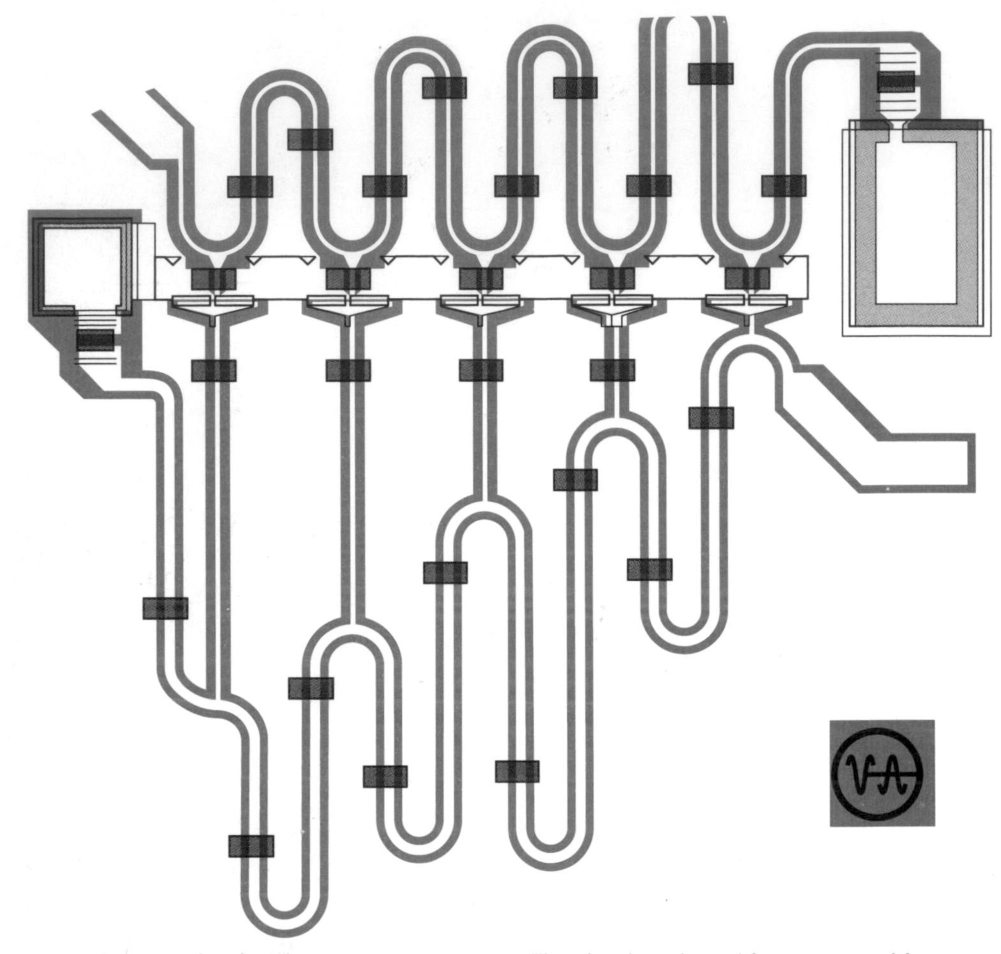

An example of a planar microwave circuit. This distributed amplifier consists of five MESFETs connected by means of coplanar waveguide (Chapter 9). The entire amplifier is fabricated monolithically on a gallium arsenide chip only a few millimeters on a side. (Artist's rendering based on a photograph of original circuit, courtesy of Varian Research Center, Palo Alto, Calif.)

ELECTROMAGNETICS FOR ENGINEERS

Steven E. Schwarz
University of California, Berkeley

SAUNDERS COLLEGE PUBLISHING
A Division of Holt, Rinehart and Winston, Inc.

Philadelphia Chicago Fort Worth San Francisco
Montreal Toronto London Sydney Tokyo

Text typeface: Times Roman
Compositor: General Graphic Services
Acquisitions Editor: Robert Argentieri
Managing Editor: Carol Field
Project Editor: Marc Sherman
Copy Editor: Linda Davoli
Manager of Art and Design: Carol Bleistine
Art and Design Coordinator: Doris Bruey
Text Designer: Caliber Design Planning, Inc.
Cover Designer: Lawrence R. Didona
Text Artwork: GRAFACON
Director of EDP: Tim Frelick
Production Manager: Bob Butler

Printed in the United States of America

ELECTROMAGNETICS FOR ENGINEERS

ISBN 0-03-006517-8

Library of Congress Catalog Card Number: 89-043041

0 1 2 3 039 9 8 7 6 5 4 3 2 1

PREFACE

One day I mentioned to a colleague that I was working on an introductory electromagnetics text. "Hmmm," he said, with a pleasant smile. "Hasn't that been done already?"

Well yes, one must admit, it has, and Maxwell's equations have not changed. And yet, looking back at some of the excellent older books, one does not feel inclined to use them in an introductory course today. Some things *have* changed. New applications have appeared, such as planar microwave technology and fiber-optic communications, and changing applications lead us to adjust our emphasis on various topics. Furthermore, the undergraduate curriculum itself is changing, and with it the introductory electromagnetics course must change.

Most of us agree that electrical engineering students should be introduced to electromagnetic technology. However, with more and more subjects entering the curriculum, it is difficult to require a traditional one-year introduction for all students. Those students who do not specialize in electromagnetics will probably only have time for a one-semester course. But unless we can afford the luxury of parallel courses, this one-semester introduction will have to be the first course for specialists as well. And here lies the problem. We must design a one-semester first course that satisfies the needs of non-specialists (who are probably in the majority), while at the same time providing a firm foundation for those who choose further study.

The ideal course for non-specialists is not the first half of a two-course sequence intended for specialists. The specialists' course is likely to go into considerable detail, even though the applications of the details do not appear until the second semester of the course. It may also present topics in strict logical order, not leaving a topic until it is completely covered. In that case, electrostatics and boundary-value problems may take up most of the first semester! Clearly such a course will not serve one-semester non-specialists very well. Their need is primarily for topics that are relevant to the rest of their studies. They will be especially interested in transmission lines; they will want to know something of high-frequency circuit technology; they will be interested in fiber optics. On the other hand, there must be no dilution in the quality of the presentation, in order to provide a sound basis for further study. This does not pose a contradiction. The course we propose contains the fundamentals found in any traditional first course. The difference is mainly in the emphasis and the order in which subjects are studied.

An unusual (although by no means unprecedented) feature of the proposed course is that it begins with transmission lines. As we all know, electromagnetics can be a difficult course, and it can be a shock for the student who finds himself suddenly at sea in the unfamiliar waters of vector analysis. However if we begin with the circuit theory of transmission lines, we do not need vector analysis right away, and the course begins more gradually, with a gentle transition from familiar circuit theory. In addition, transmission lines are probably the topic of greatest interest to most students, and by beginning with them, attention is captured at the start. Furthermore, the idea of *waves* is one of the most essential; it is good to introduce it early, with reinforcement coming later, as other types of waves are discussed. Our discussion of transmission lines also reviews sinusoidal analysis and phasors, which will then be familiar when used to describe sinusoidal fields later in the course.

The remainder of the outline is conventional, but we do not loiter. Electrostatic boundary-value problems, which we feel are of interest primarily to specialists, are mentioned but not studied in detail, with considerable saving of time. In this way we are able to get out of statics and into dynamics fairly quickly, so that by the end of the semester we can introduce plane waves, TEM waves, hollow metal guides, microstrip, and optical fiber guides. An outline of the topics, with the lecture hours we have used, is as follows:

Introduction	1
Transmission lines	6.5
Review of vector analysis	1.5
Electrostatics	6
Magnetostatics	4.5
Electrodynamics	4
Skin effect	2.5
Boundary conditions	1
Electromagnetic waves	2.5
Poynting's theorem	1.5
Reflection at normal incidence	1
Oblique incidence	3
TEM waves	1.5
Hollow metal waveguides	2
Microstrip	1
Optical fiber guides	1

Assuming a 45-hour semester, this leaves 4.5 hours for examinations and review.

The subject of electromagnetics is not easy to learn. This is partly because the concepts are abstract and require three-dimensional visualization; and partly because the main mathematical tool, vector calculus, is often unfamiliar. Thus pedagogical technique is especially important. The present book is intended primarily as an aid to learning. It does not attempt to be encyclopedic in its coverage; on the contrary, the available space is used to explain concepts at whatever length is necessary to help the reader understand. Each major concept is followed by one or more worked examples; exercises with answers are frequently inserted to let the reader test his or her knowledge. Essential to the learning process are the numerous problems at the end of each chapter, which have three levels of difficulty: no asterisk for the least difficult, one asterisk for moderate difficulty, and two asterisks for the most difficult. Two-color printing makes the book more approachable and helps to clarify the more complicated figures. Although most readers will have completed a course in vector calculus, that knowledge is likely to be rusty; therefore the various operators are discussed in some detail at the points where they appear. By means of these features we hope to reduce the mechanical difficulties encountered by readers, so they can grasp the essentials of the subject.

In writing this book the author has been greatly influenced by *Fields and Waves in Communication Electronics,* by S. Ramo, J.R. Whinnery, and T. Van Duzer. This superbly written book, in its earlier and current versions, has played a major role in defining the field of electromagnetics as we know it today. Professors Whinnery and Van Duzer have also contributed advice and useful suggestions; so have Professors D.J. Angelakos, K.K. Mei, and C.W. Turner. Further thanks are due to the following eminent reviewers: Lewis Fitch, Clemson University; Lloyd S. Riggs, Auburn University; Captain Randy Jost, Air Force Institute of Technology; Robert Samuels, Catawba Valley Technical College; John Buck, Georgia Institute of Technology; Sharadbabu Laxpati, University of Illinois, Chicago; Mac Van Valkenburg, University of Illinois, Urbana; Richard Schubert, California State College, Fullerton; Hai-Sup Lee, Pennsylvania State University; W. Mack Grady, University of Texas, Austin; Jerry Rogers, Mississippi State University; Robert Samuels, Iowa State University; Dennis Malone, SUNY Buffalo; and Susan Schneider. And appreciation is expressed to Ms. Carol Block, for her very skillful assistance with the manuscript.

Finally, the author wishes to thank the Department of Electronic and Electrical Engineering of King's College, University of London—over which presides the spirit of James Clerk Maxwell, professor 1860–65—for its hospitality and inspiration, as portions of this book were being written.

Steven E. Schwarz
Berkeley, California
July 1989

CONTENTS

INTRODUCTION

Electromagnetics is the oldest and most fundamental branch of electrical engineering. Long before there were integrated circuits, before radio, even before the telephone and the electric motor, the great minds of the nineteenth century began to define the principles of electricity and magnetism. This quality of being fundamental, of being the basis of so much else, is still one of the attractions of the field.

Because of its fundamental nature, electromagnetics belongs not only in engineering, but also in physics. The distinction lies mainly in the sort of problems one chooses to solve. Physicists may be interested in relativity, radiation from particles, or the properties of solids, while engineers tend to emphasize waves, antennas, and high-frequency devices and circuits. In practice, electromagnetic engineering and electromagnetic physics are so closely associated that it is difficult, and probably unwise, to attempt to separate the two. For instance, particle accelerators used in high-energy physics rely heavily on microwave engineering. On the other hand, there is no good way to bypass the fundamentals of electromagnetics; one cannot rely on "simplified models" or plug-in formulas. For engineers as much as for physicists, the study of electromagnetics must be based on the basic physical theory.

Electromagnetics by nature is highly abstract. Of course, this is true of electrical engineering generally. One cannot see voltages and currents; one must imagine them, and describe them by means of mathematics. However, voltages and currents are only scalar quantities. The variables of electromagnetics, on the other hand, are *fields,* which are described by three-dimensional vectors. Because of this higher dimensionality, one must develop one's powers of abstract visualization, and furthermore, one must rely on the descriptions of mathematics. Math, in fact, will be our indispensable basic tool, and for success, a good background in vector calculus is required. Sometimes this subject, studied a long time ago and never since applied, has been filed away under "useless knowledge" and forgotten. Well, here is your chance to use it! If you feel your knowledge of vector calculus has gotten rusty, it will pay to take out your old math textbooks and polish up your knowledge. There is no denying that electromagnetics is a challenging, intellectual subject, but students who come to it with strong mathematical backgrounds usually find they can do very well.

Some people find electromagnetics such an interesting and congenial subject that they make it a career. Electromagnetics specialists are needed in most kinds of electrical engineering work. Communications systems, especially, involve electromagnetics, and specialists are needed in the design of antennas, high-frequency circuits, and so forth. As technology steadily advances to higher speeds and higher frequencies, new opportunities in electromagnetics continue to appear. For example, optoelectronics, an emerging technology of great importance, relies heavily on electromagnetics, not only in the actual optical guiding, but also in the design of high-speed circuits needed to transmit the information at one end of an optical fiber and to receive it at the other.

Because it is so basic, electromagnetics is important to non-specialists as well. Transmission lines, such as the ubiquitous coax line, are used in almost every laboratory, every day. Circuit designers routinely make use of electromagnetic components such as capacitors and inductors. Even Kirchhoff's laws, which form the basis of circuit design, are low-frequency approximations that stem from Maxwell's equations. As frequency increases, Kirchhoff's laws become less valid, and the circuit designer must understand what happens and take electromagnetic effects into account. The theory of semiconductor devices also involves electromagnetics, which describes the fields that control the motion of charge carriers. Many other examples can be given; in each area of electrical engineering, one finds that at the fundamental level, electromagnetic principles are involved. Thus for innovative work—work that goes beyond established techniques and present limits—a knowledge of electromagnetics is of tremendous value.

Finally, in our overview of electromagnetics, we should mention its inherent aesthetic appeal. There are some scholarly subjects, such as plane geometry, that are self-contained, logical, elegant, free of clumsy contrivances and approximations, and united by a few graceful principles from which the entire edifice springs. Classical mechanics and special relativity possess these qualities; so, according to some people, does the game of chess. Electromagnetics, too, belongs to this elite group of beautiful intellectual constructs. Not everyone cares for this sort of beauty, but if you do, you may find in electromagnetics a new appreciation of science, and of mankind's ability to understand.

CHAPTER 1

Transmission Lines

Electrical signals are often transmitted through conductors from one place to another. Inside a small circuit this can be done by simply connecting a wire between the two places in question. However, when the transmission distance is long, more complex electromagnetic effects begin to appear. In that case, one usually connects the two points with a special elongated circuit called a *transmission system*.

The transmission systems most often encountered belong to the family called *transmission lines*. For example, coaxial cables, which are familiar in every electronic laboratory, belong to this family; so do the transmission lines used to connect TV antennas to TV sets. As frequency increases, the need for transmission lines becomes greater. Hence in modern high-speed computers, different internal components have to be connected with each other by transmission lines. It is even sometimes necessary to use a transmission line to connect different parts of a single integrated circuit! In this case, of course, the transmission line is built into the IC.

Transmission-line theory can be developed from more than one point of view. In this chapter we shall develop what is known as the *lumped-circuit theory,* in which the line is represented by a circuit model, consisting of conventional lumped-circuit elements. The model is assumed to obey Kirchhoff's laws, and its behavior is analyzed by means of conventional circuit theory. Although mathematically simple, this approach is quite adequate for most work with transmission lines. Using the model, we can then describe the action of the transmission lines in the most important situations. Primary emphasis will be placed on the case of sinusoidal excitation, and we shall see how phasor analysis can be used to describe sinusoidal waves traveling on a transmission line. We shall also consider the rather different situation that arises when a non-sinusoidal excitation, such as a short rectangular pulse, is applied to a transmission line.

Chapter 2 will build further on the lumped-circuit model. There we shall develop the techniques commonly used in transmission-line analysis. By the end of Chapter 2, we should be familiar with the idea of waves and with the use of phasors in describing them, and in fact we should have a good working knowledge of transmission-line theory.

This, however, will not be the end of the transmission-line story. In Chapter 3 the subject will change, and we will move on to the three-dimensional theory of electromagnetic fields. Many important new physical ideas will be introduced, summarized finally in the four brilliant equations known as *Maxwell's equations*. Then, armed with the powerful field theory of electromagnetics, we shall be able to return to the subject of transmission-line waves, discussing them now in terms of their fields. This point of view will add further insight to our knowledge of transmission lines, and since it is more general than the lumped-circuit theory, it will allow us to describe more complex transmission systems such as waveguides. However, we shall not find that the results of Chapter 1 are overturned. On the contrary, the full electromagnetic theory simply confirms the validity of the lumped-circuit model, for most cases of practical interest.

1.1 THE LUMPED-CIRCUIT THEORY

In this chapter our approach to transmission-line analysis will resemble what is usually done with ordinary electronic circuits. To predict the operation of a transmission line, we shall first replace it with a simple model. We then analyze the model, using conventional circuit analysis. As is always the case, the model is not precisely correct. Hence there are always some aspects of the real device's operation that a theory based on a model cannot describe. In using such a theory, one should be aware that it will not be correct under the most general conditions. However in everyday work a fairly limited range of conditions is normally encountered, and for this range the lumped-circuit transmission-line theory works extremely well.

Let us begin by imagining that our transmission line consists of two parallel conductors, not necessarily identical, as shown in Fig. 1.1(a). For the moment we shall not concern ourselves with the exact shape of the conductors; however, they are assumed to be uniform and parallel. We next imagine the line to be broken up into small sections of length h, as shown in Fig. 1.1(b). The wires in each section will contain a certain amount of inductance, which we call L_h, and between them will be a certain capacitance C_h. This leads us to the final model, shown in Fig. 1.1(c). Both L_h and C_h are proportional to h, the length of each section. Hence, we can write

$$\begin{aligned} hL &= L_h \\ hC &= C_h \end{aligned} \qquad (1.1)$$

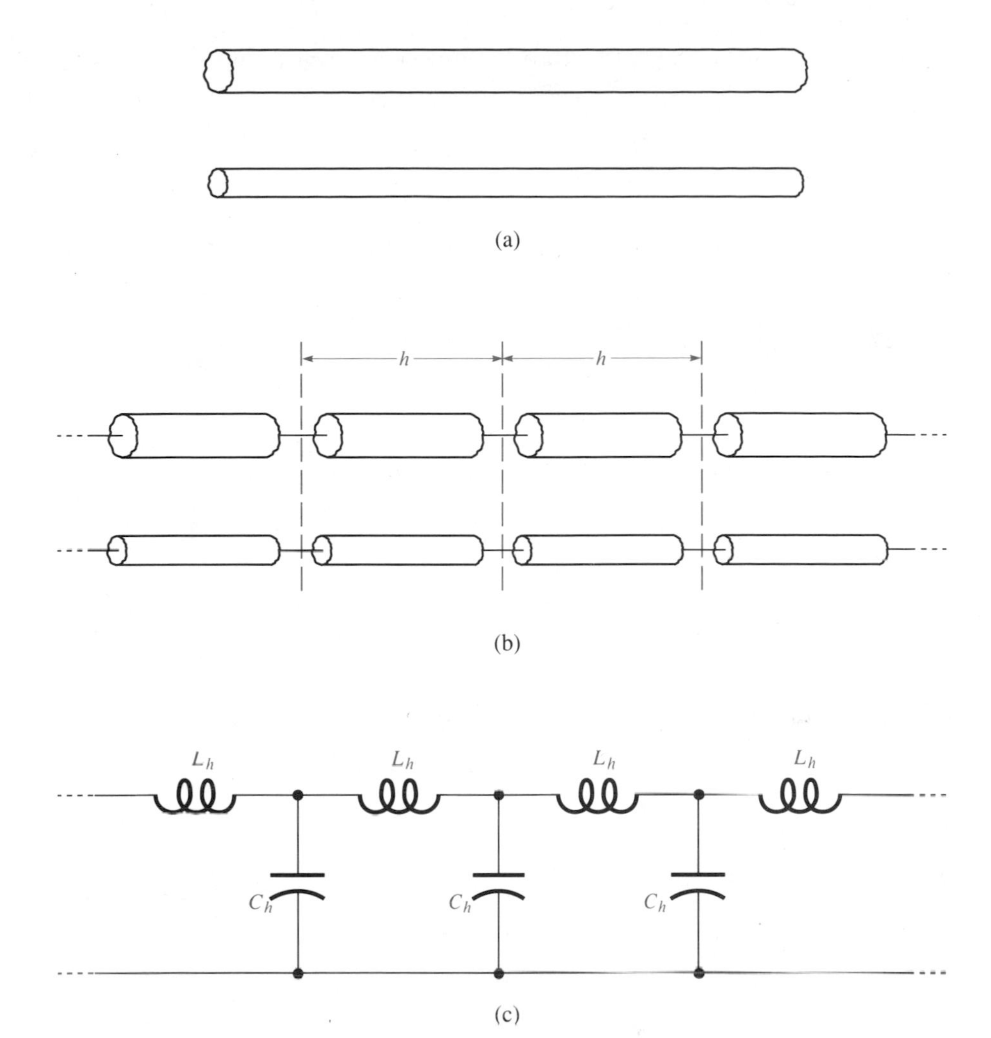

Figure 1.1 Development of the lumped-circuit model for the transmission line. (a) The line consists of two parallel conductors of arbitrary shape, which are not in contact. (b) Conductors divided into sections of arbitrary small length h. (c) Individual sections represented by lumped inductors and capacitors.

where L and C are respectively the inductance and capacitance per unit length of line. We shall refer to the model shown in Fig. 1.1(c) as the *ideal lossless transmission line*.

Let us next define voltages and currents at different places in the line, as shown in Fig. 1.2. Here one node in the long line has been arbitrarily numbered as the Nth node. The current through the Nth inductor obeys

$$L_h \frac{di_N}{dt} = v_N - v_{N+1} \tag{1.2}$$

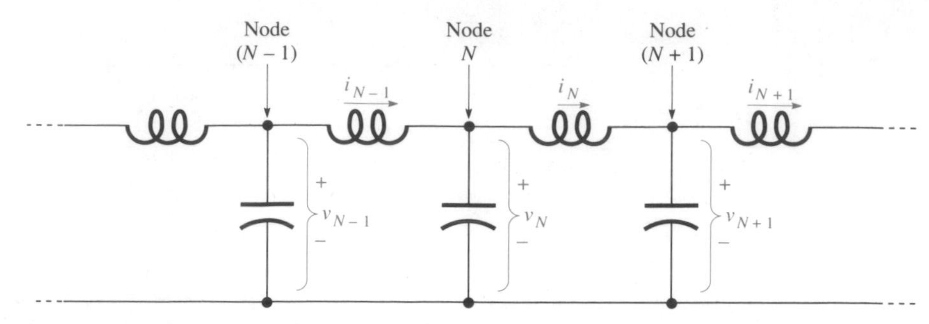

Figure 1.2 Definitions of currents and voltages for the lumped-circuit transmission-line model.

But since $L_h = hL$, (1.2) can be rewritten

$$L \frac{di_N}{dt} = -\frac{v_{N+1} - v_N}{h} \tag{1.3}$$

Now let node N be at the position z. Then node $N + 1$ is at position $z + h$, and so forth. Defining $i(z)$ as the current between z and $z + h$, we have

$$L \frac{d}{dt} i(z) = -\frac{v(z + h) - v(z)}{h} \tag{1.4}$$

We now make use of the definition of the derivative

$$\frac{df}{dz} = \lim_{h \to 0} \frac{f(z + h) - f(z)}{h} \tag{1.5}$$

Since h is an arbitrary small distance, we can let h approach zero, and from (1.4) and (1.5) we have

$$L \frac{\partial}{\partial t} i(z) = -\frac{\partial}{\partial z} v(z) \tag{1.6}$$

(In (1.6) we have written the derivatives using the notation of partial differentiation. This is a minor mathematical point, intended to emphasize that the differentiation with respect to time occurs at a fixed position z, and the differentiation with respect to z occurs at a fixed time t.)

From Kirchhoff's current law we have

$$C_h \frac{dv_N}{dt} = i_{N-1} - i_N \tag{1.7}$$

Proceeding similarly, we arrive at the result

$$C \frac{\partial}{\partial t} v(z) = -\frac{\partial}{\partial z} i(z) \tag{1.8}$$

Equations (1.6) and (1.8) are picturesquely known as the *telegraphist's equations*. Differentiating (1.6) with respect to z and (1.8) with respect to t, and combining, we obtain

$$\frac{\partial^2 v}{\partial t^2} - \frac{1}{LC}\frac{\partial^2 v}{\partial z^2} = 0 \tag{1.9}$$

This is a one-dimensional form of the *wave equation*, one of the most important equations of science and engineering. As will be seen, systems that obey this equation can be used to transmit information in the form of waves.

1.2 WAVES ON THE IDEAL LOSSLESS LINE

The phenomenon of wave motion is of very widespread importance in science and engineering. Waves of many different kinds exist: water waves, electromagnetic waves, mechanical vibrations, and even "Schroedinger waves" describing particle states in quantum mechanics. We all have an intuitive idea of what waves are, from familiar examples like waves on water, but it is not so easy to state in a few words just what a wave is. Roughly speaking, a wave is a disturbance that moves away from its source as time passes. However, that description by itself is insufficient since it might apply to a thrown baseball, which is not a wave. A key additional point is that in wave propagation it is the *disturbance* that moves, while the medium through which the wave passes is almost motionless. For example, a water wave can move all the way across an ocean, but no individual water molecule moves across the ocean. In fact, the individual water molecules only bob up and down a bit, and hardly move sideways at all. For some kinds of waves, such as electromagnetic waves, an actual physical medium is not even required. As we shall see, electromagnetic waves propagate very nicely through a vacuum.

Let us imagine that the voltage on a transmission line has the form

$$v(z, t) = f(z - Ut) \tag{1.10}$$

Here we have written $v(z, t)$ to emphasize that the voltage is a function both of time and position. The factor U is a constant. (It will turn out to be the wave's velocity.) The function f can be any reasonable function of a single variable. The purpose of writing $v(z, t)$ as we have in (1.10) is to make the waveform move *as a unit* in the positive-z direction as time passes. We recall that if $f(x)$ is any function of x, then $f(x - x_o)$ is the same function, shifted to the right a distance x_o along the x axis. If instead of $f(x - x_o)$ we write $f(x - Ut)$, then the function is shifted to the right a distance Ut. This distance increases as time increases, so the function is displaced steadily further out the x axis. The displacement is by a distance Ut, which means that the velocity of the motion is U.

For example, $f(x)$ might be the function shown in Fig. 1.3(a). Then at $t = 0$, the voltage $v(z, t)$ is as seen in Fig. 1.3(b). However, let t increase to

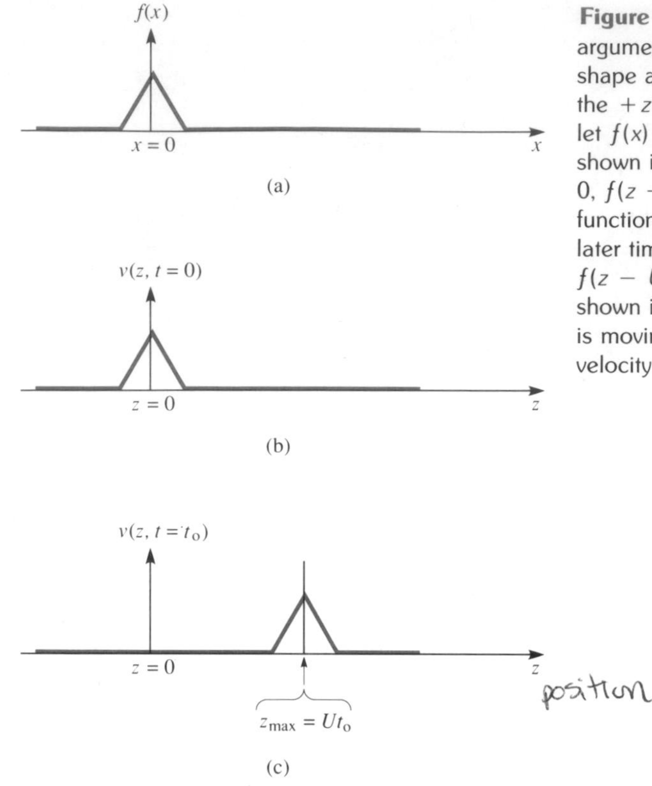

Figure 1.3 Any function of the argument $(z - Ut)$ keeps its shape and moves as a unit in the $+z$ direction. For example, let $f(x)$ be the triangular function shown in (a). Then at time $t = 0$, $f(z - Ut) = f(z)$ is the function of z shown in (b). At a later time t_o, $f(z - Ut) = f(z - Ut_o)$ is the function of z shown in (c). Note that the pulse is moving to the right with velocity U.

time t_o. Then the argument of f in (1.10) has changed to $z - Ut_o$, and therefore the function has moved to the right a distance Ut_o, as shown in Fig. 1.3(c).

A convenient way of keeping track of the motion is to locate a particular point on the waveform, for instance its maximum, and keep track of its position. In Fig. 1.3, f is a function that has its maximum when its argument is zero. Thus $v(z, t)$ is maximum where $z - Ut_o = 0$, and z_{max}, the position at which v is maximum, is given by

$$z_{max} = Ut_o \tag{1.11}$$

As time passes, the voltage waveform moves to the right by the distance stated in (1.11). Clearly this distance is proportional to time, so as time increases the waveform moves steadily to the right. The position of the maximum is always given by (1.11), and the velocity of the wave is U. Note that we are simply keeping track of the waveform's position by tracing the motion of one convenient point. However, the entire waveform moves as a unit with velocity U.

The function (1.10) is thus seen to be a wave with some especially simple properties. The disturbance keeps its size and shape, and simply moves as a unit at a constant velocity. The propagation is said to be *undistorted,* because a pulse with a certain shape that enters the line will emerge from the other end, at a later time, with its shape unchanged. One expects that undistorted propagation will be ideal for communication, since the wave that arrives contains exactly the same information that was transmitted at an earlier time. Not all transmission lines exhibit undistorted propagation. However, waves on the ideal lossless transmission line do have this useful property.

EXAMPLE 1.1

Suppose $v(z, t) = f(z + Ut)$, where $f(x)$ is the function of one variable shown in Fig. 1.3(a). Find the velocity of the wave in this case.

Solution The function $f(x)$ is maximum when its argument is zero. Thus, at time t_o the position of the maximum is given by $z_{max} + Ut_o = 0$, or $z_{max} = -Ut_o$. We see that in this case an undistorted wave moves to the *left* with velocity U.

To show that undistorted waves can propagate on an ideal transmission line, we need only verify that the wave (1.10) satisfies the wave equation (1.9). Differentiating (1.10), we see that $\partial^2 v/\partial t^2 = U^2 f''(z - Ut)$, where f'' is the second derivative of f with respect to its argument. Similarly $\partial^2 v/\partial z^2 = f''(z - Ut)$. Substituting into (1.9) we see that the wave equation is indeed satisfied, provided that

$$U = \frac{1}{\sqrt{LC}} \tag{1.12}$$

In similar fashion we can show that the leftward-traveling wave $v(z, t) = f(z + Ut)$ is also a solution.

The reader should not be confused about *what* is traveling at the velocity U. Electrons in the transmission line do *not* move with that velocity. What is propagating is an electrical disturbance. One way of visualizing the transmission-line wave is to imagine that initially we create a current in only one of the inductors in Fig. 1.2, let us say the one at the left. After a time this current charges the first capacitor, and v_{N-1} increases. But the increase of v_{N-1} creates a current i_{N-1} through the second inductor, which after a time charges the second capacitor, causing v_N to increase, and so on. Thus, the disturbance propagates down the line.

SINUSOIDAL WAVES

An especially important special case is that in which $f(x)$ is a sinusoidal function. For instance let

$$f(x) = A \cos kx \tag{1.13}$$

where A and k are constants. Then (1.10) becomes

$$v(z,t) = f(z - Ut) = A \cos k(z - Ut) \tag{1.14}$$

The constant k has not yet been specified. However, we note that at a fixed position, v varies sinusoidally in time with an angular frequency kU. It is usual to denote angular frequency by the symbol ω. Thus

$$k = \omega/U \tag{1.15}$$

This constant k is known as the *propagation constant*. In terms of k and ω,

$$v(z,t) = A \cos(kz - \omega t) \tag{1.16}$$

We already know that this wave can propagate on the transmission line, since it is of the form (1.10). At the time $t = 0$, the wave is as shown in Fig. 1.4(a). At a later time t_o it has moved to the right a distance Ut_o, as shown in Fig. 1.4(b).

The *wavelength* of the wave is defined as the distance between maxima at any fixed instant of time. (Imagine that we photograph the wave as it travels, obtaining a still picture of the wave frozen in time. The wavelength is then the distance between adjacent maxima, or equivalently between adjacent minima, in this picture.) Since the wavelength is a non-time-varying quantity, we may for convenience evaluate it at time $t = 0$. Since this particular function is maximum when its argument $(kz - \omega t)$ is zero, we see from (1.16) that at $t = 0$ there is a maximum at $z = 0$. The next maximum occurs when $kz = 2\pi$, or $z = 2\pi/k$. Thus the wavelength λ is given by[1]

$$\lambda = \text{wavelength} = \frac{2\pi}{k} \tag{1.17}$$

Since $\omega = kU$, we find that frequency, wavelength, and velocity are related by

$$f\lambda = U \tag{1.18}$$

If we stand at the position $z = 0$, we know, from (1.16), that $v(0,t) = A \cos(-\omega t) = A \cos(\omega t)$. However, if we stand at another position z_o, we find that $v(z_o, t) = A \cos(kz_o - \omega t) = A \cos(\omega t - kz_o)$. The voltage at z_o is thus a sinusoid of the same frequency, but with a phase lag of $\phi = kz_o$. This lag is explained by the time it takes for the wave to propagate from $z = 0$ to $z = z_o$.

The sinusoidal wave (1.16) is most conveniently described in complex notation. For all voltages and currents we define phasors according to the rule

$$f(t) = \text{Re}\,[\mathbf{f}\, e^{j\omega t}] \tag{1.19}$$

---

[1]The wavelength defined here is properly called the *guide wavelength;* that is, the distance between maxima for waves being guided on the transmission line. Because the wave on the line can be slowed by the presence of dielectric materials, the guide wavelength can be less than the wavelength of electromagnetic waves in free space at the same frequency. (The latter is known as the *free-space wavelength.*) In this chapter and the next the symbol λ always represents the guide wavelength.

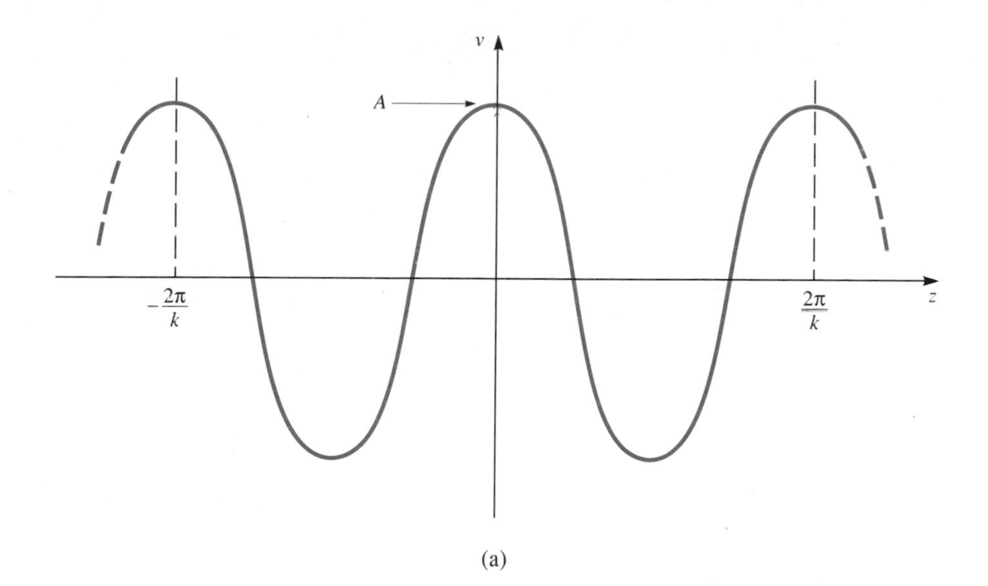

(a)

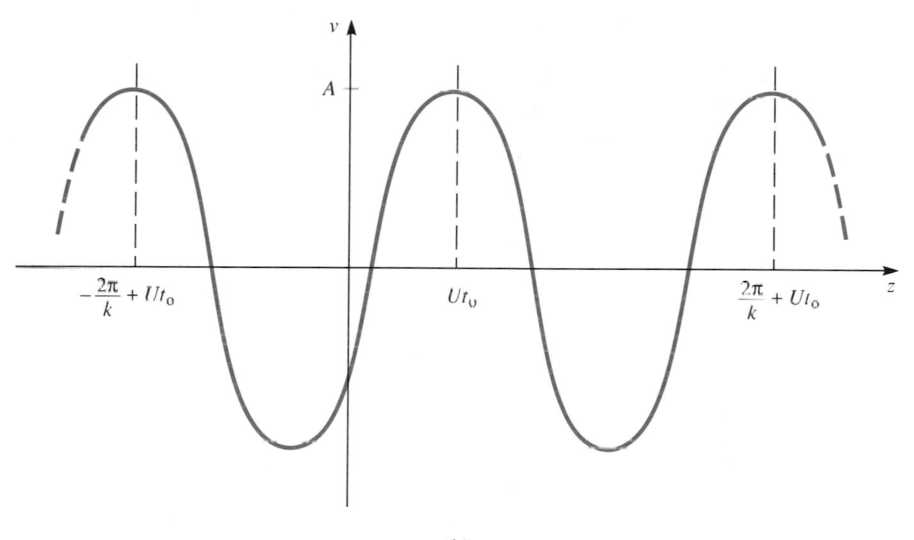

(b)

Figure 1.4 An important special case is that in which the function f is a sinusoid. Fig. (a) shows the function $v(z, t) = A \cos(kz - \omega t)$ as it appears if photographed with a flash camera at time $t = 0$. In (b) it is seen at the later time t_o.

where **f** is the phasor representing the sinusoid $f(t)$. Suppose a certain sinusoidal wave traveling in the $+z$ direction is given by $v(z, t) = A \cos(kz - \omega t)$. Then we can write

$$\begin{aligned} v(z, t) &= A \cos(kz - \omega t) = A \cos(-kz + \omega t) \\ &= \text{Re}\,[Ae^{-jkz}\, e^{j\omega t}] \end{aligned}$$

Comparing with (1.19), we see that the phasor representing this positive-going wave is

$$\mathbf{v}^+(z) = Ae^{-jkz} \tag{1.20}$$

In Example 1.1 we saw that a function $f(z + Ut)$ represents a wave moving to the left. Thus the function $A \cos k(z + Ut) = A \cos(kz + \omega t)$ is a sinusoidal wave moving to the left. For this wave

$$\begin{aligned} v(z,t) &= A \cos k(z + Ut) = A \cos(kz + \omega t) \\ &= \text{Re}\,[Ae^{jkz}\, e^{j\omega t}] \end{aligned}$$

Comparing with (1.19), we see that the phasor representing the negative-going wave is

$$\mathbf{v}^-(z) = Ae^{jkz} \tag{1.21}$$

In order to distinguish phasors from their corresponding time functions, phasors will be printed in **boldface** type.

The reader will recall from earlier study of phasors that if a certain phasor $\mathbf{v}$ represents a sinusoidal voltage $v(t)$, then the amplitude of the sinusoidal voltage is equal to the absolute value of $\mathbf{v}$, written $|\mathbf{v}|$. The absolute value can be found using $|\mathbf{v}| = \sqrt{\mathbf{v}\mathbf{v}^*}$. Thus, for example, if the phasor is $\mathbf{v} = Ae^{j\phi}$ (where A and ϕ are real numbers), the amplitude of the sinusoidal voltage $v(t)$ is $[Ae^{j\phi}\, Ae^{-j\phi}]^{1/2} = A$. If the phasor is given by $\mathbf{v}(z) = Ae^{-jkz}$, then the amplitude of the sinusoidal voltage at the position z is $[Ae^{-jkz}\, Ae^{jkz}]^{1/2} = A$. In either case, the amplitude of the sinusoid is simply equal to the magnitude, or "length," of the phasor.

EXAMPLE 1.2

A certain wave is described by

$$v(z,t) = V_o \cos(\omega t - kz + \phi)$$

where V_o and ϕ are real numbers. Find the phasor representing this wave.

Solution To find the phasor, we must express the wave in the form of (1.19). To do this we observe that

$$\begin{aligned} v(z,t) &= V_o \,\text{Re}\, e^{j(\omega t - kz + \phi)} \\ &= \text{Re}\,[(V_o e^{j\phi}\, e^{-jkz})\, e^{j\omega t}] \end{aligned}$$

Thus, comparing with (1.19), we see that the required phasor is

$$\mathbf{v}(z) = V_o e^{j\phi}\, e^{-jkz}$$

This result strongly resembles (1.20). However e^{-jkz} is now preceded by the factor $V_o e^{j\phi}$. This factor depends on V_o, the amplitude of the wave, and on ϕ, its phase.

The wave under discussion has angular frequency ω and wavelength

$2\pi/k$, and moves to the right with velocity $f\lambda = \omega\lambda/2\pi = \omega/k$. In fact, the only difference between this wave and the one of (1.16) is the addition of the constant phase shift ϕ. At any fixed position of observation, this wave *leads* the wave of (1.16) by the phase angle ϕ.

EXERCISE 1.1

Find the phasor representing a leftward-moving wave with amplitude 2 V, ordinary frequency 1 GHz, velocity 1.5×10^8 m/sec, and phase angle $-30°$ at $z = 0$.

Answer $\mathbf{v} = 2e^{-j30°}\, e^{j(41.9)z}$ V (where z is in meters).

EXAMPLE 1.3

Suppose that at a certain point z_o on an ideal lossless line the phasor representing a positive-going wave is $\mathbf{v}_o$ (where $\mathbf{v}_o$ may be complex). Find the phasor representing the voltage at all other places on the line. The propagation constant k is a given number.

Solution From Example 1.2 we know that this rightward-moving wave is represented by a phasor of the form $\mathbf{v}^+(z) = Ae^{j\phi}\, e^{-jkz}$ where A and ϕ are real numbers. We are given the value of $\mathbf{v}^+$ at the position $z = z_o$. Using this information we write

$$\mathbf{v}^+(z = z_o) = Ae^{j\phi}\, e^{-jkz_o} = \mathbf{v}_o$$

Thus $Ae^{j\phi} = \mathbf{v}_o\, e^{jkz_o}$, and the voltage elsewhere on the line is

$$\mathbf{v}^+(z) = Ae^{j\phi}\, e^{-jkz} = \mathbf{v}_o\, e^{jk(z_o - z)}$$

If the wave had been negative-going, we would have found by a similar procedure

$$\mathbf{v}^-(z) = \mathbf{v}_o\, e^{jk(z - z_o)}$$

The complex number $\mathbf{v}^+(z)$ can be regarded as a vector in the complex plane. It is important to note that the length of this vector remains constant as the position z varies. This means that *the amplitude of the sinusoidal voltage is the same at all positions*. Physically, this is because we are presently dealing with a lossless line. Since the line has no resistance, there is no way energy can be lost, and the amplitude of the wave stays constant as the wave travels. However, the phase of the voltage on the line does vary with position. If we were to attach an oscilloscope to the line at the position z_o, we would observe a sinusoidal voltage on the screen. Suppose that the phase of this sinusoid is zero. If we were now to move the oscilloscope to the right by a distance Δz, we would observe a sinusoid with the same amplitude but with a phase angle $-k\Delta z$. The phase at the second position

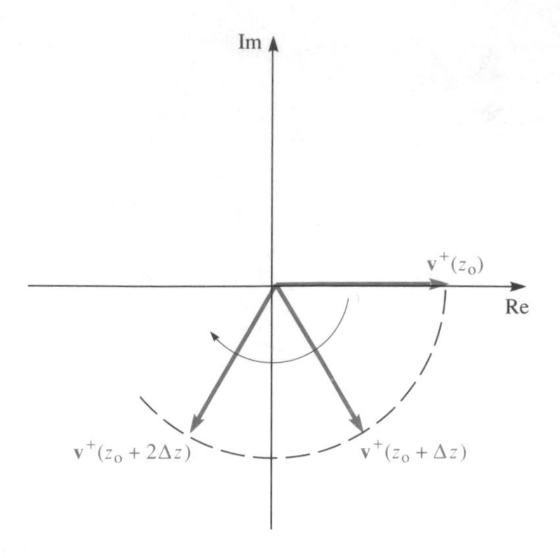

lags the phase at z_o because of the time it takes the wave to propagate the intervening distance. The sketch shows the way the phasor $\mathbf{v}^+(z)$ varies with position. For the rightward-going wave, the rotation of the phasor is clockwise. For a leftward-going wave, the phasor would rotate in the counterclockwise direction.

1.3 CHARACTERISTIC IMPEDANCE

Until now we have been considering the time-varying voltages on transmission lines. However, each wave involves currents as well as voltages. Consider the positive-going sinusoidal wave

$$v^+ = A\cos(\omega t + \phi - kz) \tag{1.22}$$

whose phasor is

$$\mathbf{v}^+(z) = Ae^{-jkz}\,e^{j\phi} \tag{1.23}$$

Since the transmission line is a linear circuit, we can assume that the current will also be sinusoidal with the same frequency. Equation (1.6) cxpressed in phasor form is

$$Lj\omega\,\mathbf{i}(z) = -\frac{\partial}{\partial z}\,\mathbf{v}(z)$$

from which we have

$$\begin{aligned} Lj\omega\,\mathbf{i}^+(z) &= jkAe^{-jkz}\,e^{j\phi} \\ &= jk\mathbf{v}^+(z) \end{aligned}$$

Thus we find that for this positive-going wave,

$$\frac{\mathbf{v}^+(z)}{\mathbf{i}^+(z)} = \frac{\omega L}{k} \tag{1.24}$$

Thus the ratio $\mathbf{v}^+/\mathbf{i}^+$ is a constant, independent of position. This ratio is known as the *characteristic impedance* of the transmission line, and is given the symbol Z_o:

$$\frac{\mathbf{v}^+}{\mathbf{i}^+} \equiv Z_o \tag{1.25}$$

Expressing k in (1.24) by means of (1.15) and (1.12), we find that

$$Z_o = \sqrt{L/C} \tag{1.26}$$

For an ideal line, the characteristic impedance is a real number and like other impedances is given in ohms. Its value is determined by the shape and size of the line and the materials used in its construction. Values in the range 50–300 ohms are typical. The fact that the characteristic impedance is *real* for an ideal line indicates that the phase of the voltage sinusoid is the same as that of the current sinusoid at every point on the line.

For a wave traveling in the negative direction, for example that of (1.21), we repeat the same steps, and find that

$$\frac{\mathbf{v}^-}{\mathbf{i}^-} = -Z_o \tag{1.27}$$

where Z_o is still given by (1.26). Note the minus sign in this equation, which differs from (1.25).

POWER TRANSMITTED BY A SINGLE WAVE

An important property of waves is their ability to carry power from one place to another. Although this notion is familiar, it is quite remarkable. For example, power is generated in the sun and brought to us by means of electromagnetic waves. Yet there is no connection of any sort between the sun and the earth. Power is delivered to us across 93 million miles of vacuum!

Transmission-line waves can carry power in a similar fashion. Figure 1.5(a) shows a positive-going transmission-line wave entering a load. The power entering the load is the time average of $v(t)i(t)$, or in terms of phasors

$$P_L^+ = \frac{1}{2}\,\mathrm{Re}\,[\mathbf{v}^+(\mathbf{i}^+)^*] = \frac{1}{2}\frac{|\mathbf{v}^+|^2}{Z_o} \tag{1.28}$$

The power entering the load is the same as the power being carried by the wave, which is given by (1.28).

In Fig. 1.5(b) we have a negative-going wave entering a load. Since we always take the reference direction of current as toward the right, the power entering this load is $-vi$, and the time-averaged power is (using 1.27)

$$P_L^- = -\frac{1}{2}\,\mathrm{Re}\,[\mathbf{v}^-(\mathbf{i}^-)^*] = \frac{1}{2}\frac{|\mathbf{v}^-|^2}{Z_o} \tag{1.29}$$

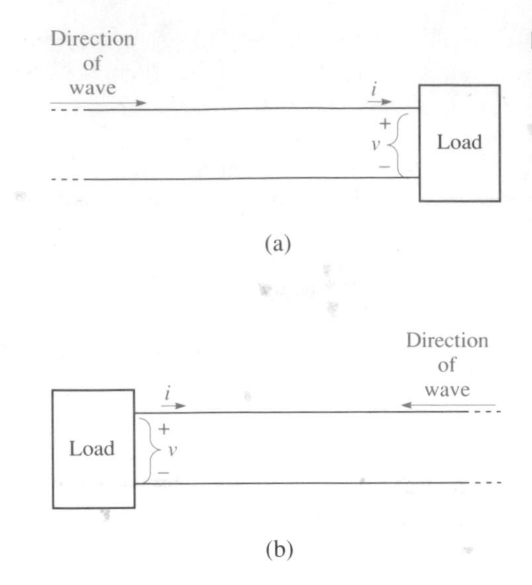

Figure 1.5 The reference direction for the current of the wave is always toward the right, regardless of which way the wave is moving. In (a), a wave is moving to the right and entering a load. Here $i = v/Z_o$, and the instantaneous power entering the load is $(v)(i)$. In (b) a wave is moving to the left and entering a load. Note that for the leftward-moving wave, $i = -v/Z_o$. However, the power entering the load is now $(v)(-i)$.

Thus the minus sign in (1.27) is consistent with the physically reasonable idea that the rightward-going wave carries power to the right, and the negative-going wave to the left.

1.4 REFLECTION AND TRANSMISSION

Until now we have assumed that our transmission lines were infinitely long, but of course in reality they must end, as, for example, in Fig. 1.5(a). In most cases the load (or "termination") is some device to which we are trying to supply power through the line. For instance, Fig. 1.5(a) could represent the line leading down from a TV antenna on the roof; the load would be the television set.

Often one wants all the power carried by a wave to be dissipated in the load, but this is never quite the case. In practice, when a positive-going wave is incident on a load, a reflected wave, negative-going, is generated, as shown in Fig. 1.6. In this figure the load is modeled as an impedance Z_L. The incident wave from the left has phasor $\mathbf{v}_i^+(z)$ and there is a reflected wave, negative-going, whose phasor is $\mathbf{v}_r^-(z)$. We choose the position of the load to be $z = 0$. Suppose we know the incident wave. How do we find the reflected wave?

To do this, we apply Kirchhoff's laws at $z = 0$. The voltage at this point is the sum of the voltages of the two waves:

$$\mathbf{v}(z = 0) = \mathbf{v}_i^+(z = 0) + \mathbf{v}_r^-(z = 0)$$

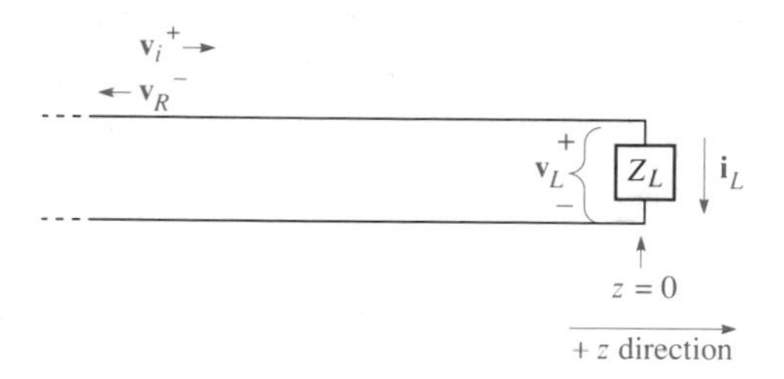

Figure 1.6 Reflection of a sinusoidal wave by a load impedance Z_L located at $z = 0$.

The current in the line at $z = 0$ is the sum of the currents of the two waves, given by (1.25) and (1.27):

$$\mathbf{i}(z = 0) = \mathbf{i}_i^+(z = 0) + \mathbf{i}_r^-(z = 0)$$

$$= \frac{1}{Z_o} [\mathbf{v}_i^+(z = 0) - \mathbf{v}_r^-(z = 0)] \qquad (1.30)$$

(Note the important minus sign that enters from (1.27).) Now the line voltage at $z = 0$ is the same as the load voltage $\mathbf{v}_L$, and for current continuity we must also have $\mathbf{i}(z = 0) = \mathbf{i}_L$. Moreover, the property of the load is that $\mathbf{v}_L/\mathbf{i}_L = Z_L$. Thus

$$\frac{\mathbf{v}(z = 0)}{\mathbf{i}(z = 0)} = Z_o \frac{\mathbf{v}_i^+(z = 0) + \mathbf{v}_r^-(z = 0)}{\mathbf{v}_i^+(z = 0) - \mathbf{v}_r^-(z = 0)} = Z_L \qquad (1.31)$$

We assume that the incident wave $\mathbf{v}_i^+$ is known. Solving for the phasor of the reflected wave, we obtain the important result

$$\mathbf{v}_r^-(z = 0) = \frac{Z_L - Z_o}{Z_L + Z_o} \mathbf{v}_i^+(z = 0) \qquad (1.32)$$

The ratio of the reflected and incident phasors at the position of the load is a complex number known as the load's *reflection coefficient* ρ_o:

$$\rho_o \equiv \frac{\mathbf{v}_r^-(z = 0)}{\mathbf{v}_i^+(z = 0)} = \frac{Z_L - Z_o}{Z_L + Z_o} \qquad (1.33)$$

EXAMPLE 1.4

Find the reflection coefficient when the line is terminated in (a) an open circuit; (b) a short circuit; and (c) a resistance whose value is nZ_o.

Solution For an open circuit we have $Z_L = \infty$, and from (1.33) we have $\rho_o = 1$. This means that the amplitude of the reflected wave is equal to that of the incident wave, consistent with our knowledge that no power can be dissipated in an open circuit. Let us investigate the distribution of the voltage on the line in this case. Suppose the incident wave is $\mathbf{v}_i^+(z) = Ae^{-jkz}$, where for simplicity we assume that A is real. The phasor representing the total voltage on the line is then

$$\mathbf{v}(z) = \mathbf{v}_i^+(z) + \mathbf{v}_r^-(z)$$

As in Example 1.3, we note that $\mathbf{v}_r^-(z) = \mathbf{v}_r^-(z = 0)e^{jkz} = \rho_o A e^{jkz}$. Thus

$$\begin{aligned}\mathbf{v}(z) &= Ae^{-jkz} + \rho_o A e^{jkz} \\ &= Ae^{-jkz} + Ae^{jkz} \\ &= 2A \cos kz\end{aligned}$$

We observe that there are certain positions on the line, given by $z = \frac{1}{2}(2N + 1)\pi/k$, where N is an integer, at which $\mathbf{v}$ is zero. The voltage as a function of time is

$$v(z, t) = \text{Re}\,[\mathbf{v}e^{j\omega t}] = 2A \cos kz \cos \omega t$$

Therefore at those positions where $\mathbf{v}$ vanishes, the actual voltage $v(t)$ is zero *at all times*. Note that this is quite different from the case of a single traveling wave, such as $v(z,t) = A\cos(kz - \omega t)$. In a traveling wave, there are positions where the voltage vanishes, but these positions *move,* at the velocity of the wave. In the present case, however, what we have is a sum of two traveling waves, of equal magnitude, moving in opposite directions, and the places of zero voltage *stand still*. This gives rise to what is known as a *standing wave,* which will be discussed further in Chapter 2.

For the short-circuit load we have $Z_L = 0$, and we find that $\rho_o = -1$. The reflected wave again has the same amplitude as the incident wave (because no power can be dissipated in a short circuit). The minus sign means that the phase of the reflected wave differs by 180° from the case of an open-circuit load. The total voltage phasor on the line in this case is

$$\begin{aligned}\mathbf{v} &= Ae^{-jkz} - Ae^{jkz} \\ &= -2jA \sin kz\end{aligned}$$

Again we have a standing wave; this time the nulls are at $z = N\pi/k$. The spacing between the nulls is the same as with the open-circuited line, but their positions are different. In particular, there is a null at $z = 0$; this is physically necessary, as the voltage must always be zero at a short circuit.

For the resistive load $R_L = nZ_o$, we have

$$\rho_o = \frac{n - 1}{n + 1}$$

We note that when $n = 1$ (i.e., $R_L = Z_o$), the reflected wave vanishes. This is the situation often desired in practice, with all incident power transmitted to the load and none reflected. One then speaks of the line as being *terminated in its characteristic impedance*. However, this ideal situation is never quite achieved in practice, since it is impossible to build a perfect load resistance with exactly the right value.

Finally we note that for all other positive values of n, either greater or less than one, the absolute value of ρ_o is always *less* than one. This is physically reasonable, as we would not expect the reflected wave to carry back more power than arrives with the incident wave.

EXERCISE 1.2

A line has a characteristic impedance of 50 ohms and carries a sinusoidal wave with an amplitude of 1V moving to the right. This wave is incident on a load whose impedance is $50(1 + j)$ ohms. Find the power carried by the reflected wave.

Answer 2 mW.

Let us now take a more general point of view and consider the situation shown in Fig. 1.7. The section of line on the left has characteristic impedance Z_{o1}, and that on the right Z_{o2}. Let there be an incident wave $\mathbf{v}_i^+$ coming from the left. In this case we expect to find a negative-going reflected wave $\mathbf{v}_r^-(z)$, and in addition there will be a positive-going *transmitted* wave $\mathbf{v}_t^+(z)$ moving off to the right. Furthermore there may be some power P_L dissipated in the impedance Z_L.

Again we let the position of the impedance Z_L correspond with $z = 0$. The voltage at this point, if we approach from the left, is $\mathbf{v}_i^+(z = 0) + \mathbf{v}_r^-(z = 0)$; but if we approach from the right, it is $\mathbf{v}_t^+(z = 0)$. Thus

$$\mathbf{v}_i^+(z = 0) + \mathbf{v}_r^-(z = 0) = \mathbf{v}_t^+(z = 0) \tag{1.34}$$

Now from Kirchhoff's current law we must also have

$$\mathbf{i}_i^+(z = 0) + \mathbf{i}_r^-(z = 0) = \mathbf{i}_t^+(z = 0) + \mathbf{i}_L \tag{1.35}$$

where $\mathbf{i}_L$ is the current through Z_L, as shown in the figure. Moreover,

$$\mathbf{i}_L = \frac{\mathbf{v}_t^+(z = 0)}{Z_L} \tag{1.36}$$

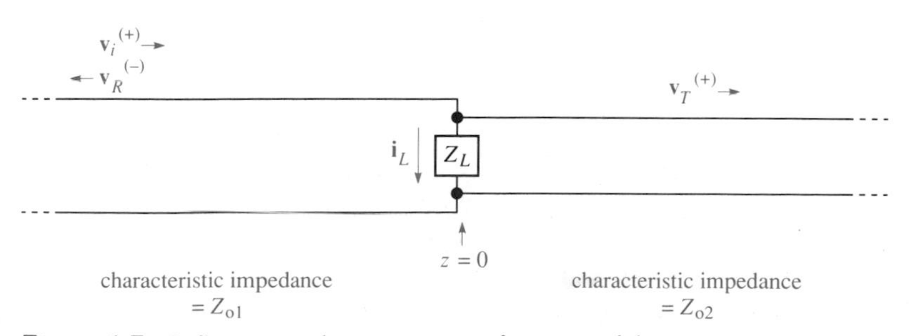

Figure 1.7 Reflection and transmission of a sinusoidal wave.

The currents $\mathbf{i}_i^+$ $\mathbf{i}_r^-$, and $\mathbf{i}_t^+$ can be expressed in terms of $\mathbf{v}_i^+$, $\mathbf{v}_r^-$, and $\mathbf{v}_t^+$ by means of (1.25) and (1.27). This leaves us with three unknowns: $\mathbf{v}_r^-(z = 0)$, $\mathbf{v}_t^+(z = 0)$, and $\mathbf{i}_L$. We also have three equations available, (1.34), (1.35), and (1.36). Solving, we find that

$$\frac{\mathbf{v}_r^-(z = 0)}{\mathbf{v}_i^+(z = 0)} = \rho_o = \frac{Z_{o2}Z_L - Z_{o1}Z_L - Z_{o1}Z_{o2}}{Z_{o2}Z_L + Z_{o1}Z_L + Z_{o1}Z_{o2}} \tag{1.37}$$

This can be rewritten

$$\rho_o = \frac{Z_\| - Z_{o1}}{Z_\| + Z_{o1}} \tag{1.38}$$

where by definition $Z_\|$ is the parallel combination of Z_L and Z_{o2}:

$$Z_\| \equiv \frac{Z_L Z_{o2}}{Z_L + Z_{o2}} \tag{1.39}$$

Now comparing (1.39) with (1.33), we see that the reflection is the same as if a line with characteristic impedance Z_{o1} were terminated in a load $Z_\|$. This is sensible, because the incident wave is reflected by the parallel combination of Z_L and Z_{o2}, the input impedance of the line going off to the right.

Continuing with our algebra, we now use (1.34) to find the transmitted wave:

$$\tau \equiv \frac{\mathbf{v}_t^+(z = 0)}{\mathbf{v}_i^+(z = 0)} = \frac{2Z_\|}{Z_\| + Z_{o1}} \tag{1.40}$$

The ratio τ is known as the *transmission coefficient*.

EXAMPLE 1.5

In Fig. 1.7, consider the special case in which $Z_{o1} = Z_{o2}$ and Z_L is purely resistive and equal to Z_{o1}. The incident wave has amplitude V_o. Find the power dissipated in Z_L.

Solution The power dissipated in Z_L is

$$P_L = \frac{1}{2} \operatorname{Re} \frac{|\mathbf{v}_t^+(z = 0)|^2}{Z_L} \tag{1.41}$$

In this case $Z_\| = Z_{o1}/2$. Hence from (1.40) $\mathbf{v}_t^+(z = 0) = \frac{2}{3}\mathbf{v}_i^+(z = 0)$, and

$$P_L = \frac{1}{2}\left(\frac{4}{9}\right) \frac{|\mathbf{v}_i^{(+)}(z = 0)|^2}{Z_{o1}} \tag{1.42}$$

However $|\mathbf{v}_i^+(z)|$ is simply the amplitude of the incident wave, which is independent of position and equal to V_o. Thus

$$P_L = \frac{2V_o^2}{9Z_{o1}} \tag{1.43}$$

This can be compared with the incident power, which is $\frac{1}{2}V_o^2/Z_{o1}$. The power in the transmitted wave must be identical with (1.43) because the power in the transmitted wave is also $\frac{1}{2}\text{Re}(|\mathbf{v}_t^+|^2/Z_{o1})$. Hence from energy conservation the reflected power must be $V_o^2/18Z_{o1}$.

EXERCISE 1.3

For the case of Example 1.5, find the reflected power directly using Equation (1.38).

Answer $V_o^2/18Z_{o1}$.

Until now we have assumed that the incident wave $\mathbf{v}^+$ was given, and we have calculated the reflected and transmitted waves. However, in the real world one does not usually know, *a priori,* the value of $\mathbf{v}^+$, as this is rather difficult to measure. One is more likely to know, for instance, the voltage across one end of the line, which can be measured by an oscilloscope. It is easy to fall into the error of thinking that the phasor representing the input voltage to the line is the same as $\mathbf{v}^+$, but this is not true, as is illustrated by the following example.

EXAMPLE 1.6

An ideal sinusoidal voltage source with phasor $\mathbf{v}_S$ is connected to the input of a coaxial transmission line of length l_1, and the voltage at the other end of the line is measured with an oscilloscope, as shown in Fig. 1.8. The input impedance of the oscilloscope is infinitely large. Find the phasor representing the voltage measured by the oscilloscope.

Solution Let the wave traveling in the $+z$ direction be $\mathbf{v}^+(z) = V_1 e^{-jkz}$, and let the wave going in the $-z$ direction be $\mathbf{v}^-(z) = V_2 e^{jkz}$, where V_1 and V_2 are constants to be determined. Note that $\mathbf{v}^+$ is *not* equal to $\mathbf{v}_S$, since the voltage at the voltage source is not just $\mathbf{v}^+$, but is the sum of $\mathbf{v}^+$ and $\mathbf{v}^-$.

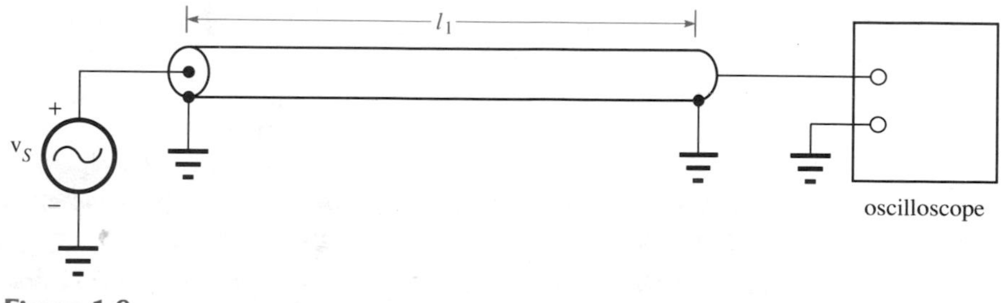

Figure 1.8

The voltage phasor at any position z is

$$\mathbf{v}(z) = \mathbf{v}^+(z) + \mathbf{v}^-(z) = V_1 e^{-jkz} + V_2 e^{jkz}$$

Let us choose $z = 0$ at the oscilloscope. Then $\mathbf{v}^+(z = 0) = V_1$, $\mathbf{v}^-(z = 0) = V_2$, and using (1.33)

$$\frac{\mathbf{v}^-(z = 0)}{\mathbf{v}^+(z = 0)} = \frac{V_1}{V_2} = \rho_o = 1$$

Thus the total voltage of the line is

$$\mathbf{v}(z) = \mathbf{v}^+(z) + \mathbf{v}^-(z) = V_1(e^{-jkz} + e^{jkz})$$
$$= 2V_1 \cos kz$$

The boundary condition at $z = -l_1$ is $\mathbf{v}(z = -l_1) = \mathbf{v}_S$. Setting $z = -l_1$ in the above equation, we have

$$\mathbf{v}(z = -l_1) = \mathbf{v}_S = 2V_1 \cos kl_1$$

Thus

$$V_1 = \frac{\mathbf{v}_S}{2 \cos kl_1}$$

and the voltage at the oscilloscope is

$$\mathbf{v}(z = 0) = 2V_1 \cos(0) = \frac{\mathbf{v}_S}{\cos kl_1}$$

It is interesting to observe that the amplitude of the voltage at the oscilloscope is larger than $|\mathbf{v}_S|$, and might even be infinite, if the line is a quarter-wavelength long. In reality, however, the voltage source would not be ideal, but would have some finite source resistance. This would change the result and prevent infinite voltages from appearing. (See Problem 1.14.)

1.5 MORE GENERAL TRANSMISSION LINES

Until now we have assumed that our transmission line is composed purely of capacitance and inductance. However, in any real line, some resistance will be present. Our model can be generalized to include this possibility, as shown in Fig. 1.9. Comparing with Fig. 1.1(c), we see that the inductors have acquired some series resistance. Furthermore, a leakage conductance now exists in parallel with the shunt capacitors. The inductance, resistance, capacitance, and leakage conductance of a small section of line of length h are represented by L_h, R_h, C_h, and G_h, respectively. We define $L = L_h/h$, $R = R_h/h$, $C = C_h/h$, $G = G_h/h$, so that L, R, C, and G are respectively the inductance, resistance, capacitance, and conductance per unit length.

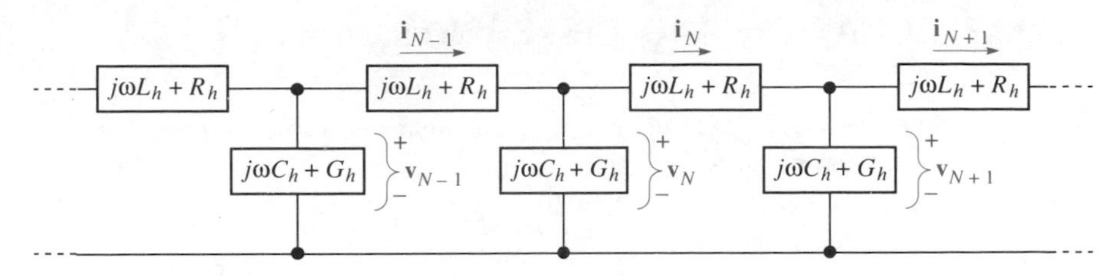

Figure 1.9 Lumped-circuit model of a more general transmission line, containing series resistance and shunt conductance.

The presence of resistance changes the physical situation in important ways. In particular, distortionless propagation is no longer possible. However, it can be shown that waves with sinusoidal time dependence can still propagate. To show this, we can adapt our earlier derivation of the wave equation to this more general case. From Kirchhoff's laws in their phasor form, we have

$$\begin{aligned} \mathbf{i}_N(j\omega L_H + R_h) &= \mathbf{v}_N - \mathbf{v}_{N+1} \\ \mathbf{v}_N(j\omega C_h + G_h) &= \mathbf{i}_{N-1} - \mathbf{i}_N \end{aligned} \tag{1.44}$$

Proceeding as before, we obtain the phasor form of the telegraphist's equations,

$$\begin{aligned} \mathbf{i}(z)(j\omega L + R) &= -\frac{\partial \mathbf{v}}{\partial z} \\ \mathbf{v}(z)(j\omega C + G) &= -\frac{\partial \mathbf{i}}{\partial z} \end{aligned} \tag{1.45}$$

which we can combine into a more general form of the wave equation

$$\frac{\partial^2 \mathbf{v}}{\partial z^2} - (j\omega C + G)(j\omega L + R)\mathbf{v} = 0 \tag{1.46}$$

The two solutions of this equation are

$$v^{\pm} = A^{\pm} e^{\mp jkz} \tag{1.47}$$

where A^+ and A^- are arbitrary constants describing the waves' amplitude and phase, and

$$k = -j(j\omega C + G)^{1/2}(j\omega L + R)^{1/2} \tag{1.48}$$

The two solutions, one with the upper and one with the lower signs, arise from the two possible signs of the square root, and simply correspond to the waves moving toward the right (upper signs) and toward the left (lower signs). Note that we can recover the case of the ideal lossless line by inserting $R = 0$, $G = 0$. In that case (1.48) becomes $k = \omega\sqrt{LC}$, in agreement with (1.15) and (1.12).

1.6 ATTENUATION AND DISPERSION

From (1.48) we observe that the propagation constant of a lossy transmission line can be a complex number. Let us now consider the physical significance of complex k. To do this, we separate k into real and imaginary parts, by defining

$$k \equiv \beta - j\alpha \tag{1.49}$$

where β and α are real numbers. Then (1.47) becomes (considering only the positive-going wave)

$$\mathbf{v}^+ = Ae^{-j\beta z}\, e^{-\alpha z} \tag{1.50}$$

Now we can visualize the propagation of the wave. The final factor, involving α, gives rise to *attenuation*. That is, as the wave travels to the right, its amplitude becomes smaller exponentially. This is physically reasonable: Currents are passing through the resistances of the line, and thus power is being lost. (From this reasoning, we see that α must be a *positive* number. Otherwise the wave would grow *larger* as it traveled.) We refer to α as the *attenuation constant*. The first factor, involving β, determines the wavelength. To increase the phase of the wave by 2π radians, we must travel a distance λ such that $\beta\lambda = 2\pi$. Thus the wavelength is now given by

$$\lambda = \frac{2\pi}{\beta} \tag{1.51}$$

Note that the phase of the wave changes by β radians per unit length. Thus we refer to β as the *phase constant*.

The actual voltage (as a function of time) is obtained from the phasor (1.50) using (1.19):

$$v(z, t) = A \cos(\beta z - \omega t)e^{-\alpha z} \tag{1.52}$$

(For convenience we have assumed that A is real.) Using this equation we can discuss the velocity of the wave. Let us locate the position at which the cosine function is maximum. This occurs when the argument of the cosine is zero. Hence $z_{\max}$, the position of a maximum, is given by

$$\beta z_{\max} = \omega t$$

As t increases, the maximum moves to the right with velocity

$$\frac{dz_{\max}}{dt} \equiv U_P = \frac{\omega}{\beta} \tag{1.53}$$

The velocity U_P found in this way is known as the *phase velocity*. For the special case of the ideal lossless line $U_P = 1/\sqrt{LC}$.

EXAMPLE 1.7

Suppose that a transmission line, if made of lossless materials, has $Z_o =$ 50 ohms and $U_P = c/3$ (where c is the velocity of light). However, since it is made of real materials, it has a series resistance of 1000 ohms/m. Find the attenuation coefficient α. How far must the wave travel to decrease in amplitude by $1/e$? The frequency is 1 GHz.

Solution From (1.12) and (1.26) we find that $C = 200$ pF/m, $L = 500$ nH/m. With resistance present, we have, from (1.48)

$$k = \frac{1}{j}\sqrt{(j\omega L + R)(j\omega C)}$$

$$= \sqrt{\omega^2 LC - j\omega RC}$$

$$= \sqrt{3948 - j1257} = 63.6 - 9.88j$$

Thus from (1.49), $\alpha = 9.88\ \text{m}^{-1}$. (Note that α is positive, as expected.) Since the wave decays as $e^{-\alpha z}$, the distance L_A required for it to decrease by $1/e$ is determined by $\alpha L_A = 1$, or

$$L_A = 1/\alpha = 0.10 \text{ meter}$$

The distance L_A so defined is called the *attenuation length*.[2]

EXERCISE 1.4

Find the wavelength and phase velocity of the wave in Example 1.7.

Answer 9.88 cm; 9.88×10^7 m/sec.

In Example 1.5 we notice that both the real and imaginary parts of k vary with frequency. This variation is referred to as *dispersion*. An important consequence is that the phase velocity is different for waves of different frequency. This means that a signal containing many frequencies tends to become "dispersed"; that is, some parts of the signal arrive sooner and others later. Thus in a dispersive transmission system, the signal that arrives can be quite different from the one that was sent. (Imagine a telephone in which the high-frequency parts of the voice arrive first, and the low-frequency parts later, and you will have the idea.) In most cases such distortion of the signal is undesirable, but it is usually present to some degree.

[2]The units of the attenuation coefficient α are sometimes given as "*nepers* per meter," instead of simply "per meter." An exponent of one neper means an attenuation of $1/e$.

Figure 1.10 shows a graph of ω as a function of β, for an arbitrarily chosen transmission system. Until now we have talked of β being a function of ω, but it is usual to plot ω on the vertical axis, and β on the horizontal. The resulting graph, known as a *dispersion diagram* or *ω–β diagram,* is useful for studying the velocity of the waves. As we see from (1.53), the phase velocity at any frequency is equal to the slope of a line drawn from the origin to the corresponding point on the graph. We note that for this particular transmission system, there is *no* value of β for frequencies below $\omega = 1.35 \times 10^{10}$ radians/sec. This indicates that waves are not propagated at frequencies below this value, which is known as the *cutoff frequency.* When ω has exactly the value 1.35×10^{10} radians/sec, the line from the origin to the curve is vertical, indicating that the phase velocity at this frequency is infinitely large. As frequency increases above the cutoff frequency, the phase velocity for this particular transmission system steadily decreases. However, the dispersion curve remains above the line $\omega = kc$ (where c is the velocity of light). Thus U_P is greater than c for all frequencies. That is just a property of this one particular system, however. In general, U_P can be either greater or less than c.

It may seem remarkable that the phase velocity can exceed c, let alone become infinitely large. This might appear to violate Einstein's postulate that messages cannot travel faster than the speed of light. Actually there is no contradiction, because in general, the information in a wave does *not* travel at the phase velocity. Information travels at a different velocity, known

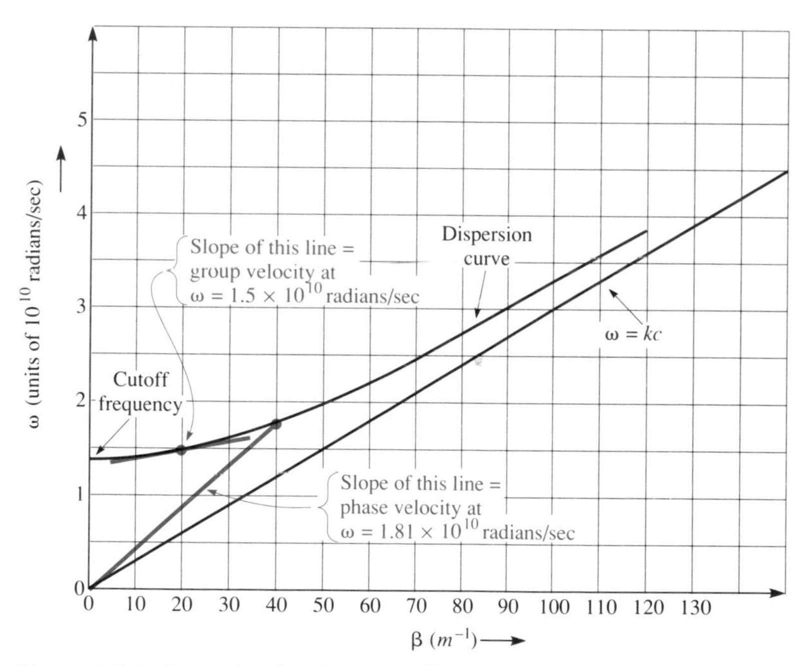

Figure 1.10 Example of a dispersion diagram.

as the *group velocity* U_G, whose value is given by

$$U_G = \frac{d\omega}{d\beta} \tag{1.54}$$

(the derivative being taken at the frequency of interest). The subject of group velocity is fairly complex, and a derivation of (1.54) is beyond the scope of our present discussion. However, we note that according to (1.54), the group velocity is equal to the slope of the *tangent* to the ω–β curve at the frequency in question. In Fig. 1.10, we see that at the cutoff frequency (where U_P is infinite) the group velocity is actually zero, meaning the information cannot be sent at this frequency at all. At higher frequencies the group velocity steadily increases, but always remains *less* than the velocity of light.

As a final point, we note that it is quite possible for U_P and/or U_G to be negative. If U_P is negative, the physical meaning is simply that the wave maxima are moving to the left. (For example, the negative-going wave (1.21).) If the *group* velocity is negative, however, it means that information is being carried from right to left. Some transmission systems can transmit waves in which the signs of the two velocities are opposite. Such waves are known as *backward waves*.

EXAMPLE 1.8

Consider an unusual transmission line whose model is shown in Figure 1.11. The capacitive reactance per unit length is $X_c = -10^{14}/\omega$ ohms/m and the shunt inductive susceptance B_L is $-10^{12}/\omega\ \text{ohms}^{-1}/\text{m}$. (Hence hX_c is the reactance of a small section of length h, etc.) Find the phase and group velocities at 1 GHz.

Solution From (1.48)

$$\begin{aligned} k &= -j\sqrt{(-10^{14}j/\omega)(-10^{12}j/\omega)} \\ &= 10^{13}/\omega \end{aligned}$$

In this case k turns out to be real, and from (1.49), $\beta = \text{Re}(k) = 10^{13}/\omega$.

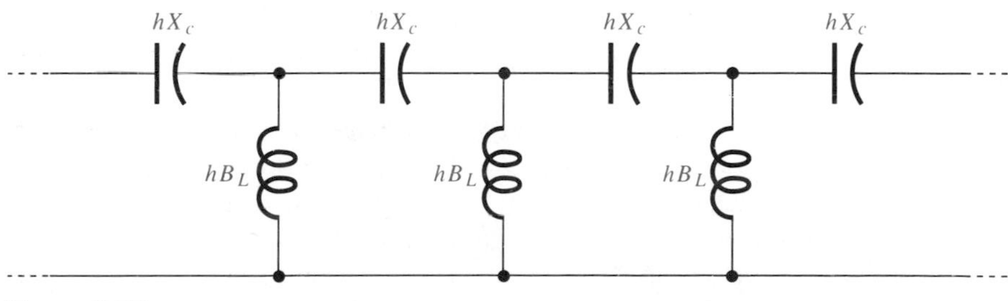

Figure 1.11

Hence

$$U_P = \frac{\omega}{\beta} = 10^{-13}\omega^2$$

$$U_G = \frac{d\omega}{d\beta} = \frac{d}{d\beta}\left(\frac{10^{13}}{\beta}\right) = -\frac{10^{13}}{\beta^2} = -10^{-13}\omega^2$$

At 1 GHz, $U_P = 3.95 \times 10^6$ m/sec, $U_G = -3.95 \times 10^6$ m/sec. For this transmission system the phase and group velocities turn out to be equal but opposite in sign. The wave is thus a backward wave.

From the result $U_G = -10^{-13}\omega^2$, one might think that at sufficiently high frequencies U_G would become greater than c. However, this is a physical impossibility. One must remember that Fig. 1.11 is only a *model* of some actual physical structure, and circuit models are only approximate equivalents, with a limited range of validity. In this case we can be sure that the circuit model becomes unrealistic by the time frequencies on the order of 8.7 GHz are reached.

1.7 NON-SINUSOIDAL WAVES

In the last few sections we have been concentrating on waves with sinusoidal time variation. This is a very important case, especially in analog systems. However, in digital systems it is common to transmit signals that are not sinusoidal—rectangular pulses, for example. Thus it is important to see how transmission-line theory applies to the non-sinusoidal case. In this section we shall be considering only the lossless ideal transmission line of Fig. 1.2.

It has already been seen that the ideal transmission line exhibits distortionless propagation. Thus waves of arbitrary form retain their shapes as they propagate. Although (1.25) was derived for the sinusoidal case, it can be shown that a similar result holds for non-sinusoidal waves. Let $v^+(z,t)$ be the voltage associated with a wave moving in the $+z$ direction, and let $i^+(z,t)$ be the current associated with the same wave. Then

$$\frac{v^+(z,t)}{i^+(z,t)} = Z_o \tag{1.55}$$

where $Z_o = \sqrt{L/C}$ as before. For a negative-going wave, $v^-(z,t)/i^-(z,t) = -Z_o$.

Non-sinusoidal waves are also reflected from discontinuities in the line. Suppose, for example, a rectangular pulse is incident on a short circuit as shown in Fig. 1.12(a). At the short the voltage must always vanish. This is accomplished by the generation of a negative reflected wave, as shown in Fig. 1.12(b). The incident and reflected waves add to give zero voltage at the position of the short. At a later time (c) the incident wave has been

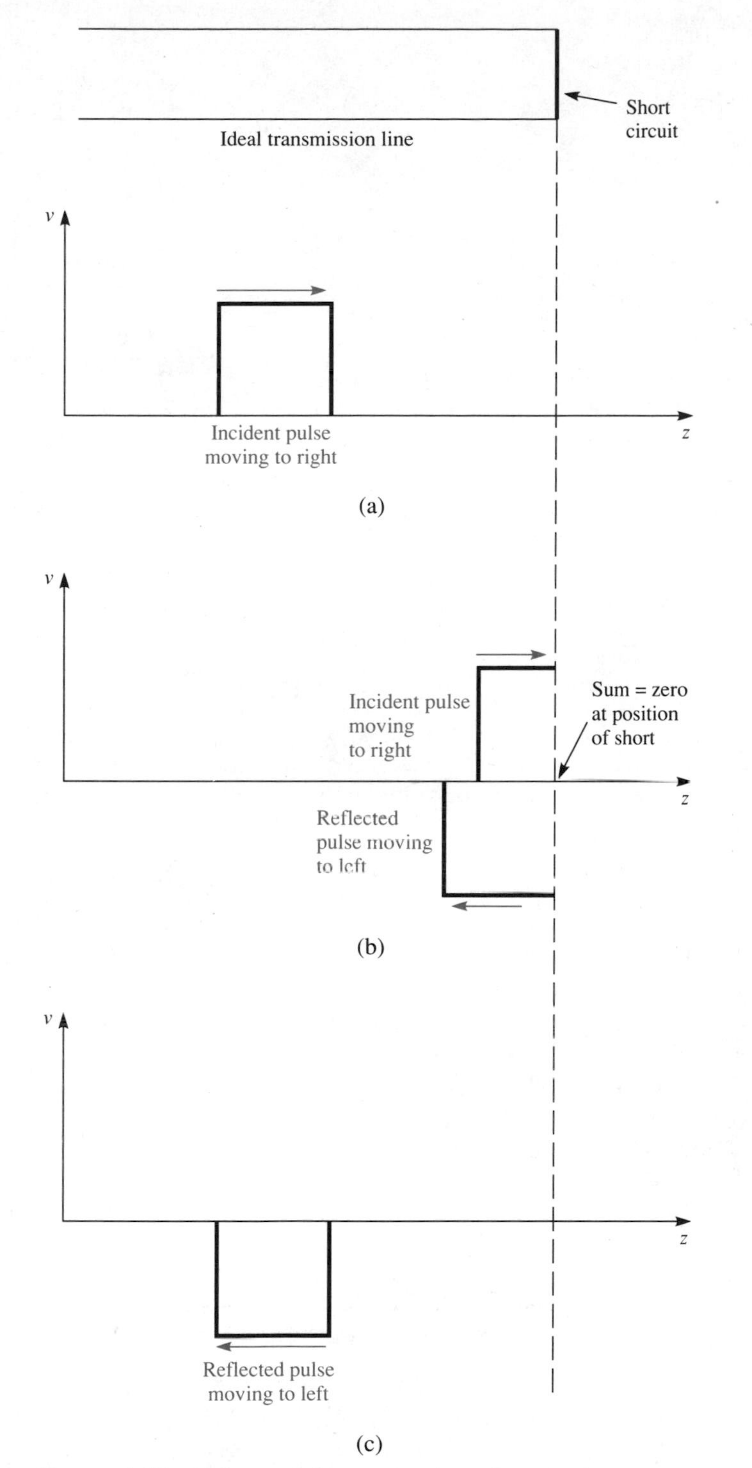

Figure 1.12 Reflection of a rectangular pulse at a short circuit. (a) shows the incident pulse moving to the right. In (b) it is striking the short-circuit termination; note that the sum of the incident and reflected voltages must always be zero at that position. In (c) the reflected pulse is moving to the left.

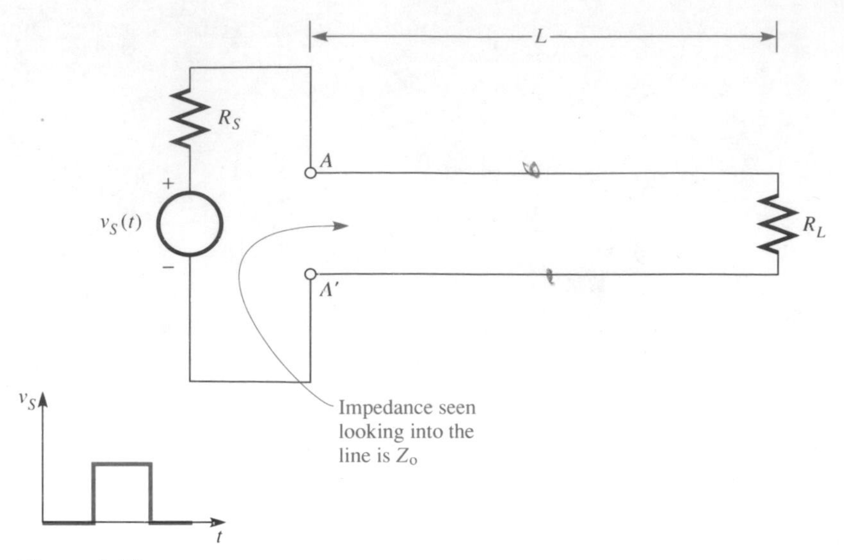

Figure 1.13

entirely reflected and there is only the reflected wave going back toward the left. Since in this case the reflected pulse has the same shape as the incident pulse, we can write

$$v^-(z_2, t + T) = \rho_o v^+(z_1, t) \tag{1.56}$$

where T is simply the time delay required for the incident wave to travel from position z_1 to the end and return to position z_2. In (1.56) ρ_o is (as before) the reflection coefficient; for the example just considered $\rho_o = -1$. It is easily shown that for *resistive* loads (including short- and open-circuit loads), reflection and transmission coefficients are still given by (1.38)–(1.40). However, if the circuit contains inductors or capacitors, the reflected pulse no longer has the same shape as the incident pulse. Our definition of the reflection coefficient (ratio of reflected amplitude to incident amplitude) then no longer makes any sense.

Multiple reflections of pulses back and forth are common. In Fig. 1.13 we have a pulsed voltage source, with source resistance R_S, connected by an ideal transmission line to a load R_L. Let $v_S(t)$ be a rectangular pulse of magnitude V_o. We can build up the total voltage on the line at any time by superimposing the incident wave and all its reflections. From (1.55) the input resistance of the transmission line, looking to the right from terminals AA', is simply Z_o.[3] This input impedance forms a voltage divider with R_S, so the

[3]The ratio of v to i is Z_o in this case because there is only *one single wave* on the line—the one that is being launched. Eventually, reflected waves may arrive, but not until the initial wave has traveled all the way to the end of the line and returned.

amplitude of the wave launched down the line is not V_o, but rather

$$V_1 = \frac{Z_o}{Z_o + R_S} V_o \tag{1.57}$$

When the incident pulse arrives at the load it is reflected, with a reflection coefficient given by (1.38),

$$\rho_1 = \frac{R_L - Z_o}{R_L + Z_o} \tag{1.58}$$

This gives rise to a reflected wave with amplitude $V_2 = \rho_1 V_1$. When this reflected wave arrives back at the signal source, it is again reflected. As usual when using superposition, the ideal voltage source is treated as a short circuit for the reflection process. The next reflection coefficient is

$$\rho_2 = \frac{R_S - Z_o}{R_S + Z_o} \tag{1.59}$$

and a new reflected wave with amplitude $V_3 = \rho_2 V_2$ moves off to the right. Since all the reflection coefficients have absolute values less than one, the reflections will gradually die away.

The preceding discussion is not restricted to pulses that are short compared with the transit time L/V. The longer the pulse, the more likely it is that the incident, reflected, and re-reflected waves will overlap each other. In such cases two or more waves can be simultaneously present at a point, and the total voltage at that point will be the sum of all the waves that are present. One special case is reached when the input pulse is *infinitely* long; that is, v_S simply increases from zero to V_o and then stays constant forever. In this case there will, in general, be an infinite number of infinitely long pulses moving back and forth. At first the voltage at any point will vary as one wave after another arrives, but since the reflections become smaller, the line voltage will ultimately settle down to some constant value.

EXAMPLE 1.9

Suppose in Fig. 1.13, $R_L = R_S$ and $v_S(t)$ is an infinitely long pulse (i.e., a step function) of amplitude V_o. Find the voltage across the line after a very long time.

Solution The initial pulse moving to the right has amplitude

$$V_1 = \frac{Z_o}{R_S + Z_o} V_o$$

The first reflected wave has amplitude $\rho_o V_1$, where

$$\rho_o = \frac{R_S - Z_o}{R_S + Z_o}$$

The second reflected wave has amplitude $\rho_o^2 V_1$, and so on. The total voltage at any point is the sum of all the voltages that have had time to arrive at that point. If we wait a very long time, the total voltage V_T is given by the infinite series

$$\begin{aligned} V_T &= V_1 + \rho_o V_1 + \rho_o^2 V_1 + \rho_o^3 V_1 + \cdots \\ &= V_1(1 + \rho_o + \rho_o^2 + \rho_o^3 + \cdots) \\ &= V_1 \left(\frac{1}{1 - \rho_o} \right) \end{aligned}$$

where in the last step an algebraic identity (easily proved by long division) has been used. Inserting the values of V_1 and ρ_o we find that

$$V_T = \frac{V_o}{2}$$

This interesting result shows that after a long time the voltage across the line simply results from the voltage divider of the two resistors, as though the transmission line were not there.

This example illustrates the relationship between transmission-line theory and ordinary circuit theory. What is happening physically is that the capacitance of the cable is being charged up by the remainder of the circuit, consisting of the voltage source, R_S, and R_L. In ordinary circuit theory we might neglect the finite transit time of waves through the line, and simply model it as a capacitance C_T. Then we would conclude that the charging time would be on the order of $(R_S \| R_L)C_T$, and that after a long time C_T would be charged to $V_o/2$. However, the details of the charging behavior—the individual waves bouncing back and forth on the line—would not be seen using ordinary circuit theory.

1.8 POINTS TO REMEMBER

1. In this chapter we have surveyed several different types of waves on transmission lines. It is important that these different cases not be confused. When approaching a transmission-line problem, the student should begin by asking, "Are the waves in this problem sinusoidal, or rectangular pulses? Is the line ideal, or does it have losses?" Then the proper approach to the problem can be taken.

2. The *ideal lossless line* supports waves of any shape (sinusoidal or non-sinusoidal), and transmits them without distortion. The velocity of these waves is $(LC)^{-1/2}$. The ratio of voltage to current is $Z_o = \sqrt{L/C}$, provided that only one wave is present. Sinusoidal waves are best treated using phasor analysis. (A common error is that of attempting to analyze *non*-sinusoidal waves with phasors. Beware! This makes no sense at all.)

3. When the line contains series resistance and/or shunt conductance, it is said to be *lossy*. Lossy lines no longer exhibit undistorted propagation; hence a rectangular pulse launched on such a line will not remain rectangular, instead evolving into irregular, messy shapes. However, *sinusoidal* waves, because of their unique mathematical properties, do continue to be sinusoidal on lossy lines. The presence of loss changes the velocity of propagation and causes the wave to be attenuated (become smaller) as it travels.

4. For lines other than the simple ideal lossless line, the velocity of propagation usually is a function of frequency. This velocity, the speed of voltage maxima on the line, is properly called the *phase velocity* U_P. The change of U_P with frequency is called *dispersion*. The velocity with which information travels on the line is not U_P, but a different velocity, known as the *group velocity* U_G. The phase velocity is given by $U_P = \omega/\beta$. However $U_G = d\omega/d\beta$.

5. Examples of *non-sinusoidal* waves are short rectangular pulses, and also infinitely long rectangular pulses, which are the same as step functions. Problems involving sudden voltage steps differ from sinusoidal problems, just as in ordinary circuits, problems involving transients differ from the sinusoidal steady state. Pulse problems are usually approached by superposition; that is, one tracks the pulses that propagate back and forth, adding up the waves to obtain the total voltage at any place and time.

6. All kinds of waves are reflected at discontinuities in the line. If the line continues beyond the discontinuity, a portion of the wave is transmitted as well. The reflected and transmitted waves are described by the *reflection coefficient* ρ_o and the *transmission coefficient* τ. For sinusoidal waves there is a simple formula (1.33) giving the reflection coefficient for any load impedance Z_L. For non-sinusoidal waves, the same formula can be used, but only if the load impedance is purely resistive. Otherwise the reflected wave has a different shape from the incident wave, and a reflection coefficient cannot be meaningfully defined.

7. In the case of non-sinusoidal waves, it is sometimes necessary to add up the contributions of many reflected waves bouncing back and forth on the line. However, for sinusoidal steady-state problems, it is only necessary to consider *two* waves, one moving to the right and the other to the left.[4]

---

[4]It is not necessary to consider the possibility of there being more than two waves, because sinusoids have the remarkable property that the sum of two or more sinusoids of the same frequency is always another sinusoid. Suppose we did have two waves moving to the right, so that $\mathbf{v} = V_1 e^{-jkz} + V_2 e^{-jkz}$. Then we can write $\mathbf{v} = V_3 e^{-jkz}$, where $V_3 = V_1 + V_2$.

REFERENCES

An entire book devoted to transmission lines is:

1. John L. Stewart, *Circuit Analysis of Transmission Lines*. New York: Wiley, 1958.

Treatments at a level slightly more advanced than this book will be found in the following books:

2. S. Ramo, J. R. Whinnery, and T. Van Duzer, *Fields and Waves in Communication Electronics*. New York: Wiley, 1984.
3. N. N. Rao, *Elements of Engineering Electromagnetics*. Englewood Cliffs, N.J.: Prentice-Hall, 1987.

PROBLEMS

Sections 1.1–1.4

1.1 Sketch the function

$$f(x) = \frac{1}{1 + (x - x_o)^2}$$

for (a) $x_o = 0$ and (b) for $x_o = 2$.

1.2 The function $f(x)$ is a pulse described by

$$\begin{aligned} f(x) &= Ax \qquad \text{for } -1 \leq x \leq 1 \\ &= 0 \qquad \text{otherwise} \end{aligned}$$

where x is in cm and A is a given positive constant.

(a) Sketch the function $f(z - Ut)$ as a function of z at the times $t = 0$, $t = 0.025$ nsec and $t = 0.05$ nsec. (One nsec $= 10^{-9}$ sec.) Assume $U = 2 \times 10^{10}$ cm/sec.

(b) Sketch $f(z - Ut)$ as a function of time at $z = 0$, $z = -2$ cm, and $z = 2$ cm.

(c) Repeat parts (a) and (b) for the function $f(z + Ut)$.

1.3 The phasor representing a sinusoidal wave is $2e^{j60°}\, e^{-jkz}$ volts where $k = 1\ \text{cm}^{-1}$. The ordinary frequency is 4 GHz.

(a) Draw the phasor in the complex plane at $z = 0$, $z = 1$ cm, and $z = 2$ cm.

(b) Sketch the voltage as a function of position at $t = 0$ and at $t = 6.25 \times 10^{-11}$ sec.

(c) Repeat parts (a) and (b) with the phasor changed to $2e^{j60°}\, e^{jkz}$ volts.

1.4 A wave with phasor $\mathbf{v}^+(z)$ propagates in the $+z$ direction on an ideal transmission line. Let the phasor representing the voltage at $z = 0$ be called $\mathbf{v}^+(z = 0)$. Show that the phasor representing the wave at other positions is

$$\mathbf{v}^+(z) = \mathbf{v}^+(z = 0)e^{-jkz}$$

*1.5 The phasor representing a certain wave is

$$\mathbf{v} = 3(2 - 3j)e^{j(0.6)z} \text{ volts}$$

where z is in cm. The ordinary frequency f is 2 GHz. (1 GHz = 10^9 Hz.)
(a) Find the wavelength.
(b) Find the earliest time after $t = 0$ at which the voltage is maximum at $z = 0$.

*1.6 A wave propagates in the $+z$ direction on an ideal transmission line with velocity $0.7c$ and frequency 5 GHz. At $z = 0$ the phasor representing this wave is $1 - 2j$ volts. Find the amplitude and phase angle of the wave at $z = -2$ cm.

*1.7 A sinusoidal wave moves in the $+z$ direction with velocity 1.8×10^{10} cm/sec. An observer stationed at $z = 5$ cm observes voltage maxima of 80 mV (millivolts) at $t = 0.15$ nsec, $t = 0.35$ nsec, $t = 0.55$ nsec, and so forth. Find the phasor representing this wave.

*1.8 Repeat Exercise 1.1 if the problem is changed so that the phase angle of the sinusoid observed at $z = 1$ cm is $-30°$.

*1.9 A sinusoidal wave, with phasor Ae^{-jkz} (where A is a real number) propagates on a transmission line.
(a) An ac voltmeter is used to measure the amplitude of the sinusoidal voltage as a function of position. That is, the voltmeter is connected across the line at various positions and the amplitude (half the peak-to-peak voltage) is measured. Sketch this amplitude as a function of the voltmeter's position.
(b) Now an additional sinusoidal wave with phasor Ae^{jkz} is added, so both waves propagate simultaneously on the line. Show that the amplitude of the sinusoidal voltage is now $2 \cos kz$. Explain this result qualitatively by sketching the phasors representing the two waves at different positions and adding them graphically.

*1.10 An ideal transmission line is terminated in a pure reactance, so that $Z_L = jX$. Prove that regardless of the value of X, *all* incident power is reflected.

*1.11 An ideal transmission line, with characteristic impedance of 50 ohms, works at a frequency of 10 GHz. In an effort to avoid reflections, the line has been terminated in a 50-ohm resistance. However, the load unintentionally contains a series inductance L, as a result of which 5% of the incident power is reflected. Find L.

*1.12 A transmission line with characteristic impedance Z_{01} feeds two transmission lines, with characteristic impedances Z_{02} and Z_{03}, connected in parallel, as shown in Fig. 1.14. Lines 2 and 3 are terminated in their characteristic impedances, and a sinusoidal signal is applied to the input. Find the fractions of the incident power transmitted into each of the two lines, and the fraction reflected.

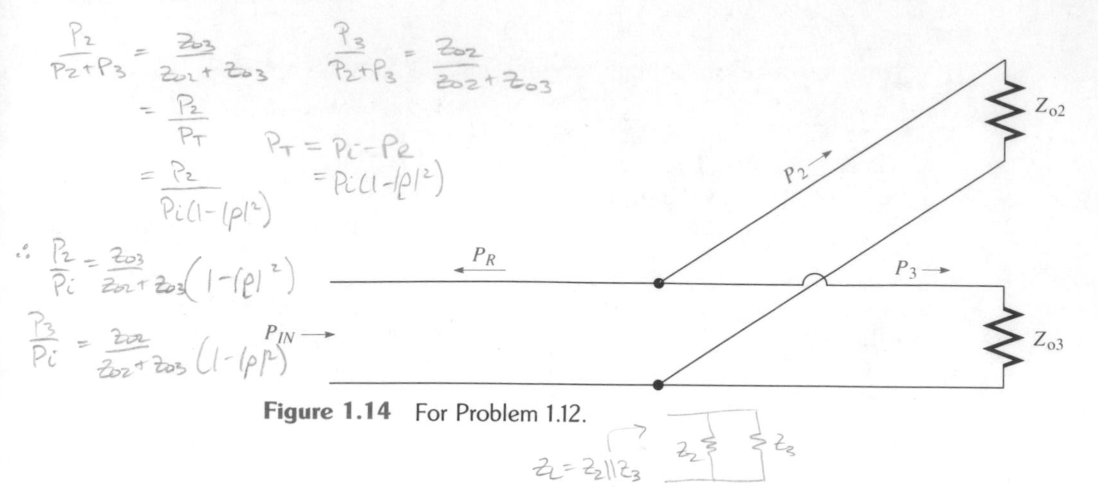

Figure 1.14 For Problem 1.12.

*1.13 The transmission line shown in Fig. 1.8 has characteristic impedance Z_o. Its input is connected to an ideal sinusoidal voltage source, the amplitude of which is $|\mathbf{v}_S|$. The other end of the line is connected to an oscilloscope, which we assume has an input resistance equal to $3Z_o$. The propagation constant k and the length of the line l_1 are related by $kl_1 = 1.75$. Find the amplitude indicated by the oscilloscope.

**1.14 The circuit of Fig. 1.8 is given a more realistic form by adding a source resistance R_S in series with the sinusoidal voltage source $\mathbf{v}_S$. The characteristic impedance of the transmission line is Z_o, and the impedance of the oscilloscope is infinitely large. Find the phasor representing the voltage measured by the oscilloscope as a function of the line length kl_1.

**1.15 An ideal sinusoidal voltage source with amplitude V_o and frequency ω is connected to a loop of coaxial cable as shown in Fig. 1.15. (Note the connections carefully.) The propagation constant of the cable is k,

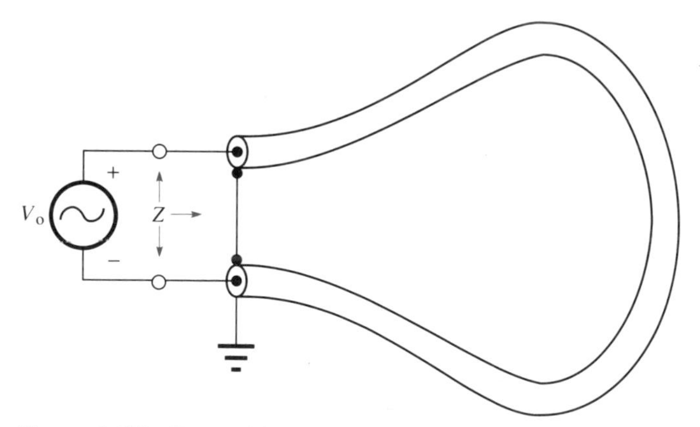

Figure 1.15 For Problem 1.15.

its total length is L, and its characteristic impedance is Z_o. Find the impedance that appears across the voltage source. Check your answer to see if it is reasonable in the limit $kL \to 0$.

Sections 1.5–1.6

**1.16 Often the series resistance and shunt conductance in a transmission line are small, so that $R \ll \omega L$ and $G \ll \omega C$. In this case we may obtain approximate values of α, β, and Z_o correct to first order in the losses. (That is, terms containing R^2, G^2, and RG, and higher powers are discarded.) Show using the binomial expansion that for a low-loss line

$$\alpha \cong \frac{R}{2Z_{oo}} + \frac{GZ_{oo}}{2}$$

$$\beta \cong \omega\sqrt{LC} \tag{1.60}$$

$$Z_o \cong Z_{oo}\left[1 + j\left(\frac{G}{2\omega C} - \frac{R}{2\omega L}\right)\right]$$

where $Z_{oo} = \sqrt{L/C}$.

1.17 A transmission line has attenuation constant α m^{-1}. By how many decibels does the power decrease for each meter of propagation?

*1.18 Suppose the resistance per unit length in a transmission line has a *small* value R', so that $\alpha l \ll 1$, where l is the length of the line. In that case, a good approximation to P_R, the time-averaged power lost in R', can be found by assuming that the current is the same as if the line were lossless. The loss is then found by integrating i^2R' over the length of the line:

$$P_R \cong \text{time avg}\left[R'\int_0^l i^2(z,t)\,dz\right] \tag{1.61}$$

A wave with complex amplitude V_i travels in the $+z$ direction. Use (1.61) to show that the approximate power lost in a length l of the line is

$$P_R \cong \frac{R'|V_i|^2 l}{2Z_o^2} \tag{1.62}$$

**1.19 Another way of calculating power loss is through the use of (1.50) and (1.60) (Problem 1.16). If the power entering a line of length l is P_{in}, the power leaving is $P_{\text{out}} = P_{\text{in}}e^{-2\alpha l}$. (Can you prove this?) The power dissipated in the line is then $P_R = P_{\text{in}} - P_{\text{out}}$. Show that this method agrees with (1.62) if $\alpha l \ll 1$.

1.20 From Eq. (1.48) we see that k can have two different values, corresponding to the two possible signs of the square root. How do these solutions differ physically?

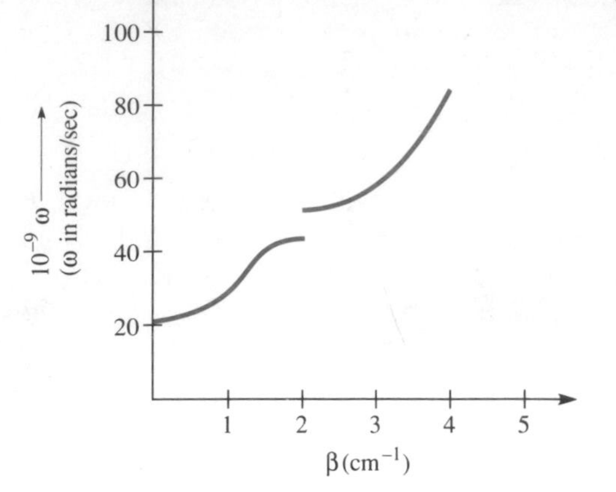

Figure 1.16 For Problem 1.21.

1.21 For a certain transmission line, the ω–β diagram is as shown in Fig. 1.16. Sketch the phase velocity and group velocity, as functions of frequency, for $\omega > 20 \times 10^9$ rps.

*1.22 For a certain unusual transmission line the propagation constant is given by

$$k = \frac{\omega}{U_o}\left(1 - \frac{\omega_o^2}{\omega^2}\right)^{1/2}$$

where U_o and ω_o are constants.

(a) Sketch α and β as functions of frequency, assuming that waves move in the $+z$ direction.

(b) For $\omega > \omega_o$, sketch U_P and U_G as functions of frequency.

(c) For $\omega < \omega_o$, find a mathematical expression for $v(z, t)$ using (1.50) and (1.19). (The constant A is arbitrary.)

(d) Using the result of part (c), sketch $v(z, t)$ as a function of z, for several fixed values of t. Is this a "wave" in the usual sense?

**1.23 Show that the group and phase velocities are related by

$$U_G = \frac{d}{d\beta}(\beta U_P) = \frac{U_P}{1 - \dfrac{\omega}{U_P}\dfrac{dU_P}{d\omega}} \tag{1.63}$$

*1.24 A wave propagating on a lossy transmission line has the form

$$v(z, t) = Ae^{-\alpha z}\cos(\omega t - \beta z)$$

where A, α, and β are positive real numbers.

(a) Show that at $t = 0$, the maximum closest to the origin is located at $z = -\frac{1}{\beta}\tan^{-1}\left(\frac{\alpha}{\beta}\right)$.

(b) Show that the distance between maxima at any time is $2\pi/\beta$.

(c) Show that the velocity of a maximum is ω/β.

Section 1.7

1.25 At one end of an ideal lossless transmission line of length L is an ideal 1-V voltage source in series with a switch. The other end of the transmission line is short-circuited. The switch is open until $t = 0$, at which time it is closed.

(a) Sketch the voltage at the center of the line as a function of time, for times between 0 and $5L/U$ (where U is the phase velocity).

(b) Repeat the previous problem with the termination changed from a short circuit to an open circuit.

*1.26 In Fig. 1.13, let $R_S = R_L$. The voltage source produces a short rectangular pulse. (Duration of pulse $\ll L/U_P$.) Sketch the voltage at the center of the line, as a function of time, for (a) $R_L = 3Z_o$ and (b) $R_L = Z_o/3$.

*1.27 Repeat the preceding problem (parts (*a*) and (*b*)) if the input pulse is infinitely long (i.e., a step function).

**1.28 An infinitely long rectangular voltage pulse is launched on a transmission line as shown in Fig. 1.17. The load at the end of the line is a capacitor C. Write Kirchhoff's current law and solve to find the voltage across C as a function of time. (Note that the reflected voltage pulse is not rectangular.) The capacitor is initially uncharged.

**1.29 In Fig. 1.18, the switch is closed at $t = 0$. The switch then remains closed, producing an infinitely long pulse. The transmission line, of length l, terminates in an open circuit. The time it takes for a wave to

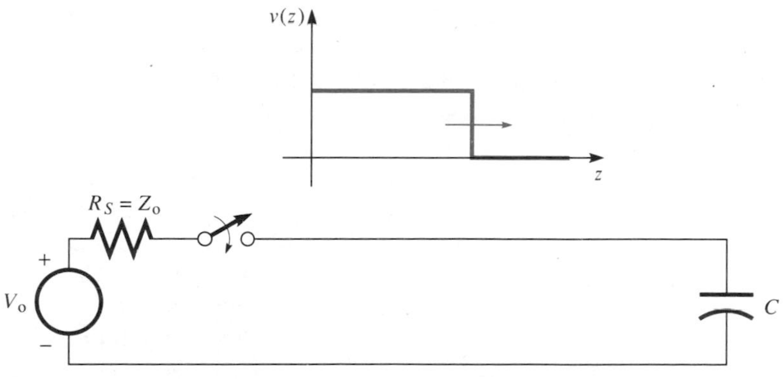

Figure 1.17 For Problem 1.28.

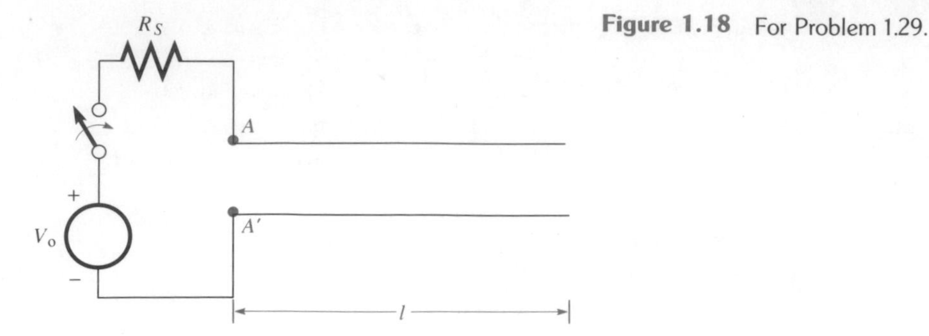

Figure 1.18 For Problem 1.29.

go from left to right and back to left on the line is T. Assume that $R_S \gg Z_o$.

(a) Find the voltage at AA' after a time NT has passed, where N is an integer. (Use the algebraic identity

$$1 + x + x^2 + \cdots + x^{n-1} = \frac{1 - x^n}{1 - x}$$

which may be checked by long division.)

(b) Find the voltage at AA' after a very long time has passed.

(c) Sketch the voltage at AA' as a function of time.

(d) Show that the "time constant" for charging the line is simply $R_S Cl$, where Cl is the total capacitance of the line.

*1.30 At time $t = 0$, the voltage on a lossless transmission line is zero everywhere, and the current is

$$\begin{aligned} i(t = 0) &= I_o \qquad \text{for } -10 \text{ cm} < z < 10 \text{ cm} \\ &= 0 \qquad \text{elsewhere} \end{aligned}$$

The characteristic impedance of the line is 50 ohms, and its phase velocity is $0.67c$; let $I_o = 10$ mA. Make a sketch of the line voltage versus position at times $t = 10^{-10}$ sec and $t = 7.5 \times 10^{-10}$ sec.

**1.31 (a) The power $P(z, t)$ transmitted by a transmission line is equal to the product of $W(z, t)$, the energy stored per unit length, and U_E, the velocity at which energy flows down the line. Show that this statement is physically reasonable. (Note, however, that U_E could be different from U_P or U_G.)

(b) In circuit theory it is shown that the energy stored in a capacitor charged to a voltage v is $\frac{1}{2}Cv^2$; also that the energy stored in an inductor is $\frac{1}{2}Li^2$. Assume that a wave (not necessarily sinusoidal) travels in the $+z$ direction on a lossless line with characteristic impedance Z_o. Let the energy per unit length stored in the capacitors be $W_C(z, t)$ and that stored in the inductors be $W_L(z, t)$. (Hence $W = W_C + W_L$.) Show that $W_C = W_L$.

(c) Using the results of parts (a) and (b), show that $U_E = U_P$.

CHAPTER 2

Transmission-Line Techniques

In the previous chapter, a variety of transmission-line waves were considered. Both sinusoidal and non-sinusoidal waves were discussed, on lines that were either ideal or lossy. Having introduced the general types of waves, we shall now consider some practical aspects of transmission-line technology. To avoid dealing with a large number of different cases, we shall confine our attention now to the case of greatest general interest, that of the sinusoidal steady state. Thus our main mathematical tool will be phasor analysis. For simplicity, we shall also confine our attention primarily to the case of lossless transmission lines.

As we have seen, a transmission line can be used to transfer power into a resistive load. If the resistance of this load is equal to the characteristic impedance of the line, no power is reflected by the load; all the power is then absorbed. This is usually what is wanted, since in most cases the goal is to transfer power into the load efficiently. In practice, however, lines are not always terminated by well-matched loads, and we must see what can be done to improve efficiency under such conditions. This brings us to the general subjects known as "impedance transformation" and "impedance matching." Then we shall go on to describe several types of transmission lines used in actual practice. Finally, we shall introduce the subject of transmission-line measurements, which are necessary to put the theory to practical use.

2.1 STANDING-WAVE RATIO

Suppose that we launch on an ideal, lossless transmission line, a wave $\mathbf{v}^+(z) = V_i e^{-jkz}$ moving to the right. Let this wave be incident on a load located at $z = 0$, so that there is a reflected wave, $\mathbf{v}^-(z) = V_r e^{jkz}$, returning to the left. The total phasor voltage, as a function of position, is

$$\mathbf{v}(z) = V_i e^{-jkz} + V_r e^{jkz}$$

Clearly $\mathbf{v}^+(z = 0) = V_i$ and $\mathbf{v}^-(z = 0) = V_r$. Using the definition of the reflection coefficient (1.33) we have

$$V_r = \mathbf{v}^-(z = 0) = \rho_o \mathbf{v}^+(z = 0) = \rho_o V_i$$

Thus

$$\mathbf{v}(z) = V_i(e^{-jkz} + \rho_o e^{jkz}) \tag{2.1}$$

Here ρ_o, the reflection coefficient of the load, is a complex number which can be expressed in the form

$$\rho_o = |\rho_o| e^{j\phi_R} \tag{2.2}$$

The amplitude of the voltage at any point is obtained from the general expression $|\mathbf{v}| = \sqrt{\mathbf{v}\mathbf{v}^*}$:

$$\begin{aligned} |\mathbf{v}(z)| &= |V_i| \, [1 + \rho_o e^{2jkz} + \rho_o^* e^{-2jkz} + |\rho_o|^2]^{1/2} \\ &= |V_i| \, [1 + |\rho_o|^2 + 2 \operatorname{Re} (\rho_o e^{2jkz})]^{1/2} \\ &= |V_i| \, [1 + |\rho_o|^2 + 2|\rho_o| \operatorname{Re} (e^{2jkz + j\phi_R})]^{1/2} \\ &= |V_i| \, [1 + |\rho_o|^2 + 2|\rho_o| \cos (2kz + \phi_R)]^{1/2} \end{aligned} \tag{2.3}$$

The dependence of $|\mathbf{v}|$ on position is illustrated in Fig. 2.1, for the case of $|V_i| = 1$ V, $|\rho_o| = 0.5$, $\phi_R = 45°$. The physical meaning of Fig. 2.1 can be understood in terms of the interference of the two waves. At some positions, the voltages of the incident and reflected waves add in phase, increasing the total voltage at those points. At other places, the interference between the two waves is destructive, and they tend to cancel each other. At any position, the voltage on the line is a sinusoidal function of time, with the amplitude given in the figure. If one moves along the line, one sees the amplitude regularly increase and decrease, as the cosine function in (2.3) varies. This variation of the amplitude should not be confused with the traveling waves. The positions of the voltage maxima and minima are *stationary*. Since the amplitude envelope stands still, the phenomenon is referred to as a *standing wave*. Note that in the special case $\rho_o = 0$, the reflected wave vanishes. In that case there is only a single traveling wave traveling to the right, and the interference between incident and reflected waves disappears. The amplitude then becomes independent of position.

Let us now determine the maximum voltage that appears anywhere on the line. From (2.3) it is evident that the amplitude is largest when $2kz + \phi_R = 2n\pi$ (where n is an integer). At these positions,

$$\begin{aligned} |\mathbf{v}|_{\max} &= |V_i| \, (1 + |\rho_o|^2 + 2|\rho_o|)^{1/2} \\ &= |V_i| \, (1 + |\rho_o|) \end{aligned} \tag{2.4}$$

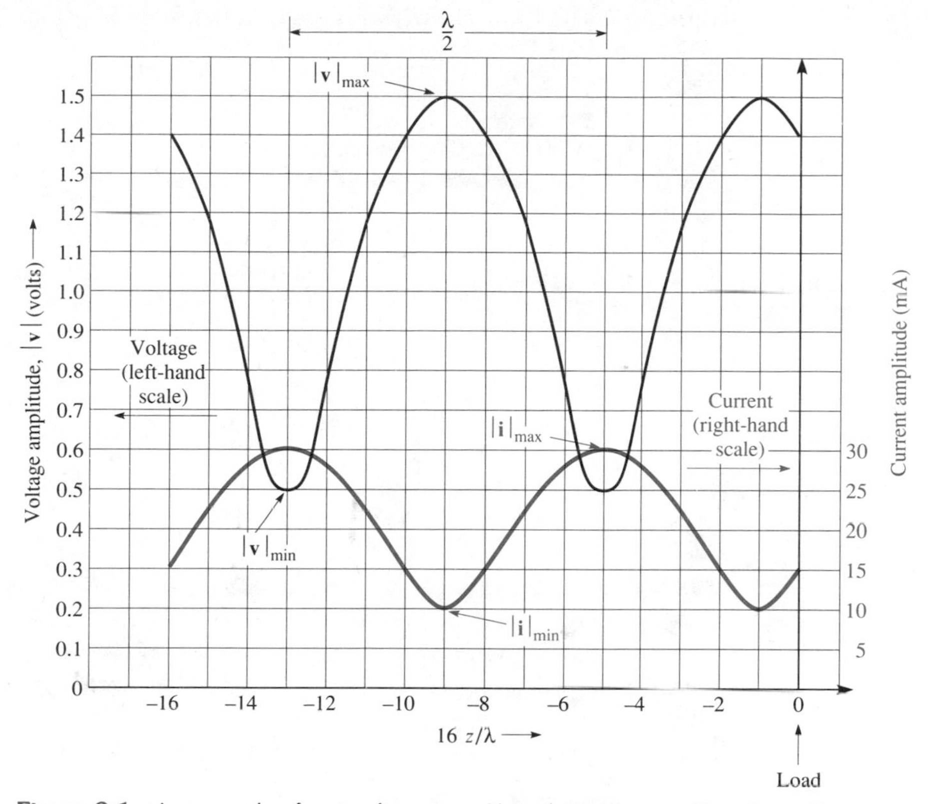

Figure 2.1 An example of a standing wave. Note that at any position, the voltage is a sinusoidal function of time. What is plotted here is the *amplitude* of the sinusoidal voltage at each position. The amplitude distribution shown here does not move; hence the name "standing wave."

Similarly, the minimum voltage occurs when $2kz + \phi_R = (2n + 1)\pi$:

$$|\mathbf{v}|_{\min} = |V_i|\,(1 - |\rho_o|) \tag{2.5}$$

The *standing-wave ratio* (often abbreviated SWR) is defined as the ratio of maximum to minimum voltages:

$$\text{SWR} = \frac{|\mathbf{v}|_{\max}}{|\mathbf{v}|_{\min}}$$

$$= \frac{1 + |\rho_o|}{1 - |\rho_o|} \tag{2.6}$$

It is clear that the SWR is always greater than or equal to 1. Larger values of SWR indicate larger values of $|\rho_o|$, or in other words, a larger reflected wave. The case of SWR = 1 occurs when the line is terminated in its characteristic impedance, and there is no reflected wave.

From (2.3) we note that $|\mathbf{v}|$ is a periodic function of position on the line. Maximum voltage appears whenever

$$2kz + \phi_R = 2n\pi \tag{2.7}$$

where n is an integer. Voltage minima occur when $2kz + \phi_R = (2n + 1)\pi$. The distance Δz between two adjacent maxima (n changes by one) is found as follows:

$$2kz_{n+1} + \phi_R = 2(n + 1)\pi$$

$$2kz_n + \phi_R = 2(n)\pi$$

Subtracting the second equation from the first,

$$\Delta z = z_{n+1} - z_n = \frac{\pi}{k} = \frac{\pi}{\left(\dfrac{2\pi}{\lambda}\right)}$$

$$= \frac{\lambda}{2} \tag{2.8}$$

Thus voltage maxima (and also voltage minima) repeat every *half* wavelength along the line, and not every full wavelength, as one might expect.

It is easily shown (Problem 2.4) that positions of maximum total voltage are the same as the positions of *minimum* total current, and vice versa. The total current amplitude $|\mathbf{i}(z)|$ is shown along with the voltage in Fig. 2.1. (The characteristic impedance of the line is assumed to be 50 ohms.)

EXAMPLE 2.1

A transmission line is terminated in a pure inductance, which presents a load impedance jZ_o. Find the distance from the load to the nearest voltage maximum. The frequency is 1 GHz and the phase velocity is 0.67c.

Solution From (1.33)

$$\rho_o = \frac{jZ_o - Z_o}{jZ_o + Z_o} = j = (1)e^{j(\pi/2)}$$

Thus $\phi_R = \pi/2$ radians. From (2.7), maxima occur when

$$z = \frac{\pi}{k}\left(n - \frac{1}{4}\right)$$

We note that the value of z we are seeking is a *negative* number, because the position of the load has been taken as $z = 0$ and the incident wave is coming from the left. Maxima occur at $-(\pi/4k)$, $-(5\pi/4k)$, and so on. The maximum closest to the load is the one corresponding to $n = 0$:

$$z_{\max} = -\frac{\pi}{4k} = -\frac{U_P}{8f} = -2.5 \text{ cm}$$

EXERCISE 2.1

Find the position of the voltage maximum closest to the load when the load is a pure capacitance with impedance $-jZ_o$. The frequency is 1 GHz and the wave velocity is 0.67c.

Answer −7.5 cm.

2.2 IMPEDANCE

We have already seen that the input impedance of a reflection-free line is simply Z_o. However, this will no longer be true if a reflected wave is present. Let us suppose that a load is located at $z = 0$, and we wish to find the input impedance at the point z (where z is a negative number). This impedance is

$$Z(z) = \frac{\mathbf{v}(z)}{\mathbf{i}(z)} = \frac{\mathbf{v}^+(z) + \mathbf{v}^-(z)}{\mathbf{i}^+(z) + \mathbf{i}^-(z)} \tag{2.9}$$

where $\mathbf{v}^+$ and $\mathbf{i}^+$ are the voltage and current associated with the incident wave (moving in the $+z$ direction) and $\mathbf{v}^-$ and $\mathbf{i}^-$ are associated with the reflected wave (moving in the $-z$ direction). Note the difference between the quantities $Z(z)$ and Z_o. The symbol $Z(z)$ represents the ratio of the voltage phasor to the current phasor at the position z. The characteristic impedance Z_o, on the other hand, is a constant number that depends only on the dimensions and construction of the transmission line.

From (1.20), (1.25), and (1.33).

$$\begin{aligned}
\mathbf{v}^+(z) &= V_i e^{-jkz} \\
\mathbf{i}^+(z) &= \mathbf{v}^+(z)/Z_o \\
\mathbf{v}^-(z) &= \rho_o V_i e^{jkz} \\
\mathbf{i}^-(z) &= -\mathbf{v}^-(z)/Z_o
\end{aligned} \tag{2.10}$$

where V_i is the complex amplitude of the incident wave. Substituting (2.10) into (2.9) we find that

$$Z(z) = Z_o \frac{e^{-jkz} + \rho_o e^{jkz}}{e^{-jkz} - \rho_o e^{jkz}} = Z_o \frac{1 + \rho_o e^{2jkz}}{1 - \rho_o e^{2jkz}} \tag{2.11}$$

This can be written in another form by substituting expression (1.33) for ρ_o. Since we are interested in finding the input impedance a distance l before the load, we also substitute $l = -z$, obtaining

$$Z(l) = Z_o \frac{Z_L \cos kl + jZ_o \sin kl}{Z_o \cos kl + jZ_L \sin kl} \tag{2.12}$$

EXAMPLE 2.2

Find the input impedance at a distance l from (a) a short-circuit load and (b) an open-circuit load.

Solution For the short-circuit load, $Z_L = 0$ and from (2.12)

$$Z(l) = jZ_o \tan kl \tag{2.13}$$

while for the open-circuit load $Z_L = \infty$ and

$$Z(l) = -jZ_o \cot kl \tag{2.14}$$

We note that in both these cases the input impedance can have any absolute value between $-\infty$ and $+\infty$, but is always purely imaginary. Physically this must be the case, since if Z had any real part, it would be possible for the line to accept power; but with an open- or short-circuit load and a lossless line, there is no place such power could be dissipated. Hence the input impedance must be purely imaginary.

With short- or open-circuit loads, $|\rho_o| = 1$, which means that the current associated with the reflected wave is as large as that of the incident wave. At current minima these two currents cancel completely and the total current is zero. The voltage at that point is maximum, and the impedance there is

$$|Z_{\text{max}}| = \frac{|\mathbf{v}_{\text{max}}|}{(0)} = \infty$$

Similarly at voltage minima the voltage associated with the two waves cancel, the total current is maximum, and

$$|Z_{\text{min}}| = \frac{(0)}{|\mathbf{i}_{\text{max}}|} = 0$$

EXERCISE 2.2

Show that regardless of what load is connected, the input impedance is periodic with period $\lambda/2$. (That is, show that $Z(l) = Z(l + \lambda/2)$.)

2.3 THE SMITH CHART

Expression (2.12) is an important result, but it is rather clumsy to use. Fortunately there is a convenient graphical technique that can be used in place of (2.12). This technique is based on a special diagram which has several names, but is most often called (after a paper published in 1939 by P. H. Smith) the Smith chart. Not only is the Smith chart a useful tool for computation; it also provides a valuable intuitive feeling for impedance variations on a transmission line.

First let us define a general reflection coefficient, $\rho(z)$, which is equal to the ratio of the reflected to the incident voltage phasors at any position z on the line:

$$\begin{aligned}\rho(z) &= \frac{\mathbf{v}^-(z)}{\mathbf{v}^+(z)} \\ &= \frac{\mathbf{v}^-(z=0)e^{jkz}}{\mathbf{v}^+(z=0)e^{-jkz}} \\ &= \rho_o e^{2jkz}\end{aligned} \tag{2.15}$$

where ρ_o, defined by

$$\rho_o \equiv \rho(z=0) \equiv \frac{\mathbf{v}^-(z=0)}{\mathbf{v}^+(z=0)} \tag{2.16}$$

is the reflection coefficient of the load. Alternatively, in terms of l, where $l = -z$ is the distance (a positive number) between the point of observation and the load, we have

$$\rho(l) = \rho_o e^{-2jkl} \tag{2.17}$$

We can now display $\rho(z)$ in the complex plane, as shown in Fig. 2.2. Since we expect $|\rho|$ to be less than or equal to one, the vector representing

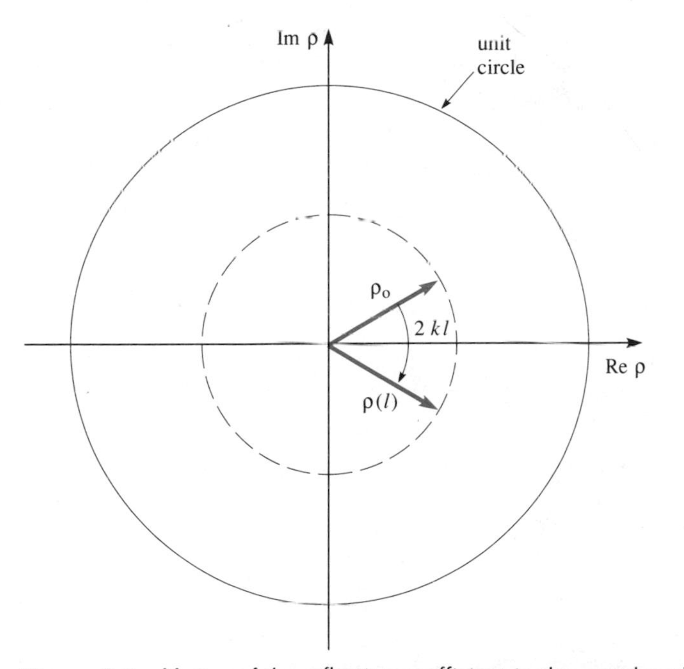

Figure 2.2 Motion of the reflection coefficient in the complex plane. At the load $\rho = \rho_o$. At a position a distance l before the load the reflection coefficient has the same magnitude, but is rotated clockwise through an angle $2kl$.

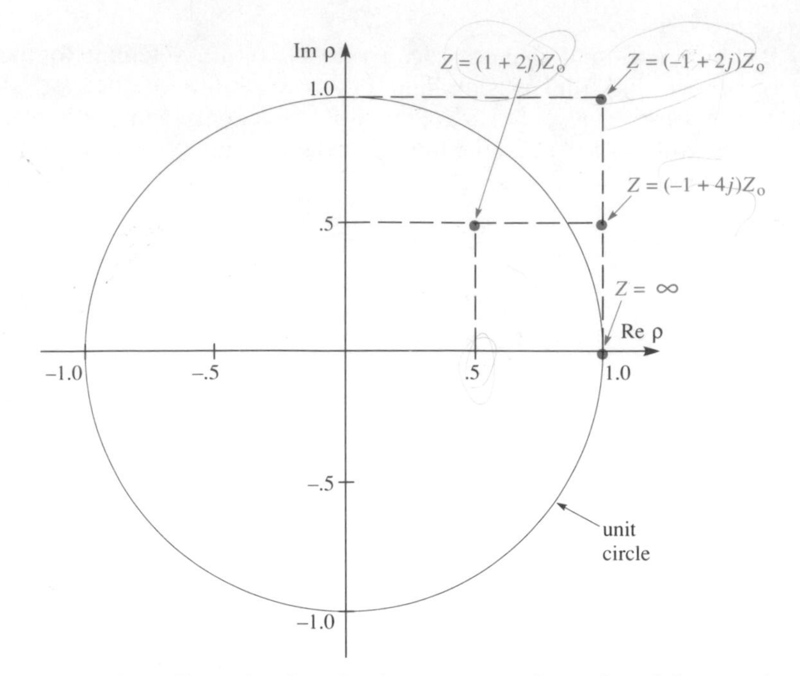

Figure 2.3 To each value of ρ there corresponds a value of the impedance Z.

$\rho(z)$ will lie inside the unit circle.[1] Let us assume that ρ_o, the load's reflection coefficient, is as shown in the figure. Now from (2.15) we see that the reflection coefficient at any position is found by simply rotating the arrow representing ρ_o through the proper angle. As l increases (i.e., as one moves farther from the load) $\rho(z)$ rotates *clockwise* in the complex plane. The angle between ρ_o and $\rho(l)$ is $2kl$. The length of the vector $\rho(l)$ remains constant.

The next step in developing the Smith chart is to find $Z(z)$ in terms of $\rho(z)$. This is easily done. Dividing top and bottom of (2.11) by e^{-jkz} and using (2.15) we find that

$$Z = \frac{1 + \rho}{1 - \rho} Z_o \tag{2.18}$$

According to (2.18), to each value of ρ there corresponds a certain value of Z. This can be represented graphically as shown in Fig. 2.3. Here we have simply indicated the value of Z corresponding to each value of ρ, as given by (2.18). For example, the point $\rho = 0.5 + 0.5j$ corresponds to $Z = (1 + 2j)Z_o$. Since (2.18) is simply a mathematical relationship, it gives a value of Z for any value of ρ, whether it is physically realizable or not. As it happens, two of the points graphed in Fig. 2.3 lie outside the unit circle, and we find that they correspond to values of Z with negative real part. Such

[1]The absolute value of the reflection coefficient will be less than or equal to one for any *passive* load; that is, for any load incapable of amplification. If the load were a transistor, we might have $|\rho| > 1$.

input impedances can never be obtained with loads composed of passive circuit elements. On the other hand, reflection coefficients lying inside the unit circle always correspond to values of impedance with positive real part.

Although Fig. 2.3 is a graphical representation of (2.18), it is a clumsy one, since it is impossible to note the values of Z for every point in the plane. Instead we can use the method of Fig. 2.4. Let us separate Z into its real and imaginary parts according to

$$Z = R + jX \tag{2.19}$$

Then in Fig. 2.4 we draw lines that are contours of constant R or constant X. It is customary for all values on the Smith chart to be given in units of Z_o. Thus all points on the curve $R = 3$ have impedances whose real part is $3Z_o$. The ratios R/Z_o, X/Z_o, and Z/Z_o are called the *normalized* resistance, reactance, and impedance, respectively. In this book we shall indicate normalized quantities with a "prime" for the sake of clarity; for example, R/Z_o

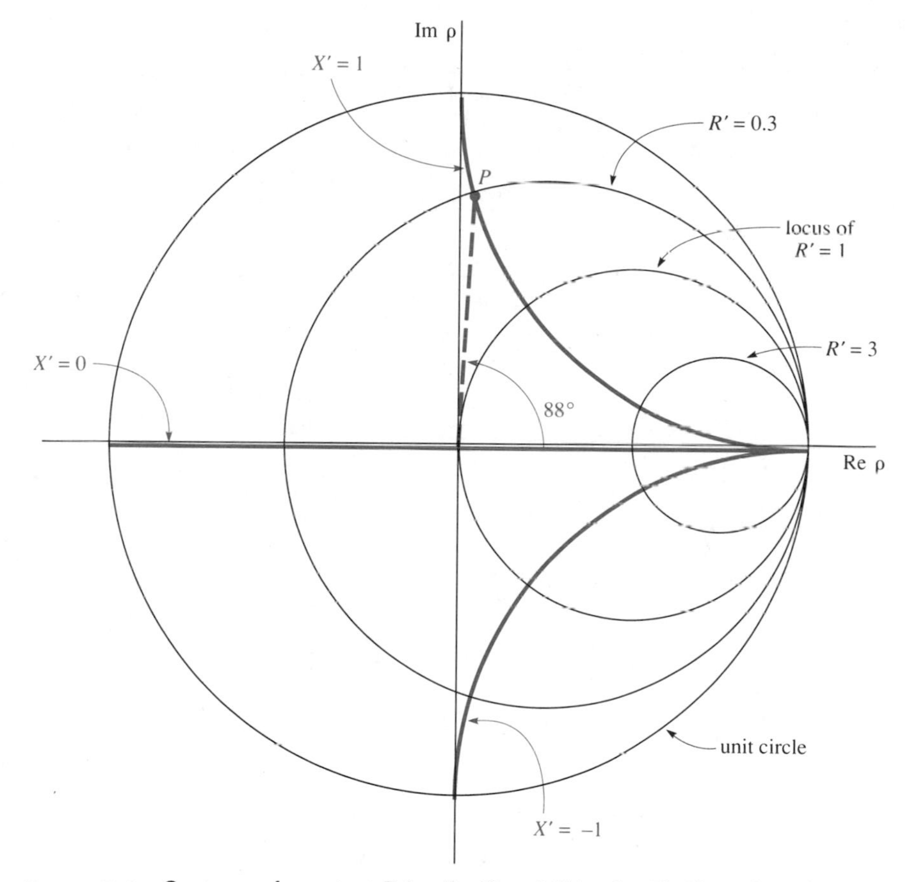

Figure 2.4 Contours of constant R (= Re Z) and X (= Im Z). The value of Z corresponding to any value of ρ can be found by locating the point representing ρ and interpolating between these contours.

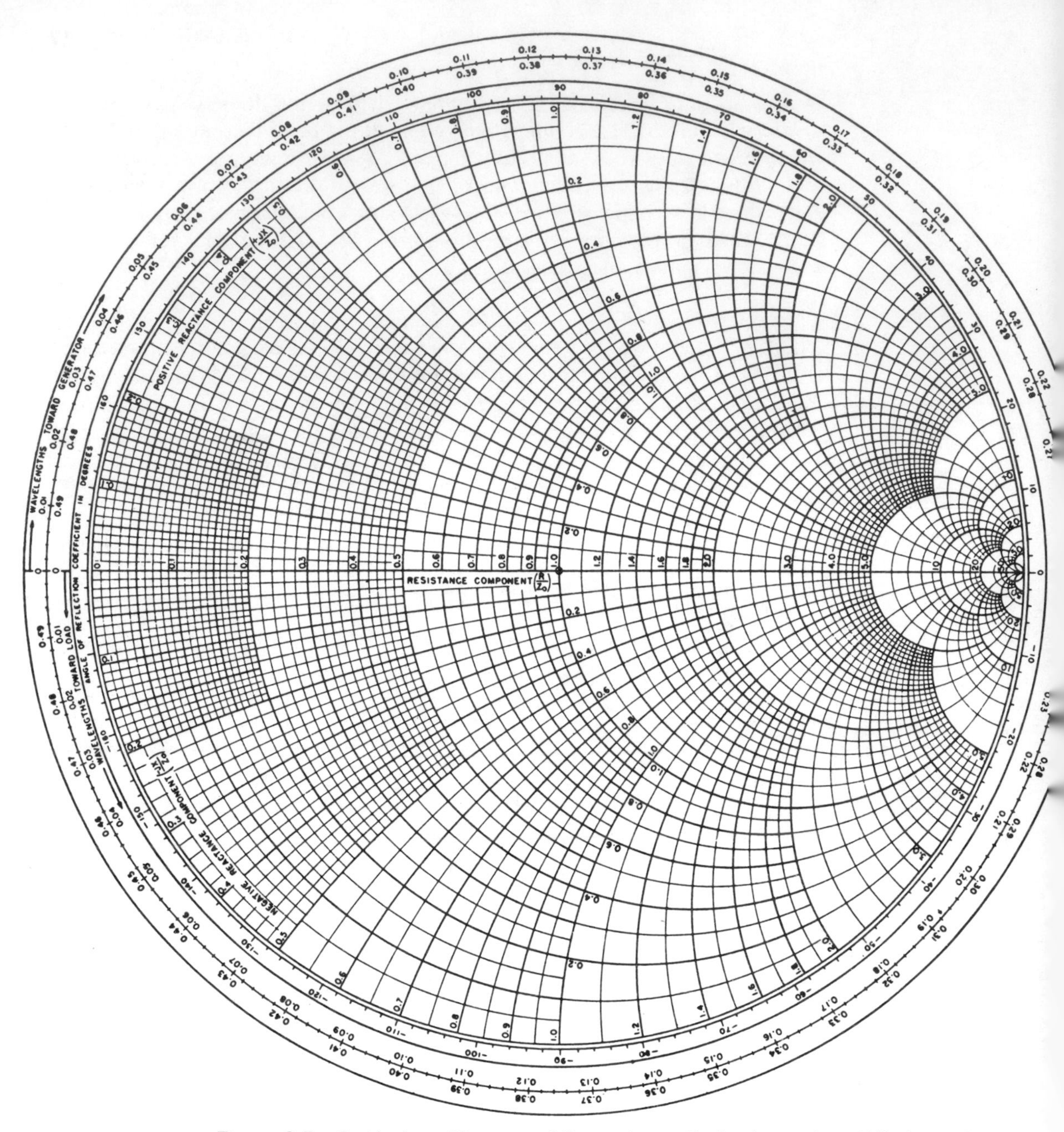

Figure 2.5 Smith chart. [Courtesy of Secure Image Technologies, Inc., Hillside, N.J.]

will be called R'. However, in the working world and on Smith charts the prime is usually omitted.

To find the value of ρ corresponding to a certain value of Z, we simply locate the point on the Smith chart corresponding to Z's value of R' and X'. The x and y coordinates of that point then give the real and imaginary parts of ρ. For example, let us find the value of ρ corresponding to $Z =$

$(0.3 + j)Z_o$. This value is located where the contour $R' = 0.3$ crosses the contour $X' = 1.0$; it is point P in Fig. 2.4. Instead of reading off the x and y coordinates of P, it is usually easier to read off its polar coordinates, obtaining ρ in the form $|\rho|e^{j\theta}$. From Fig. 2.4 we determine that $\theta = 88°$, while $|\rho| = 0.74$. (To find the value of $|\rho|$, we measure the distance from the origin to point P with a ruler. We then divide this by the measured radius of the unit circle, and the result is $|\rho|$.)

To find values that lie between the constant-R' and constant-X' curves, one must interpolate. This is much easier to do if more constant-R' and constant-X' curves are provided than the few given in Fig. 2.4. When the constant-R' and constant-X' curves are provided in detail, we arrive at the full Smith chart—Fig. 2.5.

EXAMPLE 2.3

The impedance at the input of a certain lossless transmission line (characteristic impedance, 72 ohms) is measured and found to be $45 - 100j$ ohms. What is the standing-wave ratio on this line?

Solution The point on the Smith chart representing the input impedance is shown in Fig. 2.6. (Remember that on the Smith chart, impedances are normalized to the characteristic impedance. Thus we have located the point $R' = 45/72 = 0.63$, $X' = -100/72 = -1.39$.) We now find the absolute value of the reflection coefficient. This is done by dividing the distance from the

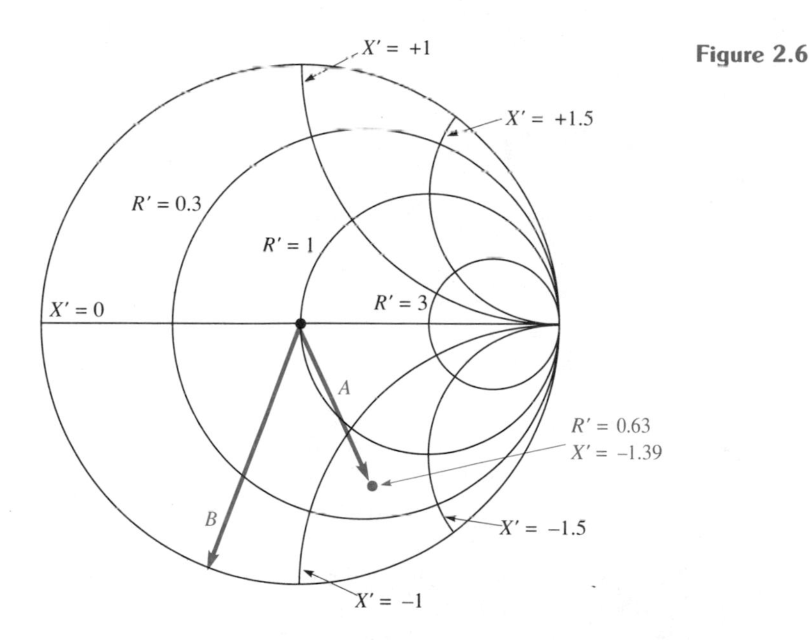

Figure 2.6

origin to the point just located (A on the diagram) by the radius of the Smith chart (B on the diagram); we find that $|\rho| = 0.66$. From (2.17) we know that the absolute value of ρ is the same everywhere on a lossless line; hence $|\rho| = |\rho_o|$, and from (2.6) we find that

$$\text{SWR} = \frac{1 + 0.66}{1 - 0.66} = 4.9$$

EXAMPLE 2.4

Find the impedance at a point 0.4λ distant from a load of impedance $(1 + j)Z_o$.

Solution Using the Smith chart, we first locate the value of ρ_o corresponding to the impedance $(1 + j)Z_o$ as Point A in Fig. 2.7. We then rotate the vector from the origin to ρ_o clockwise to find $\rho(l)$. The angle through which we rotate is, from (2.17)

$$2kl = 2(2\pi/\lambda)l = 4\pi l/\lambda$$

We note that when $l = \lambda/2$, the angle is 2π; thus *a distance of* $\lambda/2$ *corresponds to a full revolution around the Smith chart*. In this example the distance l is 0.4λ, or, equivalently, $0.8(\lambda/2)$. Thus the angle of rotation is $0.8 \times 360°$, or $288°$. The impedance at the point $\rho(l)$ is then read off the Smith chart at Point B and found to be $(0.43 + 0.34j)Z_o$.

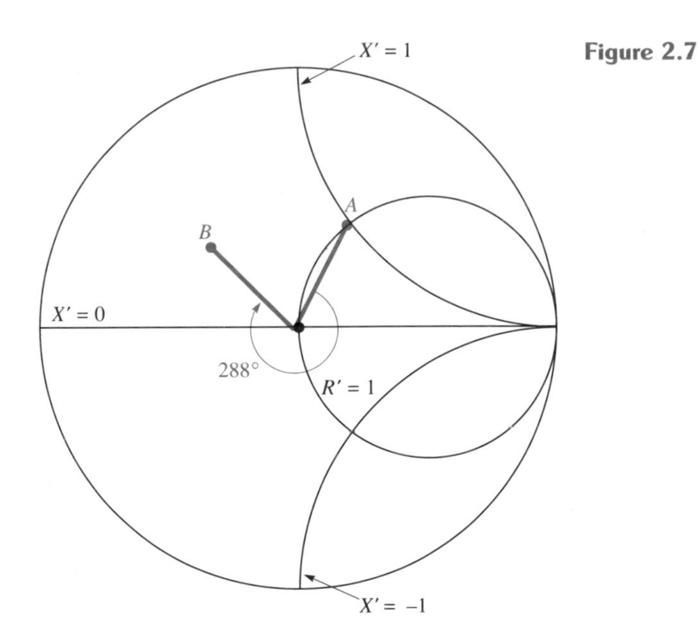

Figure 2.7

As we have seen, the impedance is a periodic function of position along the line, with period $\lambda/2$. As we travel along the line, the point on the Smith chart representing ρ crosses the horizontal axis at distance intervals of $\lambda/4$. At these positions both ρ and the impedance are *real*. Two different real values of impedance are alternately obtained, one larger than Z_o and one less than Z_o, corresponding to $\rho = +|\rho_o|$ and $\rho = -|\rho_o|$. If we give the two real values of Z the names R_{max} and R_{min}, then from (2.18) we see that

$$R_{max} = \frac{1 + |\rho_o|}{1 - |\rho_o|} = \frac{Z_o^2}{R_{min}} \tag{2.20}$$

The positions at which the impedance is real are those that make expression (2.11) a real number. These positions are the same positions already located as being voltage maxima and minima. From (2.7), voltage maxima are located where $2kz + \phi_R = 2n\pi$ (where n is an integer). Substituting this value of kz into (2.11) we obtain $Z = R_{max}$. At voltage minima $2kz + \phi_R = (2n + 1)\pi$, and $Z = R_{min}$. It has already been remarked that positions of maximum voltage are positions of minimum current, and vice versa. The properties of standing waves are summarized in Table 2.1.

Table 2.1 PROPERTIES OF STANDING WAVES

Assuming incident wave with amplitude V_i and reflection coefficient $\rho_o = \lvert\rho_o\rvert e^{j\phi_R}$:	
Maximum voltage:	$\lvert V_i\rvert\,(1 + \lvert\rho_o\rvert)$
Minimum voltage:	$\lvert V_i\rvert\,(1 + \lvert\rho_o\rvert)$
Standing-wave ratio:	$\dfrac{1 + \lvert\rho_o\rvert}{1 - \lvert\rho_o\rvert}$
Positions of voltage maxima:	$2kz + \phi_R = 2n\pi$ (n an integer)
Positions of voltage minima:	$2kz + \phi_R = (2n + 1)\pi$ (n an integer)
Positions of current maxima:	same as voltage minima
Positions of current minima:	same as voltage maxima
Positions at which impedance is real:	at voltage maxima and minima
Impedance at voltage maxima:	$R_{max} = \dfrac{1 + \lvert\rho_o\rvert}{1 - \lvert\rho_o\rvert} Z_o$
Impedance at voltage minima:	$R_{min} = \dfrac{1 - \lvert\rho_o\rvert}{1 + \lvert\rho_o\rvert} Z_o = \dfrac{Z_o^2}{R_{max}}$
$Z(l) = Z_o \dfrac{Z_L \cos kl + jZ_o \sin kl}{Z_o \cos kl + jZ_L \sin kl}$	where l is the distance (a positive number) between observation point and load.

EXERCISE 2.3

A lossless transmission line is terminated in a load with impedance $(1 - 2j)Z_o$. At what distances from the load do we find current maxima?

Answer 0.188λ, 0.688λ, 1.188λ, and so on.

Often it is more convenient to work with the admittance $Y (= 1/Z)$ than with the impedance. Let us define the characteristic admittance $Y_o = 1/Z_o$. The normalized admittance $Y' = Y/Y_o$ is then related to the normalized impedance $Z' = Z/Z_o$ by $Y' = 1/Z'$. In terms of normalized impedance, (2.18) becomes

$$Z' = \frac{1 + \rho}{1 - \rho}$$

We observe that if ρ is replaced by $-\rho$, Z' is replaced by $1/Z'$, that is by Y'. In other words, to find the normalized admittance at any point on the line, we first locate the point on the Smith chart representing ρ in the usual way. We then find the point $-\rho$ by reflecting through the origin. Finally we read off the real and imaginary contours at the point $-\rho$: The complex number thus obtained is Y'.

EXAMPLE 2.5

Find the admittance at a distance of $\lambda/16$ from a load with normalized admittance $(0.5 + 2j)$.

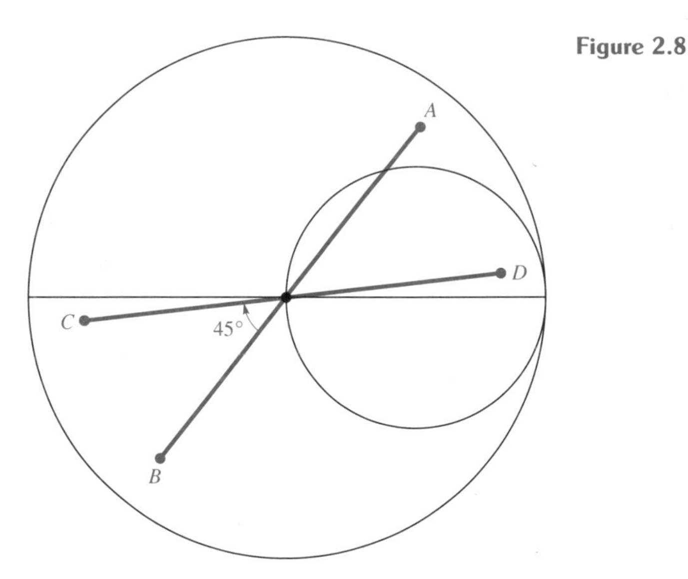

Figure 2.8

Solution Referring to Fig. 2.8, the load admittance is at point A. The load impedance is at point B. The impedance at $\lambda/16$ before the load is found by rotating point B clockwise 45° to point C. Finally the admittance at that point is found by reflecting point C through the origin to obtain point D. Reading the contours at point D gives us the desired normalized admittance, which is approximately $(8.2 + 4.1j)$.

Evidently the above procedure could be simplified. It is sufficient to simply rotate A, the load admittance point, 45° clockwise to obtain the desired admittance point D.

2.4 IMPEDANCE-MATCHING TECHNIQUES

Often it is necessary to adjust the input impedance of a transmission system to some desired value. For example, suppose a certain radio transmitter is designed to deliver radio-frequency power into a 50-ohm purely-resistive load. If a different impedance is presented to the transmitter's output terminals, it will not work properly and may even be damaged. In practice, however, the output of a transmitter is usually connected to a transmission line, at the end of which is an antenna. Now if the impedance of the antenna happens to be 50 ohms (real), there is no problem. A transmission line with characteristic impedance of 50 ohms, of any length, can be used. However, what if the impedance of the antenna is 100 ohms? In that case there is no length of 50-ohm line that will give a real input impedance of 50 ohms, and we must consider how the required input impedance can be obtained.

In practice it is usually desirable for the SWR to be as near as possible to unity on long transmission lines, because large values of SWR tend to increase power loss in the line (see Problem 2.7). Moreover, if the SWR is not unity, the driving-point impedance of a line in general has a complex value which depends on the length of the line. On the other hand, if the SWR can be made close to unity, the driving-point impedance will be nearly real and equal to Z_o for lines of any length, which is much more convenient. To reduce the SWR on the line to a value near unity, one can place a suitable matching network between the main transmission line and the load. For example, in the case of a 50-ohm transmitter driving a 100-ohm antenna, the network should have the property that when loaded by 100 ohms, it has an input impedance of 50 ohms. A line with $Z_o = 50$ ohms, of any length, can then be used to connect the transmitter to the input of the matching network, and the SWR on this line will be one.

In ordinary circuit analysis, a 100-ohm load can be matched to a 50-ohm source by means of an ideal transformer with a turns ratio of $1:\sqrt{2}$. However, conventional transformers, consisting of two coils wound on a high-permeability core, do not act much like ideal transformers at frequencies above 100 MHz or so. This is because at high frequencies, the capacitances between turns have low impedances and hence carry sizable currents. The problem, then, is to design suitable matching networks using components such as transmission lines, which work well at high frequencies. The design of high-frequency matching networks is a fairly extensive subject, and only a few of the more important techniques can be considered here.

QUARTER-WAVE TRANSFORMER

We have already noted that moving along a line a distance of $\lambda/4$ takes us halfway around the Smith chart. Thus if the reflection coefficient at the load is ρ_1, the reflection coefficient at a point $\lambda/4$ before the load is $\rho_1 e^{-j\pi} = -\rho_1$. If the normalized impedance corresponding to ρ_1 is Z_1', we know from (2.18) that the normalized impedance corresponding to $-\rho_1$ is $1/Z_1'$. Thus a line of length $\lambda/4$, with characteristic impedance Z_o, can be used as a *quarter-wave transformer*. That is, when it is loaded with an impedance Z_1, its normalized load impedance is Z_1/Z_o, its normalized input impedance is Z_o/Z_1, and its non-normalized input impedance is Z_o^2/Z_1.

EXAMPLE 2.6

A transmitter specified to work into a 50-ohm (real) load is to be connected by a transmission line to an antenna, the impedance of which is 100 ohms (real). Design an appropriate quarter-wave transformer.

Solution The input impedance of the quarter-wave section of a transmission line must be 50 ohms. Hence

$$\frac{Z_o^2}{100} = 50$$

$$Z_o = (100 \times 50)^{1/2} = 70.7 \text{ ohms}$$

The characteristic impedance of the line should thus be made equal to the geometric mean of the input and load impedances. If the actual distance between transmitter and antenna is greater than $\lambda/4$, any length of 50-ohm line can be inserted between the transmitter and the transformer, and any length of 100-ohm line between the transformer and the antenna, as shown in Fig. 2.9. The SWR will be unity on both the 50-ohm and 100-ohm lines (although not in the 70.7-ohm line composing the transformer).

Although elegant, the quarter-wave transformer has some limitations as a matching tool. Its main drawback is that with conventional transmission lines, such as coaxial cable, only a few standard values of characteristic impedance are available. (In planar technologies, such as microstrip, this is not a problem, because one can create whatever characteristic impedance

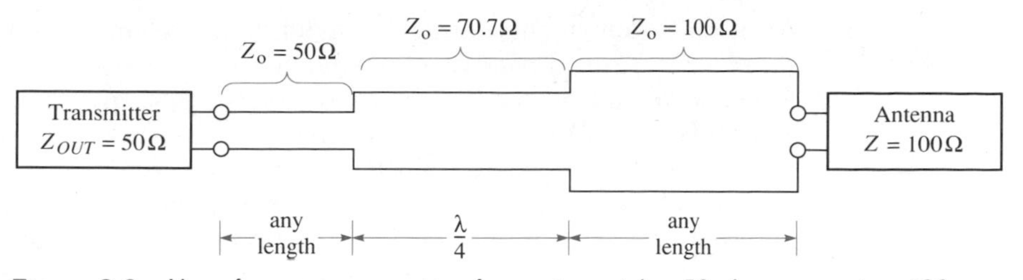

Figure 2.9 Use of a quarter-wave transformer to match a 50-ohm source to a 100-ohm load.

is needed, at least within a certain range.) Another drawback of the quarter-wave transformer is that it is difficult to adjust, for example if the frequency of operation is changed. In that case we would need a section of line of adjustable length, which can be quite inconvenient, mechanically.

It often happens that the load has a complex impedance, while the input of the matching network must present a purely real impedance. In such a case one could insert an additional length of transmission line between source and load. We know that points at which the impedance is real occur at intervals of $\lambda/4$ on the transmission line, so some additional length less than $\lambda/4$ will suffice to provide a real impedance. A quarter-wave transformer can then be used to convert this real impedance to the desired value. However, this would again require insertion of a section with adjustable length. Often it is more convenient to "tune out" the reactance of the load by means of a "backshort," that is, with a short circuit placed beyond the load at an adjustable distance x, as shown in Fig. 2.10. The input admittance of the section of line to the right of the load is purely imaginary and, depending on the value of x, can take any value between $-\infty$ and $+\infty$.[2] Thus x can be adjusted to make the input susceptance of the shorted section equal in magnitude and opposite in sign to the susceptance of the load. In that case the two susceptances cancel and what is left is just the conductance of the load. Once the admittance has been made real, a quarter-wave transformer can

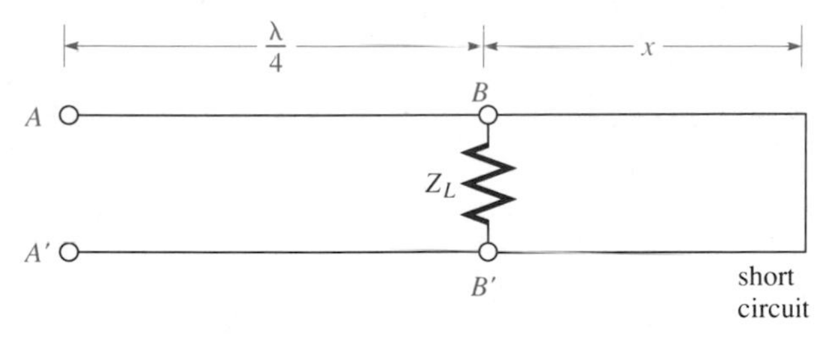

Figure 2.10 Use of a backshort to remove the susceptance of a load.

[2] An infinite range of input reactance is in fact only possible if the line is lossless. However even with normally lossy lines, quite a large range can be obtained.

be used to adjust the normalized admittance to unity. One might wonder whether it is any easier to add an adjustable backshort than it would be to insert a section of line of adjustable length. Actually, this depends on the particular kind of transmission line being used. An important case is that of hollow metal waveguide, which, though not properly a transmission line, does obey the same equations insofar as impedance matching is concerned. In hollow waveguide it is mechanically most convenient to construct short-circuited sections of adjustable length.

EXAMPLE 2.7

In Fig. 2.10, $Z_L = 50 + 50j$ ohms, and the characteristic impedance of all lines is 50 ohms. What must x be, to make the impedance at terminals AA' purely resistive?

Solution The normalized impedance of the load is $1 + j$. Using the Smith chart, we reflect this point through the origin and find that the load admittance is $Y'_L = G'_L + jB'_L = 0.5 - 0.5j$. Thus we wish the admittance looking to the right from BB' to be $+0.5j$. Again using the Smith chart, we locate the point representing the admittance of the short ($G' = \infty$, $B' = \infty$) at the extreme right of the chart.[3] As x increases the admittance looking rightward from BB' moves clockwise on the chart, reaching $+0.5j$ when $x = 0.324\lambda$. With this value of x, the total admittance at BB' will be real. The admittance at AA', $\frac{1}{4}$ wavelength away, must then also be real.

The reader should note that this calculation has to be done with admittances, not with impedances. Choosing x to make the impedance of the shorted section equal to $-j$ leads to an incorrect result. This is because the load and the shorted section are in *parallel;* hence their admittances add, but their impedances do not.

Using a backshort allows us to remove the imaginary part of a load admittance. However, we are still left with the unchanged real part of the load admittance, which may differ from the desired value. Another approach, which allows us to correct both the real and the imaginary parts of the impedance, makes use of a single shorted stub, as shown in Fig. 2.11. Both b, the position of the stub, and d, its length, must be chosen in order to provide the desired real impedance at the position of the stub. This is done as follows. The load admittance (not impedance) is rotated clockwise on the Smith chart until a point is reached whose real part equals the desired value.

[3] A short circuit clearly has $R = 0$, $X = 0$, and thus the point representing its *impedance* is at the extreme *left* of the chart. The point representing its admittance is found by reflecting through the origin.

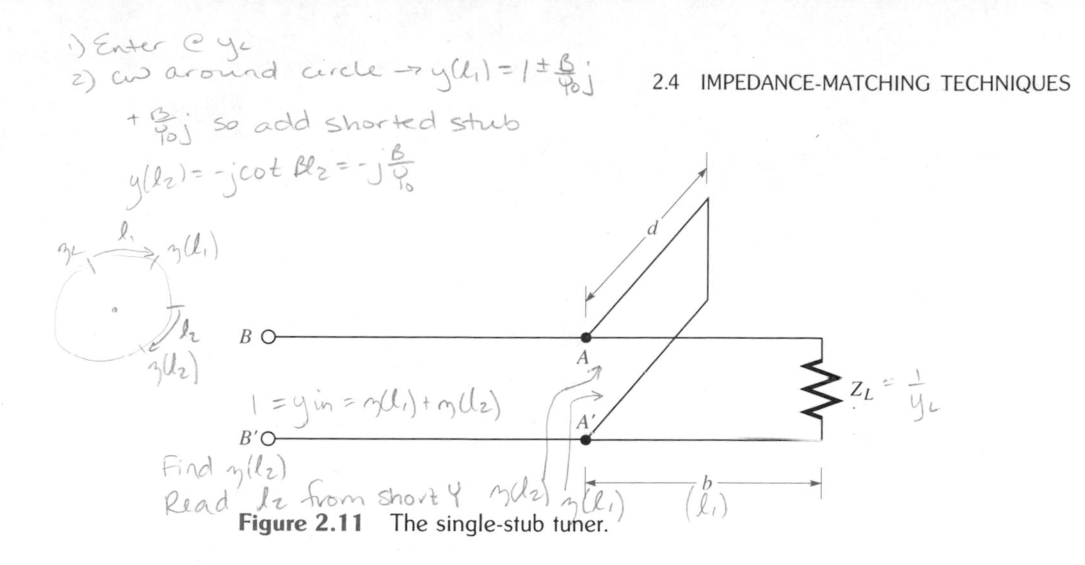

Figure 2.11 The single-stub tuner.

This determines b. The imaginary part of the admittance is then tuned out by adjusting d to provide an equal and opposite susceptance, just as with the backshort.

EXERCISE 2.4

In Fig. 2.11, let the normalized load impedance be $Z_L/Z_o = 2 + j$. Determine the lengths b and d required to make the normalized impedance at AA' equal to unity.

Answer $b = 0.197\lambda$, $d = 0.125\lambda$.

EXAMPLE 2.8

Figure 2.12(a) shows a more sophisticated impedance matching scheme known as a *double-stub tuner*. Here the lengths h_1 and h_2 are fixed, while b and a are to be adjusted in order to produce a normalized impedance of unity at AA'. Let $Z_L/Z_o = 2 + j$, $h_2 = 0.1\lambda$, $h_1 = 0.25\lambda$. Find b and a.

Solution On the Smith chart, Fig. 2.12(b), point L represents the admittance of the load $(0.4 - 0.2j)$ and point B is this admittance after the rotation produced by h_2 $(0.43 + 0.35j)$. The stub connected at BB' can add or subtract any susceptance, and thus moves the admittance point anywhere along the $G' = 0.43$ contour. We wish to move it to a point which, when reflected through the origin (the effect of $h_1 = 0.25\lambda$) will place us on the $G' = 1$ contour. A convenient way to do this is to reflect the $G' = 1$ contour through the origin, resulting in the dashed circle in the figure. Point C, where the dashed curve meets the $G' = 0.43$ contour, is the desired total admittance (load plus stub) looking right from BB'. The normalized admittance at point C is $0.43 + 0.50j$. Going from B to C requires an addition to the admittance

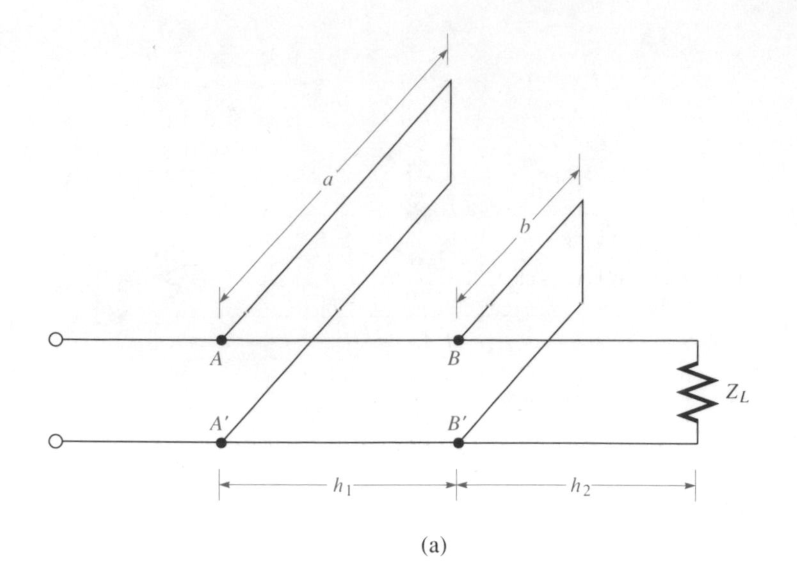

(a)

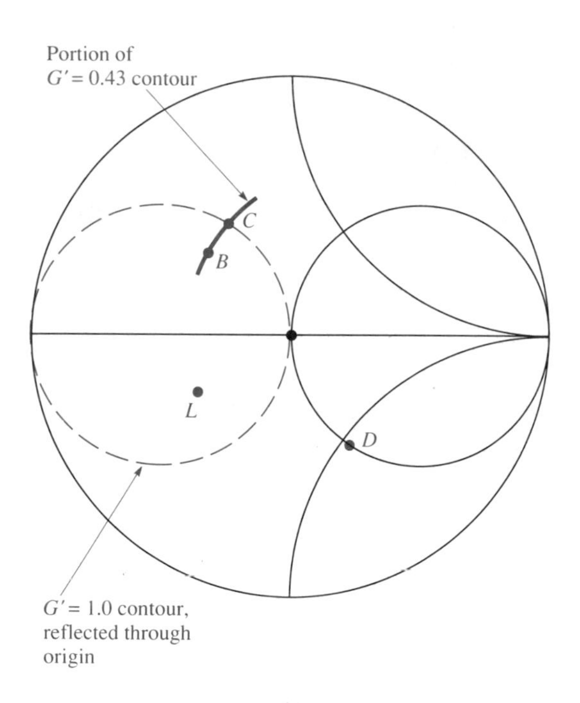

(b)

Figure 2.12 The double-stub tuner: (a) circuit arrangement (b) design technique using Smith chart.

of $0.15j$, which is provided by a stub with length $b = 0.272\lambda$. The quarter-wave section h_1 rotates this point halfway around the Smith chart to point D, where the admittance is $1.00 - 1.15j$. Finally, the susceptance $-1.15j$ must be tuned out by means of the stub of length a, just as in Fig. 2.11. The required parallel susceptance is $+1.15j$, which is provided by the second stub if we make $a = 0.386\lambda$.

The reader may wonder why one would use this double-stub tuner, and not the simpler single-stub tuner of Fig. 2.11. The reason is that in Fig. 2.11 the stub has to be placed at a particular point along the line, which may be inconvenient. A double-stub tuner can be placed at almost any point where the main transmission line can conveniently be interrupted. Coaxial double-stub tuners are available commercially. They consist of the quarter-wave section (h_1) with connectors at each end, and two convenient adjustable-length sections of line for the stubs. These devices are very handy for obtaining unity SWR when driving a partially reactive load.

2.5 PRACTICAL TRANSMISSION LINES

Two important types of transmission line are encountered. The first class includes commercially available cables such as coaxial line and the various parallel-conductor lines. These are used at frequencies ranging from dc to a few gigahertz, over distances ranging from a few centimeters to many kilometers. The other class includes the planar lines used in microwave integrated-circuit (MIC) technology. These are generally fabricated by the circuit builder himself as components of a planar circuit, and typically are no more than a few centimeters in length.[4] *Microstrip* is the most common transmission line of this kind.

COAXIAL LINE

Coaxial line consists of a cylindrical center conductor surrounded by a cylindrical metal shield, as shown in cross section in Fig. 2.13. Commercial cables for use at relatively low frequencies often use metal braid for the outer conductor, rather than solid metal, so that the resulting cable is flexible. The outer diameter is in the range 0.3–1.5 cm. The space between the con-

[4]Microwave integrated circuits are not always "integrated" in the sense that digital ICs are. If a circuit contains transmission lines more than, say, 1 mm long, it is probably too large to be fabricated on a semiconductor chip. Instead the transmission lines are built on a large dielectric substrate, and semiconductor devices are added as discrete components. The result is known as a *hybrid* planar circuit. At high frequencies, or at lower frequencies when special design techniques are used, it does sometimes become possible to put all components on a semiconductor chip. The result is then known as a *monolithic* microwave integrated circuit, or MMIC.

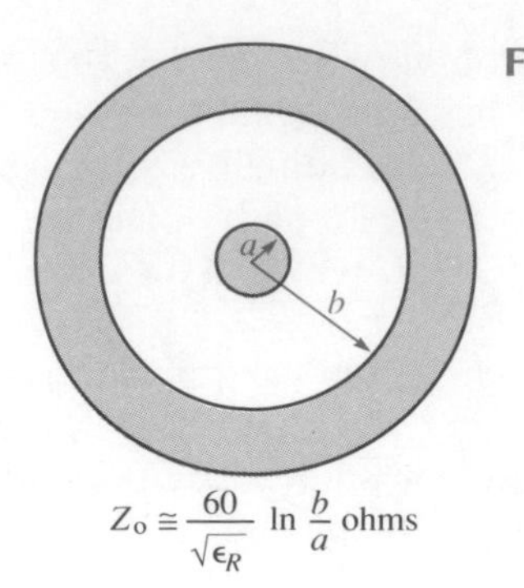

Figure 2.13 The coaxial transmission line.

ductors is filled with soft dielectric material, such as polyethylene or rubber, and the characteristic impedance is usually either 50 or 75 ohms. Special coaxial lines with smaller diameter (a few millimeters) can be used at frequencies as high as 30 GHz, but as with all transmission lines, attenuation due to resistance in the conductors becomes large at the higher frequencies. The velocity of propagation is given approximately by

$$U_P = \frac{c}{\sqrt{\epsilon_R}} \tag{2.21}$$

where ϵ_R is the relative dielectric permittivity of the material separating the two conductors (a constant which may be found in tables) and c is the velocity of light. When the dielectric is polyethylene, the ratio c/U_P is about 1.5. Since U_P is nearly independent of frequency, the line is to a first approximation nondispersive and nondistorting. The characteristic impedance is $\sqrt{L/C}$, but to use this formula one must first determine L and C, the inductance and capacitance of the line per unit length. The relationships between these quantities and the line dimensions will be considered later. For now, we merely note that for an ideal coaxial line, Z_o is given by the formula in Fig. 2.13.

A great advantage of coaxial line arises from the fact that in a well-designed system the current in the outer conductor flows on its *inner* surface. The finite thickness of the outer conductor then acts as a shield. This tends to eliminate pickup of outside signals and to prevent loss of energy by radiation, and also makes the line insensitive to its supports and any other objects that may be near. These advantages account for the very frequent use of coax as a general-purpose transmission line.

PARALLEL-CONDUCTOR LINES

Instead of placing one conductor inside the other, the two conductors can be placed side by side, as in Fig. 2.14. In practice the two conductors must be physically supported in some way. Sometimes they are spaced by periodic dielectric separators ("ladder line"); another method is to separate them by

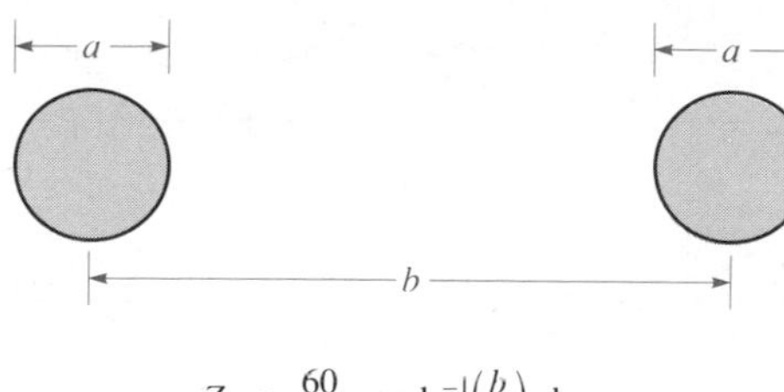

$$Z_o \cong \frac{60}{\sqrt{\epsilon_R}} \cosh^{-1}\left(\frac{b}{a}\right) \text{ohms}$$

Figure 2.14 Parallel-conductor transmission line.

a thin uniform dielectric membrane (''twin-lead''). For two unsupported parallel conductors in a vacuum, the phase velocity would be c, and the characteristic impedance would be given by the formula in Fig. 2.14. However, the presence of the dielectric supports causes the phase velocity to be slightly less than c, and causes Z_o to be slightly less than that of Fig. 2.14.

Parallel-conductor lines are often cheaper than coax and have somewhat lower attenuation. They may also be specified when a higher characteristic impedance, in the range of 300–500 ohms, is required. Another, rather subtle consideration that may enter is that these lines are what are known as *balanced* lines; that is, the two conductors are equivalent. Coax, by comparison, is an *unbalanced* line, because the conductors are not identical. When a signal source and load are symmetrical, it is usually desirable to connect them with a symmetrical line. If coax is used instead, one may find that currents are excited on the *outside* of the coax's outer conductor. This leads to unpredictable operation and radiation losses. Although this is a subtle point, the distinction between balanced and unbalanced transmission lines can be quite important in practice.

MICROSTRIP

An important trend in microwave technology is the increasing importance of planar circuits, known as microwave integrated circuits or MICs. In MICs, handmade wire connections are replaced by patterns in a metal film lying on the surface of a dielectric. This has advantages in terms of cost, durability, and reproducibility; moreover modern photolithographic techniques allow such structures to be scaled down to small dimensions, so that they can be used at high microwave frequencies. Transmission lines are important parts of such circuits, and thus must also be built in planar form. The most common choice is microstrip, as illustrated in Fig. 2.15. (Figure (a) is a cross-sectional view in the plane perpendicular to the propagation direction.) Microstrip consists of a uniform metal layer, the *groundplane*, on one side of a slab of dielectric, and a narrow line of metallic conductor on the other side. It can be fabricated by first plating metal uniformly on both faces, and then etching away most of the metal from the upper face, to leave a conductor of the desired size. In operation current flows through the narrow conductor and

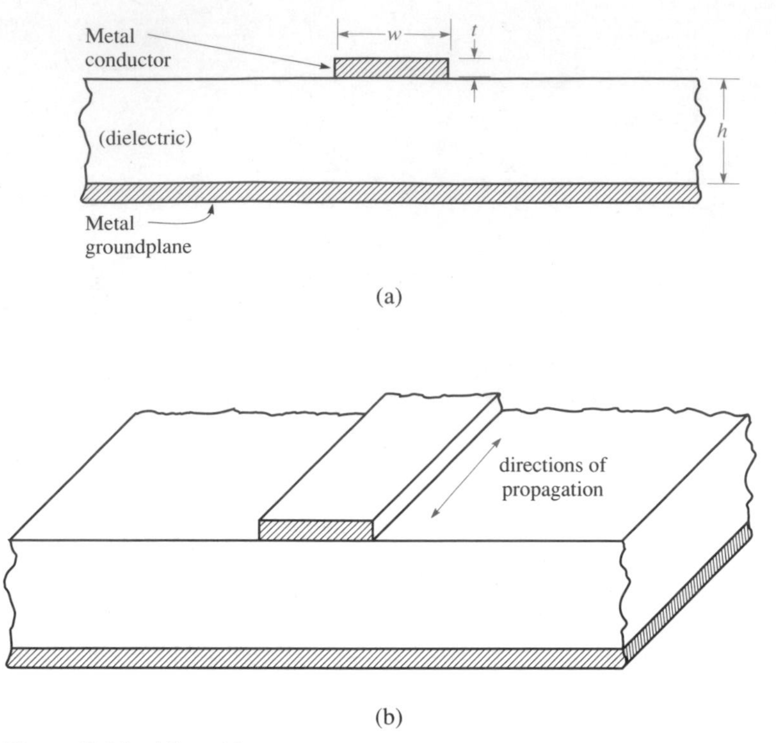

Figure 2.15 Microstrip.

returns by flowing through the groundplane, which acts as the second conductor.

Strictly speaking, microstrip is not a true transmission line. However, at low frequencies (say, below 10 GHz) its behavior differs only slightly from transmission-line theory. Thus it is often treated as though it were a true transmission line, and the terminology of transmission lines (for example, "characteristic impedance") is commonly used. At higher frequencies, or when high accuracy is required, the true nature of the microstrip wave becomes more important. This subject will be discussed further in Chapter 9. Figure 2.16 shows a typical MIC based on microstrip.

2.6 MEASUREMENTS ON TRANSMISSION LINES

In practical work it is often necessary to measure the reflection coefficient on a transmission line. This measurement allows one to determine whether power is being transferred into the load or reflected back. It can also be

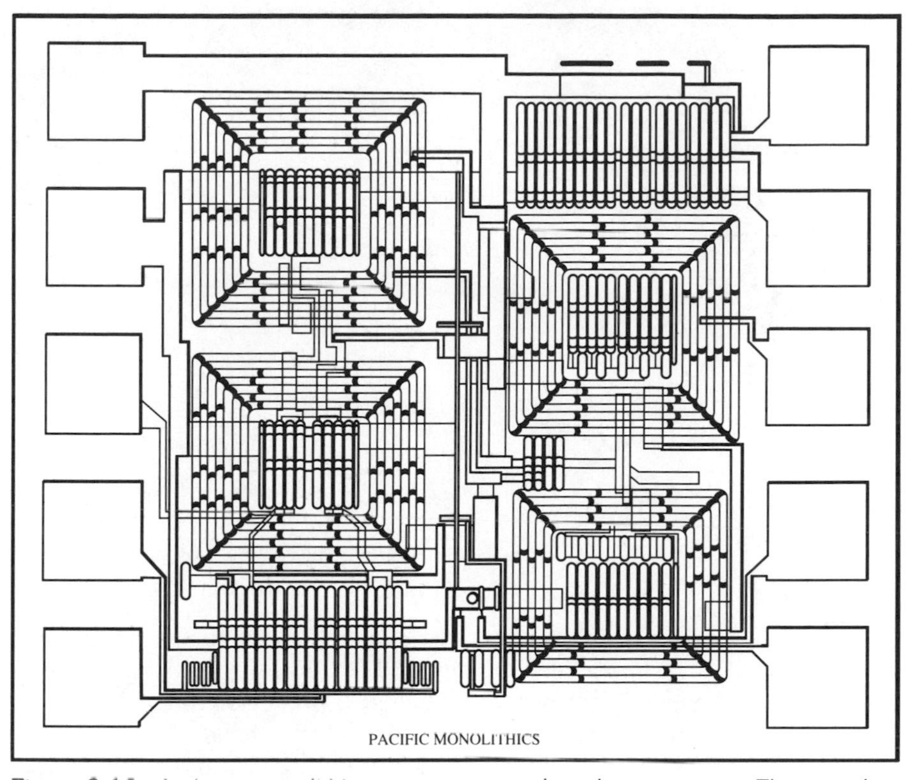

Figure 2.16 A planar monolithic microwave circuit based on microstrip. This complete low-noise downconverter for reception of satellite television (3.7–4.2 GHz band) is only 1 mm square. The four large square structures are spiral inductors, rectangular spirals of microstrip used in planar circuits in place of three-dimensional coils. [Adapted from a photo by Pacific Monolithics, Sunnyvale, Calif.]

used to determine the impedance of an unknown load, as well as for other purposes.

A traditional (and inexpensive) means of measuring ρ is by means of *probe* measurements. A probe is a sensitive detector that is very weakly coupled to a transmission line, in order to measure the voltage or current at the position of the probe. The coupling to the line must be weak, so that the probe will have a negligible effect on the line being measured.

For coaxial line one can use the *slotted line* shown in Fig. 2.17. A conducting probe is inserted into a longitudinal slot in a section of coaxial line. (There is no solid dielectric material between the conductors of this special coax section; it is what is known as an ''air line.'') The probe is not long enough to touch the center conductor; however it couples to it by capacitance. The amplitude of the high-frequency voltage appearing on the

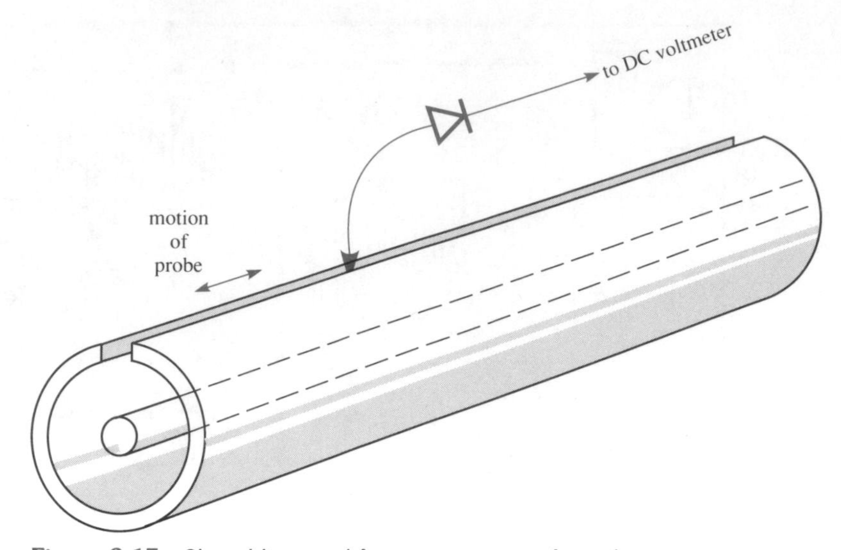

Figure 2.17 Slotted line used for measurement of standing-wave ratio on coax.

probe is then measured by a rectifying diode or other square-law device.[5] The probe can be moved back and forth along the guide by means of a mechanical positioner called a *probe carriage*. In addition to keeping the probe inserted at a constant depth, the probe carriage also has a scale or mechanical gauge that accurately reads out the position of the probe. Thus measurements of radio-frequency (rf) amplitude versus position can be made.

Figure 2.18 shows a typical set of probe measurements. (A real set of experimental data would exhibit some random errors—"scatter"—in the points, but in Fig. 2.18 ideal data has been assumed.) Note that Fig. 2.18 differs from Fig. 2.1 because the signal from the probe is proportional to the *square* of the amplitude. (The value of C, the constant of proportionality, is not important.) From these measurements one can determine λ, the SWR, and ρ_o, the reflection coefficient of the load (both magnitude and phase angle). In taking the data of Fig. 2.18, experimental points have been plotted at position intervals of $\frac{1}{16}$ wavelength. In addition, it is useful to locate the positions of maxima and minima and their corresponding probe voltages, and these points are also shown in the figure. For instance, the usual method of finding λ is to observe the spacing between two voltage minima.[6]

---

[5]Bolometers are often used for this purpose. A bolometer is a temperature-sensitive resistor. Minute rf currents from the probe heat the bolometer and the change of its dc resistance is measured. The change in its resistance is proportional to the square of the amplitude of the rf voltage.

[6]Minima are slightly easier to locate than maxima because the meter can be set on a more sensitive scale. An interesting and rather subtle advantage of locating minima is that when the effect of the probe on the wave is taken into account, it is found that maxima are shifted, but minima are not. See *Technique of Microwave Measurements*, C.G. Montgomery (Ed.), Vol. 11 of the M.I.T. Radiation Laboratory Series, New York: McGraw-Hill, 1947.

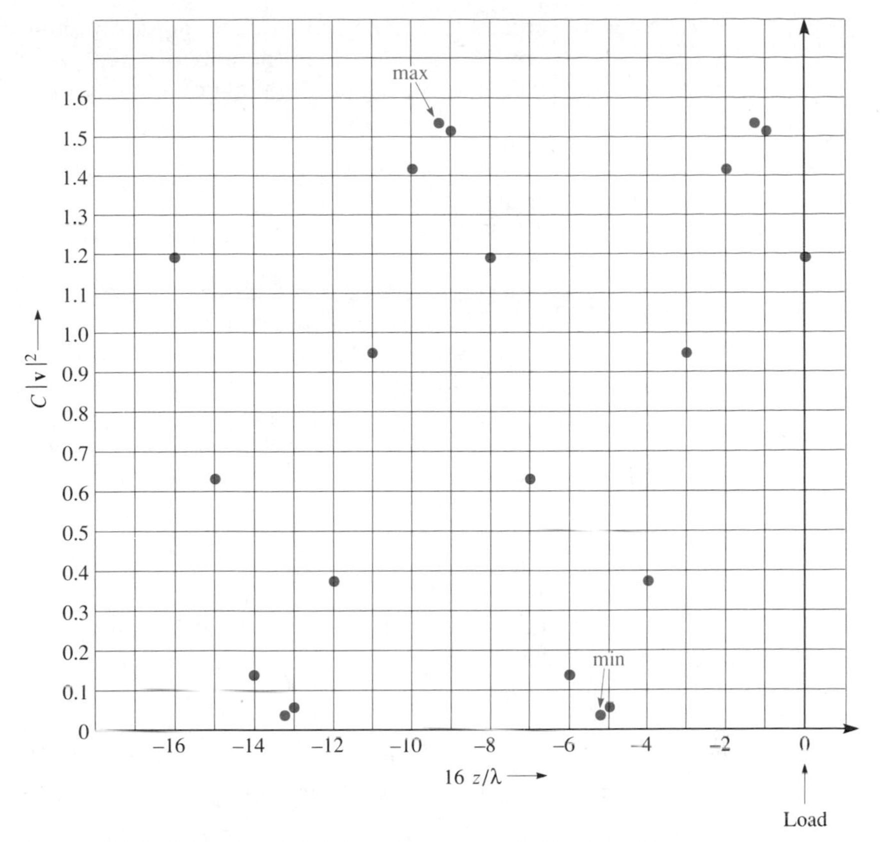

Figure 2.18 Experimental data points obtained with slotted line.

EXERCISE 2.5

Find the SWR from the probe measurements of Fig. 2.18. (Take the value of $C|v|^2$ at the minimum to be 0.032.)

Answer 6.9.

In present-day work, probe measurements are becoming obsolete. Instead, one uses a piece of modern electronic equipment known as a *network analyzer*. These versatile instruments measure both transmission and reflection coefficients. For example, a reflection measurement is made by simply connecting the analyzer to the input of the transmission line. The instrument then automatically measures the reflection coefficient (both magnitude and phase) and displays this information as a function of frequency.

Information is obtained much more quickly and easily than with probe measurements. The only disadvantage of the network analyzer is its cost, which can be substantial, particularly in the higher frequency ranges.

2.7 POINTS TO REMEMBER

1. When both an incident and a reflected wave are simultaneously present on a transmission line, a *standing wave* is said to be present. This means that a stationary pattern of voltage maxima and minima is present. The ratio of the maximum voltage to the minimum voltage is called the *standing-wave ratio* (SWR). The positions of the voltage maxima are determined by the phase angle of the load's reflection coefficient, and the spacing between each pair of adjacent maxima is $\lambda/2$ (and not λ, as one might think). Positions of maximum voltage are positions of minimum current, and vice versa.

2. The impedance $Z(z)$ at any point on a line is defined as the ratio of the voltage phasor to the current phasor at the point z. If a standing wave is present, the impedance will be a periodic function of position along the line, with period $\lambda/2$. Note that this impedance is different from the "characteristic impedance" Z_o, which is a constant that depends only on the construction of the line.

3. The general reflection coefficient $\rho(z)$ is defined as the ratio of the phasor representing the incident wave to that representing the reflected wave. The general reflection coefficient also varies with position and is related to the impedance according to

$$Z = \frac{1 + \rho}{1 - \rho} Z_o$$

The value of the reflection coefficient at the position of the load is called ρ_o; this number is determined by the nature of the load, and for passive loads has magnitude less than or equal to one. At a distance l before the load the reflection coefficient is given by

$$\rho(l) = \rho_o e^{-2jkl}$$

4. The *Smith chart* is a graphic device for finding Z when ρ is known, and vice versa. It can also be used to find the impedance at one point on a line when the impedance at another point is known. It can also be used as an admittance chart.

5. *Impedance-matching techniques* are used to transform the impedance of a line to a desired value. Often it is desired to reduce the SWR of an improperly terminated line to unity. Some techniques useful for this purpose are the *quarter-wave transformer,* the *adjustable backshort,* and *single-* and *double-stub tuners.*

6. At low frequencies commercial transmission lines with fixed values of characteristic impedance are generally used. *Coaxial cable* is the most familiar line of this kind. At higher frequencies, one also encounters planar transmission lines, such as *microstrip*. These are often fabricated as components of planar circuits, and can have any desired value of characteristic impedance within a certain range.

7. Measurements of SWR or reflection coefficient can be made by means of *probe measurements*. A modern electronic instrument that performs these measurements much more quickly is the *network analyzer.*

REFERENCES

1. J. L. Stewart, *Circuit Analysis of Transmission Lines.* New York: Wiley, 1958.
2. S. Ramo, J. R. Whinnery, and T. Van Duzer, *Fields and Waves in Communication Electronics.* New York: Wiley, 1984.
3. N. N. Rao, *Elements of Engineering Electromagnetics.* Englewood Cliffs, N.J.: Prentice-Hall, 1987.
4. P. H. Smith, "Transmission-Line Calculator." *Electronics,* **12,** 29–31 (Jan. 1939); "An Improved Transmission-Line Calculator," *Electronics,* **17,** 130 (Jan. 1944).

For information on microstrip see:

5. T. C. Edwards, *Foundations for Microstrip Circuit Design.* New York: Wiley, 1981.
6. K. C. Gupta, R. Garg, and I. J. Bahl, *Microstrip Lines and Slotlines.* Dedham, Mass.: Artech House, 1979.

A general reference on measurements is:

7. T. S. Laverghetta, *Handbook of Microwave Testing.* Dedham, Mass.: Artech House, 1981.

Computer software:

8. Charles H. Roth, Jr. (Dept. of Electrical and Computer Engineering, Univ. of Texas), "TLS: Transmission Line Simulator." Available for Macintosh from Kinko's Academic Courseware Exchange, 255 West Stanley Ave., Ventura, CA 93001. Telephone (800) 235-6919 (in California (800) 292-6400).

PROBLEMS

In these problems the symbol λ always represents the guide wavelength.

Section 2.1

2.1 The voltage on a transmission line with an unknown load is measured at various positions. The largest amplitude measured is 1.3 V and the smallest is 0.9 V. Find the magnitude of the load's reflection coefficient.

*2.2 A 50-ohm line is terminated in a load $Z_L = 40 - 60j$ ohms. Find the standing-wave ratio and the position of the voltage maximum nearest the load. The wavelength is 3 cm.

*2.3 Show that the ratio of maximum current to minimum current on a line is also given by equation (2.6).

*2.4 Show that a position on the transmission line where the total voltage is maximum is also a position of minimum current, and vice versa.

*2.5 Show that at a position where the absolute value of the impedance is maximum or minimum, the impedance is real.

*2.6 The power transmitted on a transmission line is

$$P = \text{time avg } [v(t)i(t)] = \frac{1}{2} \text{Re } (\mathbf{vi}^*)$$

Show that regardless of SWR this power is independent of position.

**2.7 The purpose of this problem is to demonstrate that the power lost in the resistance of a transmission line can be reduced by reducing the SWR. A transmission line of length L is terminated by a load, which has reflection coefficient ρ_o. The transmission line has resistance R_1 per unit length, and R_1 is small, so that $\alpha L \ll 1$. The line is very long in terms of wavelengths. It is required that a power P_L be dissipated in the load. Use the method of Problem 1.18 to show that P_R, the power lost in the line resistance R_1, depends on ρ_o according to

$$P_R \cong \frac{R_1 P_L}{Z_o} \frac{1 + |\rho_o|^2}{1 - |\rho_o|^2} L$$

Section 2.2

2.8 We wish to construct a load with impedance $-23j$ ohms at the frequency 100 MHz. To do this we use the shortest possible length of a coaxial cable with characteristic impedance 50 ohms and phase velocity 2×10^8 m/sec, terminated in a short circuit. What is the length of the cable?

*2.9 Show, using equation (1.45), that the impedance along a lossy transmission line varies according to

$$\frac{dZ}{dz} = (j\omega C + G)Z^2 - (R + j\omega L) \qquad \textbf{(2.22)}$$

**2.10 Apply equations (2.22) (preceding problem) to the case of a lossless line. Integrate the equation and recover equation (2.12).

2.11 On a graph of the complex plane, locate the points representing the reflection coefficient of
(a) a short circuit

(b) an open circuit
(c) a resistive termination with resistance Z_o.

2.12 On a graph of the complex plane, locate the points representing the reflection coefficient
(a) of a pure inductance with reactance jZ_o
(b) of a pure capacitance with reactance $-2jZ_o$
(c) at the input of a line one-eighth wavelength long, terminated in a pure inductance with reactance jZ_o.

2.13 Suppose the impedance $Z(z)$ at point z on a line is known. Show that the generalized reflection coefficient $\rho(z)$ at that point is

$$\rho(z) = \frac{Z - Z_o}{Z + Z_o} \tag{2.23}$$

*2.14 Show that if the impedance $Z(z)$ at any point on a line has a positive real part, the reflection coefficient satisfies $|\rho(z)| \leq 1$.

2.15 Show that if the generalized reflection coefficient $\rho(z)$ at any point z lies on the unit circle in the complex plane, the real part of $Z(z)$ vanishes.

Section 2.3

2.16 Find the reflection coefficient corresponding to $Z = (2 - 2j)Z_o$
(a) using equation (2.23) (Problem 2.13)
(b) using the Smith chart.
Do the two results agree?

2.17 Find the impedance corresponding to a reflection coefficient of $0.5e^{j30°}$.
(a) using equation (2.18)
(b) using the Smith chart.
Do the two results agree?

2.18 Use the Smith chart to find the normalized impedance corresponding to reflection coefficients of
(a) 0
(b) -1
(c) j
(d) $(0.4)e^{j160°}$
(e) $-0.6 - 0.2j$

*2.19 Find the normalized impedance at a point 0.6λ distant from a load
(a) whose reflection coefficient is $0.5 - 0.3j$
(b) whose normalized impedance is $0.5 - 0.3j$
(c) whose normalized admittance is $0.5 - 0.3j$.

2.20 The normalized impedance measured at a certain point on a transmission line is $3 + j$. If the load is known to be purely resistive, what two values may it have? At what distances (in wavelengths) might it be located?

*2.21 A transmission line is terminated in a load $Z_L = (1.3 - 0.7j)Z_o$. Find the distance from the load to the nearest
(a) voltage maximum
(b) current maximum
(c) current minimum
(d) place where the impedance is real.

2.22 Find the (non-normalized) admittance at a point on a transmission line where $\rho = (0.6)e^{-j100°}$. The characteristic impedance of the line is 72 ohms.

*2.23 Find the normalized admittance at a distance of $\lambda/8$ before a load with admittance $Y_L = (2 + 4j)Z_o^{-1}$.

*2.24 Immediately preceding a 75-ohm resistive load is a quarter-wave section of line with characteristic impedance 50 ohms. Before this comes a quarter-wave section of 100-ohm line, before that another quarter-wave section of 50-ohm line, and so on in alternation until there are a total of six quarter-wave sections (three of each kind). What is the impedance at the input? Assume that the transmission lines are all lossless and otherwise ideal.

Section 2.4

*2.25 In Fig. 2.19, let $Z_L = 50 + 30j$ ohms. The characteristic impedance of all lines is 50 ohms. Find the smallest distance x that makes the impedance at AA' equal to 50 ohms, purely resistive.

2.26 In Fig. 2.10, all lines have characteristic impedance 50 ohms, and $x = 0.8\lambda$. Find the value of the impedance Z_L required to make the impedance at AA' equal to 50 ohms, regardless of the distance between points A and B.

*2.27 In Fig. 2.20, the stub is open-circuited at its end, and is in series with the line. Let $Z_L = (0.8 - 3j)Z_o$. Find the smallest b and d (in units of λ) required to give unity SWR at AA'.

*2.28 In Fig. 2.11, $Z_L = (0.5 - j)Z_o$, where Z_o is the characteristic impedance of all the lines. Find the smallest lengths b and d (measured in units of λ) that make the SWR equal to unity, to the left of AA'.

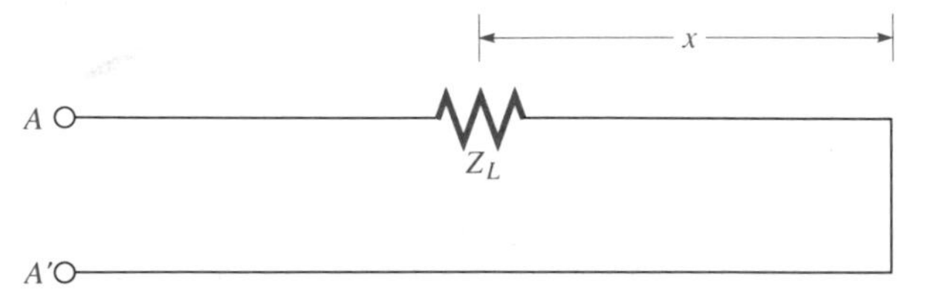

Figure 2.19 For Problem 2.25.

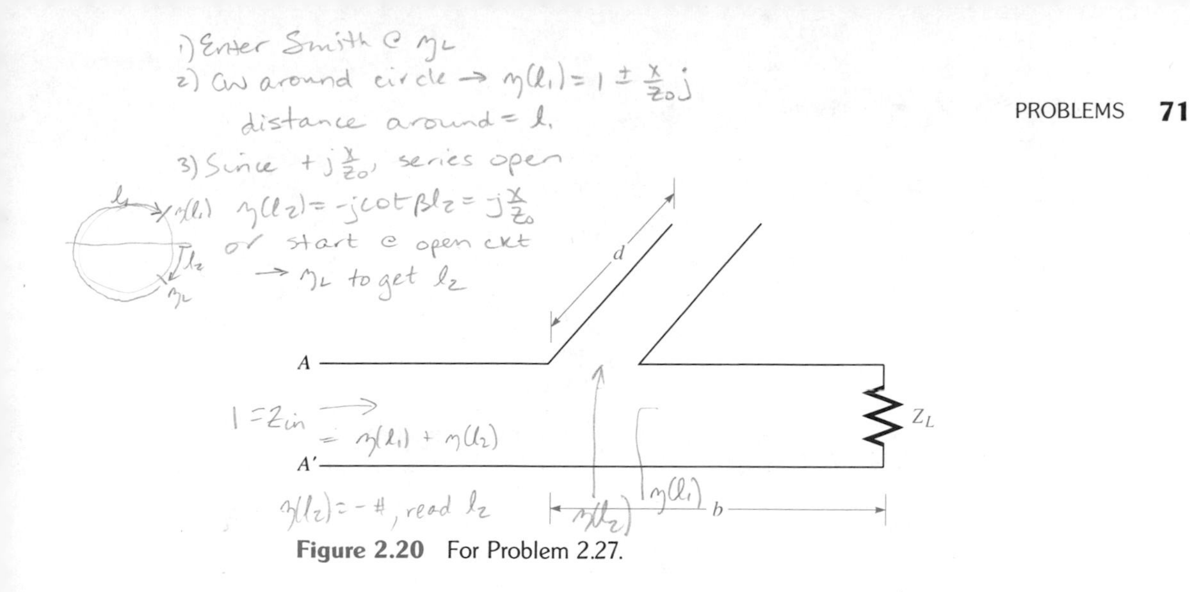

Figure 2.20 For Problem 2.27.

*2.29 In Exercise 2.4, suppose that due to unavoidable mechanical errors, the stub d is made 5% longer than intended. What is the resulting SWR at BB'?

**2.30 Suppose that a transmission line has a series resistance R per unit length, and that this resistance is small, so that equation (1.60) (in Problem 1.16) can be used. The characteristic impedance of the line is 50 ohms. Let R be 0.5 ohms/m, $\lambda = 1.3$ m, and let the length of a shorted stub be 0.6 m. Find the input impedance of the stub.

**2.31 A certain transmission line, with characteristic impedance 50 Ω, is slightly lossy, so α (in equation (1.49)) is not equal to zero. A short-circuited stub of length l is required to have an input reactance of 1000 ohms, inductive.

(a) Find the resistance that appears in series with this reactance, in terms of the product αl. Assume that the stub is made as short as possible, and that $\alpha l \ll 1$.

(b) The stub is connected in series with a lumped capacitor C, in order to obtain series resonance. Let us define a quality factor Q for the resulting circuit by $Q = (R\omega C)^{-1}$, where ω is the resonant frequency and R is the resistance found in part (a). Find Q in terms of αl.

**2.32 The double-stub tuner of Fig. 2.12(a) is to be used to eliminate reflections from a load impedance $Z_L = (1.6 - j)Z_o$. Let $h_2 = 0.1\,\lambda$, and $h_1 = 0.25\,\lambda$. Find b and a to produce SWR = 1 at the input.

**2.33 Suppose the double-stub tuner of Fig. 2.12(a) has $h_1 = 0.25\,\lambda$. The position of point B (in Fig. 2.12(b)) depends on Z_L and h_2. For some positions of B, it may not be possible to correct the SWR to unity.

(a) Determine the region on the Smith chart in which B can lie, if SWR is to be corrected to unity.

(b) Assuming neither Z_L nor h_1 can be changed, can h_2 be changed to correct the difficulty? What is the largest change that may be required?

Section 2.6

*2.34 A transmission line is terminated in a normalized load impedance of $1 + 0.5j$. Probe measurements are made with a slotted line, using a square-law detector for which V_M, the measured voltage, is proportional to the square of the amplitude of the voltage on the transmission line. Sketch the expected appearance of V_M as a function of distance from the load (measured in wavelengths).

2.35 Figure 2.21(a) illustrates an impedance measurement using a network analyzer. The instrument has a screen, Fig. 2.21(b), with the markings of a Smith chart, on which it projects a spot of light representing the reflection coefficient at the plane AA'. All transmission lines have a

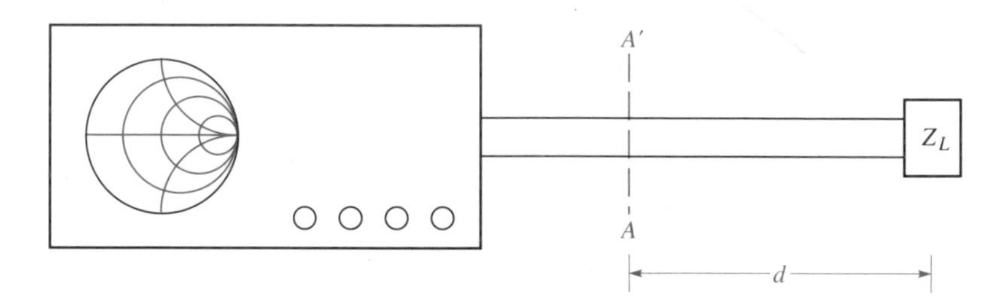

(a)

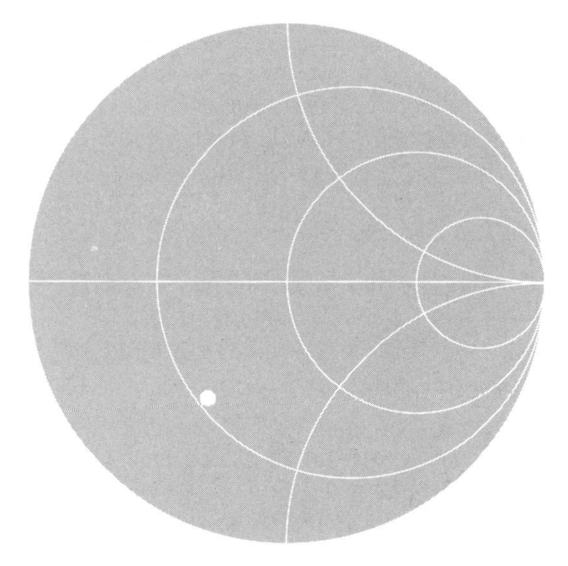

(b)

Figure 2.21 For Problem 2.35.

characteristic impedance of 50 ohms, the frequency is 1 GHz, and d is 3.2 guide wavelengths.

(a) If the display is as in Fig. 2.21(b), find $|\rho_o|$, the SWR, and Z_L.

(b) Sketch the new position of the display dot if the length d is increased by 10%.

(c) Suppose the frequency is swept continuously between 0.9 GHz and 1.1 GHz. Sketch the curve swept out by the display dot.

(d) Suppose the frequency is fixed at 1 GHz, but the line between AA' and the load attenuates the wave by 3 dB (round trip attenuation from AA' to load and back to AA'). Where is the display spot in this case?

COMPUTER EXERCISE 2.1

In this exercise we shall design the single-stub tuner of Fig. 2.11 by a numerical technique. Let the load to be matched have normalized admittance $Y_L' = Y_1 + jY_2$, and let $\theta = kb$.

(a) Verify that θ must satisfy

$$Y_1 + 2Y_2 \sin\theta \cos\theta - \cos^2\theta - (Y_1^2 + Y_2^2)\sin^2\theta = 0$$

(b) Using a library equation-solving subroutine, construct a program for numerical solution of this equation, for arbitrary given values of Y_1 and Y_2. Note that the equation has an infinite number of solutions. For the tuner to have minimum size, you should find the smallest positive solution.

(c) Add program steps for finding $\psi = kd$.

(d) Test your program for the case $Y_1 = 1.1$, $Y_2 = 0$. By inspection of the Smith chart, what should θ be, approximately?

(e) Find θ and ψ for $Y_1 = 0.8$, $Y_2 = -1.4$. Check your results by means of the Smith chart.

CHAPTER 3

Fields and Field Operators

Central to the subject of electromagnetics is the notion of a *field*. In this chapter we shall begin with a review of vector analysis, after which, in section 3.2, we shall introduce scalar and vector fields. The remainder of the chapter is then devoted to the important mathematical operations used in describing fields. Line and surface integration are reviewed in section 3.3. Finally, in section 3.4, we shall introduce the general subject of differential field operators. However, the details of the individual operators—gradient, divergence, curl, and Laplacian—will be left for presentation later, as they are needed.

It should be noted that the subject of fields and field operators belongs fundamentally to mathematics, and that a full mathematical treatment contains many subtleties. Our discussion here, however, will be directed toward electromagnetic applications, and will not attempt to deal with points that are unlikely to arise in engineering work. Many excellent books on mathematical analysis are available for those who wish a rigorous treatment of the subject.

It is assumed that the reader is familiar with the rectangular coordinate system. However, many problems in electromagnetics possess cylindrical or spherical symmetry, and effort can often be saved through use of cylindrical coordinates or spherical polar coordinates. A review of the cylindrical and spherical coordinate systems will be found in Appendix A.

3.1 REVIEW OF VECTOR ANALYSIS

VECTORS

Vectors are used to represent quantities such as velocity that have both magnitude and direction. Electric and magnetic fields also possess both magnitude and direction, and thus are also conveniently represented by vectors.

A vector can be described by a set of scalars. The number of scalars needed to represent a vector depends on the dimensionality of the space. For two-dimensional problems, a vector is represented by a set of two scalars, while for three-dimensional problems a set of three scalars is needed to describe each vector. Most of the time we shall be dealing with rectangular coordinates. In that case the scalars are simply the rectangular components of the vector. For example, suppose a certain vector $\vec{V}_1$ is given by

$$\vec{V}_1 = 2\vec{e}_x + 3\vec{e}_y + 4\vec{e}_z$$

where $\vec{e}_x$, $\vec{e}_y$, and $\vec{e}_z$ are the unit vectors in the x, y, and z directions. The three scalars 2, 3, 4 are the rectangular components of the vector and describe it completely. A graphical representation of $\vec{V}_1$ is shown in Fig. 3.1. If the tail of the vector is pictured at the origin, its tip is located at the point $x = 2$, $y = 3$, $z = 4$. The *magnitude* or *absolute value* of the vector $\vec{V}_1$ with the symbol $|\vec{V}_1|$, can be considered to be the vector's *length*. From the pythagorean theorem, it is clear that

$$|\vec{V}_1| = \sqrt{(2)^2 + (3)^2 + (4)^2} = 5.385$$

The direction of $\vec{V}_1$ is also determined when its rectangular coordinates have been given, as is seen from the figure.

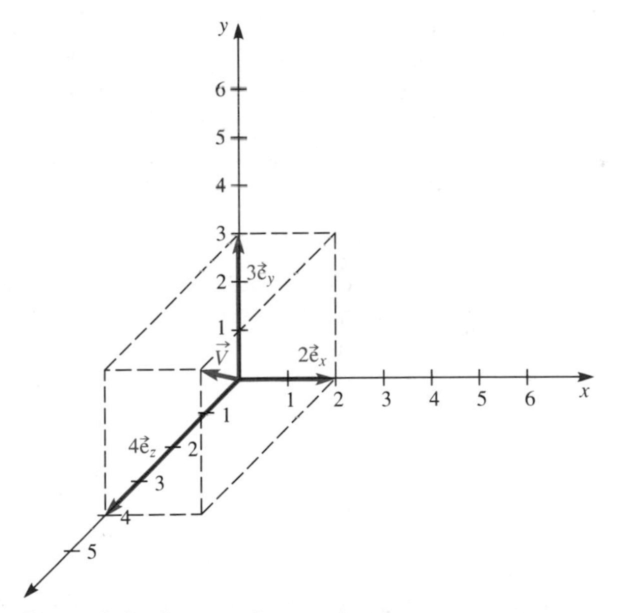

Figure 3.1 Rectangular coordinates.

EXERCISE 3.1

Find the angle that the vector $\vec{V}_1$ makes with the x axis.

Answer 68.20°.

THE RADIUS VECTOR

The vector describing the location of a point is known as the *radius vector* $\vec{r}$. If the point P is located at the position x, y, z, the radius vector describing P is

$$\vec{r} = x\vec{e}_x + y\vec{e}_y + z\vec{e}_z \tag{3.1}$$

From Fig. 3.1 we can see that $\vec{r}$ is simply the vector directed from the origin to the point P.

Often we need the *unit vector* in the direction of $\vec{r}$. This unit vector $\vec{e}_r$ is obtained by dividing $\vec{r}$ by its length $|\vec{r}|$. Thus

$$\vec{e}_r = \frac{x\vec{e}_x + y\vec{e}_y + z\vec{e}_z}{\sqrt{x^2 + y^2 + z^2}} \tag{3.2}$$

VECTOR ALGEBRA

Vectors are added by adding their individual components. That is, $(a_x\vec{e}_x + a_y\vec{e}_y) + (b_x\vec{e}_x + b_y\vec{e}_y) = [(a_x + b_x)\vec{e}_x + (a_y + b_y)\vec{e}_y]$. Graphically this corresponds to adding the vectors head to tail, as shown in Fig. 3.2.

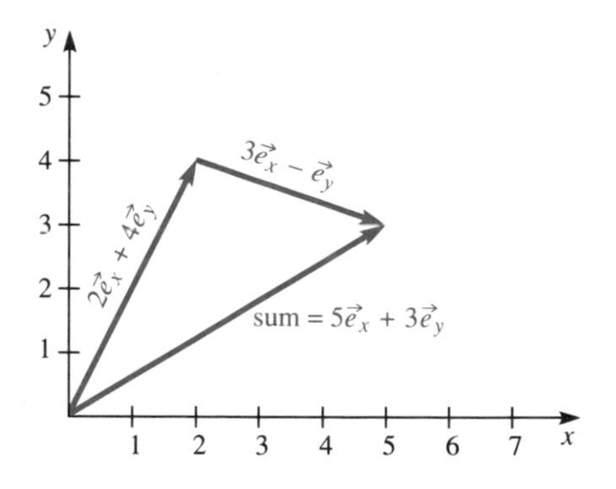

Figure 3.2 Addition of vectors in rectangular coordinates.

EXERCISE 3.2

Let $\vec{V}_1 = 3\vec{e}_x - 2\vec{e}_y - 4\vec{e}_z$ and $\vec{V}_2 = -2\vec{e}_x - 3\vec{e}_y + 5\vec{e}_z$, and let $\vec{V}_3 = \vec{V}_1 - \vec{V}_2$. Find $|\vec{V}_3|$.

Answer 10.34.

We shall be interested in two types of vector product, the *dot product* and the *cross product*. The dot product of the two vectors $\vec{A}$ and $\vec{B}$ is written $\vec{A} \cdot \vec{B}$, and is a *scalar*. It has the following significance:

$$\vec{A} \cdot \vec{B} = |\vec{A}|\,|\vec{B}| \cos\psi \tag{3.3}$$

where ψ is the angle between the vectors. (If they point in the same direction, $\psi = 0$; if in opposite directions, $\psi = 180°$.) Alternatively, $\vec{A} \cdot \vec{B}$ can be found by the following method. Let $\vec{A} = a_x\vec{e}_x + a_y\vec{e}_y + a_z\vec{e}_z$, $\vec{B} = b_x\vec{e}_x + b_y\vec{e}_y + b_z\vec{e}_z$. Then

$$\vec{A} \cdot \vec{B} = a_x b_x + a_y b_y + a_z b_z \tag{3.4}$$

EXERCISE 3.3

Let $\vec{V}_1 = 3\vec{e}_x - 2\vec{e}_y - 4\vec{e}_z$, $\vec{V}_2 = -2\vec{e}_x - 3\vec{e}_y + 5\vec{e}_z$. Find $\vec{V}_1 \cdot \vec{V}_2$.

Answer -20.

EXERCISE 3.4

Find the angle between the vectors $\vec{V}_1$ and $\vec{V}_2$ in the preceding exercise.

Answer 127°.

On the other hand, the *cross product* of the vectors $\vec{A}$ and $\vec{B}$, which is written $\vec{A} \times \vec{B}$, is a *vector*. The direction of the vector $\vec{A} \times \vec{B}$ is perpendicular to the plane containing $\vec{A}$ and $\vec{B}$, and its magnitude is given by

$$|\vec{A} \times \vec{B}| = |\vec{A}|\,|\vec{B}| \sin\psi \tag{3.5}$$

where ψ is the angle between the vectors. Alternatively, $\vec{A} \times \vec{B}$ can be found by the following prescription:

$$\vec{A} \times \vec{B} = \begin{vmatrix} \vec{e}_x & \vec{e}_y & \vec{e}_z \\ a_x & a_y & a_z \\ b_x & b_y & b_z \end{vmatrix}$$

$$= \vec{e}_x(a_y b_z - a_z b_y) + \vec{e}_y(a_z b_x - a_x b_z) + \vec{e}_z(a_x b_y - a_y b_x) \quad \textbf{(3.6)}$$

Note that $\vec{B} \times \vec{A} = -(\vec{A} \times \vec{B})$. (However, $\vec{B} \cdot \vec{A} = \vec{A} \cdot \vec{B}$.)

EXERCISE 3.5

Find the cross product of $\vec{V}_1$ and $\vec{V}_2$, as defined in Exercise 3.3.

Answer $-22\vec{e}_x - 7\vec{e}_y - 13\vec{e}_z$.

EXERCISE 3.6

Show that the volume of the solid shown in Fig. 3.3 is equal to $\vec{V}_3 \cdot (\vec{V}_1 \times \vec{V}_2)$.

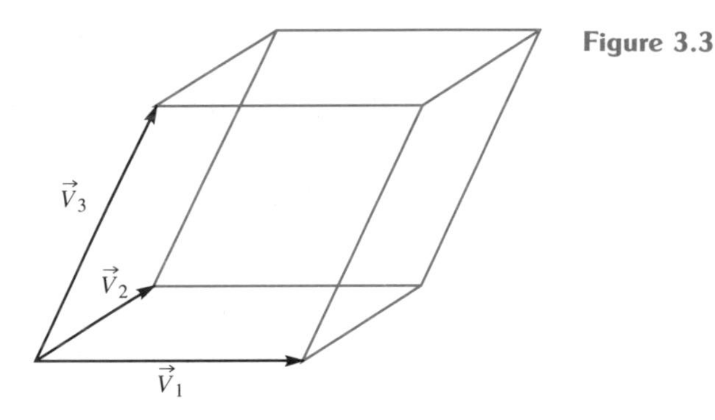

Figure 3.3

VECTOR IDENTITIES

There are a number of relationships between the operations of vector algebra, known as *vector identities*. These need not be memorized, but it is useful to remember that they exist, so that they can be looked up when needed. Two of the most useful vector identities are:

$$\vec{A} \cdot (\vec{B} \times \vec{C}) = \vec{B} \cdot (\vec{C} \times \vec{A}) = \vec{C} \cdot (\vec{A} \times \vec{B}) \quad \textbf{(3.7)}$$

$$\vec{A} \times (\vec{B} \times \vec{C}) = \vec{B}(\vec{A} \cdot \vec{C}) - \vec{C}(\vec{A} \cdot \vec{B}) \quad \textbf{(3.8)}$$

Some other vector identities will be found in the endpapers (inside the covers) of this book.

3.2 SCALAR AND VECTOR FIELDS

There are many situations in which a physical quantity depends upon position. For example, let us think about temperature in the atmosphere. We can begin by setting up a rectangular coordinate system, so that each position is designated by its coordinates x, y, and z. At each position, the temperature will have some value, which we designate as $T(x, y, z)$. The temperature is then said to be a scalar function of position, or, equivalently, we say that the temperature is a *scalar field*. The important point is that to each position there corresponds a single, well-defined value of temperature. Later we shall make use of this concept with electrical quantities. For example, the electrostatic potential is a scalar field.

The notion of a *vector field* is similar; the difference is only that in a vector field, a certain *vector* is associated with each position in space. For example, let us think of winds in the atmosphere. At each position the vector $\vec{U}$, representing the wind velocity (both magnitude and direction) has some value. We would then say that $\vec{U}(x, y, z)$ is a vector function of position, or equivalently, that $\vec{U}$ forms a vector field.

The dependence of $\vec{U}$ on position can be specified in various ways. However, it will be remembered that to describe $\vec{U}$, *three* scalars (e.g., its x, y, and z components) must be specified. Some examples:[1]

1. $\vec{U} = 3\vec{e}_x + 2\vec{e}_y + 5\vec{e}_z$

In this case $\vec{U}$ is a constant; it has the same value at all positions.

2. $\vec{U} = 3\vec{e}_x + (x^2 + 2y + yz)\vec{e}_z$

Here $\vec{U}$ is a two-dimensional vector; its y component is always zero. Its x component is constant, but its z component depends on y as well as on x and z.

3. $\vec{U} = f(x, y, z)\vec{e}_x + g(x, y, z)\vec{e}_y + h(x, y, z)\vec{e}_z$

This represents the most general case. The functions f, g, h can be any three different functions of the three position variables.

4. $\vec{U} = \vec{e}_x \times (x\vec{e}_x + y\vec{e}_y + z\vec{e}_z)$

Here the vector $\vec{U}$ is given as a vector function of other vectors.

---

[1]These are illustrative examples of vector fields. For physical reasons it may not be possible for these particular functions to represent wind velocity.

The thing to be remembered is that at each point in space, the vector field $\vec{U}$ is specified to have some particular value (both magnitude and direction). For example, the magnetic field of a bar magnet is a vector field. We could illustrate this field by drawing an arrow at each point, indicating the direction of the magnetic field at that point. The length of the arrow could indicate the magnitude of the field at the point. This has been done in Fig. 3.4(a). Of course, we cannot indicate $\vec{U}$ at *every* point by this means; only at a selection of points. A more common graphical way of illustrating the same field is that of Fig. 3.4(b). Here the direction of the field is indicated by the continuous "field lines." The direction of the field at a point is always

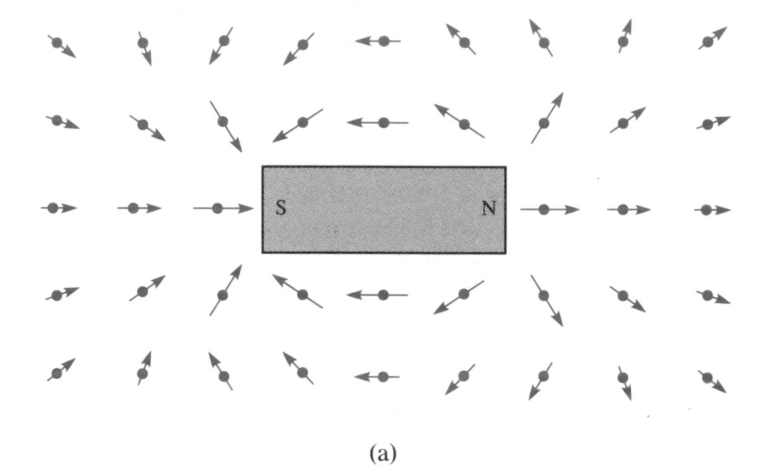

(a)

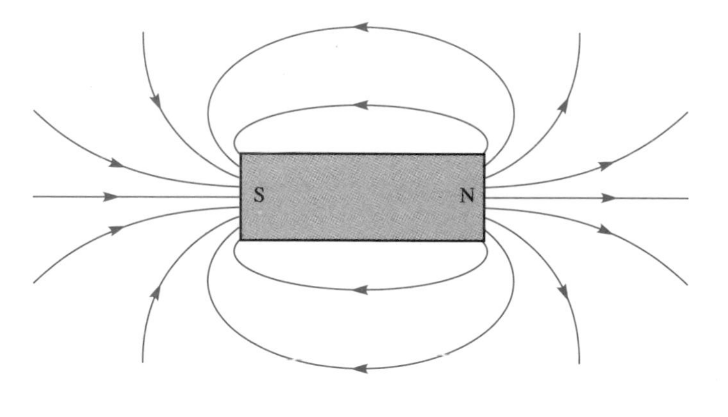

(b)

Figure 3.4 Two ways of drawing a vector field. In (a), the lengths of the arrows indicate the field strength at each point. In the more common method (b), the side-to-side spacing of the lines indicates the field strength: the closer together the lines, the stronger the field.

in the direction of the field line through that point. The strength of the field is indicated by the side-to-side spacing of the lines. The closer together the lines are, the stronger the field.

EXERCISE 3.7

Let $\vec{A} = \vec{r} \times \vec{e}_y$, where $\vec{r}$ is the "radius vector" defined by $\vec{r} = x\vec{e}_x + y\vec{e}_y + z\vec{e}_z$. Find $\vec{A}$ at the following points: (a) $(x_o, 0, 0)$; (b) $(0, y_o, 0)$; (c) $(0, 0, z_o)$.

Answer (a) $x_o\vec{e}_z$; (b) 0; (c) $-z_o\vec{e}_x$.

3.3 LINE AND SURFACE INTEGRALS

Now that vector fields have been introduced, we are ready for various types of calculations based on vector fields. In this section we shall discuss functions of vector fields obtained by integrating them. In section 3.4 we shall discuss functions obtained by differentiating vector fields.

LINE INTEGRALS

A line integral can be calculated whenever a path has been specified through a known vector field. The notation

$$\int_P \vec{V} \cdot \vec{dl}$$

is read "The line integral of the field $\vec{V}$ along the path P." The meaning is as illustrated in Fig. 3.5. The path is broken up into very small straight line segments. Each of these is a small vector whose direction is parallel to the original path. We call the first of these little vectors $\vec{dl}_1$, the second $\vec{dl}_2$, and so on until we arrive at the end of the path.

Now let us extract one of these little vectors, for instance $\vec{dl}_1$ and look at it by itself as in Fig. 3.5(c). The vector field $\vec{V}$ has some value at the position of $\vec{dl}_1$. To form this part of the line integral, we take the dot product of $\vec{dl}_1$ with the value of $\vec{V}$ that exists at this point. We add this to the dot product of $\vec{dl}_2$ with the value of $\vec{V}$ existing at the position of $\vec{dl}_2$, and proceed in this way until we have added up the contributions from all the segments of the path. The dot product of two vectors is a *scalar,* so each term being added is a scalar, and so is their sum, the line integral.

From (3.3) we know that

$$\vec{dl}_1 \cdot \vec{V} = |\vec{dl}_1|\,|\vec{V}| \cos \theta_1$$

where θ_1 is the angle between $\vec{dl}_1$ and $\vec{V}$. We can think of $\vec{V} \cos \theta_1$ as the *component* of the vector $\vec{V}$ in the direction of $\vec{dl}_1$. If $\vec{V}$ is parallel to $\vec{dl}_1$, the

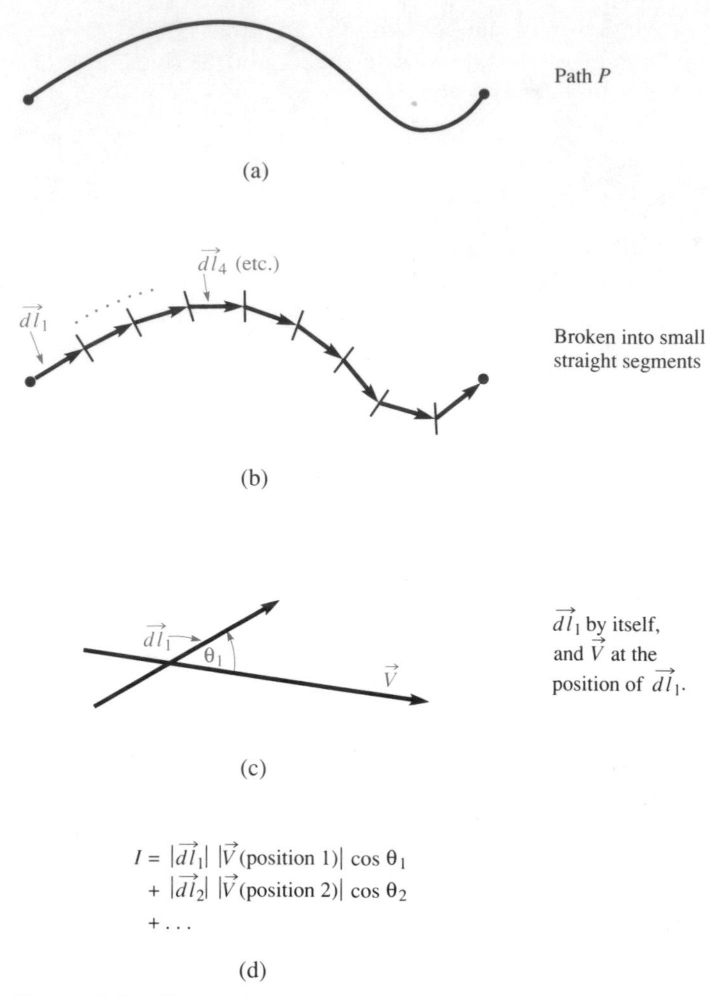

Figure 3.5 The line integral.

contribution of this segment to the line integral will be maximized. If $\vec{V}$ happens to be perpendicular to $\vec{dl}_1$, the contribution from this segment will be zero.

EXAMPLE 3.1

Let the vector field $\vec{V}$ be given by $\vec{V} = V_o\vec{e}_x$, where V_o is a constant. Find the line integral

$$I_1 = \int_{P_1} \vec{V} \cdot \vec{dl}$$

where the path P_1 is the closed path shown in Fig. 3.6.

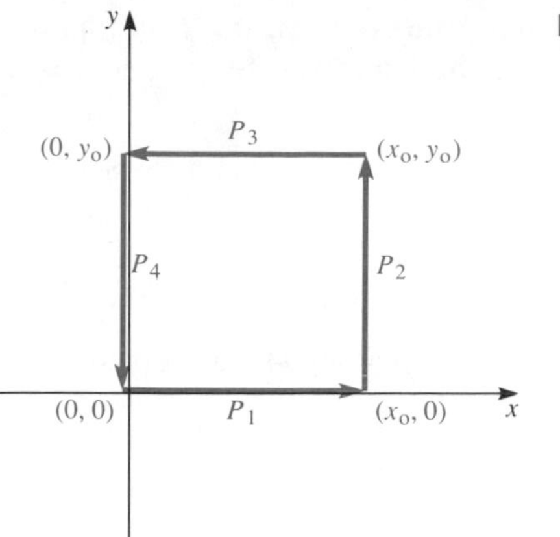

Figure 3.6

Solution It is convenient to break the path up into the four parts P_1, P_2, P_3, P_4, as indicated in the figure. On segments P_2 and P_4, the field $\vec{V}$ is always perpendicular to $\vec{dl}$; hence P_2 and P_4 make no contribution to the integral.

For segment P_1, $\vec{dl} = dx\,\vec{e}_x$. Thus

$$\int_{P_1} \vec{V} \cdot \vec{dl} = \int_{x=0}^{x=x_o} (V_o\vec{e}_x) \cdot (dx\,\vec{e}_x) = V_o x_o$$

where we have used the fact that $\vec{e}_x \cdot \vec{e}_x = 1$. On P_3 each little arrow $\vec{dl}$ points to the *left,* and hence $\vec{dl} = -dx\,\vec{e}_x$. Hence

$$\int_{P_3} \vec{V} \cdot dl = \int_{x=0}^{x=x_o} (V_o\vec{e}_x) \cdot (-dx\,\vec{e}_x) = -V_o x_o$$

The contribution of P_3 is negative because the direction of travel on this part of the path is opposite to the direction of $\vec{V}$. The entire integral is

$$I_1 = \int_{P_1} + \int_{P_2} + \int_{P_3} + \int_{P_4} = 0$$

This particular vector field happens to have the property that its line integral around *any* closed path is zero. A field with this property is known as a *conservative field.* It is easy to show that for a conservative field the line integral from point A to point B has the same value regardless of which path is chosen to go from A to B.

EXAMPLE 3.2

Let the vector field $\vec{V}$ be given by $\vec{V} = V_o\vec{e}_x$. Find the line integral of $\vec{V}$ over the semicircular path of Fig. 3.7(a).

Solution Let us consider the contribution of the path segment located at the angle θ (Fig. 3.7(b)). The length of this segment is $a\ d\theta$, and from trigonometry

$$\vec{dl} = |dl| \cos\phi \vec{e}_x + |dl| \sin\phi \vec{e}_y$$

However, $\phi = \theta - 90°$; hence

$$\vec{dl} = |dl| \sin\theta \vec{e}_x - |dl| \cos\theta \vec{e}_y$$

$$= a\ d\theta(\sin\theta \vec{e}_x - \cos\theta \vec{e}_y)$$

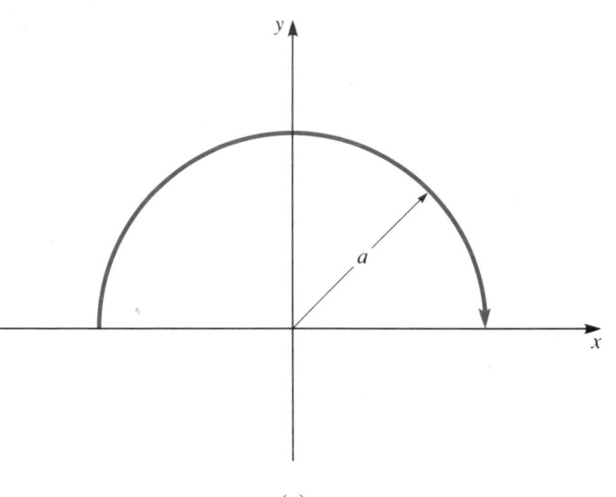

(a)

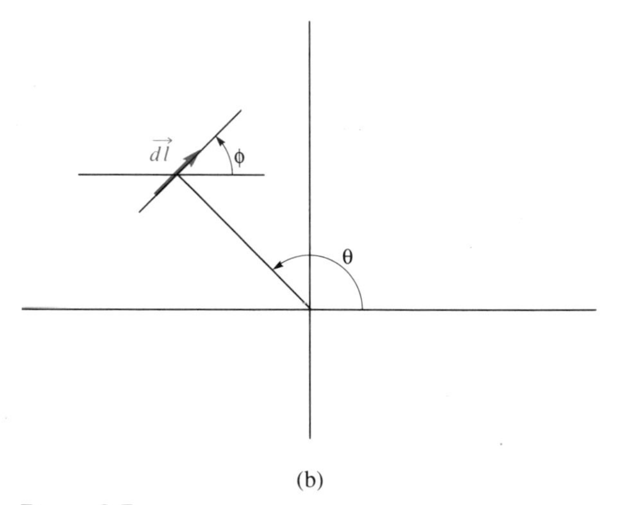

(b)

Figure 3.7

The integral is obtained by adding the contributions from all the sections $d\theta$:

$$I = \int_{\theta=0}^{\theta=180^\circ} (V_o \vec{e}_x) \cdot (\sin\theta\, \vec{e}_x - \cos\theta\, \vec{e}_y)\, a\, d\theta$$

$$= aV_o \int_0^{180^\circ} \sin\theta\, d\theta = 2aV_o$$

Computational note: When integrating over dx or $d\theta$, one should always integrate from the lower limit to the higher (e.g., from $x = 0$ to $x = 1$, not from $x = 1$ to $x = 0$), regardless of the direction of the path. The direction of the path is taken care of by the dot product.

SURFACE INTEGRALS

Surface integration amounts to adding up normal components of a vector field over a given surface S. Let this surface be broken up into small elements. To each bit of surface we assign a vector $\vec{dS}$. Its magnitude, $|\vec{dS}|$, is equal to the area of the surface element, and its direction is *perpendicular* to the surface at that point. (If S is a *closed* surface,[2] the vector $\vec{dS}$ is by convention directed *outward*. Otherwise physical reasoning must be used to choose which of the two normal directions is that of $\vec{dS}$.) We then use the value of the vector field $\vec{V}$ at the position of the surface element, and take its dot product with $\vec{dS}$; the result is a small (differential) scalar. The sum of these scalar contributions over all the surface elements is the surface integral, which is itself a scalar. The notation for the surface integral is

$$\int_S \vec{V} \cdot \vec{dS}$$

We recall that $\vec{V} \cdot \vec{dS} = |\vec{V}|\, |\vec{dS}| \cos\theta$, where θ is the angle between the vectors $\vec{V}$ and $\vec{dS}$. Hence $\vec{V} \cdot \vec{dS}$ is equal to the product of the area $|\vec{dS}|$ and the *normal component* of $\vec{V}$. This provides us with a physical interpretation of the surface integral: it can be thought of as the flow of the vector field through the surface S. For example, suppose the vector field $\vec{V}$ represents the flow of water at any point. Let the direction of $\vec{V}$ be the direction of flow, and let its magnitude $|\vec{V}|$ be the rate of flow per unit area (gallons per second per square meter). Then at points where the flow is parallel to the surface, $\vec{V} \cdot \vec{dS}$ is zero and there is no contribution to the integral. When $\vec{V}$ is perpendicular to $\vec{dS}$, the contribution to flow across the surface is maximum. The integral $\int_S \vec{V} \cdot \vec{dS}$ then represents the rate at which water is going across the entire surface, in gallons per second. If the surface S is closed, this integral will have to turn out to be zero, because (assuming

---

[2]A closed surface is one that is continuous, with no breaks, such as, for example, a balloon.

that the water is uniform and incompressible) the amount of water inside the volume bounded by S must be constant. Hence, the vector field describing water flow must have the property that its surface integral over any closed surface vanishes. A vector field with this property is said to be *solenoidal*. Of course, not all vector fields have this property.

EXAMPLE 3.3

Let the surface S_1 lie in the plane $z = 0$, and be a square with sides of length a. Let the vector field $\vec{V}$ have the constant value $2\vec{e}_x - 3\vec{e}_y - 4\vec{e}_z$. Evaluate $\int_{S_1} \vec{V} \cdot \vec{dS}$. Let the normal to S_1 be toward the $+z$ direction.

Solution In this case $\vec{dS} = |\vec{dS}|\vec{e}_z$, and $\vec{V} \cdot \vec{dS} = -4|\vec{dS}|$. The integral of $|\vec{dS}|$ is the total area of S_1. Hence $\int_{S_1} \vec{V} \cdot \vec{dS} = -4a^2$.

EXAMPLE 3.4

Let $\vec{V}$ be the radius vector $x\vec{e}_x + y\vec{e}_y + z\vec{e}_z$, and let S_2 be a sphere of radius b centered at the origin. Find $\int_{S_2} \vec{V} \cdot \vec{dS}$.

Solution The radius vector has the property that it points radially outward from the origin to the point (x, y, z). Hence, $\vec{V}$ is perpendicular to S_2 over the entire spherical surface. Hence, $\vec{V} \cdot \vec{dS} = |\vec{V}|\ |\vec{dS}| = (\sqrt{x^2 + y^2 + z^2})\ |\vec{dS}| = b\ |\vec{dS}|$. The integral is obtained by summing $|\vec{dS}|$ over the entire area of the sphere $(4\pi b^2)$, giving $\int_{S_2} \vec{V} \cdot \vec{dS} = 4\pi b^3$.

EXAMPLE 3.5

Again let $\vec{V}$ be the radius vector $x\vec{e}_x + y\vec{e}_y + z\vec{e}_z$, and now let the surface S_3 be defined by $z = c$, $-d < x < d$, $-d < y < d$. Find $\int_{S_3} \vec{V} \cdot \vec{dS}$. The normal to the surface is directed in the $+z$ direction.

Solution The situation is illustrated in Fig. 3.8. The vector $\vec{V}$ is not perpendicular to S_3, except at one point on the z axis: The cosine of the angle between $\vec{V}$ and $\vec{dS}$ is $c/\sqrt{x^2 + y^2 + c^2}$. Also $|\vec{V}| = \sqrt{x^2 + y^2 + c^2}$, and $|\vec{dS}| = dx\,dy$. Thus

$$\int_{S_3} \vec{V} \cdot \vec{dS} = \int_{x=-d}^{x=d} \int_{y=-d}^{y=d} \sqrt{x^2 + y^2 + c^2}\ \frac{c}{\sqrt{x^2 + y^2 + c^2}}\ dx\,dy$$

$$= 4\,d^2c$$

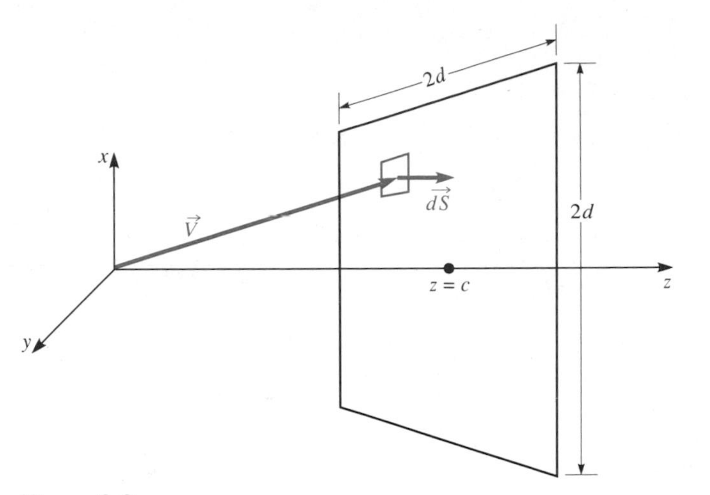

Figure 3.8

3.4 INTRODUCTION TO DIFFERENTIAL OPERATORS

In the last section we considered two operations on the vector field that involve integration. Now we shall consider operations involving differentiation. Just as in ordinary calculus, differential operations on the vector field are somewhat simpler than integral operations (in the sense that less ingenuity and labor are required to perform them). However, the differential operations have peculiar names and symbols, and some of them are difficult to visualize physically. As a result, they are sometimes seen as an obstacle in the study of electromagnetics. In the following chapters, we shall be introducing the important differential operators as they are needed. However, at this point it should be useful to present the general ideas, so they will not be unfamiliar when we meet them later.

An operator acts *on* a vector field *at* a point, to produce some function of the vector field. It is like a function of a function. Let us imagine, as a simple starting point, a function $f(x)$ of the single variable x. Let O be an operator. Then what happens when O acts on $f(x)$? The result is written $O[f(x)]$, and means that first f acts on x, and then O acts on f. For example, let $f(x) = x^2$, and let the operator O be $\left(\dfrac{d}{dx} + 2\right)$. Then

$$O[f(x)] = \frac{d}{dx}(x^2) + 2(x^2) = 2x(x + 1)$$

EXERCISE 3.8

Let $f(x) = 2x^2 + 3x + 4$, and let the operator O be $\left(3\frac{d^2}{dx^2} - 2\frac{d}{dx} + 1\right)$. Find the value of $O[f(x)]$ evaluated at $x = 2$.

Answer 8.

In the case of vector fields, the operator O acts on $\vec{V}(x,y,z)$ just as above it acted on $f(x)$. Now, however, O is acting on a vector function of three variables instead of a scalar function of one variable. Also, an important point to note is that *$O[\vec{V}(x,y,z)]$ can be either a scalar or a vector,* depending on what operator O happens to be.

EXAMPLE 3.6

Let the operator O be the *length operator,* defined by $O(\vec{A}) = \sqrt{\vec{A} \cdot \vec{A}}$. Let the vector field $\vec{V}$ be defined by $\vec{V} = 3y\vec{e}_x + z\vec{e}_y$. Evaluate $O(\vec{V})$ at the point $x = 1$, $y = 2$, $z = -2$.

Solution The length of the vector $\vec{V}$ is

$$O(\vec{V}) = \sqrt{\vec{V} \cdot \vec{V}} = \sqrt{9y^2 + z^2}$$

The value of $O(\vec{V})$ at the given position is $\sqrt{40} = 6.32$. This is an example of a *scalar* operator acting on a vector field.

EXAMPLE 3.7

The operator O_1 is defined by

$$O_1(\vec{A}) = \vec{A}\sqrt{\vec{A} \cdot \vec{A}} + 2\vec{A}$$

Let O_1 operate on the vector field of the previous example. Find $O_1(\vec{V})$ at $x = 1$, $y = 2$, $z = -2$.

Solution In the previous example we found that $\sqrt{\vec{V} \cdot \vec{V}} = \sqrt{9y^2 + z^2}$. Thus

$$O_1(\vec{V}) = (3y\vec{e}_x + z\vec{e}_y)\sqrt{9y^2 + z^2} + 6y\vec{e}_x + 2z\vec{e}_y$$

At the given position

$$O_1(\vec{V}) = (6\vec{e}_x - 2\vec{e}_y)\sqrt{40} + 12\vec{e}_x - 4\vec{e}_y$$

$$= 49.95\vec{e}_x - 16.65\vec{e}_y$$

Note that in this example the operator acts on the vector field to produce a *vector*.

Vector fields are very often specified in terms of their rectangular components, as follows

$$\vec{V}(x,y,z) = V_x(x,y,z)\vec{e}_x + V_y(x,y,z)\vec{e}_y + V_z(x,y,z)\vec{e}_z \tag{3.9}$$

Here V_x, V_y, and V_z are three scalar functions of position. When they have been specified, the vector field is fully defined. Operators can then be specified in terms of V_x, V_y, and V_z.

EXAMPLE 3.8

The *divergence operator* is a special operator, with the symbol $\nabla\cdot$, defined by

$$\nabla\cdot\vec{V} = \frac{\partial}{\partial x}V_x + \frac{\partial}{\partial y}V_y + \frac{\partial}{\partial z}V_z$$

where V_x, V_y, V_z are as defined in (3.9). Let the vector field be $\vec{V} = x^2\vec{e}_x + y\vec{e}_y + (2 + x)\vec{e}_z$. Evaluate $\nabla\cdot\vec{V}$ at the point $x = 1$, $y = -1$, $z = 2$.

Solution We have $V_x = x^2$, $V_y = y$, $V_z = (2 + x)$. Hence,

$$\frac{\partial}{\partial x}V_x = 2x, \qquad \frac{\partial}{\partial y}V_y = 1, \qquad \text{and} \qquad \frac{\partial}{\partial z}V_z = 0$$

(because the partial derivative of x with respect to z vanishes). Thus

$$\nabla\cdot\vec{V} = 2x + 1$$

and at the given point its numerical value is 3.

Clearly the divergence operator is a scalar operator.

The divergence operator seen in this last example will be of importance to us later. The other important differential operators—gradient, curl, and Laplacian—will be defined as they arise.

3.5 POINTS TO REMEMBER

1. Vectors are mathematical quantities used to describe both magnitude and direction. In three-dimensional space a vector consists of an ordered set of three scalar numbers. In rectangular coordinates these numbers are known as the vector's rectangular components. The magnitude, or "length," of the vector, is the square root of the sum of the squares of the rectangular components.

2. The *radius vector* at a certain point is the vector directed from the origin to the point. The rectangular components of the radius vector at the point (x, y, z) are simply x, y, and z.

3. Vectors are added by adding their individual components. The dot product of two vectors is a scalar. Its value is equal to the product of the magnitudes of the two original vectors and the cosine of the angle between them. The cross product of two vectors is a vector that is perpendicular to both of the original vectors. Its magnitude is equal to the product of the magnitudes of the two original vectors and the sine of the angle between them.

4. A field is an assignment of some value to each position. There are both scalar fields and vector fields. In the case of a vector field, some vector value is assigned to each position.

5. A line integral can be defined by specifying a certain path of integration through a vector field. It represents the sum of the component of the field parallel to each path element, times the length of the path element.

6. A surface integral can be defined by specifying a certain surface in the presence of a vector field. It represents the sum of the component of the field normal to each surface element, times the area of the element.

7. An operator is a mathematical operation that acts on a scalar or vector field to give a different scalar or vector field. Four different types exist (scalar-valued operators acting on scalar fields, vector-valued operators acting on scalar fields, etc.). The operators of interest to us will involve partial differentiation, as for example in the *divergence operator*.

REFERENCES

All elementary texts on electromagnetics contain some review of this material. For example, see:

1. W. H. Hayt, *Engineering Electromagnetics,* 4th ed. New York: McGraw-Hill, 1981.
2. D. K. Cheng, *Field and Wave Electromagnetics*. Reading, Mass.: Addison-Wesley, 1983.

A good mathematical reference is:

3. H. Margenau and G. M. Murphy, *The Mathematics of Physics and Chemistry*, 2nd ed. New York: Van Nostrand, 1956.

PROBLEMS

Section 3.1

3.1 Sketch the following vectors in the x-y plane:
(a) $\vec{V}_1 = 3\vec{e}_x - 2\vec{e}_y$
(b) $\vec{V}_2 = -2\vec{e}_x + \vec{e}_y$
(c) $\vec{V}_1 + \vec{V}_2$
(d) $\vec{V}_1 - \vec{V}_2$

3.2 Let $\vec{V}_1 = -\vec{e}_x - 2\vec{e}_y$, $\vec{V}_2 = 2\vec{e}_x + \vec{e}_y$.
(a) Sketch $\vec{V}_1$ and $\vec{V}_2$ in the x-y plane.
(b) Find $\vec{V}_1 + \vec{V}_2$ numerically.
(c) Add the vectors $\vec{V}_1$ and $\vec{V}_2$ graphically. Does your result agree with (b)?

3.3 Let $\vec{V}_1$ and $\vec{V}_2$ be as defined in Problem 3.2. Find:
(a) $|\vec{V}_1|$
(b) $|\vec{V}_2|$
(c) $|(\vec{V}_1 + \vec{V}_2)|$

3.4 Show, graphically or otherwise, that for any two vectors $\vec{V}_1$, $\vec{V}_2$,

$$|\vec{V}_1 + \vec{V}_2| \leq |\vec{V}_1| + |\vec{V}_2|$$

(A two-dimensional proof will be sufficient. Do you think the statement also holds for three dimensions?)

3.5 Find the unit vector in the direction from the origin toward the point ($x = -2$, $y = 3$, $z = 2$).

3.6 Let $\vec{r}_1$ and $\vec{r}_2$ be the radius vectors for two points P_1, P_2 in the x-y plane.
(a) Show, by means of a sketch, that $\vec{r}_2 - \vec{r}_1$ is the vector from P_1 to P_2.
(b) Show that

$$\vec{e}_{21} = \frac{\vec{r}_2 - \vec{r}_1}{|\vec{r}_2 - \vec{r}_1|}$$

is a unit vector in the direction from P_1 to P_2.

3.7 Let $\vec{V}_1 = 46\vec{e}_x - 27\vec{e}_y - 52\vec{e}_z$; $\vec{V}_2 = 10\vec{e}_x - 17\vec{e}_z$. Find:
(a) $\vec{V}_1 \cdot \vec{V}_2$
(b) $\vec{V}_2 \cdot \vec{V}_1$
(c) the angle between these vectors.

3.8 Prove that for any two vectors $\vec{V}_1$ and $\vec{V}_2$

$$|\vec{V}_1 - \vec{V}_2|^2 = |\vec{V}_1|^2 - 2\vec{V}_1 \cdot \vec{V}_2 + |\vec{V}_2|^2$$

3.9 Let $\vec{V}_1$ be a given vector, and let $\vec{x}$ be any vector from the origin to point P satisfying

$$\vec{x} \cdot \vec{V}_1 = 1$$

Show that the locus of possible points P is a plane perpendicular to $\vec{V}_1$.

3.10 Let $\vec{V}_1$, $\vec{V}_2$ be as defined in Problem 3.1. Find:
(a) $\vec{V}_1 \times \vec{V}_2$
(b) $\vec{V}_2 \times \vec{V}_1$

3.11 Let $\vec{r}$ be the radius vector to the point P, and let $\vec{r} \times \vec{e}_x = \vec{e}_z$. Show that the locus of P is a straight line. Describe this line.

3.12 Verify (3.8), assuming, for simplicity, that all three vectors lie in the x-y plane.

For a review of curvilinear coordinates, see Appendix A.

3.13 Let r and ϕ be cylindrical coordinates in the plane $z = 0$. Sketch the vector $-\vec{e}_r + 2\vec{e}_\phi$
(a) at a point where $\phi = 30°$, and
(b) at a point where $\phi = -70°$.

3.14 The cylindrical coordinates of point P are ($r = 10$, $\phi = -110°$, $z = 6$). Express P in rectangular coordinates.

3.15 Express the position ($x = -2$, $y = 3$, $z = -5$) in cylindrical coordinates.

3.16 In cylindrical coordinates, $\vec{V}_1 = 2\vec{e}_r - 3\vec{e}_\phi$, $\vec{V}_2 = -\vec{e}_r - \vec{e}_\phi$. Find
(a) $\vec{V}_1 \cdot \vec{V}_2$
(b) the angle between the two vectors.

3.17 Express the following positions in spherical coordinates:
(a) ($x = -70$, $y = 30$, $z = -50$)
(b) ($r = 10$, $\phi = 60°$, $z = -8$), where r, ϕ, z are cylindrical coordinates.

3.18 Make a careful three-dimensional sketch of the rectangular coordinate system, similar to Fig. 3.1. Let the $\theta = 0$ axis of a system of spherical coordinates coincide with the z axis, and the $\phi = 0$ direction with the x axis. Locate the following points:
(a) $r = 3$, $\theta = 180°$, $\phi = 0$
(b) $r = 3$, $\theta = 90°$, $\phi = 90°$
(c) $r = 3$, $\theta = 120°$, $\phi = -90°$.

3.19 Refer to a world globe or atlas. Let the polar axis of spherical coordinates point toward the north pole and let $\phi = 0$ correspond to the longitude of Greenwich (i.e., London). Find the spherical coordinates of:

(a) Sydney, Australia, and

(b) your present location.

3.20 Let $\vec{V}_1 = 2\vec{e}_r - 4\vec{e}_\theta + \vec{e}_\phi$, $\vec{V}_2 = -\vec{e}_r + 2\vec{e}_\theta + \vec{e}_\phi$. Find:

(a) $\vec{V}_1 \cdot \vec{V}_2$

(b) $\vec{V}_1 \times \vec{V}_2$.

3.21 Let $f(r,\theta,\phi) = \dfrac{\cos\theta}{r^2}$ (in spherical coordinates). Sketch the surfaces $f = 1$ and $f = -2$ in the plane $\phi = 0$. What do these surfaces look like in three dimensions?

Section 3.2

3.22 A section of water pipe is shaped like a capital letter "S." Water flows smoothly through the pipe. Sketch the vector field representing water velocity.

3.23 Consider the vector field $\vec{V}(x,y,z) = x\vec{e}_x + y\vec{e}_y + z\vec{e}_z$. Show that at any point P (except the origin) the direction of the field is the same as the direction from the origin to P. Sketch the field, using the style of Fig. 3.4(a).

3.24 Sketch (using the style of Fig. 3.4a) the vector field $\vec{V}(x,y) = \sin kx\, e^{ky}\, \vec{e}_y$ over the range $-\dfrac{\pi}{k} < x < \dfrac{\pi}{k}$, $0 < y < \dfrac{1}{k}$.

*3.25 Sketch, using the style of Fig. 3.4a, the magnitude and direction of the vector field

$$\vec{V}(x,y) = \cos kx\, e^{ky}\, \vec{e}_x + \sin kx\, e^{ky}\, \vec{e}_y$$

on the line $-\dfrac{\pi}{k} < x < \dfrac{\pi}{k}$, $y = 0$.

**3.26 Sketch the vector field (in spherical coordinates)

$$\vec{E}(r,\theta,\phi) = \frac{2\cos\theta\vec{e}_r + \sin\theta\vec{e}_\theta}{r^3}$$

in the plane $\phi = 0$. Begin with the style of Fig. 3.4(a); then construct a drawing in the style of Fig. 3.4(b).

Section 3.3

3.27 (a) Let $\vec{V}(x,y) = V_o x\vec{e}_x$. Find the line integral of $\vec{V}$ over the closed path shown in Fig. 3.6.

(b) Repeat with the field changed to $\vec{V}(x,y) = V_o y\vec{e}_x$.

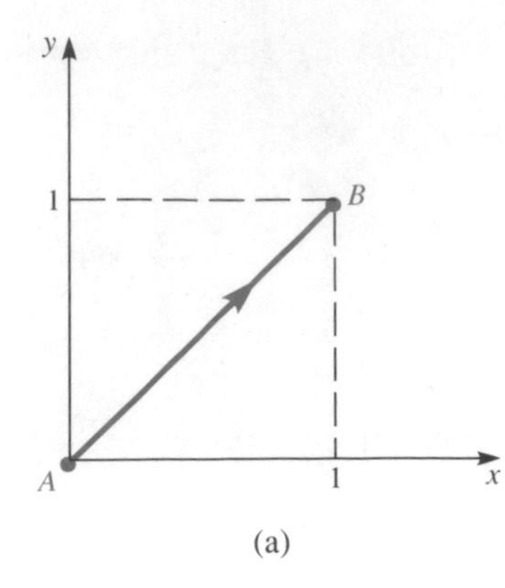

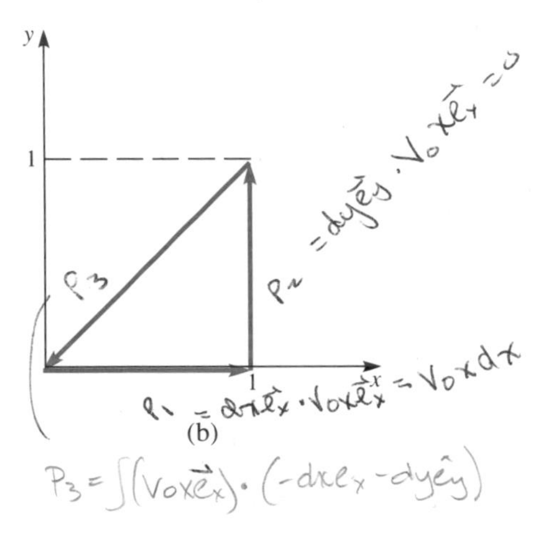

Figure 3.9 For Problem 3.29.

3.28 Find the line integral of $\vec{V}(x,y) = V_o x\vec{e}_x$ on the contour from A to B shown in Fig. 3.9(a).

3.29 Find the line integral of $\vec{V}(x,y) = V_o y\vec{e}_x$ on the contour of Fig. 3.9(a).

*3.30 Find the line integral of the field $\vec{V}(x,y) = V_o x\vec{e}_x$ on the closed triangular contour shown in Fig. 3.9(b). (*Note:* This problem is instructive because it tempts one to make a common sign error. The correct answer is zero. See the last paragraph of Example 3.2.)

*3.31 Repeat Example 3.2 with the vector field changed to $\vec{V} = V_o x\vec{e}_x$.

*3.32 Consider a parabolic path of integration that follows the curve $y = Kx^2$ from the origin to the point $(x = a, y = Ka^2)$. Find the line integral of the field $\vec{V} = Ay\vec{e}_x + B\vec{e}_y$ on this path.

*3.33 Verify the following statements:

(a) The line integral of a field from A to B is the negative of its line integral along the same path from B to A.

(b) If a field is conservative, its line integral from point A to point B is the same regardless of what path is followed.
(c) If the line integral between any two paths is independent of the path chosen, the field is conservative.

3.34 Let the surface S be the rectangle $-a < x < a$, $-b < y < b$, in the plane $z = 0$, and let the normal to S be in the $+z$ direction. Find the surface integral over S of the field $\vec{V} = Ax\vec{e}_x + By^2\vec{e}_z$.

*3.35 Consider the hemispherical surface described by $r = a$, $0 < \theta < 90°$. Let $\vec{V} = A\vec{e}_z$, where the z axis is the same as the polar axis. Find the surface integral. (Let the normal to the surface be in the direction away from the origin.)

*3.36 Repeat the preceding problem with the field changed to $\vec{V}(x,y,z) = Ax^2\vec{e}_z$.

*3.37 The field $\vec{V}$ is described in cylindrical coordinates by $\vec{V}(r,\phi,z) = r^3\vec{e}_\phi$. Let the surface S be described by $x = a$, $0 < y < b$, $-c < z < c$. The x direction coincides with the direction $\phi = 0$, and the normal to S is in the $+x$ direction. Find the surface integral of $\vec{V}$ over S.

Section 3.4

3.38 Find the divergence $\nabla\cdot\vec{V}$ of the vector field $\vec{V} = x\vec{e}_x + y\vec{e}_y + z\vec{e}_z$.

3.39 An example of a vector differential operator is the gradient operator, with the symbol ∇, defined by

$$\nabla f = \frac{\partial f}{\partial x}\vec{e}_x + \frac{\partial f}{\partial y}\vec{e}_y + \frac{\partial f}{\partial z}\vec{e}_z$$

where f is any differentiable scalar field. Find the gradient of the field $f(x,y,z) = Axy^2 \sin z$, as a function of position.

CHAPTER 4

The Electrostatic Field

Fundamentally, the science of electromagnetics deals with the ways in which charged particles interact with one another. In the early days, various electromagnetic interactions were observed in experiments, but there was no general theory to explain them. To bring order and structure into the subject, the great thinkers of the time introduced the idea of electric and magnetic fields. These are two separate vector fields that, in the usual point of view, are generated by charges and currents. The fields in turn make themselves known by exerting forces on charged particles. In reality, the electromagnetic fields are only a bookkeeping device, introduced to allow us to describe the otherwise mysterious phenomenon of action at a distance. But they are such an extremely useful tool for describing electromagnetic phenomena, and bring such beautiful order into the subject, that electromagnetics is seldom described in any other way.

In the present chapter we shall introduce the electric field. In particular we shall be concerned with situations in which the electric and magnetic fields do not vary with time; the electric field in such a case is known as the *electrostatic* field. Chapter 5 will introduce the magnetostatic field in similar fashion. The remainder of the book, however, will be devoted to the more interesting time-varying case. As we shall see in Chapter 6, the electric and magnetic fields are not really unrelated; once we consider time-varying fields, interactions between the electric and magnetic fields will be found to occur. These *electromagnetic* phenomena, described by the famous equations of James Clerk Maxwell, are responsible for electromagnetic waves, which are the basis of electronic communication in all its forms.

4.1 SOURCE OF THE ELECTROSTATIC FIELD

An electric field is set up whenever charged particles are present. In the present chapter we are dealing with electrostatics, and therefore assume that no time-varying fields exist. This assumption allows us to use simple formulas to calculate the electric field when the positions of the charges are known.

The simplest case is the field of a single point charge Q located at the origin. By experiment it is found that this field is directed away from the charge and has a magnitude proportional to r^{-2}, where r is the distance from the charge to the point where the field is measured. Thus, after introducing the appropriate constant of proportionality, we can write

$$\vec{E} = \frac{Q}{4\pi\epsilon_o r^2}\vec{e}_r \tag{4.1}$$

Here $\vec{e}_r$ is a unit vector pointing in the radial direction, that is, in the direction *away* from the charge. In the international system of units (known as SI units), which is almost universally used in practical work, Q is measured in *coulombs* and $\vec{E}$ in *volts per meter*. The constant ϵ_o is known as the *electric permittivity* of vacuum. Its value is

$$\epsilon_o = 8.854 \times 10^{-12} \text{ SI units}$$

If the region of interest does not contain vacuum, but instead is uniformly filled with some material, equation (4.1) still applies, but ϵ_o is replaced by a different number ϵ. In that case the value of ϵ is determined by the material. The permittivity of air is almost exactly equal to that of vacuum. For other common materials, ϵ can be set equal to $\epsilon_R\epsilon_o$, where ϵ_R is called the *relative permittivity*. The value of ϵ_R is typically in the range 1–100.

If several point charges are present, the total electric field is obtained by adding together the fields of the individual charges. The fields are added as *vectors:* that is, the x components add together to give the x component of the sum, and so forth. Let there be several charges Q_i located at positions $\vec{r}_i$, with respect to the origin. The vector from the ith charge to the observation point $\vec{r}$ is $(\vec{r} - \vec{r}_i)$, and the distance is $|\vec{r} - \vec{r}_i|$. Thus

$$\vec{e}_{rr_i} = \frac{\vec{r} - \vec{r}_i}{|\vec{r} - \vec{r}_i|}$$

is a unit vector directed from Q_i to the observation point, and the total field is

$$\begin{aligned}\vec{E} &= \sum_i \frac{Q_i}{4\pi\epsilon|\vec{r} - \vec{r}_i|^2}\vec{e}_{rr_i} \\ &= \sum_i \frac{Q_i}{4\pi\epsilon|\vec{r} - \vec{r}_i|^3}(\vec{r} - \vec{r}_i)\end{aligned} \tag{4.2}$$

EXAMPLE 4.1

A positive charge Q is located on the x axis at $x = b$, and an equal but opposite negative charge $-Q$ is located at $x = -b$, as shown in Fig. 4.1. Find the electric field at position P located on the y axis at $y = d$.

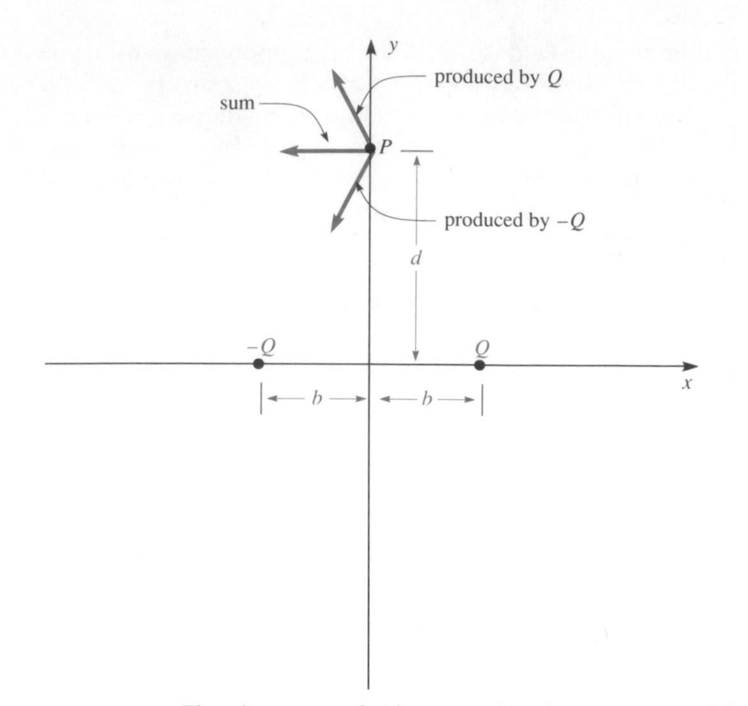

Figure 4.1 The electrostatic field at point P is the vector sum of the fields produced by the two charges Q and $-Q$.

Solution Before beginning any mathematics, it is very useful to first visualize the fields and determine which components will be present at P. To add the contributions of the two charges it is necessary to add up separately their x, y, and z components. It will save effort if we can see by inspection that some of these sums will have to be zero.

First of all, the vector from Q to P lies in the x-y plane. Therefore it has no z component, and hence from (4.2) the electric field produced by Q at P will have no z component. The same is true for the other charge, and hence it is evident that E_z (the z component of $\vec{E}$) vanishes at P.

Now let us consider E_y. The contributions of the two charges to the field at P are sketched in the figure. We see that their y components are equal and opposite and therefore add to zero. Thus before doing any calculation, we already know that $\vec{E}$ has only an x component.

The x component of any electric field is the product of $|\vec{E}|$ and the cosine of the angle made by $\vec{E}$ with the x axis. In this case for the field arising from Q that cosine is equal to $-b/\sqrt{b^2 + d^2}$. The field arising from $-Q$ makes the same cosine with the x axis. The total field is the sum of the contributions from the two charges:

$$E_x = -\frac{2Qb}{4\pi\epsilon(b^2 + d^2)^{3/2}}$$

In the real world, engineers seldom deal with single point charges. Usually a very large number of charges are present. We can, for example, consider a charged wire, with a certain amount of charge per unit length. The field of this wire can be found by imagining the wire to be composed of small charges Δq, spaced by a small distance Δl. The actual charge per unit length, τ, must then be equal to $\Delta q/\Delta l$. The field is found from (4.2),

$$\vec{E}(\vec{r}) = \sum_i \frac{\Delta q}{4\pi\epsilon} \frac{\vec{r} - \vec{r}_i}{|\vec{r} - \vec{r}_i|^3}$$

where as before, $\vec{r}_i$ is the position of the ith charge. This can be rewritten as

$$\vec{E}(\vec{r}) = \frac{1}{4\pi\epsilon} \sum_i \frac{\vec{r} - \vec{r}_i}{|\vec{r} - \vec{r}_i|^3} \frac{\Delta q}{\Delta l} \Delta l$$

If we now let $\Delta l \rightarrow 0$, keeping $\Delta q/\Delta l = \tau$, this expression becomes an integral:

line charge

$$\vec{E}(\vec{r}) = \frac{1}{4\pi\epsilon} \int \frac{\vec{r} - \vec{r}'}{|\vec{r} - \vec{r}'|^3} \tau(\vec{r}')\, dl'$$

where $\vec{r}'$ is the position of the incremental length dl'. (Note the difference between $\vec{r}$ and $\vec{r}'$. The former is the place where the field is measured, called the *field point*. The symbol $\vec{r}'$ represents the location of a bit of charge; it is called the *source point*.)

Charges can also be distributed on a surface, in which case it is convenient to consider the *surface charge density* σ, the units of which are coulombs per square meter. The evident extension of (4.2) to this case is

$$\vec{E}(\vec{r}) = \int \frac{\sigma(\vec{r}')\vec{e}_{rr'}\, dA'}{4\pi\epsilon|\vec{r} - \vec{r}'|^2}$$

$$= \int \frac{\sigma(\vec{r}')(\vec{r} - \vec{r}')\, dA'}{4\pi\epsilon|\vec{r} - \vec{r}'|^3} \qquad \textbf{(4.3)}$$

Here the integral is a surface integral taken over the surface bearing the charge, $\vec{e}_{rr'}$ is the unit vector directed from $\vec{r}'$ to $\vec{r}$, $\sigma(\vec{r}')$ is the surface charge density at the position $\vec{r}'$, and $\vec{r}$ is the position of observation, as before. The charge may also be distributed through a volume, in which case we speak of the *volume charge density* ρ, with units coulombs per cubic meter. The electric field in that case is

$$\vec{E}(\vec{r}) = \int \frac{\rho(\vec{r}')\vec{e}_{rr'}\, dV'}{4\pi\epsilon|\vec{r} - \vec{r}'|^2} \qquad \textbf{(4.4)}$$

where the integral is now over the volume containing the charge. If some of the charge were in the form of point charges, some of surface charge, and some of volume charge, we would simply add (vectorially) the contributions of (4.2), (4.3), and (4.4).

EXAMPLE 4.2

A circular disk of radius b, with uniform surface charge density σ, lies in the x-y plane with its center at the origin. Find the electric field at the point P located on the z axis, at $z = h$.

Solution The disk is seen edgewise in Fig. 4.2. Let us consider the contributions of two small areas on opposite sides of the disk, at the same distance from its center. From the diagram we see that their contributions in the x direction cancel, while those in the z direction add. By always considering pairs of areas at opposite positions, we then see that all contributions in the x and y directions are cancelled, and the field at P will contain only E_z.

To perform the integral (4.3) we shall add the contributions of rings of radius u and width du, lying in the x-y plane. The area of one of those rings is $2\pi u\,du$, its distance from P (which is $|\vec{r} - \vec{r}'|$) is $(h^2 + u^2)^{1/2}$, and the cosine of the angle between its contribution and the z axis is $h/(h^2 + u^2)^{1/2}$. Thus from (4.3)

$$E_z = \int_0^b \frac{\sigma}{4\pi\epsilon(h^2 + u^2)} \frac{h}{(h^2 + u^2)^{1/2}} 2\pi u\,du$$

Carrying out the integration we find that

$$E_z = \frac{\sigma}{2\epsilon}\left(1 - \frac{h}{(h^2 + b^2)^{1/2}}\right)$$

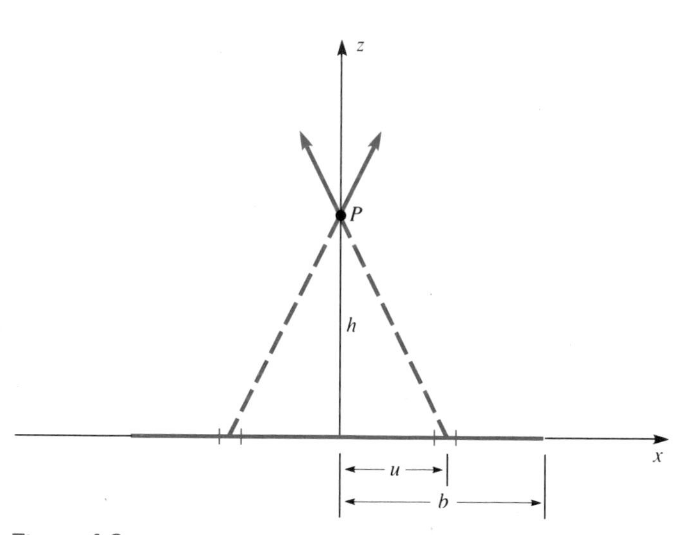

Figure 4.2

EXERCISE 4.1

A spherical surface of radius b has a uniform surface charge density σ. The charges are firmly attached to the surface and are not free to move. The surface is placed with its center at the origin, and is then cut in half, leaving only the hemisphere in the region $z > 0$. Find the electric field at the origin.

Answer $\sigma/4\epsilon$, in the $-z$ direction.

4.2 THE SOURCE EQUATION

As we have seen, the electric field can be expressed as an integral of the charges that produce it. Thus it should not be surprising that the relationship can also be turned around: that is, the charge can be expressed in terms of derivatives of the field. The equation expressing this relationship, which we shall call the *source equation,* is written:

$$\nabla \cdot (\epsilon \vec{E}) = \rho \tag{4.5}$$

The quantity $\nabla \cdot \vec{E}$ is called the *divergence* of $\vec{E}$. The symbol $\nabla \cdot$ represents the *divergence operator,* a differential operator of the kind introduced in section 3.4.

THE DIVERGENCE OPERATOR

The divergence operator is of the type that acts on a vector field to produce a scalar. Let $\vec{E}$ be a function of x, y, and z, such that

$$\vec{E}(x,y,z) = E_x(x,y,z)\vec{e}_x + E_y(x,y,z)\vec{e}_y + E_z(x,y,z)\vec{e}_z \tag{4.6}$$

Then the divergence of $\vec{E}$ is given by

$$\nabla \cdot \vec{E} = \frac{\partial E_x}{\partial x} + \frac{\partial E_y}{\partial y} + \frac{\partial E_z}{\partial z} \tag{4.7}$$

EXAMPLE 4.3

A semiconductor pn junction can be approximated as one-dimensional: that is, none of its properties are functions of the coordinates x or y. The electrostatic field in the junction is

$$\vec{E} = A(z^2 - h^2)\vec{e}_z \qquad \text{for } -h < z < h$$

$$= 0 \qquad \text{elsewhere}$$

where A and h are given constants. The permittivity of the semiconductor is ϵ. Find the charge density.

Solution In this one-dimensional case, derivatives with respect to x and y must vanish. Thus $\nabla \cdot \vec{E} = \dfrac{\partial E_z}{\partial z}$, and from the source equation we have

$$\rho = \epsilon \nabla \cdot \vec{E} = \epsilon \frac{\partial}{\partial z}[A(z^2 - h^2)] = 2A\epsilon z \qquad \text{for } -h < z < h$$

$$= 0 \qquad \text{elsewhere}$$

The electric field and charge density are sketched in Fig. 4.3. This type of pn junction is known as a *linearly graded* junction; it is made by adding charged impurities to the semiconductor in a nonuniform way, so that the resulting charge density has the linear variation shown in the figure. Note that the source equation provides a convenient way to find the charge density when the electric field is known.

EXAMPLE 4.4

Let $\vec{E}$ be the field of a single point charge Q at the origin, as given by (4.2). Show that the divergence of $\vec{E}$ is zero everywhere except at the origin.

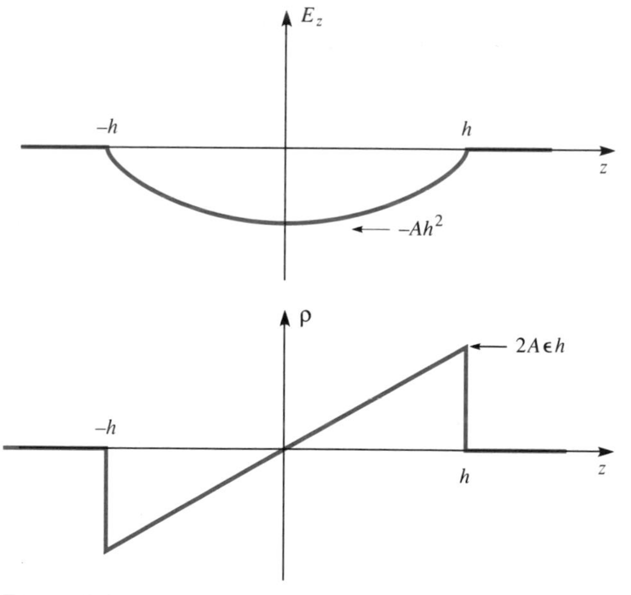

Figure 4.3

Solution Since there is only one point charge, which is located at $\vec{r}_i = 0$, (4.2) takes the form

$$\vec{E} = \frac{Q\vec{r}}{4\pi\epsilon|r|^3}$$

Noting that $\vec{r} = x\vec{e}_x + y\vec{e}_y + z\vec{e}_z$, we have

$$\vec{E} = \frac{Q}{4\pi\epsilon} \frac{x\vec{e}_x + y\vec{e}_y + z\vec{e}_z}{(x^2 + y^2 + z^2)^{3/2}}$$

the x component of which is

$$E_x = \frac{Q}{4\pi\epsilon} \frac{x}{(x^2 + y^2 + z^2)^{3/2}}$$

Taking its partial derivative with respect to x,

$$\frac{\partial E_x}{\partial x} = \frac{Q}{4\pi\epsilon} \frac{y^2 + z^2 - 2x^2}{(x^2 + y^2 + z^2)^{5/2}}$$

Now $\dfrac{\partial E_y}{\partial y}$ and $\dfrac{\partial E_z}{\partial z}$ are taken in similar fashion, and it is seen that the sum of the three derivatives is zero (except at the origin, where the derivatives are undefined).

The conclusion, that $\nabla \cdot \vec{E} = 0$ (except at the origin), is consistent with the source equation (4.5), which states that the divergence of $\vec{E}$ must vanish wherever there is no charge. A closer examination of the situation at the origin would show that the infinitely large divergence there is consistent with the infinitely large charge density of the point charge, but that is beyond the scope of the present discussion.

It sometimes happens that a vector field is expressed in cylindrical or spherical coordinates instead of in rectangular coordinates. (Curvilinear coordinate systems are reviewed in Appendix A.) If one needs to calculate the divergence, it is convenient to express the divergence operator in the same coordinate system. The rectangular form of the divergence operator was given in (4.7). In cylindrical coordinates it takes the form

$$\nabla \cdot \vec{E} = \frac{1}{r}\frac{\partial}{\partial r}(rE_r) + \frac{1}{r}\frac{\partial E_\phi}{\partial \phi} + \frac{\partial E_z}{\partial z} \tag{4.8}$$

and in spherical coordinates

$$\nabla \cdot \vec{E} = \frac{1}{r^2}\frac{\partial}{\partial r}(r^2E_r) + \frac{1}{r\sin\theta}\frac{\partial}{\partial\theta}(\sin\theta E_\theta) + \frac{1}{r\sin\theta}\frac{\partial E_\phi}{\partial\phi} \tag{4.9}$$

EXAMPLE 4.5

Let $\vec{E}$ be the electrostatic field of a point charge Q at the origin, as given by (4.2). Express this field in spherical coordinates and find its divergence by means of (4.9).

Solution From (4.2),

$$\vec{E} = \frac{Q}{4\pi\epsilon r^2}\vec{e}_r$$

From (4.9)

$$\nabla \cdot \vec{E} = \frac{1}{r^2}\frac{\partial}{\partial r}\left(\frac{Q}{4\pi\epsilon}\right) = 0$$

everywhere except at the origin, where the divergence is undefined (because the field is infinite). Note that less calculation has been required here than in the preceding example, where rectangular coordinates were used. This is because spherical coordinates are better suited to the spherical symmetry of the problem.

4.3 GAUSS' LAW

The relationship between the electrostatic field and the charges that produce it has already been expressed in two ways: (4.1) and (4.5). A third formulation of the same relationship is known as *Gauss' Law*. It follows from a general mathematical result known as the *divergence theorem*.

The divergence theorem relates the surface integral of a vector field to the divergence of the field inside the surface. The statement of the theorem begins with the free choice of any arbitrary closed surface.[1] This closed surface might be thought of as resembling a balloon: that is, it is like a thin film that can have any shape, but has no holes through which air might escape. If a vector field $\vec{V}(\vec{r})$ is present, we can construct the surface integral of $\vec{V}$ over the surface we have chosen, as described in section 3.3. This integral is written

$$\int_S \vec{V} \cdot \overrightarrow{dS}$$

Let us now construct $\nabla \cdot \vec{V}$ everywhere inside the surface S. It is a scalar function of position. We can integrate that scalar over the *volume*

[1]We assume that surfaces met in practice will be free of unpleasant mathematical peculiarities. A more general discussion of the divergence theorem will be found in mathematics texts.

enclosed by the surface S. According to the divergence theorem, this integral is related to the surface integral of $\vec{V}$ by

$$\int_V \nabla \cdot \vec{V}\, dV = \int_S \vec{V} \cdot \vec{dS} \tag{4.10}$$

where the volume integral is over the volume enclosed by the surface S. The fact that S can be chosen freely makes this a very powerful theorem.

To apply the divergence theorem to electrostatics, we make use of the source equation (4.5). This leads us to Gauss' law:

$$\int_S \epsilon\vec{E} \cdot \vec{dS} = \int_V \rho\, dV = Q \tag{4.11}$$

where Q is the total charge enclosed by the surface S. In words, the surface integral of the electric field over any surface is determined by the charge enclosed by that surface.

From (3.3) we note that

$$\text{flux} = \int_S \epsilon\vec{E} \cdot \vec{dS} = \int_S (\epsilon|\vec{E}| \cos\psi)\, dA$$

where $dA = |\vec{dS}|$ is simply a differential area, and ψ is the angle between $\vec{E}$ and the outward normal to the surface. We recognize $|\vec{E}| \cos\psi$ as the normal component of the electric field. Thus the surface integral in Gauss' law represents the normal component of $\epsilon\vec{E}$ integrated over the area of the surface. This surface integral can be thought of as the outflow of the field $\epsilon\vec{E}$ through the surface S. It is called the *flux* of the field $\epsilon\vec{E}$ through the surface S.

The divergence theorem is helpful in gaining an intuitive feeling for the divergence operator. Let us consider a small spherical surface. According to the divergence theorem, if the divergence of a vector field is zero inside the sphere, the flux of the field through the sphere must vanish; that is, just as much flux enters the surface as leaves. On the other hand, if the divergence is *positive* inside the surface, there must be a net *outflow* of flux. We sometimes refer to a region of positive divergence as a *source*. Thus, for example, a positive charge is a source of the electric field. Similarly, a region of negative divergence must be associated with an *inflow* of the field; such a region is referred to as a *sink*. A negative charge is a sink of the electric field. See Fig. 4.4.

EXAMPLE 4.6

Find the electric field at a distance r from a point charge q, by means of Gauss' law.

Solution We expect the field to be radial and spherically symmetric. Let the surface S be a sphere of radius r centered on the charge. Then the cosine

Figure 4.4 At a "source," the divergence of the field is positive, and field lines appear to originate and flow outward. At a "sink," the divergence of the field is negative; field lines appear to flow into such a region and disappear.

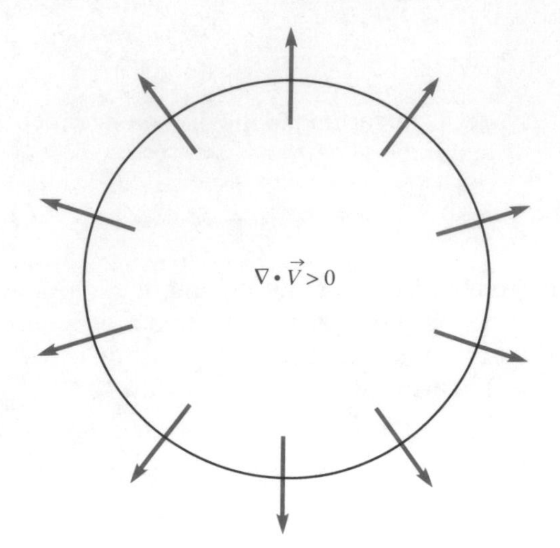

(a) Divergence is *positive:* region contains a *source*

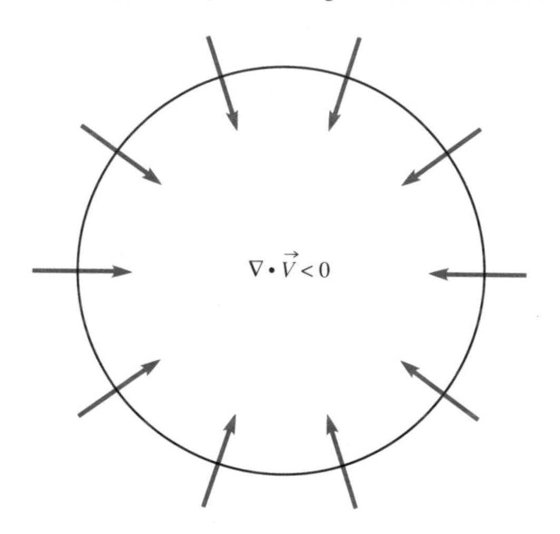

(b) Divergence is *negative:* region contains a *sink*

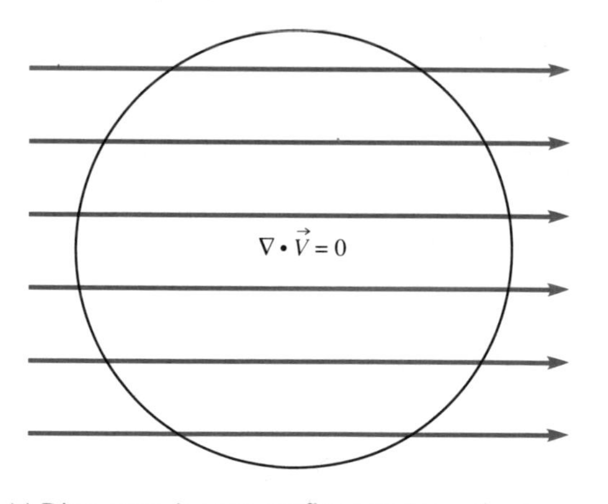

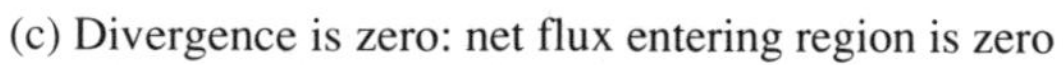

(c) Divergence is zero: net flux entering region is zero

factor in $\vec{E} \cdot \vec{ds}$ is unity, and (4.11) becomes

$$\int_S E_r \, dA = \frac{q}{\epsilon}$$

where $\vec{E} = E_r \vec{e}_r$ and $\vec{e}_r$ is the unit vector in the radial direction. Because of symmetry, E_r is the same everywhere on the sphere, and thus we find that

$$4\pi r^2 E_r = \frac{q}{\epsilon}$$

$$E_r = \frac{q}{4\pi\epsilon r^2}$$

in agreement with (4.1).

EXAMPLE 4.7

A uniform sphere of charge, with charge density ρ_o and radius b, is centered at the origin. Find the electric field at a distance r from the origin, for the cases $r > b$ and $r < b$.

Solution As in the previous example, the electric field is radial and spherically symmetric. For the case $r > b$, (4.11) becomes

$$4\pi r^2 E_r \epsilon = \frac{4\pi b^3 \rho_o}{3}$$

$$E_r = \frac{b^3 \rho_o}{3\epsilon r^2}$$

For $r < b$, the charge inside a surface of radius r is $\frac{4}{3}\pi r^3 \rho_o$, and Gauss' law becomes

$$4\pi r^2 E_r \epsilon = \frac{4\pi r^3 \rho_o}{3}$$

$$E_r = \frac{r\rho_o}{3\epsilon}$$

Inside the sphere, the field strength increases linearly with r; outside it decreases as $1/r^2$. As we expect, the two results are equal at $r = b$.

EXERCISE 4.2

A charged wire is infinitely long and has charge τ per unit length. Find the electric field at a distance r from the wire.

Answer $E_r = \tau/2\pi\epsilon r$.

THE *D* VECTOR

In addition to the electric field $\vec{E}$, it is customary to define a second electric vector $\vec{D}$, known as the *flux density* (or *displacement* or *electric induction* or—most often—simply "D") vector. The relationship between $\vec{D}$ and $\vec{E}$ is[2]

$$\vec{D} = \epsilon \vec{E} \tag{4.12}$$

where ϵ is the permittivity at the place in question.

The reasons for the introduction of a second electric vector are profound and rather subtle. If space were entirely filled with vacuum, or with constant, uniform dielectric, one vector, the E vector, would be sufficient. In the real world, however, the permittivity ϵ often varies with position. For instance, consider Fig. 4.5, which shows a slab of dielectric material placed between two given external charges, $+Q$ and $-Q$. Immediately certain difficulties arise. In this case it is found that additional charges—ones that are not intentionally placed there—appear on the interfaces between the dielectrics. As usual, electric field lines are directed from $+Q$ to $-Q$, passing through the dielectric on the way. But where the field enters the dielectric, negative charges are induced on the surface; and where the field exists, positive charges are induced on the surface. These induced charges are known as *polarization charges*. There is no net charge on the dielectric, so the sum of all the polarization charges is zero. Polarization charges appear because of small movements of charged particles inside the dielectric, in response to the forces of the fields.

Polarization charges are still charges, so they must contribute to the electric fields. Thus the symbol "ρ" in (4.4) and (4.5) must refer to the total charge, including both the "real" charges (the ones over which one has direct control) and also the polarization charges. Unfortunately one does not know exactly where the polarization charges are located until one solves the field problem. It therefore simplifies matters to introduce a second vector—the D vector—which arises only in the real charges. The source equation for $\vec{D}$ is found to take the form

$$\nabla \cdot \vec{D} = \rho_{\text{real}} \tag{4.13}$$

That is, the sources of the $\vec{D}$ field are the *real* charges (such as $+Q$ and $-Q$ in Fig. 4.5); polarization charges do not appear. By applying the divergence theorem to (4.13) we obtain

$$\int_S \vec{D} \cdot \vec{dS} = Q_{\text{real}} \tag{4.14}$$

where again, the charge on the right now includes only the real charge.

[2]Equation (4.12) implies that $\vec{D}$ and $\vec{E}$ have the same direction. In anisotropic materials, such as some crystals, $\vec{D}$ and $\vec{E}$ can have different directions. In such cases ϵ is a matrix, rather than a scalar. Anisotropic materials will not be considered here.

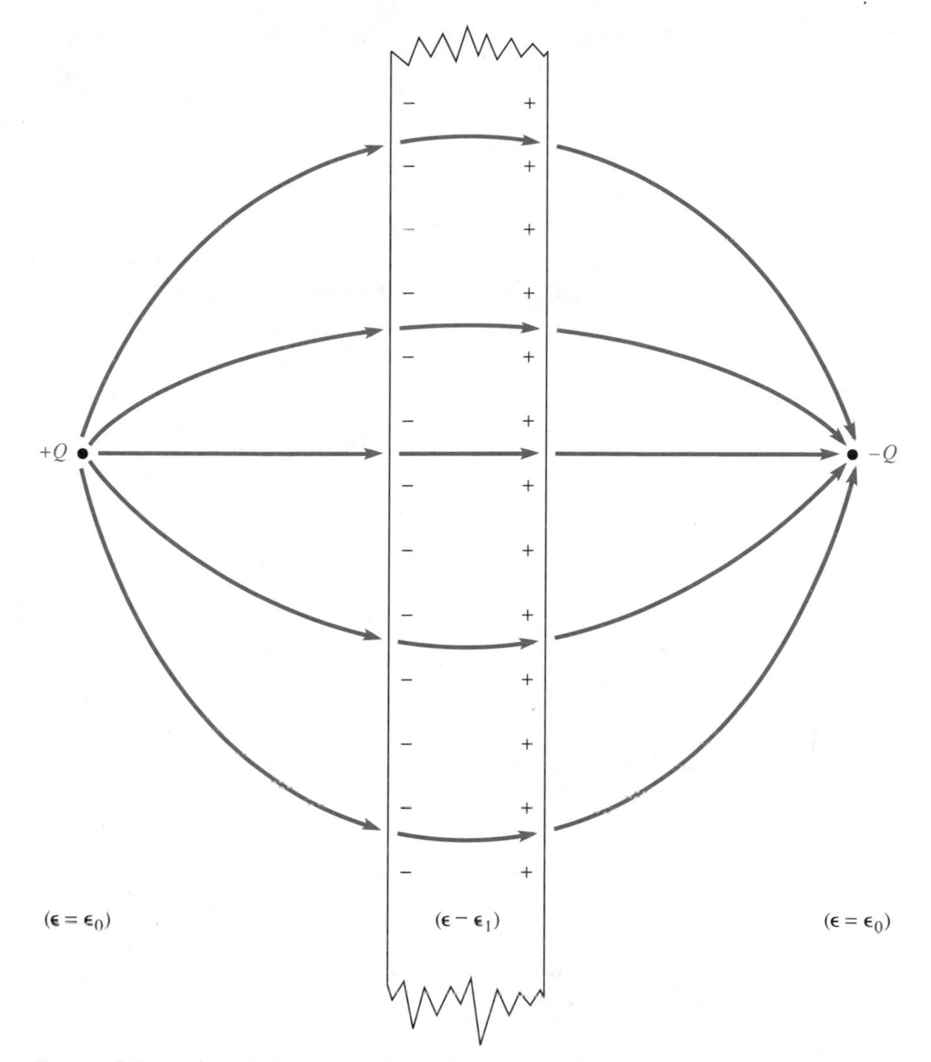

Figure 4.5 Induced charges on the surface of a dielectric.

The subtleties of real and polarization charge are really beyond the level of our present discussion. The uses of the D vector will become clear as we proceed. There are some important cases in which dielectric discontinuities arise. For instance, reflection of light from a sheet of glass is a problem of this kind.

OHM'S LAW

The term "current" usually means "real current," that is, a motion of real charges. This is the same "current" that is familiar to the reader from circuit theory. The value of a current through a wire, in amperes, is equal to the amount of real charge (in coulombs) that passes a given point each second.

Often it is more convenient to work with the *current density* vector $\vec{J}$. The direction of the current density vector is the direction in which, on the average, charge is moving. The magnitude of the current density is the amount of charge that moves across a plane perpendicular to that direction, per unit area, per unit time. Since current density is current per unit area, the total current is found by integrating current density over the entire surface through which current is flowing. As a simple example, suppose the current density in a wire has the same value J_o everywhere inside the wire, and is directed parallel to the axis of the wire. In that case, the total current through the wire will be $I = J_o \pi a^2$, where a is the radius of the wire. On the other hand, suppose the current density in the wire is still directed parallel to the wire axis, but varies with the distance r from the axis according to $|J(r)| = Ar$, where A is a constant. In that case the total current through the wire will be $I = 2\pi A a^3/3$. The general formula relating current density to the total current through a surface is

$$I = \int_S \vec{J} \cdot \vec{dS}$$

For most everyday materials, it is found that current density has the same direction as electric field, and that its magnitude is linearly proportional to that of the electric field. This relationship is known as *Ohm's law:*

$$\vec{J} = \sigma_E \vec{E}$$

Here the scalar constant of proportionality σ_E is known as the *electric conductivity*. (We use the subscript "E" to distinguish it from the surface charge density, which conventionally also has the symbol σ.) Often one encounters the *resistivity* ρ, which is defined simply by $\rho = 1/\sigma_E$.

Ohm's law is not really a "law" in the sense that it has universal validity. Perhaps it should be called "Ohm's property," since it is a property possessed by some materials, though not by all. For electrical engineers the most important materials that do not obey Ohm's law are the semiconductors: diffusion current in a semiconductor is unrelated to $\vec{E}$. Moreover, Ohm's law fails in all materials, when the electric field becomes very large. However, Ohm's law does apply, at least approximately, in many common situations. The form in which the law has been stated here is known as the "microscopic" form of Ohm's law. When it holds, the "macroscopic," or large-scale, form of Ohm's law can also be shown to hold. This is the equation $V = IR$ familiar in circuit theory.

4.4 ELECTROSTATIC ENERGY AND POTENTIAL

Experimentally, an electric field reveals its presence by exerting forces on charged particles. A point charge q placed in an electric field $\vec{E}$ experiences a force

$$\vec{f} = q\vec{E}$$

The direction of the force on a positive charge is the same as the direction of the electric field.

Suppose a particle with charge q moves through an electric field from point P_1 to point P_2. The work done on the particle is, according to the laws of mechanics

$$W = \int_{P_1}^{P_2} \vec{f} \cdot d\vec{l} = q \int_{P_1}^{P_2} \vec{E} \cdot d\vec{l} \tag{4.15}$$

If the particle is stationary at P_1 and we release it to travel to P_2, it acquires kinetic energy W. Accordingly, when it was at P_1 it must have had *potential energy* W. If it had kinetic energy W_o when it was at P_1, it arrives at P_2 with kinetic energy $W_o + W$. Either way, the potential energy at P_1 exceeds that at P_2 by

$$W_1 - W_2 = q \int_{P_1}^{P_2} \vec{E} \cdot d\vec{l} \tag{4.16}$$

It is customary to define the potential energy of a *unit* charge as the *electrostatic potential* V; that is,

$$V_1 = \frac{W_1}{q}$$

and so forth. Thus the *electrostatic potential difference* between P_1 and P_2 is

$$V_1 - V_2 = \int_{P_1}^{P_2} \vec{E} \cdot d\vec{l} \tag{4.17}$$

This quantity is also called the *voltage* between point 1 and point 2. We note that V_1 is higher than V_2 if the electric field points in the direction from P_1 toward P_2.

For the present we shall assume that the value of the line integral in (4.17) is independent of the path chosen between P_1 and P_2. In Chapter 6 we shall verify that this is indeed true, provided that the fields do not vary in time.

EXAMPLE 4.8

A point charge q is located at the origin. Find the potential difference between point P_1, located at $x = a$, $y = 0$, $z = 0$, and point P_2, located at $x = 0$, $y = b$, $z = 0$.

Solution The electric field, given by (4.1) is radial. In choosing a path of integration, we can begin at P_1 and move from the x axis to the y axis along a line that remains at a constant distance from the origin; this is the circular line designated "Segment I" in Fig. 4.6. Segment I contributes nothing to

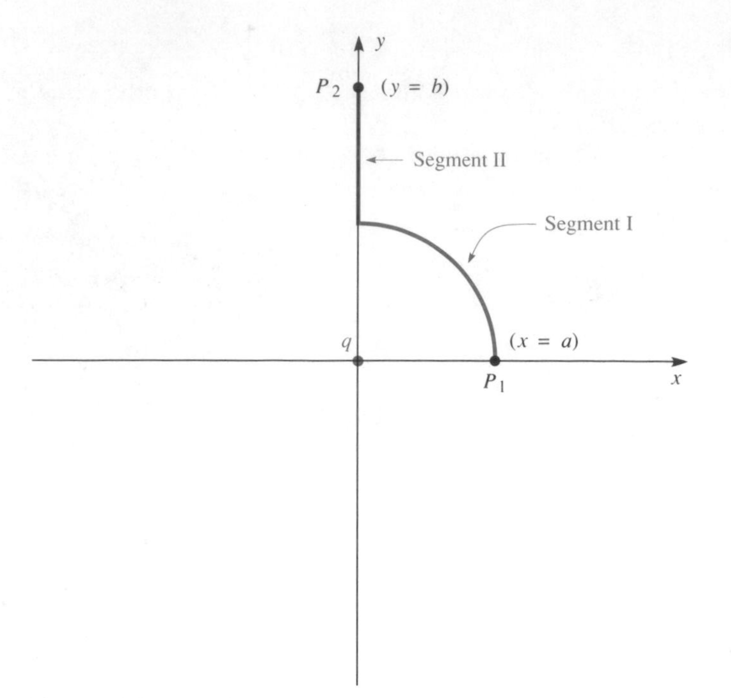

Figure 4.6

the integral in (4.17), because $\vec{E}$ is perpendicular to $\vec{dl}$ everywhere along it. As a consequence, the potential is constant everywhere on Segment I; it is said to be *equipotential*. In fact, the potential is constant everywhere on the sphere $r = a$; it is said to be an *equipotential surface*. Spheres with other radii are also equipotential surfaces, with different values of potential.

On Segment II, $\vec{e}_r = \vec{e}_y$, $\vec{dl} = dy\,\vec{e}_y$, and $r = y$. Thus on Segment II,

$$\vec{E} \cdot \vec{dl} = \frac{q\,dy}{4\pi\epsilon y^2}\,\vec{e}_y \cdot \vec{e}_y = \frac{q\,dy}{4\pi\epsilon y^2}$$

$$V_1 - V_2 = \int_a^b \frac{q\,dy}{4\pi\epsilon y^2} = \frac{q}{4\pi\epsilon}\left(\frac{1}{a} - \frac{1}{b}\right)$$

As we have seen, *differences* in potential are physically meaningful: the difference in potential between two points is the line integral of the electric field along a path from one point to the other. However, the position at which potential is said to be zero can be chosen arbitrarily. The situation is similar to that in describing altitude. One could say that at sea level the elevation is zero—or one could just as well say that at the top of Mount Everest, the elevation is zero, in which case we would say that the elevation of the sea is about −29,000 feet. It is often convenient to choose that the potential is zero when we are infinitely far away from all charges. If we do this in the last example, by letting $b \to \infty$ and setting $V_2 = 0$, we find that

$$V_1 = \frac{q}{4\pi\epsilon a} \tag{4.18}$$

where a is the distance of the observation point from the charge.

It is easily shown that if there are many charges q_i, the potential (with reference to infinity) is the sum of the potentials of the individual charges, that is,

$$V(\vec{r}) = \sum_i \frac{q_i}{4\pi\epsilon|\vec{r} - \vec{r}_i|} \tag{4.19}$$

If there is a continuous distribution of charge, with charge density $\rho(r')$ at position r', the sum of (4.19) is replaced by an integration:

$$V(\vec{r}) = \int \frac{\rho(\vec{r}\,')\,dv'}{4\pi\epsilon|\vec{r} - \vec{r}\,'|} \tag{4.20}$$

where dv' is the differential volume located at r'. For a surface charge distribution $\rho(r')\,dv'$ is replaced by $\sigma(r')\,dA'$.

EXERCISE 4.3

A charged wire has charge τ per unit length. Find the potential difference between P_1, located at a distance a from the wire, and P_2, at distance b from the wire. (Use the result of Exercise 4.2.)

Answer $V_1 - V_2 = \dfrac{\tau}{2\pi\epsilon} \ln\left(\dfrac{b}{a}\right)$.

THE GRADIENT OPERATOR

In (4.17) we have the potential difference expressed as an integral of the electric field. Thus we should also be able to express the field as some sort of derivative of the potential. To do this we shall introduce a differential operator known as the *gradient* operator. This operator will act on the potential, which is a scalar field, and will yield the electric field, which is a vector. Thus the gradient operator differs from the divergence operator, which acts on a vector field to produce a scalar.

Let us consider two points P_1 and P_2, which are quite close together. Point P_1 is located at (x_1, y_1, z_1), and P_2 is at $(x_1 + \Delta x,\ y_1 + \Delta y,\ z_1 + \Delta z)$. As we move from P_1 to P_2, the potential changes from V_1 to $V_1 + \Delta V$. This change is due to the change in x, the change in y, and the change in z. The change in V due to changing x is expressed by means of the partial derivative with respect to x; this change is

$$\frac{\partial V}{\partial x} \Delta x$$

with similar expressions for y and z. The total change is

$$\Delta V = \frac{\partial V}{\partial x} \Delta x + \frac{\partial V}{\partial y} \Delta y + \frac{\partial V}{\partial z} \Delta z \tag{4.21}$$

The small vector

$$\overrightarrow{\Delta l} = \Delta x \vec{e}_x + \Delta y \vec{e}_y + \Delta z \vec{e}_z \tag{4.22}$$

expresses the change in position as we go from P_1 to P_2.

Now let us define the gradient operator. The gradient of V is given, in rectangular coordinates, by:

$$\nabla V = \frac{\partial V}{\partial x} \vec{e}_x + \frac{\partial V}{\partial y} \vec{e}_y + \frac{\partial V}{\partial z} \vec{e}_z \tag{4.23}$$

Note that the gradient of the scalar field V is a *vector*. In terms of the gradient, (4.21) and (4.22) become $\Delta V = \nabla V \cdot \overrightarrow{\Delta l}$, and in the limit as P_1 and P_2 become very close together,

$$dV = \nabla V \cdot \overrightarrow{dl} \tag{4.24}$$

On the other hand, if we let P_1 and P_2 be close together (so that $\vec{E}$ is nearly constant), equation (4.17) becomes

$$dV = -\vec{E} \cdot \overrightarrow{dl} \tag{4.25}$$

(The minus sign appears because $dV = V_2 - V_1$.) Comparing (4.25) with (4.24) we have the important result

$$\vec{E} = -\nabla V \tag{4.26}$$

EXAMPLE 4.9

Find the electric field of a point charge q, located at the origin, by differentiating (4.18), using (4.26) and (4.23).

Solution The potential is

$$V = \frac{q}{4\pi\epsilon\sqrt{x^2 + y^2 + z^2}}$$

Thus

$$\frac{\partial V}{\partial x} = -\frac{qx}{4\pi\epsilon(x^2 + y^2 + z^2)^{3/2}}$$

$$\frac{\partial V}{\partial y} = -\frac{qy}{4\pi\epsilon(x^2 + y^2 + z^2)^{3/2}}$$

$$\frac{\partial V}{\partial z} = -\frac{qz}{4\pi\epsilon(x^2 + y^2 + z^2)^{3/2}}$$

Hence

$$\vec{E} = \frac{q}{4\pi\epsilon}\frac{x\vec{e}_x + y\vec{e}_y + z\vec{e}_z}{(x^2 + y^2 + z^2)^{3/2}}$$

$$= \frac{q}{4\pi\epsilon}\frac{\vec{r}}{|\vec{r}|^3}$$

in agreement with (4.2).

EXERCISE 4.4

Differentiate the result of Exercise 4.3, by means of (4.26) and (4.23), to recover the answer of Exercise 4.2.

A feeling for the significance of the gradient operator can be gained by inspection of (4.24). This equation describes the change in V that occurs when position changes by $\vec{dl}$. We note that dV is greatest when the vectors ∇V and $\vec{dl}$ are parallel (making the cosine factor in the dot product equal to one). Thus *the direction of the vector ∇V is the direction in which V increases most rapidly*. The magnitude of ∇V describes how strongly V changes with position: the larger $|\nabla V|$, the larger the rate of change.

On the other hand, if the displacement $\vec{dl}$ is *perpendicular* to ∇V, then, because of the dot product in (4.24), it produces no change in V at all. Thus if we move perpendicular to ∇V, potential remains constant. In other words, ∇V is always perpendicular to equipotential surfaces, and from (4.26) so is $\vec{E}$.

In particular, the boundaries of ideal metallic conductors are equipotential surfaces. This is seen from the following argument. The current density $\vec{J}$ in most conductors obeys Ohm's law:

$$\vec{J} = \sigma_E \vec{E} \tag{4.27}$$

In an ideal metal, σ_E is assumed to be infinitely large. Hence there can be no electric field inside an ideal metal; if there were, (4.27) would result in an infinitely large current density, which is impossible. Now consider two points P_1 and P_2 located inside a metal or on its surface. A path between them can be chosen that goes entirely through the metal, where $\vec{E}$ is always zero. Thus from (4.17) P_1 and P_2 must be at the same potential. The entire metal is therefore at a single potential, and its surface is an equipotential surface. Thus $\vec{E}$ is always perpendicular to the surface of an ideal metal.

The situation is illustrated in Fig. 4.7. Here a potential difference is maintained between two metal electrodes by an external 4-V voltage source. The potential of the right-hand electrode has been arbitrarily chosen to be zero. Thus, that of the left electrode is 4 V. The electric field is directed from the conductor at higher potential toward the one at lower potential. At the metal surfaces, $\vec{E}$ is always perpendicular to the surface.

Between the two metals the potential takes on intermediate values. For any intermediate value of potential, there is an equipotential surface with that potential somewhere between the metals. Three of these surfaces are indicated by dashed lines in the figure.

Just as with the divergence operator, the gradient operator can be expressed in cylindrical or spherical coordinates. These forms, which are usually more convenient in problems with cylindrical or spherical symmetry, are as follows. For cylindrical coordinates,

$$\nabla V = \frac{\partial V}{\partial r}\vec{e}_r + \frac{1}{r}\frac{\partial V}{\partial \phi}\vec{e}_\phi + \frac{\partial V}{\partial z}\vec{e}_z \tag{4.28}$$

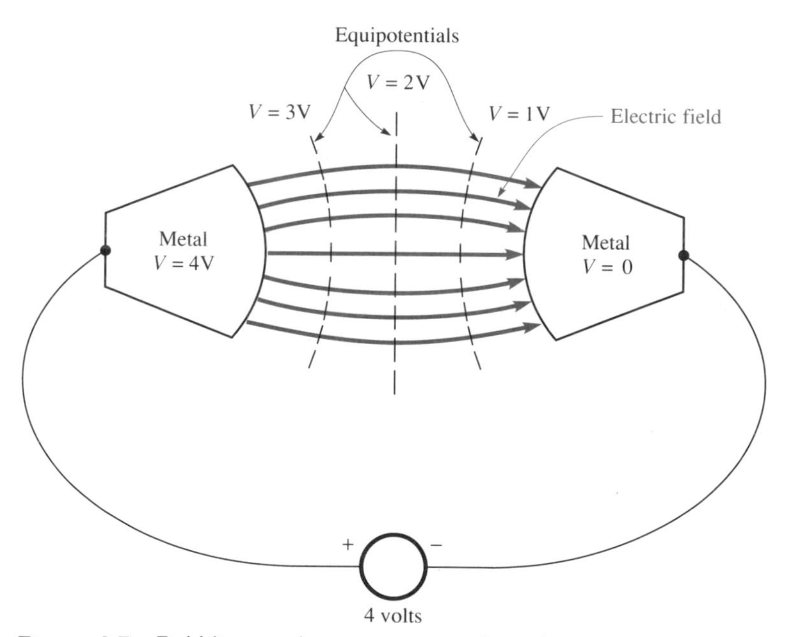

Figure 4.7 Field lines and equipotential surfaces between two metallic electrodes.

and for spherical coordinates,

$$\nabla V = \frac{\partial V}{\partial r}\vec{e}_r + \frac{1}{r}\frac{\partial V}{\partial \theta}\vec{e}_\theta + \frac{1}{r\sin\theta}\frac{\partial V}{\partial \phi}\vec{e}_\phi. \tag{4.29}$$

EXERCISE 4.5

Repeat Example 4.9 using (4.29) in place of (4.23).

4.5 CAPACITORS

Whenever two electrical conductors are brought near to one another, a capacitance exists between them. Often this effect is undesirable, placing limits on the operation of circuits at high frequencies. However, it also serves a useful purpose in capacitors, which are common circuit components.

As a first example, let us consider the two parallel metal plates shown in Fig. 4.8. If an external voltage is applied, making the potential of the

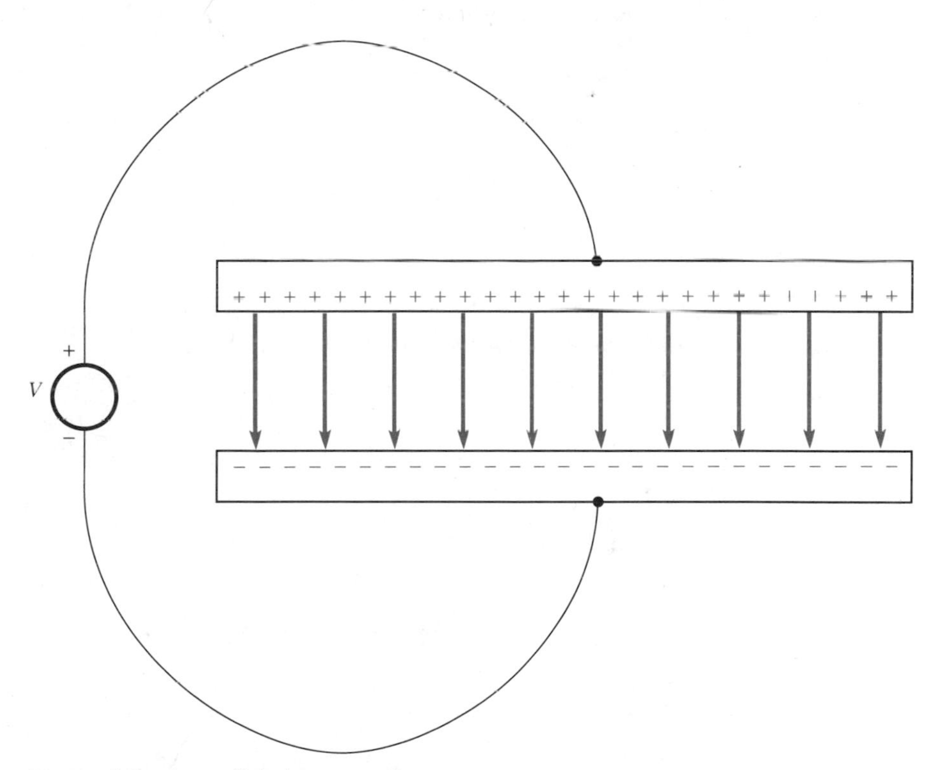

Figure 4.8 A parallel-plate capacitor.

upper conductor larger than that of the lower, an electric field is set up, directed downward. We have seen that electric fields are produced by charges, so we expect that charges will be present to create this field. Since the space between the plates is nonconductive, no charges can move into that region; hence the charges must be on the surfaces of the metals.[3] They are indicated by the plus and minus signs in the figure.

To find the charge density on the plates, we can make use of Gauss' law. Figure 4.9 is a magnified view of a portion of the upper plate. In order to apply Gauss' law, we have selected the surface of integration indicated by the dashed line. This surface is shaped like a "pillbox"; that is, it is a circular cylinder of small height.[4] Its plane faces are parallel to the surface, one inside the metal surface, the other outside. Let the area of the plane faces be ΔA, and let the charge per unit area on the surface of the metal be σ. Then the total charge inside the surface is $\sigma \Delta A$.

To perform the surface integration of $\vec{E} \cdot \vec{dS}$ in Gauss' law, we first observe that the curved sides of the pillbox contribute nothing, since the field is parallel to those surfaces. The plane face inside the metal also contributes nothing, because $\vec{E}$ vanishes inside the perfect conductor. That leaves the plane face outside the metal as the only contributor. $\vec{E}$ is perpendicular to the surface. Hence (4.11) becomes $|\vec{E}|\Delta A = \sigma \Delta A/\epsilon$, from which

$$|\vec{E}| = \frac{\sigma}{\epsilon} \tag{4.30}$$

or, equivalently, $|\vec{D}| = \sigma$. This important result is valid at the surface of any ideal metal.

Capacitance is defined as the ratio of the stored charge to the voltage applied:

$$C = \frac{Q}{V} \tag{4.31}$$

Returning to the parallel-plane capacitor of Fig. 4.8, let the plates have area A and be spaced by a distance d. The charge on one plate is $Q = \sigma A = \epsilon|\vec{E}|A$. However, from (4.17), $|\vec{E}|d = V$. Thus we have $Q = \epsilon AV/d$, and

$$C = \frac{\epsilon A}{d} \tag{4.32}$$

This result is not exact, because we have neglected "edge effects." That is, we have treated the capacitor as though it were infinitely large, so we did not have to concern ourselves with what happens at the ends. This gives a

---

[3]It can be shown that charge can reside only on the *surface* of a conductor, and not inside its bulk. This is true even of conductors with finite conductivity, provided they obey Ohm's law. Fundamentally this is because like charges repel each other and thus move apart until they arrive at the surface.

[4]Traditionally, this surface of integration is said to be "pillbox"-shaped, but no one puts pills in boxes like this any more. For modern readers, it is shaped like a can of tuna fish, or perhaps cat food.

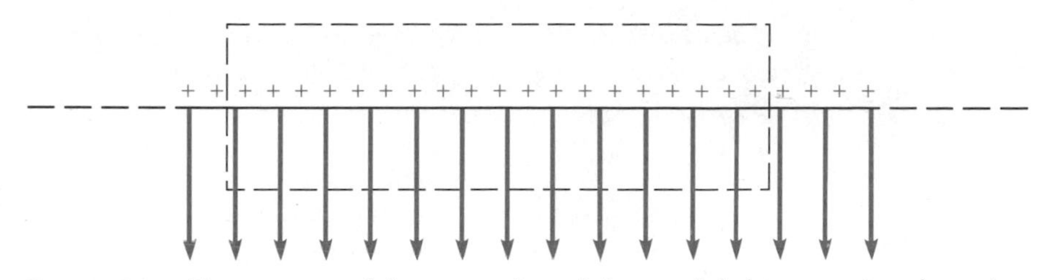

Figure 4.9 Close-up view of the upper plate of the parallel-plate capacitor shown in Fig. 4.8.

good approximation if the lateral dimensions of the capacitor are large in comparison with d. In that case most of the capacitance arises from the interior, and edge effects are insignificant. It is useful to remember equation (4.32), because it allows one to estimate capacitance, at least roughly, in many situations.

EXAMPLE 4.10

Find the capacitance between two concentric spheres having radii a and b, with $a < b$.

Solution Let us assume that the charge per unit area on the inner conductor is σ, and proceed to find the voltage. In order to use Gauss' law, we shall choose a spherical surface with radius r such that $a < r < b$. The lines of $\vec{E}$ are radial and hence perpendicular to this surface. Thus from Gauss' law,

$$4\pi r^2 |\vec{E}(r)| = \frac{4\pi a^2 \sigma}{\epsilon}$$

We find that $\vec{E}(r)$ is a function of r:

$$\vec{E}(r) = \frac{a^2\sigma}{\epsilon r^2}\vec{e}_r$$

Now using (4.17) we find that

$$V = \int_a^b \vec{E} \cdot dr = \frac{a^2\sigma}{\epsilon}\left(\frac{1}{a} - \frac{1}{b}\right)$$

The charge on the inner sphere is $4\pi\sigma a^2$. Hence the capacitance is

$$C = \frac{4\pi\epsilon ab}{b - a}$$

It is interesting to note that the capacitance remains finite if b, the radius of the outer sphere, becomes infinitely large. The limit obtained in this case is $4\pi\epsilon a$, which is often referred to as the capacity of a sphere (of radius a) in free space. This result can be useful for estimating capacitances to ground, even when objects are only roughly spherical. For example, the additional capacitance to ground of a blob of solder on a thin wire might be estimated in this way.

EXERCISE 4.6

Find the capacitance per unit length between two coaxial cylinders having radii a and b.

Answer $2\pi\epsilon/\ln(b/a)$.

4.6 LAPLACE'S AND POISSON'S EQUATIONS

When approaching an electrostatic problem, one may begin by calculating the electric field. Or alternatively, one may begin by calculating the electrostatic potential. In practice it usually turns out to be easier to calculate the potential, because it is easier to find a scalar than a vector. When the potential has been found, it is easy to find the field at any point, simply by taking the gradient of the potential. For this reason, the usual approach to electrostatic problems is to begin by calculating the potential.

The equation governing the potential is obtained from (4.5) and (4.26). We have

$$\nabla \cdot \vec{E} = \nabla \cdot (-\vec{\nabla} V) = \frac{\rho}{\epsilon} \tag{4.33}$$

We note that the gradient operator acts on the scalar field V to produce the vector field $\vec{\nabla} V$. The divergence operator then acts on that vector field to produce another scalar field. The sequence of operations amounts to a new operator of the kind that acts on a scalar field and gives another scalar field. Let us give this new operator the symbol ∇^2. Then

$$\nabla^2 V \equiv \nabla \cdot (\vec{\nabla} V) \tag{4.34}$$

and from (4.33) we have

$$\nabla^2 V = -\frac{\rho}{\epsilon} \tag{4.35}$$

The operator ∇^2 is known as the *Laplacian* operator, and equation (4.35) is *Poisson's equation*. If no charge is present, (4.35) simplifies to

$$\nabla^2 V = 0 \tag{4.36}$$

which is known as *Laplace's equation*.

The form of the Laplacian operator can be found by combining the gradient and divergence operations. Thus in rectangular coordinates

$$\begin{aligned}
\nabla^2 V &= \nabla \cdot (\vec{\nabla} V) \\
&= \nabla \cdot \left(\frac{\partial V}{\partial x} \vec{e}_x + \frac{\partial V}{\partial y} \vec{e}_y + \frac{\partial V}{\partial z} \vec{e}_z \right) \\
&= \frac{\partial^2 V}{\partial x^2} + \frac{\partial^2 V}{\partial y^2} + \frac{\partial^2 V}{\partial z^2}
\end{aligned} \tag{4.37}$$

In the same way the forms of ∇^2 applicable to other coordinate systems can be found. In cylindrical coordinates,

$$\nabla^2 V = \frac{1}{r} \frac{\partial}{\partial r} \left(r \frac{\partial V}{\partial r} \right) + \frac{1}{r^2} \frac{\partial^2 V}{\partial \phi^2} + \frac{\partial^2 V}{\partial z^2} \tag{4.38}$$

and in spherical coordinates

$$\nabla^2 V = \frac{1}{r^2} \frac{\partial}{\partial r} \left(r^2 \frac{\partial V}{\partial r} \right) + \frac{1}{r^2 \sin\theta} \frac{\partial}{\partial \theta} \left(\sin\theta \frac{\partial V}{\partial \theta} \right) + \frac{1}{r^2 \sin^2\theta} \frac{\partial^2 V}{\partial \phi^2} \tag{4.39}$$

Formulas (4.38) and (4.39) may look alarmingly complex, but actually they only represent straightforward sequences of mathematical steps. One would not of course be expected to memorize (4.38) or (4.39); one simply looks them up in a table when needed.

EXERCISE 4.7

Using (4.39), verify that the potential of a point charge at the origin satisfies Laplace's equation everywhere except at the position of the charge.

In practical work it is often necessary to find the electrostatic field produced by conductors to which external voltage sources have been applied. For example, Fig. 4.10 is a simplified diagram of an electron accelerator. Electrons from a heated cathode pass through hollow cylindrical electrodes of increasing length. Voltages applied to these electrodes accel-

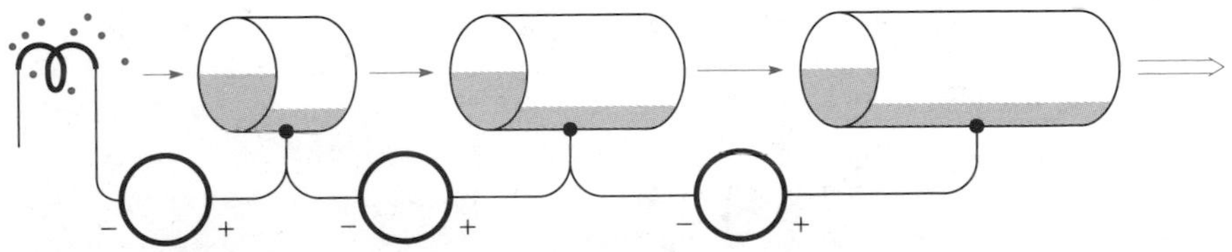

Figure 4.10 Linear electron accelerator (simplified diagram).

erate electrons to speeds approaching that of light. However, the effectiveness of the scheme depends on the details of the electrodes' electrostatic field, and hence it is necessary to find the field (or, equivalently, the potential) with a high degree of accuracy. Here we are confronted with one of the classic problems of electromagnetics, the *electrostatic boundary-value problem*. There is a theorem that assures us that electrostatic boundary-value problems do have solutions. This theorem, the *uniqueness theorem,* states that if the potential is specified everywhere on the boundaries of a region, there is one and only one solution of Laplace's equation that has the given boundary values.

The difficulty of a boundary-value problem depends strongly on the complexity of the boundaries. When the conductors have simple, highly symmetric shapes, such as planes, cylinders, or circles, we can find the potential rather easily. However, in practical work one often finds boundaries with much more complicated shapes, and the solution of boundary-value problems becomes a large subject, almost a small science in itself. Various methods, both experimental and mathematical, are available. At present we can only survey these approaches, but many detailed references are available.

EXPERIMENTAL METHODS

Laplace's equation remains true even if a medium has electrical conductivity. (This is easily seen by repeating the derivation of Laplace's equation; the presence of conductivity has no effect on the derivation.) Therefore if a set of electrodes, at given potentials, is immersed in a conducting medium, the resulting potentials in the region enclosed by the electrodes are, according to the uniqueness theorem, identical with what they would be if the electrodes were in vacuum. This is the basis of an experimental method for investigating the potential between two electrodes with complicated shapes. The electrodes are lowered into a tank of conducting liquid and a voltage V is applied between them. Then a probe can be used to measure the potential at any point in the liquid. Furthermore, it is easily shown (Problem 4.38) that the current I that will flow through the liquid between the electrodes is related to C, the capacitance between the same electrodes when they are in vacuum. The relationship between the two is $I/V = (\sigma_E/\epsilon)C$, where σ_E is the conductivity of the medium. Since I is easily measured, one can determine C. This is known as the *electrolytic tank method*.

There is also a planar version of the above technique which is suitable for two-dimensional problems. In this method one uses a sheet of resistive paper; shapes like those under study are painted onto this paper with conducting paint. Again it can be shown that the resulting two-dimensional potential distribution is the same as would exist for infinitely big cylinders of the same cross sections immersed in vacuum. The details of the potential distribution can be measured by means of a probe connected to a high-impedance voltmeter.

ANALYTIC METHODS

In a few problems the potential distribution can be calculated, as a function of position, and presented as a mathematical expression in closed form.

EXAMPLE 4.11

A spherical capacitor is composed of two concentric conducting spheres with radii a and b, with $a < b$. The potential of the outer sphere with respect to ground is zero, and that of the inner is V_o. Find the potential between the spheres as a function of position.

Solution From Example 4.10, we know that the electric field at a distance r from the spheres' center is

$$\vec{E}(r) = \frac{a^2\sigma}{\epsilon r^2}\vec{e}_r$$

where a is the radius of the inner sphere and σ is the charge density upon it. It was also found in Example 4.10 that

$$V_o = \frac{a^2\sigma}{\epsilon}\left(\frac{1}{a} - \frac{1}{b}\right)$$

which we can solve for σ, obtaining

$$\sigma = \frac{\epsilon V_o b}{a(b - a)}$$

Inserting this into the expression for $\vec{E}$, we find

$$\vec{E}(r) = V_o \frac{ab}{r^2(b - a)}\vec{e}_r$$

The potential at any radius can now be found using (4.17)

$$\begin{aligned} V_o - V(r) &= \int_a^r \vec{E}(r) \cdot \vec{e}_r \, dr \\ &= \int_a^r \frac{V_o ab}{r^2(b - a)} dr \end{aligned}$$

Carrying out the integration, we find

$$V(r) = V_o\left[\frac{a(b - r)}{r(b - a)}\right]$$

Note that this result reduces to the given potentials at the boundaries. That is, $V(a) = V_o$ and $V(b) = 0$.

A few problems can be solved analytically by the ingenious *method of images*. This method is based on the uniqueness theorem, which states that if by any means we can find a solution of Laplace's equation that satisfies the boundary conditions, this must be the one and only correct solution.

Let us consider a problem in which there is a grounded metal sheet in the plane $z = 0$, and above it a point charge Q, as shown in Fig. 4.11(a). Note that we cannot find the potential by simply using (4.20), because there are unknown induced charges on the surface of the metal. (These charges flow up onto the metal through the ground connection, under the attraction of Q.) Instead, we proceed by constructing a second situation as shown in

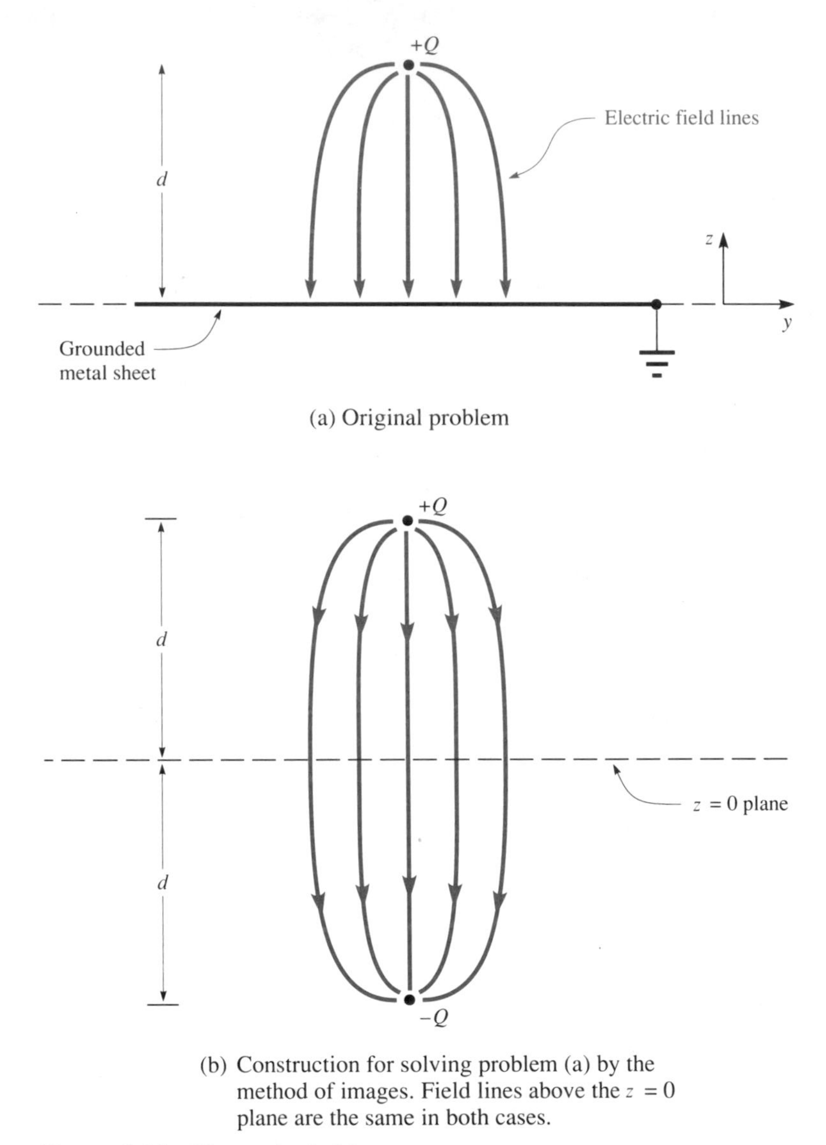

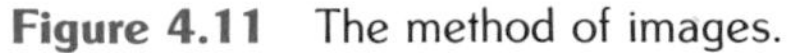
Figure 4.11 The method of images.

Fig. 4.11(b). In (b), the metal plane has been removed, and instead a charge $-Q$ has been placed at the "image" point of the original charge. Now in (b), consider the potential of the plane where the metal originally was. By symmetry, the potential in this plane is zero. That means that the field existing in the image problem, Fig. 4.11(b), possesses the requirements for being a solution of the original problem (a). That is, it is a solution of Laplace's equation, and it satisfies the boundary conditions of the original problem. Therefore, we can write down the potential of the original problem (a):

$$V(x, y, z) = \frac{Q}{4\pi\epsilon[x^2 + y^2 + (z - d)^2]^2} - \frac{Q}{4\pi\epsilon[x^2 + y^2 + (z + d)^2]^2} \tag{4.40}$$

The method of images may be extended to the case of several point charges above the grounded plane, and to a few other problems, but it is by no means a general method. Unfortunately, this is the nature of analytic solutions to boundary-value problems in general: they are said to be solved by "the method of ingenious devices," meaning that often a certain problem will have a specific clever analytic solution valid for that one problem alone. However, there is no general way of finding such solutions, even when they exist, and a good deal of experience and cleverness is required to find them. In two-dimensional problems a method called *conformal mapping* can sometimes be used. This method, based on the algebra of complex variables, can be applied, at least approximately, in a surprisingly large variety of situations, but its applicability is hardly general, and again, experience and cleverness are required in its use. Thus one must look elsewhere for methods that are straightforward and widely useful. The best bet for such a general approach is the large class of numerical methods, most of which rely on digital computation.

NUMERICAL METHODS

In three-dimensional problems with simple, highly symmetrical boundaries, the *method of separation of variables* has been a traditional choice. This method amounts to solving the partial differential equation (Laplace's equation) by expanding its solutions in an infinite series of eigenfunctions. The results are obtained in the form of infinite series that must be summed, perhaps by a computer. The accuracy of the method depends on how many terms of the series are summed. However, the requirement of simple boundaries places severe limitations on the use of this method.

A technique with extremely wide applicability is the *finite-difference* method. In this approach derivatives are replaced by nearly equivalent subtractions. For example, from the definition of the derivative,

$$\left.\frac{\partial y}{\partial x}\right|_{x=x_o} \quad \text{is approximated by} \quad \frac{y\left(x_o + \dfrac{h}{2}\right) - y\left(x_o - \dfrac{h}{2}\right)}{h}$$

where h is a small finite number. An approximation for the second derivative is obtained in similar fashion:

$$\left.\frac{\partial^2 y}{\partial x^2}\right|_{x=x_o} = \frac{\partial}{\partial x}\left(\frac{\partial y}{\partial x}\right) \cong \frac{\left.\frac{\partial y}{\partial x}\right|_{x=x_o+\frac{h}{2}} - \left.\frac{\partial y}{\partial x}\right|_{x=x_o-\frac{h}{2}}}{h}$$

$$= \frac{y(x_o + h) - 2y(x_o) + y(x_o - h)}{h^2}$$

Using this approximation, Laplace's equation (for simplicity, in two dimensions), becomes

$$\nabla^2 V|_{x_o, y_o} \cong \frac{V(x_o + h, y_o) - 2V(x_o, y_o) + V(x_o - h, y_o)}{h^2} + \frac{V(x_o, y_o + h) - 2V(x_o, y_o) + V(x_o, y_o - h)}{h^2} = 0 \quad \textbf{(4.41)}$$

As an example of the technique, let us consider the boundary-value problem shown in Fig. 4.12. Here the four walls of the box are held at potentials V_1, V_2, V_3, and V_4, by external voltage sources. We wish to find the potential inside the walls as a function of position. We begin by constructing a rectangular lattice of points, equally spaced by the distance h. The points are numbered as shown in the figure, and the potential at point ij is called V_{ij}. Laplace's equation must hold at every one of these points. Using the above approximation for the laplacian operator, the statement of Laplace's equation at point 11 is

$$\frac{V_{12} - 2V_{11} + V_1}{h^2} + \frac{V_2 - 2V_{11} + V_{21}}{h^2} \cong 0$$

or, equivalently,

$$V_{11} \cong \frac{V_1 + V_2 + V_{12} + V_{21}}{4} \quad \textbf{(4.42)}$$

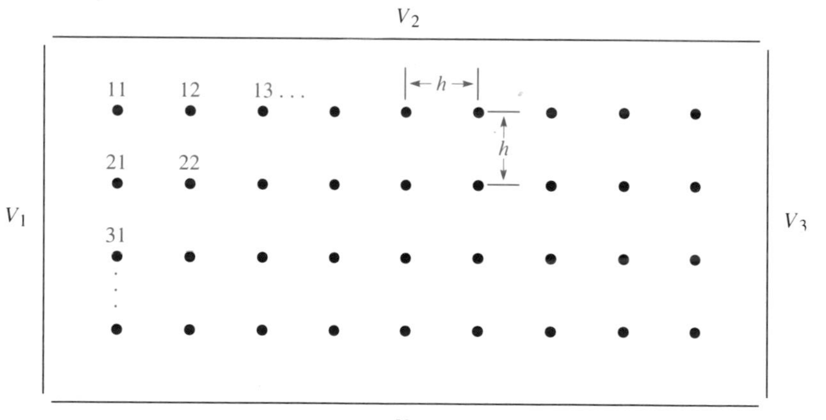

Figure 4.12 Solving an electrostatic boundary-value problem by the finite-difference method.

This equation states that the potential at the point in question is approximately equal to the average of the potentials at the four nearest-neighbor points. A similar equation can be written at each of the other lattice points. The number of equations obtained is equal to the number of lattice points, and the unknowns in these equations are the potentials at the lattice points. Thus there are as many equations as there are unknowns, and the equations can be solved simultaneously for the potentials at all the points. Of course one would not do this by hand; one uses a computer routine that solves simultaneous equations by Cramer's rule or some other method. Since each of the original equations has only five terms, most of the elements of the determinants that arise in Cramer's rule are zeros; this fact makes the solution of the simultaneous equations faster than it otherwise would be. Intuitively it is evident that the smaller h is, the more accurate the method; on the other hand, the smaller h is, the greater the number of simultaneous equations that have to be solved. The method works equally well in three dimensions. A great advantage of this sort of method is that it is an orderly procedure that works with oddly shaped boundaries; once it has been set up, it does not require intuition or cleverness. Thus it can be used as a general-purpose procedure.

One of the best approaches for solving finite-difference equations is through *iterative* methods. In these, one begins by making a guess at a solution, and then uses this guess to calculate a better guess, and so on. Details of the method will be found in Computer Exercise 4.4. Iterative methods lend themselves particularly well to formulation as computer programs, and in fact such programs are available as commercial software packages. They are efficient in terms of computer time, and can be used in very powerful ways. For instance, they can be used with a light pen, so that a designer can sketch electrodes on a screen and watch the potential distribution change as he varies his design.

The subject of numerical techniques is large and interesting. One may study the various methods mathematically to learn about their efficiency (in terms of computer time), and about their convergence properties (that is, for which problems they lead to correct solutions). For those who are mathematically inclined this field is recommended for further study.

4.7 POINTS TO REMEMBER

1. Electrostatic fields originate from charges. If the locations of all charges are known, the electric field can be found by summing (or integrating) the fields of the individual charges. Note that these fields must be added as *vectors*.

2. The divergence operator acts on a vector field to give a scalar field.

3. The source equation

$$\nabla \cdot \vec{E} = \frac{\rho}{\epsilon}$$

allows one to find the charge density when the electric field is known.

4. Gauss' law relates the surface integral of the electric field over a closed surface to the charge contained within that surface. It is a useful problem-solving tool in situations where a high degree of symmetry is present.

5. The electric flux-density vector $\vec{D}$ is related to $\vec{E}$ in most materials through $\vec{D} = \epsilon\vec{E}$, where ϵ is the electric permittivity, a constant of the medium.

6. When no time-varying fields are present, it is possible to assign to each position in space a value of electrostatic potential. The difference in potential between points A and B is given by a line integral of the electric field along any path from A to B. This potential difference is also known as a voltage. Only differences in potential have physical significance.

7. The gradient operator acts on a scalar field to give a vector field. The direction of $\vec{\nabla}\phi$ is the direction in which ϕ increases most rapidly, and the magnitude of the gradient indicates the rate of increase.

8. The electric field is related to the electrostatic potential V through $\vec{E} = -\vec{\nabla}V$.

9. In many (but by no means all) materials, the current density $\vec{J}$ is related to the electric field by $\vec{J} = \sigma_E\vec{E}$, where σ_E is the electrical conductivity of the medium. This relationship is known as Ohm's law.

10. The capacitance between two conductors is the ratio of the charge stored on each conductor to the potential difference between them.

11. The scalar Laplacian operator ∇^2 acts on a scalar field to produce another scalar field. It is the equivalent of the gradient operation followed by the divergence operation. The electrostatic potential is related to the charge density through Poisson's equation

$$\nabla^2 V = -\frac{\rho}{\epsilon}$$

(If no charge is present we have $\nabla^2 V = 0$, which is known as Laplace's equation.)

12. It sometimes happens that one must find the electrostatic potential at all positions, in the presence of a given distribution of charges and perfect conductors that are at given potentials. This is known as an electrostatic boundary-value problem. To solve it one must find a potential distribution that satisfies Poisson's equation and also reduces to the given values on the boundaries. Boundary-value problems can be solved experimentally, analytically, or by numerical methods.

REFERENCES

1. S. Ramo, J. R. Whinnery, and T. Van Duzer, *Fields and Waves in Communication Electronics*. New York: Wiley, 1984.
2. N. N. Rao, *Elements of Engineering Electromagnetics*. Englewood Cliffs, N.J.: Prentice-Hall, 1987.

For experimental and numerical methods, see:

3. P. Silvester, *Modern Electromagnetic Fields*. Englewood Cliffs, N.J.: Prentice-Hall, 1968.
4. S. V. Marshall and G. G. Skitek, *Electromagnetic Concepts and Applications*, 2nd ed. Englewood Cliffs, N.J.: Prentice-Hall, 1987.
5. V. F. Fusco, *Microwave Circuits: Analysis and Computer-Aided Design*. Englewood Cliffs, N.J.: Prentice-Hall, 1987.
6. P. Lorrain and D. Corson, *Electromagnetic Fields and Waves*, 2nd ed. New York: Freeman, 1970.

More advanced treatments of the fundamental physics include:

7. W. K. H. Panofsky and M. Phillips, *Classical Electricity and Magnetism*. Reading, Mass.: Addison-Wesley, 1955.
8. J. D. Jackson, *Classical Electrodynamics*. New York: Wiley, 1962.

With regard to dielectric materials, see:

9. R. S. Elliott, *Electromagnetics*. New York: McGraw-Hill, 1966.

PROBLEMS

Section 4.1

4.1 The average distance between the electron and the proton in an unexcited hydrogen atom is 5.3×10^{-11} m. Find the magnitude of the proton's electrostatic field at this distance. (The charge of a proton is 1.6×10^{-19} C.)

4.2 Two point charges, each with charge q, are located at $(a, 0, 0)$ and $(-a, 0, 0)$. Find the electric field vector, expressed in rectangular components, at (a) $(0,0,0)$; (b) $(0,a,0)$; (c) $(a,a,0)$.

4.3 A point charge with charge q is located at $(a,0,0)$ and one with charge $-q$ is at $(-a,0,0)$. Find the electric field vector, expressed in rectangular components, at (a) $(0,0,0)$; (b) $(0,a,0)$; (c) $(a,0,a)$.

*4.4 A line charge with charge τ per unit length lies on the x axis between $x = -a$ and $x = a$. Find the electric field at $(0,b,0)$. Find the limit of your answer as the line charge becomes infinitely long $(a \to \infty)$.

**4.5 A line charge with charge τ per unit length lies on the x axis between $x = -a$ and the origin. Find all rectangular components of the electric field at $(0, b, 0)$.

*4.6 A surface charge with the shape shown in Fig. 4.13 lies in the x-y plane. Its charge density is σ. Find the magnitude and direction of the electric field at the origin.

*4.7 An infinitely long sheet of charge with charge density σ lies in the x-y plane between $x = a$ and $x = b$. Find the electric field at $(c, 0, 0)$, where $c > b > a$.

**4.8 A line charge with charge per unit length τ is in the form of a circle of radius a in the x-y plane with its center at the origin. Find an approximate expression for the electric field at the point $(-d, 0, 0)$ where $d \ll a$. (*Suggestion:* Simplify the integral by means of the binomial expansion, retaining only terms of first order in d/a.)

*4.9 A thin wire is in the shape of a circle lying in the plane $z = 0$ with its center at the origin. Its radius is a, and it carries a total charge Q. Find the positions on the z axis where the electric field is maximum.

*4.10 A volume in the shape of a right-circular cylinder has a uniform charge density ρ. The cylinder has radius b, and its two plane faces are at $z = 0$ and $z = c$. Find the electric field at $(0, 0, d)$, where $d > c$.

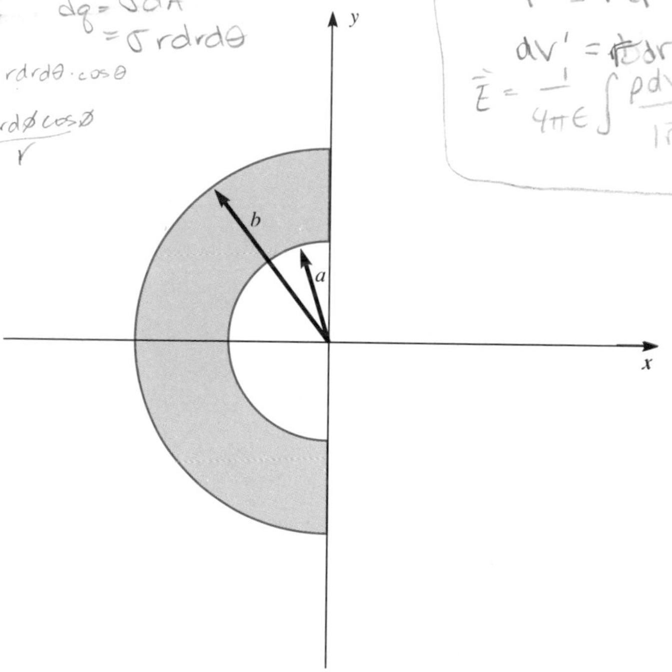

Figure 4.13 For Problem 4.6.

Section 4.2

4.11 The field inside a one-dimensional semiconductor pn junction is

$$\vec{E}(x) = -\frac{qN_D}{\epsilon}(x_n - x)\vec{e}_x$$

for $0 \le x \le x_n$. Here q is the absolute value of the charge of an electron, and N_D is the number of donor impurity atoms per unit volume. Find the charge density.

*4.12 The volume charge density in the region $-x_o \le x \le x_o$ is $\rho = Ax^3$, independent of y and z. The electric field vanishes at $x = x_o$. Find the electric field inside the charged region by integrating the source equation (4.5).

4.13 Suppose that $E_x = Ax$, where A is a constant, and $E_z = 0$. No charge is present in the region of interest. Find E_y, under the assumption that the field at the origin vanishes. Sketch the magnitude and direction of $\vec{E}$ at various points in the x-y plane.

4.14 The field of a line charge (in cylindrical coordinates) with charge τ per unit length is

$$\vec{E} = \frac{\tau}{2\pi\epsilon r}\vec{e}_r$$

Verify that $\nabla \cdot \vec{E} = 0$ everywhere except at $r = 0$.

4.15 The electric field of a certain charge distribution (expressed in spherical coordinates) is

$$\vec{E} = \frac{A}{4}r^2\vec{e}_r$$

where A is a constant. Find the charge density.

4.16 The field of an *electric dipole* (expressed in spherical coordinates) is

$$\vec{E} = \frac{qd}{4\pi\epsilon r^3}(2\cos\theta\vec{e}_r + \sin\theta\vec{e}_\theta)$$

where θ is the polar angle (See Appendix A). Make a sketch showing the magnitude and direction of this field at various positions.

*4.17 For the field of Problem 4.16, show that $\nabla \cdot \vec{E} = 0$ everywhere except at the origin.

Section 4.3

4.18 Imagine a spherical surface of radius b, centered on a point charge q. Apply Gauss' law to this surface and show that its prediction is correct.

4.19 Apply Gauss' law to the dipole field of Problem 4.16 to find the total charge on an electric dipole. Assume that the charge, if any, is located at the origin, and use a spherical surface with the origin at its center.

*4.20 An infinitely long cylindrical region of radius R contains a uniform volume charge density ρ. Find the electric field for $r < R$ and $r > R$.

4.21 Verify that the units of the flux density vector $\vec{D}$ are coulombs/meter^2 and that those of ϵ are coulombs/volt-meter.

4.22 A hollow metal sphere is made of perfect conductor of finite thickness. It carries a total charge Q. Using Gauss' law and arguments based on symmetry, investigate the following questions: (a) Is any of the charge on the inside surface? (b) Is there an electric field inside the sphere?

4.23 In a certain region the current density has constant magnitude J_o and is in a direction that makes an angle of 40° with the z axis. What is the total current through a region with area A lying in the x-y plane?

*4.24 Find the total current passing through a circular area with radius a, lying in the x-y plane and centered on the origin. The electric field is directed toward the $+z$ direction and has magnitude $|\vec{E}| = B(x^2 + y^2)$, where B is a given constant. The conductivity in the region of interest is σ_E.

Section 4.4

4.25 The electric field in a certain region is $\vec{E} = A\vec{e}_x + B\vec{e}_y$. Find the potential difference between the points $(0,0,0)$ and $(a,b,0)$: (a) by integrating first from $(0,0,0)$ to $(a,0,0)$ and thence to $(a,b,0)$, and (b) by integrating first from $(0,0,0)$ to $(0,b,0)$ and thence to $(a,b,0)$.

4.26 Sketch several equipotential surfaces for the case of (a) a point charge; (b) a system of two point charges on the x axis: $+q$ at $x = a$ and $-q$ at $x = -a$.

*4.27 Find the potential associated with the dipole field of Problem 4.16. Let the potential at infinity be zero. (Be careful of the algebraic sign.) Sketch several equipotential surfaces.

*4.28 A circular disk of radius b, with uniform surface charge density σ, lies in the x-y plane with its center at the origin. Find the electrostatic potential at a point on the z axis at a distance h from the disk.

**4.29 Suppose there were a place in an electrostatic field where, although no charge is present, the electric potential is a local minimum. (a) Show that a point charge could be trapped and held stably at such a position. (b) Show, by means of Gauss' law, that no such place can exist (*Earnshaw's theorem*).

**4.30 A line charge with charge τ per unit length lies on the y axis between $y = -c$ and $y = c$. (a) Find the potential at an arbitrary point $(x, y, 0)$ in the x-y plane. (b) Find the electric field in the x-y plane by taking the gradient of the result of part (a).

4.31 Sketch a map of the state in which you live, and dot it with a variable density of pencil dots to indicate regions of high or low population density. Now draw arrows at different points on your map representing the gradient of population density. (Indicate the magnitude of the vector by the lengths of the arrows.) Also sketch in lines of constant population density.

4.32 In a certain region the potential is given by

$$V(x, y, z) = A \cos(k_x x) \cos(k_y y) \cos(k_z z)$$

(a) Find the electric field vector as a function of x, y, and z. (b) Sketch the field in the plane $z = 0$.

*4.33 A certain potential is described (in cylindrical coordinates) by

$$V(r, \phi) = Ar^n \sin n\phi$$

in the region $0 < r < r_o$, $0 < \phi < \phi_o$. Here $n = \pi/\phi_o$. Find and sketch the electric field.

*4.34 A certain potential is given (in spherical coordinates) by

$$V(r, \theta, \phi) = \frac{A}{r^2} \cos\theta$$

Find the electric field and show that it is the same as the dipole field of Problem 4.16, when the constant A is suitably chosen.

Section 4.5

*4.35 A capacitor is made from two long strips of metal foil, each 1 cm wide and 0.005 cm thick. These are separated by strips of insulating material with thickness 0.003 cm and permittivity $3\epsilon_o$. A stack is made consisting of a layer of foil, a layer of insulator, the other layer of foil, and another layer of insulator. The stack of strips is then rolled up to make a cylinder 0.7 cm in diameter. Find the capacitance of the resulting capacitor, approximately.

*4.36 Using the capacitance of a sphere in free space, make an approximate estimate of how much additional capacitance to ground is added to a circuit, if a wire is anchored to a metal binding post. The binding post (with solder) is a conducting mass 8 mm high, 5 mm wide, and 3 mm deep.

**4.37 Consider a cylindrical capacitor like that of Exercise 4.6 with a small spacing, so that $(b - a)/a \ll 1$. Obtain an approximate expression for its capacitance per unit length by modeling it as a parallel-plate capacitor having width $2\pi[(a + b)/2]$. Show that your result agrees with the result of the exercise in the limit of small spacing.

Section 4.6

*4.38 A capacitor is modeled by placing conductors of exactly the same size and shape in a very large bath of conducting fluid with resistivity ρ_o. A voltage V is applied to the conductors and a current I is measured. It can be shown that the potential distribution is the same as if the fluid were replaced by vacuum. Find the capacitance of the vacuum capacitor in terms of the measured I.

*4.39 Use the method of images to find the force acting on a point charge q located at $(0, 0, a)$, if there is a grounded conducting plane at $z = 0$ and a point charge $-q$ at $(a, 0, a)$.

*4.40 Grounded metal sheets are located in the $x = 0$ and $y = 0$ planes, and a point charge $+Q$ is located at $(a, b, 0)$. Find the potential in the region $x > 0$, $y > 0$.

**4.41 An infinitely long, grounded metal cylinder of radius a lies centered on the z axis, and a line charge with charge τ per unit length is parallel to the cylinder, at $x = c$, $y = 0$. Show that the potential everywhere on the cylinder remains the same if the metal is taken away and a line charge $-\tau$ is placed at $x = a^2/c$, $y = 0$ and another line charge τ' is placed at $x = 0$, $y = 0$. Find τ'.

COMPUTER EXERCISE 4.1

A circular loop of wire of radius a, with charge τ per unit length, lies in the x-y plane with its center at the origin.

(a) Using a library numerical-integration subroutine, write a program for finding the electrostatic potential V at a point on the x axis at a distance x from the origin. The integral to be evaluated is that of (4.20), with $\rho\, dV$ replaced by $\tau\, dl$. Define $u \equiv x/a$, and let

$$V(u) = \frac{\tau}{\epsilon_o} f(u)$$

Your program should find the dimensionless function $f(u)$.

(b) Test your program by finding the potential at the origin. What should the result be?

(c) The electrostatic field at any distance x can be found, approximately, by means of

$$|\vec{E}(u)| \cong \left| \frac{V(x + \delta x) - V(x - \delta x)}{2\, \delta x} \right|$$

where $\delta x/x \ll 1$. Let $|\vec{E}(u)|$ be expressed in the form

$$|\vec{E}(u)| = \frac{\tau}{\epsilon_o a} g(u)$$

where $g(u)$ is a dimensionless function. Find $g(u = 2)$.

COMPUTER EXERCISE 4.2

For the case of Computer Exercise 4.1, show that the electrostatic field on the x axis at a distance x from the origin is given by

$$E_x(u) = \frac{\tau}{\epsilon_o a} \frac{1}{2\pi} \int_0^{\pi} \frac{(u - \cos\theta)\, d\theta}{(1 + u^2 - 2u\cos\theta)^{3/2}}$$

where as before $u \equiv x/a$. Let

$$E_x(u) = \frac{\tau}{\epsilon_o a} g(u)$$

(a) Write a program that performs the above integration numerically and finds $g(u)$ for given values of u. Sketch a graph of $g(u)$ over the range $0 < u < 3$. (Note the singularity at $u = 1$. You may find that your integration program converges slowly for values of u near one.)

(b) Compare $g(2)$ with the result of Computer Exercise 4.1(c).

*COMPUTER EXERCISE 4.3 FINITE-DIFFERENCE METHOD

In this exercise we shall solve an electrostatic boundary-value problem by the finite-difference method. A program for solution of simultaneous linear equations (or a matrix-inversion program) is required.

The basic idea of the finite-difference method has been outlined in section 4.6. To be specific, let us solve the boundary-value problem illustrated in Fig. 4.14(a). Here we have a two-dimensional potential distribution $V(x, y)$ inside the region bounded by four metal plates. The plates are held at given boundary voltages V_{B1}, V_{B2}, V_{B3}, V_{B4}, and as in any boundary-value problem, our job is to find the potential at all positions inside the space enclosed by the plates.

In the finite-difference approach, one selects a lattice of points, and calculates the potential, approximately, at each of these points. In Fig. 4.14(a), six points have been selected. (Once the potentials at these points have been found, potentials at other positions can be found by interpolation.) Let us call the unknown potentials at these points $V_1, V_2, \ldots, V_6$. The points are spaced in a regular rectangular lattice with horizontal spacing h_1 and vertical spacing h_2.

Let us consider any one of the six points and refer to it as point A, as shown in Fig. 4.14(b). From the discussion of section 4.6, the potential at A is related to the potentials at neighboring points B, C, D, and E by

$$\frac{V_C + V_E - 2V_A}{h_1^2} + \frac{V_B + V_D - 2V_A}{h_2^2} \cong 0$$

from which we have

$$\alpha(V_C + V_E) + (V_B + V_D) - 2(1 + \alpha)V_A \cong 0$$

where $\alpha \equiv h_2^2/h_1^2$. In this equation some of the voltages $V_B \cdots V_E$ will also be unknown, whereas others will be given boundary voltages. In this example

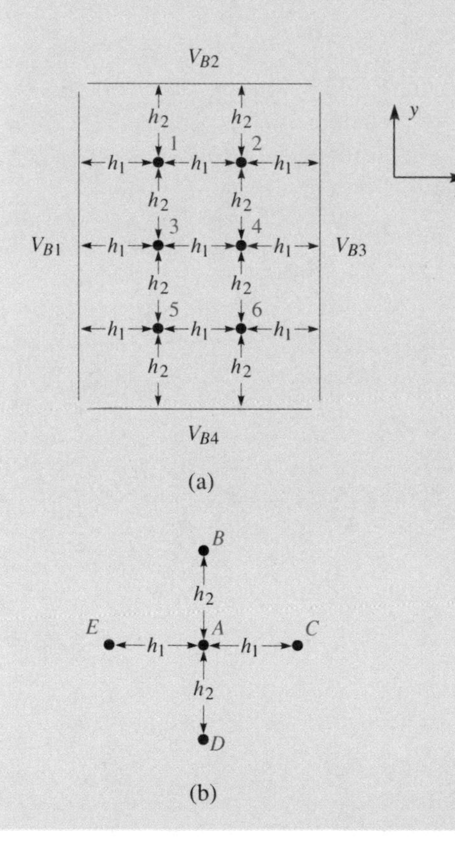

Figure 4.14

there are six unknown voltages, and six simultaneous equations of this form can be written, by letting point A be first point 1, then point 2, and so on. For instance, the equation for point 1 is

$$\alpha(V_2 + V_{B1}) + (V_{B2} + V_3) - 2(1 + \alpha)V_1 = 0$$

In this exercise, you are asked to find the potential in Fig. 4.14 with $V_{B2} = 10$ V, $V_{B1} = V_{B3} = V_{B4} = 0$, and $h_1 = h_2$. How many points are to be used is for you to decide, and depends on the computing power available. It seems clear intuitively that the larger the number of points, the finer the mesh and the greater the accuracy of the method.[5] On the other hand, if the number of unknown voltages is N, then N simultaneous equations must be solved; we note that in general the required computational effort increases as N^3. Moreover, for large N, entering the N^2 matrix elements becomes a task in itself; this task must be systematized and carried out by a subsidiary program. Using $N = 6$, as in the figure, will not give very good accuracy, but results in matrices that are small enough to enter by hand, and which can be inverted by an Hp-15C hand-held calculator. If you are using a larger computer and a higher language such as FORTRAN (with a library subroutine for solving simultaneous linear equations), you can use more points and obtain more accurate results.

*COMPUTER EXERCISE 4.4 ITERATIVE SOLUTION OF FINITE-DIFFERENCE EQUATIONS

In the previous exercise you were asked to solve a set of simultaneous finite-difference equations by means of a standard subroutine. Use of the standard subroutine is convenient, but general-purpose subroutines are rather slow. This is because they are intended for general problems in which the N equations may have N^2 nonzero coefficients. However, the equations arising from the finite-difference method turn out to be simpler than general equations could be: most of the coefficients, in fact, are zero. As a result, they can be solved by a simple iterative method, known as Liebmann's method, which is much faster than a general simultaneous-equation solver. Thus many more solution points can be used and greater accuracy obtained. The entire program can be written directly; no library subroutine is required.

As in Computer Exercise 4.3, we shall calculate the potential at a set of chosen points. In this case assume the points are equally spaced, and

[5]Actually, there is a limit to the accuracy that can be obtained. If too many points are used, the computer finds itself taking very small differences between almost-equal numbers, and roundoff errors begin to increase. However, this is only likely to be important when extremely large numbers of points are used.

since they may be numerous, it is convenient to number them by row and column, in the same way as matrix elements are numbered; this is illustrated in Fig. 4.15(a). The subscripts begin with V_{22} instead of V_{11} because we shall consider points V_{12}, V_{13}, and so on, to lie on the top conducting plate. Similarly, points V_{21}, V_{31}, and so on, lie on the left-hand plate at voltage V_{B1}. From (4.42), we know that the correct potential at each point is equal to the *average* of the potentials at the four nearest-neighbor points. For example, $V_{23} = (V_{22} + V_{24} + V_{13} + V_{33})/4$. Note that V_{22}, V_{23}, V_{24}, and V_{33} are unknowns, but V_{13} is known, as it is the given boundary potential V_{B2}. It is convenient to write down all the voltages in matrix form. For the case of Fig. 4.15(a), the voltage matrix is 7×5, because the fixed boundary voltages are included. The voltage matrix is shown in Fig. 4.15(b). Our goal is to find the 15 unknown voltages enclosed by the dashed line. The four voltages indicated by the asterisk do not enter the calculations and their values do not matter.

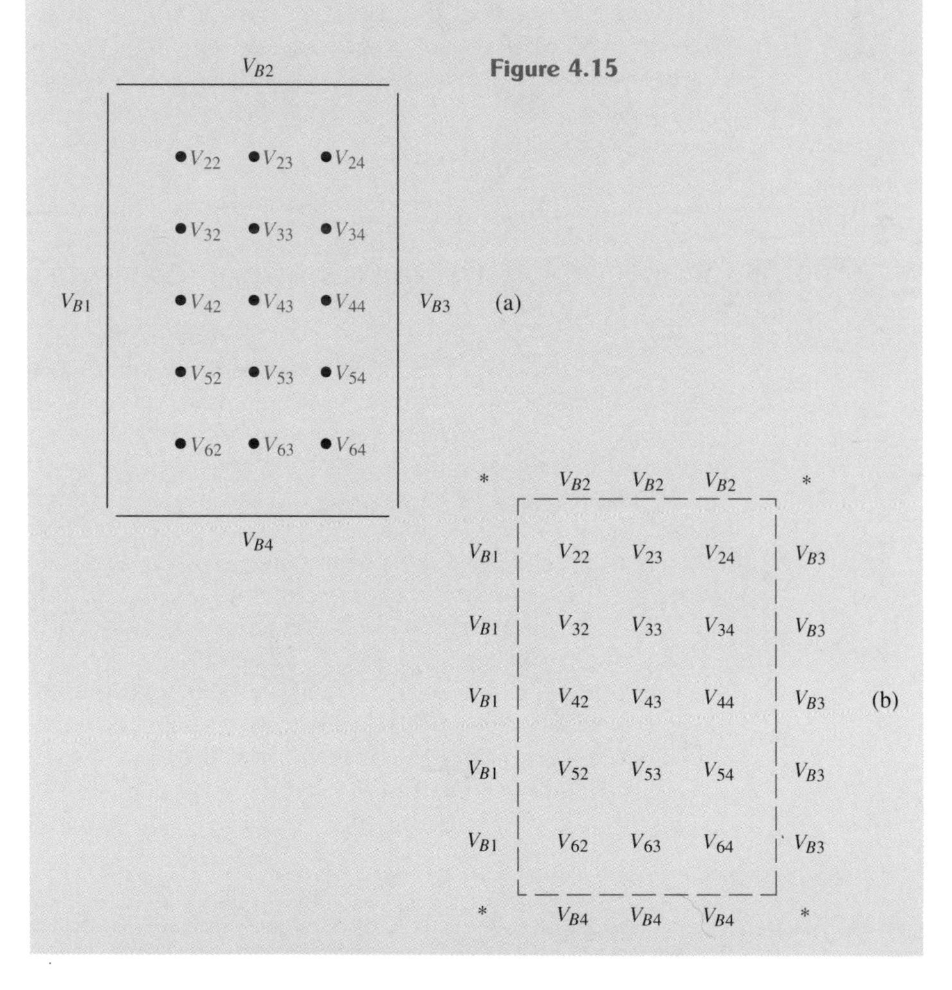

Figure 4.15

In this method we guess an initial value $V_{ij}^{(0)}$ for each voltage; for instance, we can start by setting all the $V_{ij}^{(0)}$ equal to zero.[6] We then construct the average of the four nearest-neighbor voltages, which we can call V'_{ij}:

$$V'_{ij} = \frac{1}{4}\left(V_{(i+1),j}^{(0)} + V_{(i-1),j}^{(0)} + V_{i,(j+1)}^{(0)} + V_{i,(j-1)}^{(0)}\right) \tag{4.43}$$

If all the $V_{ij}^{(0)}$ were the actual correct voltages, then $V_{ij}^{(0)}$ would be equal to V'_{ij}. But since the $V_{ij}^{(0)}$ are only arbitrary guesses, they do not equal V'_{ij}. The difference between $V_{ij}^{(0)}$ and V'_{ij} is the *residual* R:

$$R = V'_{ij} - V_{ij}^{(0)} \tag{4.44}$$

We can improve the approximation by changing one of the old $V_{ij}^{(0)}$ to a new value $V_{ij}^{(\text{new})}$ which makes the residual vanish. This can be done by applying

$$V_{ij}^{(\text{new})} = V_{ij}^{(\text{old})} + R \tag{4.45}$$

We have not yet obtained a correct solution, however, because the residuals at all the other points will still fail to vanish. We must go on to each of the other points, applying (4.45) to each point in turn. One might think that after applying (4.45) to all the points, we would finally have a correct solution, one for which the residuals vanish at every point. But this is not the case, because every time we "correct" the voltage at one point, we reintroduce residuals at the points we have already "corrected." However, after "correcting" all the points, one after another, we will have arrived at a new potential distribution that is more nearly correct than the original set of guesses. There will still be a set of nonvanishing residuals, obtained by substituting the "new" voltages (in place of the "old" voltages $V_{ij}^{(0)}$) into (4.43) and (4.44), and these new residuals should be smaller than the ones obtained from the voltages originally guessed. Then we have only to repeat the process. A sequence of repetitions will result in a sequence of approximate potential distributions, which eventually converges to the potential distribution that correctly satisfies (4.42) and the given boundary conditions.

An interesting point is that faster convergence can be obtained if instead of (4.45) we use

$$V_{ij}^{(\text{new})} = V_{ij}^{(\text{old})} + \beta R \tag{4.46}$$

where β is called an "overrelaxation factor." Its value affects the speed at which the sequence of approximations $V_{ij}^{(0)}, V_{ij}^{(1)}, V_{ij}^{(2)} \cdots$ approaches the correct value of V_{ij}. The optimum value of β is difficult to predict, because it depends on the details of the problem being solved. Generally a value

[6]*Caution:* We choose initial values only for the unknown voltages inside the dashed line in Fig. 4.15(b). Voltages such as V_{13}, V_{35}, etc. are fixed boundary voltages and cannot be varied.

between 1 and 2 will give significantly faster convergence than is obtained with $\beta = 1$. However, as β increases further, improvement ceases, and for $\beta \geq 2$ the sequence fails to converge at all.

These steps are to be followed. We begin at the first point, point 22. We use the arbitrary $V_{ij}^{(0)}$ to calculate V'_{22}. Then we apply (4.46) to obtain $V_{22}^{(1)}$. Then we construct V'_{23}, using V_{13} ($\equiv V_{B2}$), $V_{24}^{(0)}$, $V_{33}^{(0)}$, *and the new value* $V_{22}^{(1)}$. We continue with this procedure, improving the approximation by means of (4.46) at each individual point, until we have worked through all the points, and the first iteration is completed. The entire process is then repeated as many times as necessary for the V_{ij} to approach their final values. This may seem a roundabout procedure, but each step is very simple and the computation goes very quickly. An idea of how close one has come to the correct result can be gained by noting the change between values of any V_{ij} on successive iterations; this change should approach zero as V_{ij} converges to its correct value.

As usual in the finite-difference method, increasing the number of points would be expected to increase the accuracy of the result. The technique can be demonstrated with an Hp-15C calculator, but its small memory imposes a limit of about fifteen unknown voltages, as in Fig. 4.15. If you are using a larger computer you can space the points more closely and use more of them, thus obtaining more accurate results. The suggested approach to programming is to use two index registers for the indices i and j. First write a subroutine for generating $V_{ij}^{(n+1)}$ according to (4.46). Then use two loops for taking i and j through all possible values, and finally use another loop to perform successive iterations.

As a typical problem, try $V_{B2} = 10V$, $V_{B1} = V_{B3} = V_{B4} = 0$. For convenience begin with $V_{ij}^{(0)} = 0$. You may wish to experiment with different values of β between 1 and 2.

**COMPUTER EXERCISE 4.5 THE METHOD OF MOMENTS

In this exercise we shall calculate the capacitance of a parallel-plate capacitor, including edge effects. We shall use a simple form of a technique known as the *method of moments.*

The structure in question is the two-dimensional capacitor shown in Fig. 4.16. It consists of parallel metal plates with width a, spaced a distance d apart and assumed to be infinitely long in the direction perpendicular to the plane of the page. The plates are charged to $+V$ and $-V$ as shown, so that the total potential difference is $2V$. If this structure were a part of an infinitely wide parallel-plate capacitor we would expect its capacitance per

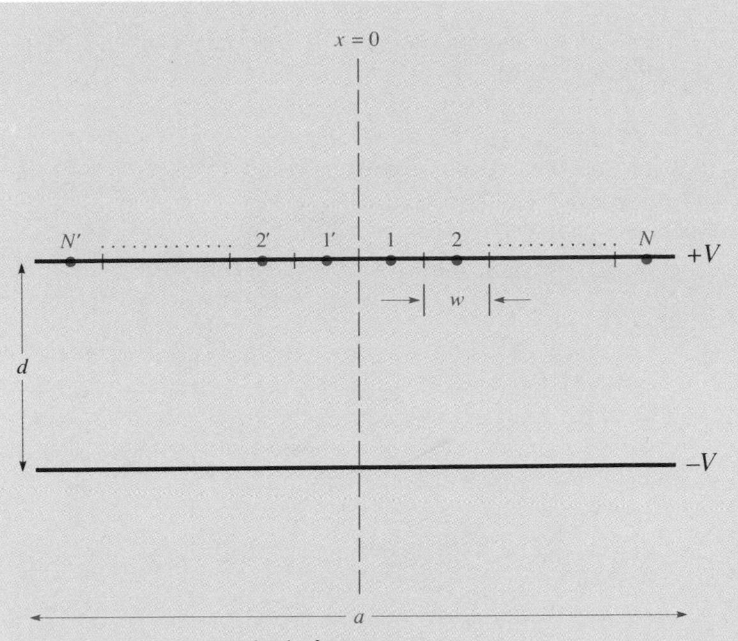

Figure 4.16 The method of moments.

unit length (perpendicular to the page) to be $\epsilon_0 a/d$. However, we shall see that the actual capacitance is larger than this value.

In this method we shall find the capacitance by calculating the distribution of charges on the plates. Let us divide each plate into $2N$ strips of width w, as shown in the figure. The numbered points are located at the centers of the strips. Thus, point M is located at $x = (M - \frac{1}{2})w$, and point M' is located at $x = -(M' - \frac{1}{2})w$. By symmetry we see that the charge on strip M' is the same as the charge on strip M. Also by symmetry, the charges on the lower plate are equal in magnitude and opposite in sign to the corresponding charges on the upper plate.

We must somehow obtain a set of equations which will allow us to solve for the N unknown charges. To do this, we first observe that the potential that exists at any point must be related to the charges that produce it through (4.20). In this case, the potentials on the capacitor plates are given, and it is the charges that we must find. Thus we shall write N equations expressing the known potentials on the strips in terms of the charges on the strips, and then solve these equations for the charges. The reader may feel that this approach is somehow "backwards." The way to think of it is that when voltage is applied to the plates by an external voltage source, such as a battery, charges must flow out of the battery to charge the capacitor. The quantity and location of these charges must turn out to be exactly right, so that the applied potentials on the plates are consistent with the charges through (4.20).

Now we must turn to the problem of calculating the potentials in terms of the charges on the strips. Let us make the approximation that the potential produced by each charged strip is approximately that of a line charge at its center, which, from Gauss' law is

$$V \cong -\frac{Q_M}{2\pi\epsilon_o} \ln R \qquad \textbf{(4.47)}$$

where R is the distance from charge to the place where potential is measured. Evidently this approximation cannot be used to find the contribution to the potential of a strip arising from the charge on that same strip, so that contribution will have to be considered separately. Thus the total potential at point K (which is in the right half of the upper plate) is

$$\begin{aligned} V_K = \sum_{\substack{M=1 \\ M\neq K}}^{M=N} -\frac{Q_M}{2\pi\epsilon_o} \ln |x_M - x_K| - \sum_{M=1}^{M=N} \frac{Q_M}{2\pi\epsilon_o} \ln (x_M + x_K) \\ + \sum_{M=1}^{M=N} \frac{Q_M}{2\pi\epsilon} \ln [(x_M - x_K)^2 + d^2]^{1/2} \\ + \sum_{M=1}^{M=N} \frac{Q_M}{2\pi\epsilon} \ln [(x_M + x_K)^2 + d^2]^{1/2} + V_{KK} \end{aligned} \qquad \textbf{(4.48)}$$

Here $x_j \equiv (j - \frac{1}{2})w$, and Q_M is the charge on the Mth strip. The first sum in (4.48) represents the contribution of all the strips in the right-hand half of the upper plate, except for the strip containing point K. The second sum is the contribution of the left half of the upper plate; the third sum is from the right side of the lower plate; and the fourth sum is from the left side of the lower plate. The final term V_{KK} is the contribution of the strip containing point K.

Now the value of V_K is actually known; it is simply V, the value at which the capacitor plate is held. There are N unknown charges and from (4.48) we obtain N different equations, one for each value of K. Thus we can solve for the unknown charges Q_K, and, by adding them, find the total charge on the plate, and hence the capacitance. Note that in the finite-difference and iterative methods we solved for the potentials, whereas in this method we solve for the charges. The latter approach is more convenient when charge is what we wish to find. Moreover, the problem space in the present problem is unbounded. If we had chosen a finite-difference method we would have had to use artificial grounded boundaries far from the capacitor, which would have required a large number of lattice points. Thus we would have had a larger number of unknowns to solve for than in the present method.

There still remains the problem of V_{KK}. We may estimate this quantity by making the assumption that charge is uniformly distributed over the narrow strip. The potential at its center is then found by summing the potentials of many parallel line charges, each with charge per unit length $\tau = \frac{Q_K}{w} dx$:

$$V_{KK} = \int_{x_K-(w/2)}^{x_K+(w/2)} - \frac{\left(\frac{Q_K}{w}\right) dx}{2\pi\epsilon_o} \ln|x - x_K|$$

$$= -\frac{Q_K}{2\pi\epsilon_o}\left[\ln\left(\frac{w}{2}\right) - 1\right] \quad \textbf{(4.49)}$$

Inserting (4.49) into (4.48) we can now solve for the N different Q_K, using a library subroutine for simultaneous linear equations. Note that w can be eliminated from the equations by using $w = a/2N$ and introducing the dimensionless parameter $Z = 2Nd/a$. A hand-held calculator can be used, but the value of N will be limited. For accurate results a larger value of N, say 30, is desirable. It is convenient to set $2\pi\epsilon_o V = 1$. (The value chosen for V does not matter, since all the Q's are proportional to V; thus $C = \Sigma Q_K/V$ will ultimately be independent of V.)

(a) First test your program by choosing a small spacing, say $d/a = 1/(2N)$, and finding the total charge on the upper plate. Compare the capacitance per unit length with the infinite-width approximation $(\epsilon a/d)(2V)$. If there are no errors, you should obtain a capacitance that is similar, but slightly larger due to edge effects.
(*Caution:* Serious roundoff errors leading to unphysical results may occur with extremely small values of d/a.)

(b) Find $Q_K(K)$ for large spacing, $d/a = 4$. Make a graph of Q_K versus K.

(c) Find and graph the capacitance as a function of the ratio d/a, for the range $0.1 < \frac{d}{a} < 4$. On the same graph plot the infinite-width approximation $\epsilon_o a/d$ for comparison.

CHAPTER 5

The Magnetostatic Field

We have now seen that distant charged particles can exert forces on one another, and that such forces can be described by means of the electric field vector $\vec{E}$. However, this is only half the picture. There is another kind of force that is experimentally observable, but which is physically different from the force described by $\vec{E}$. This is the magnetic force. In order to describe magnetic effects, we introduce another vector field $\vec{H}(x, y, z)$, which is usually called the *magnetic intensity*. Mathematically, $\vec{H}$ is a field in exactly the same sense that $\vec{E}$ is; that is, to each position (x, y, z) in space there corresponds a value of the three-dimensional vector $\vec{H}$. Physically, however, magnetic fields originate from currents and exert forces on currents, while electric fields originate from charges and exert forces on charges. In the present chapter, we shall be dealing with *magnetostatics,* that is, with situations in which $\vec{H}$ is non-time-varying, or at most slowly varying, so its rate of change can be neglected.

Magnetostatic effects appear quite different from electrostatic effects, which might lead one to think that electric and magnetic phenomena are entirely separate from one another. Actually, this impression is quite false. In Chapter 6 we shall see that as soon as time variations appear, $\vec{E}$ and $\vec{H}$ cease to be independent of each other. In fact, there is really only *one* field, the electromagnetic field; it is just a convenience to divide this field into the two parts we call $\vec{E}$ and $\vec{H}$. Only in the static case can one talk about "pure" magnetic effects that are unrelated to $\vec{E}$. In Chapter 6 we shall go on to the time-varying case, and interactions between $\vec{E}$ and $\vec{H}$ will be described.

5.1 ORIGIN OF THE MAGNETOSTATIC FIELD

As already remarked, magnetostatic fields originate from currents. In electrostatics, we saw that if the positions of all charges were known, the resulting electric field could be calculated by means of (4.2), (4.3), or (4.4). Similarly, in magnetostatics, $\vec{H}$ can be found by means of an integral involving the existing currents. Initially, let us just consider one small elemental current: we imagine a current I flowing through a short length of wire with length dl. From experiments it is known that the magnetic flux density produced at the point $\vec{r}$ (known as the *field point*) by this wire element is

$$\overrightarrow{dH} = \frac{I(\vec{dl} \times \vec{e}_{r'r})}{4\pi|\vec{r} - \vec{r}'|^2} \tag{5.1}$$

In this formula the wire element is assumed to be located at $\vec{r}'$ (the *source point*), and $\vec{e}_{r'r}$ is the unit vector directed from $\vec{r}'$ to $\vec{r}$. The quantity $|\vec{r} - \vec{r}'|$ appearing in the denominator is of course the distance from the current element to the place where $\overrightarrow{dH}$ is measured. As seen from (5.1), the mks units of $\vec{H}$ are amperes per meter.

Expression (5.1) is known as the law of Biot and Savart. It can also be written

$$\overrightarrow{dH} = \frac{I[\vec{dl} \times (\vec{r} - \vec{r}')]}{4\pi|\vec{r} - \vec{r}'|^3} \tag{5.2}$$

EXAMPLE 5.1

A current element is located at $x = 2$ cm, $y = 0$, $z = 0$, as shown in Fig. 5.1. The current has a magnitude of 150 mA and flows in the $+y$ direction. The length of the current element is 1 mm. Find the contribution of this element to the magnetic field at $x = 0$, $y = 3$ cm, $z = 0$.

Solution In this case $\vec{dl} = 10^{-3}\vec{e}_y$, $\vec{r} = (0,3,0) \times 10^{-2}$, $\vec{r}' = (2,0,0) \times 10^{-2}$, and $|\vec{r} - \vec{r}'| = \sqrt{13} \times 10^{-2}$ mks units. Taking the cross product, we have

$$\begin{aligned} \vec{dl} \times (\vec{r} - \vec{r}') &= 10^{-3}\vec{e}_y \times (-2\vec{e}_x + 3\vec{e}_y) \times 10^{-2} \\ &= 10^{-5}(2\vec{e}_z) \end{aligned}$$

(since $\vec{e}_y \times \vec{e}_x = -\vec{e}_z$ and $\vec{e}_y \times \vec{e}_y = 0$.) Thus

$$\begin{aligned} \overrightarrow{dH} &= \frac{(0.15)(2 \times 10^{-5})\vec{e}_z}{4\pi(\sqrt{13} \times 10^{-2})^3} \\ &= 5.09 \times 10^{-3}\vec{e}_z \text{ A/m} \end{aligned}$$

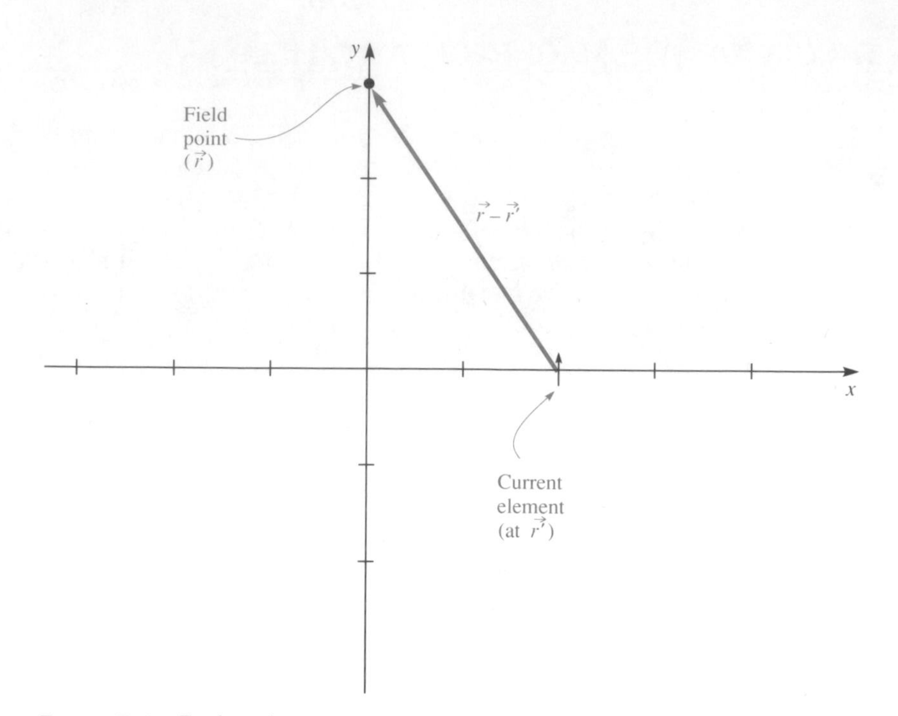

Figure 5.1 Finding the magnetostatic field of a current element using the law of Biot and Savart.

EXERCISE 5.1

In Example 5.1, the source point is moved to (2,4,0) and the direction of the current element is changed to the x direction. Now what is $\overrightarrow{dH}$ at the same field point?

Answer $-1.07 \times 10^{-2}\,\vec{e}_z$ A/m.

Inspection of (5.1) or (5.2) reveals that the H field produced by the current element "curls around" the element in the manner shown in Fig. 5.2. Because of the cross product, the direction of $\overrightarrow{dH}$ at any field point is perpendicular to the current element $\overrightarrow{dl}$ that produces it. It is also perpendicular to a line drawn from the source point to the field point. (Check that Fig. 5.2 is consistent with these statements.) The sign of $\vec{H}$ can be determined by what is known as the *right-hand rule*. If one makes a fist of the right hand and extends the thumb in the direction of the current element, the lines of $\vec{H}$ point in the direction of the curled fingers.

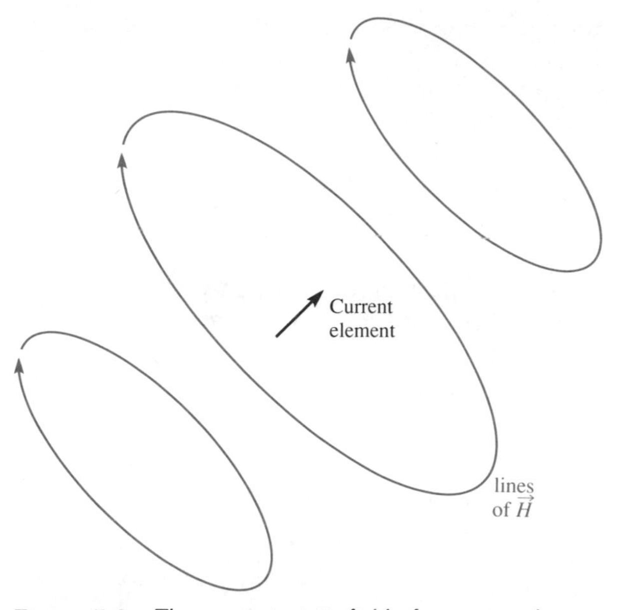

Figure 5.2 The magnetostatic field of a current element is in the form of circles around the current that produces it. The direction of the field can be found from the right-hand rule: If the thumb of the right hand points in the direction of the current, the fingers point in the direction of the field.

EXERCISE 5.2

A current element is located at the origin and is directed in the $+y$ direction. By inspection (no written mathematics) determine the direction of $\overrightarrow{dH}$ at (a) $(1,0,0)$; (b) $(-1,0,0)$; (c) $(0,0,1)$.

Answers (a) $-z$; (b) $+z$; (c) $+x$.

It may have occurred to the reader that an isolated static current element must be some kind of fiction. How does the current get into the element and where does it go when it leaves? The reader is correct. We have only introduced the current element as a convenience for discussion. In fact, direct currents can flow only in continuous closed paths. However, the principle of superposition applies, and we can find the field produced by a closed wire loop by thinking of it as a large number of current elements placed head to tail. We then add up the fields produced by the individual elements to find the total field.

An important example is the field of an infinitely long straight wire. Let this wire be located on the z axis, and let it carry a current I in the $+z$ direction. Consider a field point located at $z = 0$, at a distance R from the wire, as shown in Fig. 5.3. To find the contribution of a current element located at z, we note that

$$\vec{dl} = dz\vec{e}_z$$

$$\vec{r} = R\vec{e}_r$$

(where $\vec{e}_r$ is the unit vector in the r direction in cylindrical coordinates; that is, the unit vector perpendicular to the wire). Also,

$$\vec{r}' = z\vec{e}_z$$

$$|\vec{r} - \vec{r}'| = (R^2 + z^2)^{1/2}$$

We now use (5.1):

$$\overrightarrow{dH} = \frac{I\,dz(\vec{e}_z \times \vec{e}_{r'r})}{4\pi|\vec{r} - \vec{r}'|^2}$$

whereas before, $\vec{e}_{r'r}$ is the unit vector directed from $\vec{r}'$ to $\vec{r}$. We observe that the cross product $\vec{e}_z \times \vec{e}_{r'r}$ is a vector perpendicular to both, that is, *out* of the paper in Fig. 5.3. (The fact that it is out of the paper, and not in, can be seen from the right-hand rule.) The direction of $\overrightarrow{dH}$ at this field point is the same (outward), no matter where on the wire the source element is taken to be; in other words, all elements of the wire give contributions in the same direction. Thus we need only add up their magnitudes. The total field magnitude is

$$|\vec{H}| = \int_{z=-\infty}^{\infty} \frac{I\,dz}{4\pi(R^2 + z^2)} \frac{R}{(R^2 + z^2)^{1/2}}$$

$$= \frac{I}{2\pi R} \tag{5.3}$$

Thus the magnitude of the field is inversely proportional to distance from the wire, and its direction is in circles around the wire, as shown in Fig. 5.4(a). The sign of $\vec{H}$ is a given by the right-hand rule.

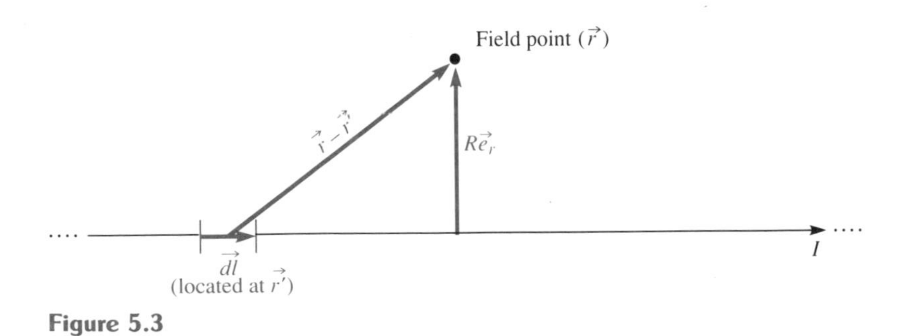

Figure 5.3

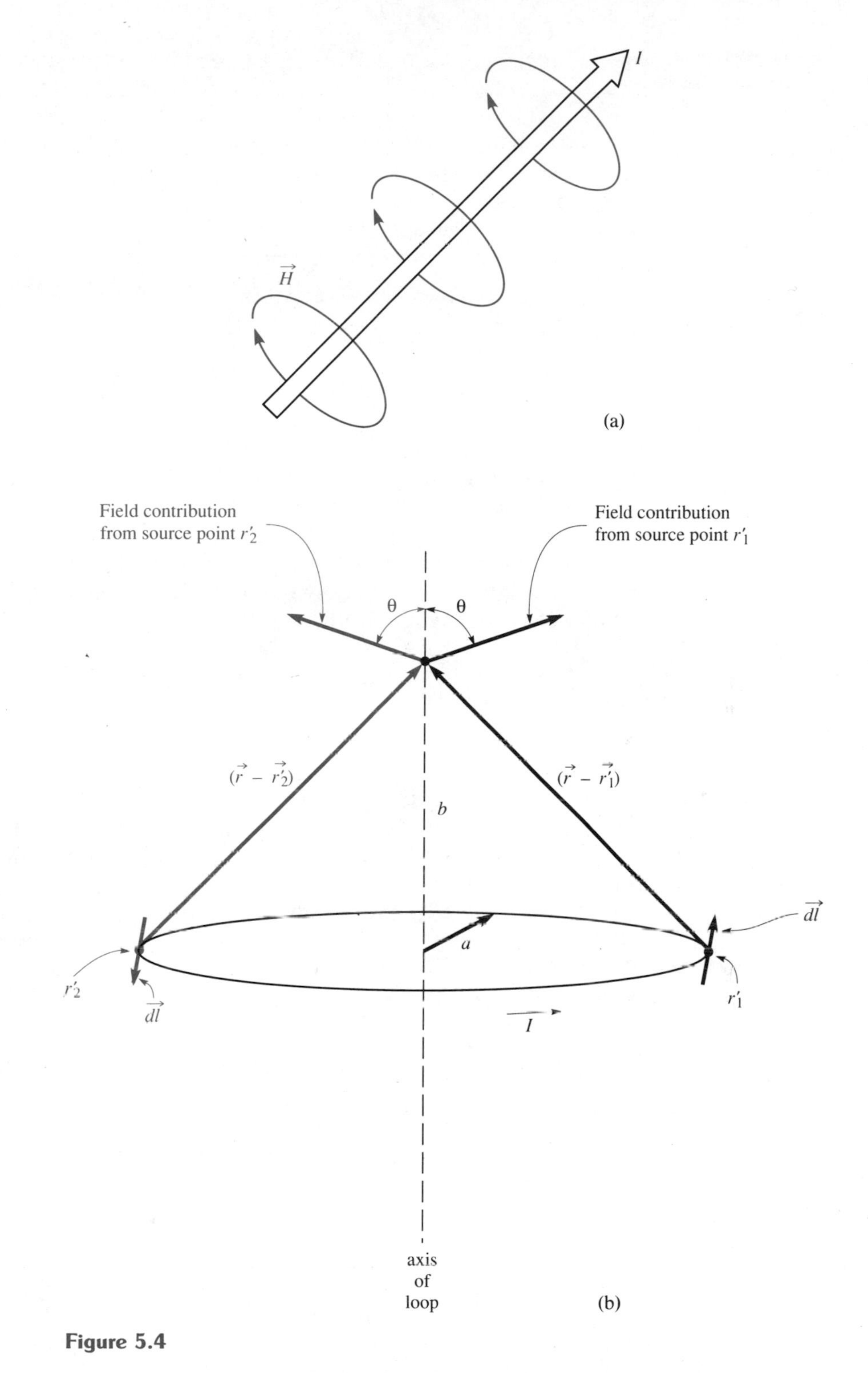

I
$\vec{H}$
(a)
Field contribution from source point r'_2
Field contribution from source point r'_1
θ
θ
$(\vec{r} - \vec{r}'_2)$
$(\vec{r} - \vec{r}'_1)$
b
a
$\vec{dl}$
r'_2
$\vec{dl}$
r'_1
I
axis of loop
(b)

Figure 5.4

EXERCISE 5.3

A current I flows in the clockwise direction around a circular loop of radius a. Find $\vec{H}$ at the center of the loop.

Answer $I/2a$, perpendicular to the plane of the loop.

The situation is slightly more complicated when the fields contributed by each element of the wire are not all in the same direction. Consider, for example, the field of a circular loop of wire of radius a, carrying a current I, lying in the x-y plane and centered at the origin. Suppose we wish to find H at the point $(0,0,b)$, which is on the axis of the loop a distance b above its center, as shown in Fig. 5.4(b). We note that $\vec{dl}$ is always perpendicular to $\vec{r} - \vec{r}'$; also, the distance from source point to field point is the same for all the source points, and is equal to $\sqrt{a^2 + b^2}$. Thus the *magnitude* of the contribution of each length dl of the loop is

$$|\vec{dH}| = \frac{I\,dl}{4\pi(a^2 + b^2)}$$

However, the directions of these contributions are all different. In each case the direction of the contribution is perpendicular to both $\vec{dl}$ and $\vec{e}_{r'r}$. The total field at the field point is the sum of all these contributions. In this case one can see that the component of the sum in the plane of the loop is zero. This is because the horizontal component of the contribution from any point on the loop is cancelled by the horizontal contribution of the point on the opposite side of the loop. (See Fig. 5.4(b).) However, the contributions in the z direction all add. The angle θ that each contribution makes with the z axis is the same, and the cosine of that angle is

$$\frac{a}{\sqrt{a^2 + b^2}}$$

Thus the total magnetic intensity at the point in question is

$$\begin{aligned}\vec{H} &= \frac{I}{4\pi}\frac{2\pi a}{a^2 + b^2}\frac{a}{\sqrt{a^2 + b^2}}\vec{e}_z \\ &= \frac{Ia^2}{2(a^2 + b^2)^{3/2}}\vec{e}_z\end{aligned}$$

5.2 MAGNETIC FLUX DENSITY AND MAGNETIC FLUX

In electrostatics, a second vector field $\vec{D}$ was introduced to supplement $\vec{E}$. Similarly in magnetostatics it is usual to introduce a second magnetic vector $\vec{B}$, which we assume is related to the intensity $\vec{H}$ by the equation[1]

$$\vec{B} = \mu \vec{H} \tag{5.4}$$

$\vec{B}$ is known as the magnetic *flux density* vector. Its mks unit is the tesla (T). (Sometimes this unit is called the weber per square meter (Wb/m^2).) The constant μ describes the effects of atoms in the vicinity; its value depends on what material is present. When the surrounding medium is vacuum, the value of μ is called μ_o, given by

$$\mu_o = 4\pi \times 10^{-7} \text{ mks units} \tag{5.5}$$

Except for ferromagnetic materials, most materials have values of μ very nearly equal to μ_o. For ferromagnetic materials, μ is much larger than μ_o.

When a field with flux density $\vec{B}$ passes normally through an area A, we say that a *flux* Φ, with magnitude $|B|A$, passes through the area A. Magnetic flux is measured in *webers*. (This is why the units of $\vec{B}$ are sometimes called webers per square meter.) If the field is not perpendicular to the area A, but makes an angle θ with the normal, the flux depends on the normal component of $\vec{B}$: that is, $\Phi = |B|A \cos\theta$. Most generally, if the angle of $\vec{B}$ varies over the surface, the flux is given by

$$\Phi = \int_S \vec{B} \cdot \vec{dS} \tag{5.6}$$

One can visualize the notion of "flux" by thinking in terms of fluid flow. If a vector field $\vec{V}$ is thought of as representing the direction and flow rate of some fluid, then the flux of that vector through a surface S, which is written

$$\int_S \vec{V} \cdot \vec{dS}$$

represents the total amount of flow through that surface. Magnetic flux, measured in webers, is the total flow of the flux density $\vec{B}$ through a surface. Similarly electric flux, written

$$\int_S \vec{D} \cdot \vec{dS}$$

[1] Equation (5.4) assumes that the directions of $\vec{B}$ and $\vec{H}$ are the same, and that their magnitudes are linearly proportional. There are some cases in which these simplifying assumptions are inappropriate, but we shall not deal with such cases in this book.

is the total flow of the electric flux density $\vec{D}$ through the surface S. As we have seen, the electric flux through a surface is related to the charge enclosed by that surface, through Gauss' law. Magnetic flux has a different significance, as will be seen in the next chapter. There we shall see how voltages are produced by time-varying magnetic flux.

EXERCISE 5.4

A spherical surface with radius a has its center at the origin. Half the surface is removed, leaving a hemisphere in the region $x > 0$. There is a magnetic field $\vec{B}$ in the x direction. Find the flux through the hemispherical surface.

Answer $\pi a^2 B$.

5.3 MAGNETIC FORCE

How can one know if a magnetostatic field is present? We can tell because it exerts forces on moving charges and on currents. If a current element $I\vec{dl}$ is located in a magnetic field $\vec{B}$, it experiences a force

$$\vec{dF} = I\vec{dl} \times \vec{B} \tag{5.7}$$

This force, known as the *Lorentz force,* is in a direction perpendicular to the current element and also perpendicular to $\vec{B}$. It is largest when $\vec{B}$ and the wire are perpendicular. The Lorentz force plays an important role in everyday life: it is the force that drives electric motors.

EXAMPLE 5.2

Two wires are parallel and separated by a distance b. They each carry a current I, in the same direction. Find the magnitude and the direction of the force per unit length exerted on one wire by the magnetic field of the other.

Solution For convenience let us assume that the currents flow in the $+z$ direction, and that wire #1 is located at $x = 0$, $y = 0$ and wire #2 is at $x = b$, $y = 0$. From (5.3) and (5.4) the magnitude of $\vec{B}$ at either wire is $\mu I/2\pi b$. From the right-hand rule, its direction at wire #1 is $-y$, while its direction at #2 is $+y$.

The force on every element of a wire is the same. The magnitude of the force acting on a length L is $IL(\mu I/2\pi b)$. The direction of the force on wire #1 is $(+\vec{e}_z) \times (-\vec{e}_y) = +\vec{e}_x$, while that on wire #2 is $(+\vec{e}_z) \times (+\vec{e}_y) = -\vec{e}_x$. Thus we see that the currents *attract*.

Sometimes one is interested in the motion of an individual charged particle, such as an electron, as it moves through a magnetic field. Since a current in a wire is simply a flow of moving electrons, one expects the force on a single moving particle to resemble expression (5.7). From experiment the force acting on a charge q, moving with velocity $\vec{v}$, is found to be

$$\vec{F} = q(\vec{v} \times \vec{B}) \tag{5.8}$$

As before, the force is perpendicular to the direction of charge motion, and is also perpendicular to $\vec{B}$. The magnetic force (5.8) can be used, for example, to deflect a beam of electrons in a cathode-ray tube.

5.4 THE CURL OPERATOR

We must now introduce another vector operator, known as the *curl operator*. This operator differs from the others we have seen in that it acts *on a vector field* to produce *another vector field*. Because vectors are more complicated than scalars, the curl operator is more complicated than—for example—the divergence operator, which acts on a vector field to produce a scalar field. However, the curl operator plays a vital role in electromagnetics. Thus it is desirable to work with it a bit, until it becomes familiar.

Let $\vec{B}(x,y,z)$ be any vector-valued function of position. The expression for the curl of $\vec{B}$, written curl $\vec{B}$ or $\nabla \times \vec{B}$, is, in rectangular coordinates

$$\nabla \times \vec{B} = \left(\frac{\partial B_z}{\partial y} - \frac{\partial B_y}{\partial z}\right)\vec{e}_x + \left(\frac{\partial B_x}{\partial z} - \frac{\partial B_z}{\partial x}\right)\vec{e}_y + \left(\frac{\partial B_y}{\partial x} - \frac{\partial B_x}{\partial y}\right)\vec{e}_z \tag{5.9}$$

where B_x, B_y, B_z are the rectangular components of $\vec{B}$. This rather complicated formula does not have to be memorized; it can be looked up in tables of vector operators, such as the one in the endpapers of this book. The operator can also be written in the form of a determinant (which helps one to remember it):

$$\nabla \times \vec{B} = \begin{vmatrix} \vec{e}_x & \vec{e}_y & \vec{e}_z \\ \dfrac{\partial}{\partial x} & \dfrac{\partial}{\partial y} & \dfrac{\partial}{\partial z} \\ B_x & B_y & B_z \end{vmatrix}$$

Sometimes one finds the field $\vec{B}$ expressed in cylindrical or spherical coordinates instead of rectangular. Expressions for the curl operator in those coordinate systems will also be found in the endpaper table.

The physical significance of the curl operator is that it describes the "rotation" or "vorticity" of the field $\vec{B}$ at the point in question. Let us imagine we have a tank of flowing water, such that the flow velocity at any point x, y, z is equal to $\vec{B}(x,y,z)$. Let us also imagine that we probe the moving fluid with a tiny paddlewheel mounted on a shaft. The paddles are not tilted with respect to the shaft (i.e., the shaft lies in the plane of each paddle); therefore, water flowing *parallel to the shaft* does *not* cause the paddlewheel to turn. Now if this paddlewheel is placed in the flowing water field $\vec{B}$, its resulting rotation is proportional to curl $\vec{B}$. But $\nabla \times \vec{B}$ is a vector; it has direction as well as magnitude. The full analogy is this: the component of $\nabla \times \vec{B}$ *in the direction of the shaft* is proportional to the velocity of the paddlewheel.[2] Some examples of fields with zero or nonzero curl are given in Fig. 5.5. In Fig. 5.5(a), the field is independent of position and points in the y direction. Thus there is as much upward force on the right side of the wheel as on the left, and we conclude that the z component of $\nabla \times \vec{B}$ is zero. (The z component is the component being measured, because the shaft of the paddlewheel is perpendicular to the page, therefore in the z direction.) In (b), the field is changing its direction at the point in question; mathematically, $\frac{\partial B_y}{\partial x}$ is negative. This makes the wheel turn *clockwise*. From (5.9) we see that negative $\frac{\partial B_y}{\partial x}$ implies that the z component of $\nabla \times \vec{B}$ is negative. (There is no B_x in this example, so the $\frac{\partial B_x}{\partial y}$ term makes no contribution to $(\nabla \times \vec{B})_z$.) In (c), $\frac{\partial B_y}{\partial x}$ is positive; the wheel rotates counterclockwise, indicating that $(\nabla \times B)_z$ is positive. In (d) the field is directed in the x direction, and its strength varies as a function of y: $\frac{\partial B_x}{\partial y}$ is negative. The force on the lower side of the wheel is greater than the force on its upper side, and it turns counterclockwise, which, as we have just seen, means that $(\nabla \times \vec{B})_z$ is positive. This agrees with (5.9), where negative $\frac{\partial B_x}{\partial y}$ implies a positive z component of curl $\vec{B}$. Finally in (e), B_x varies as a function of x: $\frac{\partial B_x}{\partial x}$ is negative. However, in this case the forces on the upper and lower sides of the wheel are equal, and it does not turn. Again this agrees with (5.9). Since in that equation $\frac{\partial B_x}{\partial x}$, $\frac{\partial B_y}{\partial y}$, and $\frac{\partial B_z}{\partial z}$ do not appear, these derivatives (with the same coordinate in both numerator and denominator) make no contribution to the curl.

Mathematically one finds the curl of any given vector field by simply applying (5.9), or the corresponding formula for another coordinate system. This process is straightforward, but care is required.

---

[2]One may still ask toward which of the two possible directions along the shaft the curl vector points. The direction along the shaft can be found from the right-hand rule. If the fingers of the right hand point the way the paddlewheel turns, the sign of the component of the curl along the shaft is in the direction of the thumb.

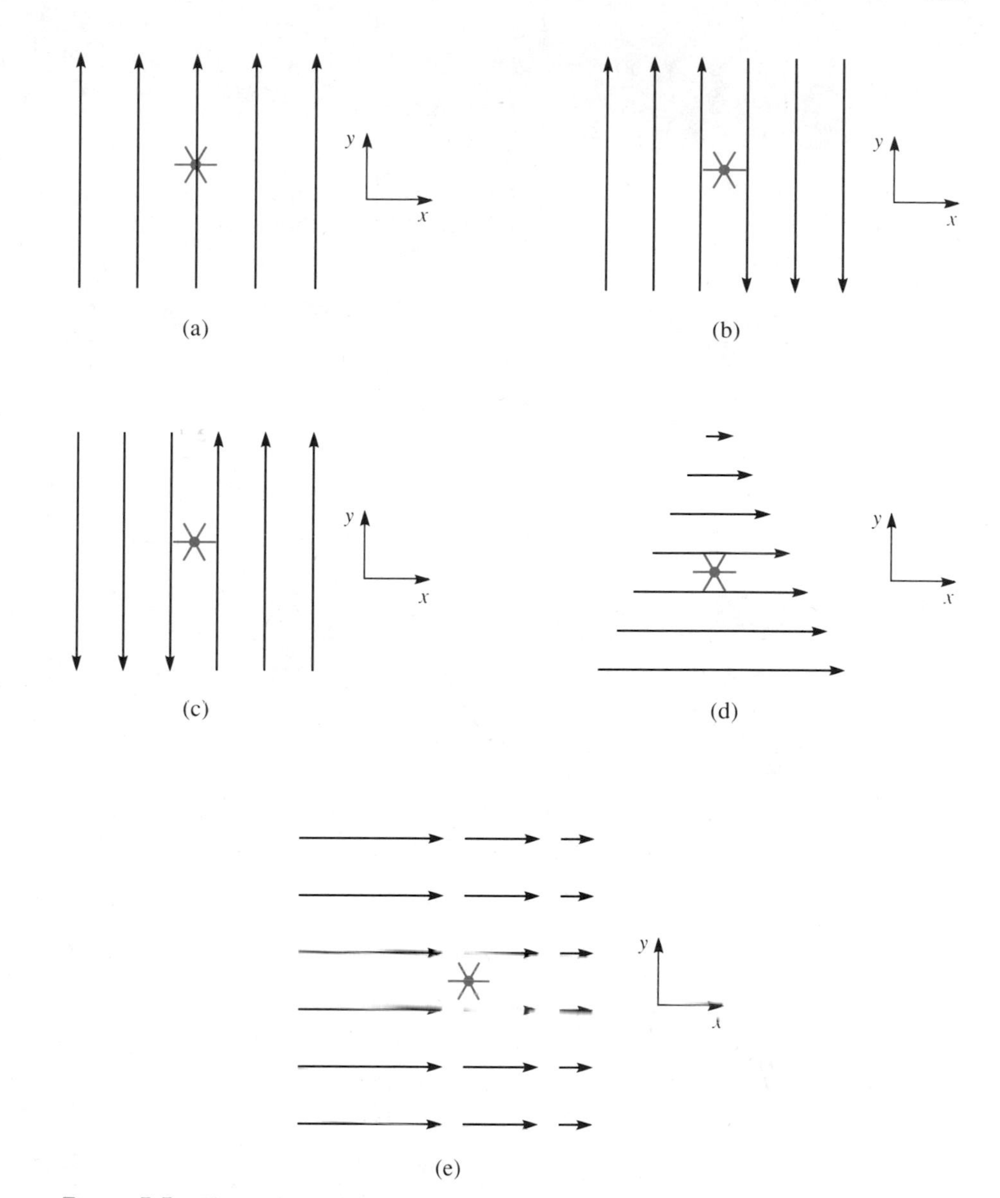

Figure 5.5 Physical significance of the curl operator.

EXAMPLE 5.3

A certain vector field $\vec{V}(x, y, z)$ is described by

$$\vec{V}(x, y, z) = Az^2 \vec{e}_y + Cx^2 \vec{e}_z$$

where A and C are given constants. Find (a) the value of $\vec{V}$ at (0, 2, 2); (b) the value of $\vec{V}$ at (2, 2, 0); (c) the value of $\nabla \times \vec{V}$ at (0, 2, 2); (d) the value of $\nabla \times \vec{V}$ at (2, 2, 0).

Solution The vector $\vec{V}$ takes on a value that depends on position; in this case on x and z (but not y). From the given expression we see that

$$\vec{V}(0,2,2) = 4A\vec{e}_y$$

$$\vec{V}(2,2,0) = 4C\vec{e}_z$$

The curl of $\vec{V}$ is also a vector field; that is, the vector $\nabla \times \vec{V}$ is another function of position. In this case $V_y = Az^2$ and $V_z = Cx^2$. Thus from (5.9),

$$\nabla \times \vec{V} = -2Az\vec{e}_x - 2Bx\vec{e}_y$$

At (0, 2, 2) $\nabla \times \vec{V} = -4A\vec{e}_x$, and at (2, 2, 0) $\nabla \times \vec{V} = -4B\vec{e}_y$.

EXAMPLE 5.4

Let

$$\vec{E} = \frac{A\vec{e}_r}{r}$$

where $r = (x^2 + y^2)^{1/2}$ is the distance from the z axis, $\vec{e}_r$ is the radial unit vector in cylindrical coordinates, and A is a given constant. Find $\nabla \times \vec{E}$.

Solution We can proceed by converting the expression for $\vec{E}$ into rectangular coordinates and using (5.9). We note that

$$\vec{e}_r = \frac{\vec{r}}{r} = \frac{x\vec{e}_x + y\vec{e}_y}{(x^2 + y^2)^{1/2}}$$

Thus

$$E_x = \frac{Ax}{x^2 + y^2} \qquad E_y = \frac{Ay}{x^2 + y^2}$$

Differentiating, we find that

$$\frac{\partial E_x}{\partial y} = -\frac{2Axy}{(x^2 + y^2)^2}$$

$$\frac{\partial E_y}{\partial x} = -\frac{2Axy}{(x^2 + y^2)^2}$$

while all other derivatives appearing in (5.9) vanish. Thus

$$\nabla \times \vec{E} = \left(\frac{\partial E_y}{\partial x} - \frac{\partial E_x}{\partial y}\right)\vec{e}_z = 0$$

We can also proceed more directly by using the expression for the curl in cylindrical coordinates, which (from the table in the endpapers) is

$$\nabla \times \vec{E} = \left(\frac{1}{r}\frac{\partial E_z}{\partial \phi} - \frac{\partial E_\phi}{\partial z}\right)\vec{e}_r + \left(\frac{\partial E_r}{\partial z} - \frac{\partial E_z}{\partial r}\right)\vec{e}_\phi + \left(\frac{1}{r}\frac{\partial (rE_\phi)}{\partial r} - \frac{1}{r}\frac{\partial E_r}{\partial \phi}\right)\vec{e}_z$$

The given $\vec{E}$ field has only an r component and its value depends only on r. The derivative $\frac{\partial E_r}{\partial r}$ does not appear in the cylindrical curl expression, and all the derivatives that do appear vanish, in this case. Thus it is immediately seen that $\nabla \times \vec{E} = 0$.

EXERCISE 5.5

The field $\vec{U}(r, \theta, \phi)$ is described in spherical coordinates by

$$\vec{U} = \frac{2p\cos\theta}{r^3}\vec{e}_r + \frac{p\sin\theta}{r^3}\vec{e}_\theta + \frac{A\sin\theta}{r}\vec{e}_\phi$$

where p and A are constants. Find $\nabla \times \vec{U}$.

Answer $\frac{2A}{r^2}\cos\theta\vec{e}_r$ (except at the origin, where the fields become infinite and derivatives fail to exist).

MAGNETIC VECTOR POTENTIAL

The curl operator can be used to derive the magnetic field B from a suitable potential function. The idea is similar to that of calculating the electric field $\vec{E}$ from its potential V. However, in keeping with the generally more complicated (shall we say, more ''vectorial'') nature of the magnetic field, the usual potential function for $\vec{B}$ is a vector function of position, rather than a scalar, as in electrostatics. This vector function is the *magnetic vector potential* $\vec{A}(x, y, z)$. The field is found by taking the curl of the magnetic vector potential:

$$\vec{B} = \nabla \times \vec{A}$$

This formula is analogous to (4.26). To find the magnetic potential, we require an additional formula analogous to (4.20). For the case of magneto*statics*, this formula is

$$\vec{A}(\vec{r}) = \oint \frac{\mu\, I(\vec{r}\,'')\, \vec{dl}\,'}{4\pi|\vec{r} - \vec{r}\,''|} \tag{5.10}$$

Using (5.10) we can in principle find $\vec{A}$ at any field point $\vec{r}$ by integrating the contributions of all the current elements of a circuit located at source points $\vec{r}''$. An interesting point about this integral is that the contribution to $\vec{A}$ of each differential current element $\vec{dl}'$ points in the same direction as the current element that produces it.

The magnetic potential can be useful as a computational tool for finding the magnetic field, much in the same way as the electric potential is useful. However, the magnetic vector potential is a bit more complicated to use. One case in which it is quite convenient is that in which all currents producing the field flow in the same direction—say, the x direction. In that case (5.10) becomes a scalar equation and has exactly the same form as (4.20); one can find the magnetic potential by analogy with the electric potential in the same geometry. The magnetic potential is also useful as a conceptual tool; some theorems are easiest to derive in terms of $\vec{A}$. Actually it is not in statics, but in dynamics that the magnetic vector potential finds its most important uses. It plays an important role in the theory of antennas, as we shall see in Chapter 10.

THE MAGNETOSTATIC CURL EQUATION

We can now make use of the curl operator to write a basic equation of magnetostatics. We have already seen in section 5.1 that it is possible to find the magnetic field if the currents that produce it are known. Now we shall go in the other direction, writing an expression that can be used to find the current when the magnetic field is known. This expression is

$$\nabla \times \vec{H} = \vec{J} \tag{5.11}$$

Here $\vec{H}$ is the magnetic intensity, and $\vec{J}$ is the current density. (Current density is the current per unit area passing through a plane perpendicular to the flow.) Equation (5.11) contains the same physics as (5.1), and in fact (5.1) can be derived from (5.11). Note that like (5.1), (5.11) is valid *only for magnetostatics*. When time variations appear in the next chapter, an additional term will appear on the right side of (5.11). Equation (5.11) is analogous to the electric source equation $\nabla \cdot \vec{D} = \rho$, in that it relates the field to its causative agent.

5.5 STOKES' THEOREM AND AMPÈRE'S CIRCUITAL LAW

We shall now proceed to develop an important law of magnetostatics, known as *Ampère's circuital law*. This law plays a role similar to that of Gauss' law in electrostatics, in that it provides a convenient method of calculating the field, in those cases where its use is appropriate. The law of Biot and Savart (5.1) can in principle be used to find the field when currents are known, but it is too clumsy to be useful in solving problems by inspection; Ampère's circuital law is often more convenient.

To arrive at Ampère's law we shall make use of a mathematical theorem known as *Stokes' theorem*. The author must confess to the reader that Stokes' theorem is rather difficult to master, but it is worth the effort. It is not just useful in magnetostatics; it has many applications in electromagnetics and in other subjects. Before proceeding, one may wish to review section 3.3, on line and surface integrals.

The statement of Stokes' theorem is as follows.[3] Let S be any unbroken surface, and let L be the closed path around its border. Then the line integral of a vector field $\vec{V}$ around the path L is related to the surface integral of $\nabla \times \vec{V}$ over the surface:

$$\int_S (\nabla \times \vec{V}) \cdot \vec{dS} = \oint_L \vec{V} \cdot \vec{dl} \tag{5.12}$$

The direction of integration around L is related to the direction of $\vec{dS}$ by the right-hand rule.

We note that Stokes' theorem is very powerful because one can choose *any* surface. It doesn't matter *which* surface one chooses. It can be flat or curved, just so long as it is continuous (has no holes in it). Moreover (5.12) applies to *any* physically reasonable vector field. (Infinitely large fields, which presumably do not exist physically, are excluded.) Stokes' theorem is a mathematical tool, like the divergence theorem, which in a way it resembles. Its statement (5.12) is certainly simple enough, but practice is required to gain a "feeling" for its use.

EXAMPLE 5.5

Let $\vec{v}$ be a uniform field in the x direction, given by $\vec{v} = A\vec{e}_x$ (where A is a constant). Verify Stokes' theorem using a square surface in the x-y plane bounded by the points (0, 0), (1, 0), (1, 1), (0, 1).

Solution The curl of $\vec{v}$ is obviously zero, since all its derivatives vanish. Therefore the left side of (5.12) vanishes. The line integral on the right is

$$\int_{(0,0)}^{(1,0)} \vec{v} \cdot \vec{dl} + \int_{(1,0)}^{(1,1)} \vec{v} \cdot \vec{dl} + \int_{(1,1)}^{(0,1)} \vec{v} \cdot \vec{dl} + \int_{(0,1)}^{(0,0)} \vec{v} \cdot \vec{dl}$$

The second and fourth of these integrals vanish because $\vec{v}$ is perpendicular to $\vec{dl}$ on those segments. Thus the right side of (5.12) is

$$A + 0 + (-A) + (0) = 0$$

in agreement with Stokes' theorem.

---

[3]As usual, we sidestep mathematical refinements and state the theorem in a form suitable for practical use. Readers interested in more general cases should consult works on mathematical analysis.

EXAMPLE 5.6

Repeat the previous example with $\vec{v} = Ky\vec{e}_x$, where K is a constant.

Solution In this case $\nabla \times \vec{v} = -K\vec{e}_z$, which is constant over the surface in question and normal to it. Thus (taking $\vec{dS}$ to be in the $+z$ direction, so that $\vec{dS} = \vec{e}_z\, dA$)

$$\int_S (\nabla \times \vec{v}) \cdot dS = -K \int_S \vec{e}_z \cdot \vec{e}_z\, dA = -K \int_S dA = -K$$

Proceeding to evaluate the right side of (5.12) as in the previous example,

$$\oint_L \vec{v} \cdot \vec{dl} = (0) + (0) + K \int_{(1,1)}^{(0,1)} y\vec{e}_x \cdot \vec{dl} + (0)$$

In the remaining integral we set $y = 1$, because we are integrating along the $y = 1$ contour from $x = 1$ to $x = 0$. On this contour $\vec{dl} = -\vec{e}_x dx$. Thus

$$\oint_L \vec{v} \cdot \vec{dl} = -K \int_0^1 dx = -K$$

in agreement with Stokes' theorem.

In the preceding two examples the surface of integration was planar, but this need not be the case. For instance, imagine a rigid metal ring over which is stretched a rubber membrane. The ring acts as the line L that bounds the rubber surface S. Now let us stretch the rubber so that it bulges out into, say, a hemisphere. From the point of view of Stokes' theorem, the rubber surface, although now a hemisphere (or any other shape), is still bounded by the same metal ring. If the surface is shaped like a glove, it is bounded by the line that forms the opening at the wrist, as shown in Fig. 5.6.

EXAMPLE 5.7

Let $\vec{V} = Kz\vec{e}_y$, and let the surface of integration be a hemisphere of radius a, with center at the origin, occupying the region $x > 0$. Verify Stokes' theorem.

Solution In this case $\nabla \times \vec{V} = -K\vec{e}_x$. Refer to Fig. 5.7 on p. 162. Choosing $\vec{dS}$ in the direction outward from the center of the hemisphere,

$$\int_S (\nabla \times \vec{V}) \cdot \vec{dS} = -K \int_S \vec{e}_x \cdot \vec{dS} = -K \int_0^{\pi/2} \cos\theta\,(2\pi a^2 \sin\theta)\, d\theta$$
$$= -\pi a^2 K$$

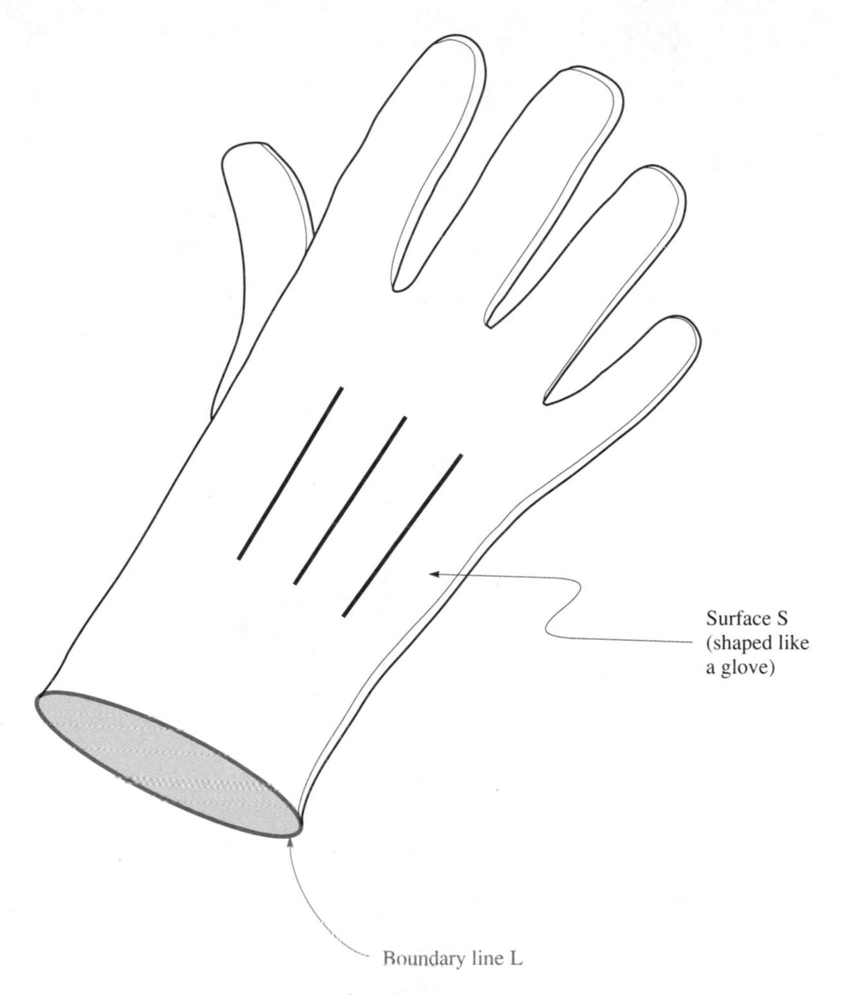

Figure 5.6 The glove-shaped surface is bounded by the line L.

As for the right side of (5.12), we use $\vec{dl} = -a\cos\phi\, d\phi\, \vec{e}_y - a\sin\phi\, d\phi\, \vec{e}_z$, $z = a\cos\phi$. (Note the important minus sign that enters because $\vec{dl}$ points in the $-\phi$ direction.) Thus:

$$\oint_L \vec{V} \cdot \vec{dl} = K \oint z\vec{e}_y \cdot \vec{dl} = K \int_0^{2\pi} (a\cos\phi)(-a\cos\phi\, d\phi)$$
$$= -Ka^2 \int_0^{2\pi} \cos^2\phi\, d\phi = -\pi a^2 K$$

Thus the surface integral is found to be equal to the line integral, in agreement with Stokes' theorem. Note that we might have integrated over a *planar* surface, a circular disc in the y-z plane, and, according to Stokes' theorem, the surface integral would also have turned out to be $-\pi a^2 K$. Moreover, the

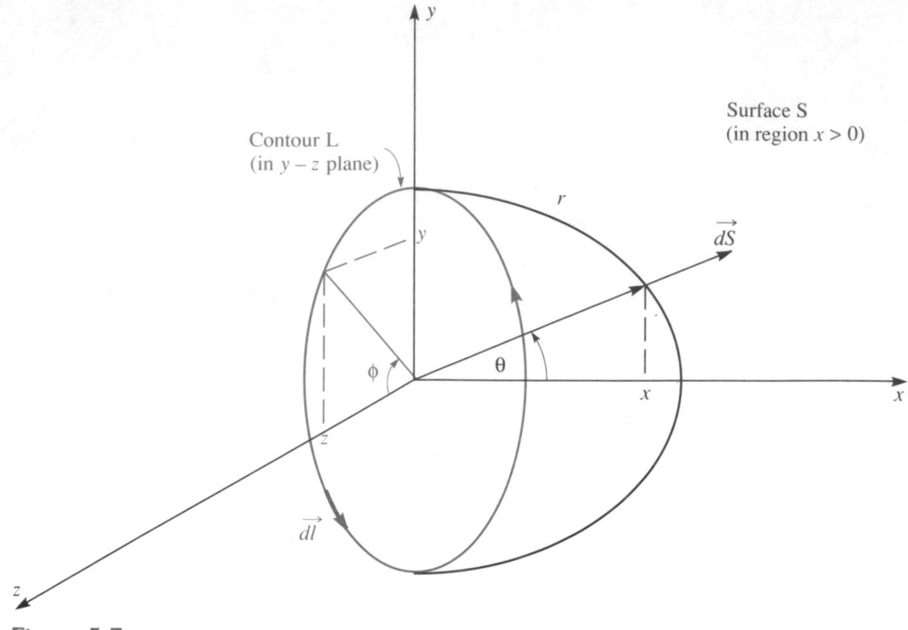

Figure 5.7

surface integral would turn out to be $-\pi a^2 K$ for *any* arbitrary non-planar surface bounded by the same L, such as, for example, Fig. 5.6. Of course performing a surface integration over a glove-shaped surface would be a difficult task, which would have to be carried out numerically. Using Stokes' theorem would save all this effort. One can get the answer by simply evaluating the line integral around L; or one can evaluate the surface integral on the planar disc.

AMPÈRE'S CIRCUITAL LAW

Let us apply Stokes' theorem to the magnetic intensity vector $\vec{H}$. Choosing any surface S bounded by the border line L, we have from (5.12):

$$\int_S (\nabla \times \vec{H}) \cdot \vec{dS} = \oint_L \vec{H} \cdot \vec{dl} \qquad \textbf{(5.13)}$$

Substituting (5.11), we have

$$\int_S \vec{J} \cdot \vec{dS} = \oint_L \vec{H} \cdot \vec{dl} \qquad \textbf{(5.14)}$$

which is Ampère's circuital law. It states that the surface integral of the current density passing through S is equal to the line integral of $\vec{H}$ around the path L. The surface integral of the current density is equal to I_S, the total current passing through S. Thus Ampère's law may also be written

$$\oint_L \vec{H} \cdot \vec{dl} = I_S \tag{5.15}$$

EXAMPLE 5.8

Use Ampère's circuital law to find the magnetic field outside an infinitely long straight wire carrying a current I.

Solution As in similar problems using Gauss' law, we take advantage of the symmetry to observe that the lines of H are circles centered on the wire, as in Fig. 5.4. The direction of $\vec{H}$ is related to the direction of I by the right-hand rule. Let us apply Ampère's law to a circular path of radius R, centered on the wire, as shown in Fig. 5.8. Let the surface S be a planar disc perpendicular to the wire. The current passing through S, which is called I_S in (5.15), is clearly equal to I. To perform the line integral we make use of the fact that $\vec{H}$ is everywhere parallel to $\vec{dl}$. Thus

$$\oint \vec{H} \cdot \vec{dl} = 2\pi R|\vec{H}|$$

From (5.15) we then have

$$2\pi R|\vec{H}| = I$$

$$|\vec{H}| = \frac{I}{2\pi R}$$

$$|\vec{B}| = \frac{\mu I}{2\pi R}$$

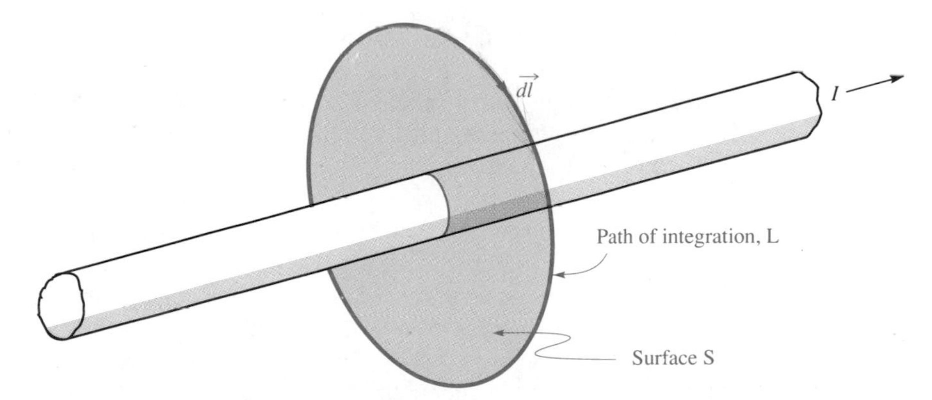

Figure 5.8 Illustrating Ampère's circuital law.

in agreement with (5.3), which was found from the law of Biot and Savart. We note that it is easier to obtain the result from Ampère's law. However, like Gauss' law, Ampère's law is not a general method. It is only useful in certain cases, where the direction of $\vec{H}$ can be guessed at the start.

EXERCISE 5.6

Find the magnetic intensity *inside* an infinitely long straight wire of radius a, carrying a current I. Assume the current density in the wire is constant.

Answer $\dfrac{Ir}{2\pi a^2}$, where r is the distance from the wire axis.

5.6 BOUNDARY CONDITIONS

It often happens that magnetic or electric fields are present at the boundary of two different materials. This raises the question of how the fields on one side of the boundary are related to those on the other side. Are they the same, or is there a change in the fields as one crosses the boundary? The rules that relate fields on opposite sides of a boundary are known as *boundary conditions*. This is a fairly extensive subject, because there are many different cases, involving different types of boundaries. However, we shall see that there are general methods that are useful for deriving boundary conditions. Once one becomes familiar with the derivations, it is easy to remember the various boundary conditions, or to rederive them.

As our first example, let us assume that magnetic fields are present at the boundary of two materials having different values of μ. Ampère's law can be used to obtain a relationship between the fields on the two sides of the boundary. Such a situation is shown in Fig. 5.9. The field $\vec{H}_1$, inside medium 1 at some position close to the boundary, can be decomposed into its components tangential to the boundary and normal to the boundary. We call these components H_{T1} and H_{N1}, respectively. Just across the boundary in medium 2, the corresponding components are H_{T2} and H_{N2}. We shall now show that for any materials with finite conductivity, $H_{T1} = H_{T2}$.

We begin by choosing a path of integration L, as shown in Fig. 5.10. This path is very narrow, so that part L_1 lies in medium 1 but is very close to the boundary, and part L_3 lies inside medium 2, but is also very close to the boundary. Sides L_1 and L_3 have length l. When we construct the line integral of $\vec{H}$ around this path, parts L_2 and L_4 contribute nothing because their lengths are (in the limit as the path becomes arbitrarily narrow) zero. The contribution to the line integral from L_1 is $H_{T1}l$ and from L_3 it is $-H_{T2}l$.

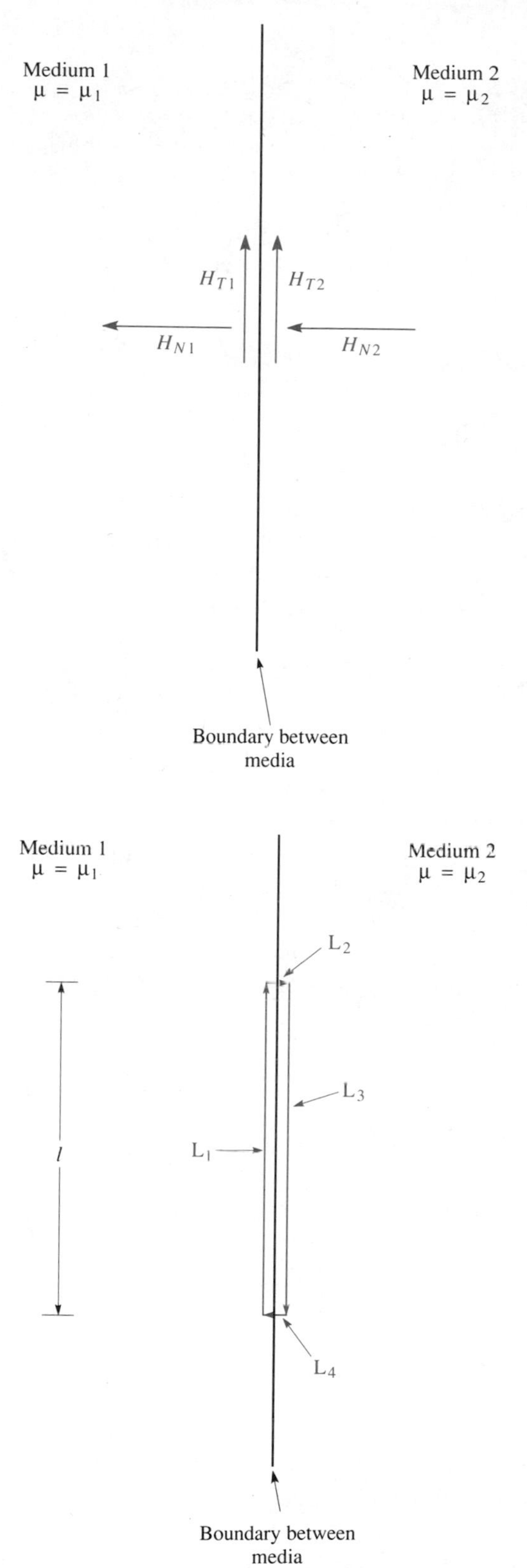

Figure 5.9 Boundary conditions at the interface between two materials with different magnetic permeability.

Figure 5.10 Information about fields tangential to an interface is obtained using a long, thin path for line integration.

Thus

$$\oint_L \vec{H} \cdot dl = (H_{T1} - H_{T2})l \tag{5.16}$$

Now consider the surface integral of the current density over the surface enclosed by L. Since (in the limit) the area of this surface is zero, we must have

$$\int_S \vec{J} \cdot \vec{dS} = I_S = 0 \tag{5.17}$$

From (5.15), (5.16), and (5.17) we then conclude that

$$H_{T1} = H_{T2} \tag{5.18}$$

which is the desired result.[4] One says that the tangential component of $\vec{H}$ is *continuous,* meaning "the same on both sides of the boundary." Note that the above argument fails to tell us anything about the *normal* components of $\vec{H}$ on the two sides, and these in general are not the same. Later we shall see that $\mu_1 H_{N1} = \mu_2 H_{N2}$.

The proof we have just seen demonstrates one of the two common types of boundary conditions. The distinguishing features of this type are: (a) we are finding a property of the *tangential* fields, and (b) we make use of the line integral of the field in question. (In the above proof we used (5.15), which is derived from the curl equation (5.11).) The technique for finding boundary conditions on a tangential field involves the use of a long, narrow path of line integration, which straddles the boundary. In the next chapter we shall apply the same method to the tangential component of the electric field. Note that we could *not* have used this method (or any other method) to show that the tangential component of $\vec{B}$ is continuous. That is because there is no expression like (5.15) for the line integral of $\vec{B}$.

EXERCISE 5.7

Suppose $|\vec{H_2}|$, the magnitude of $\vec{H_2}$ close to the boundary, is equal to $2|\vec{H_1}|$, where $\vec{H_1}$ is the field just across the boundary. $\vec{H_1}$ makes an angle of 60° with the normal to the boundary. What angle does $\vec{H_2}$ make with the normal?

Answer 25.7°.

---

[4]This argument could fail if the current density on the interface were infinitely large, since the product of an infinitely large current density and an infinitely small area might fail to vanish. An infinitely large current density cannot exist in any normal material. However, in a superconductor, with infinite conductivity, a current could flow entirely on the metal surface; that is, the current density, as a function of depth, would be a delta function. In such a case the integral (5.17) would fail to vanish, and H_T would be discontinuous. In an ordinary metal with finite conductivity, current may flow in a thin—but not infinitely thin—sheet along the surface. In that case the integral does vanish when the path is made infinitely narrow, and (5.18) is correct.

A second magnetic boundary condition is needed in order to find the relationship between the *normal* components of $\vec{B}$ on either side of a boundary. To obtain this second condition, we must use an additional physical law describing magnetic fields. From experiments it has been found that in general

$$\nabla \cdot \vec{B} = 0 \tag{5.19}$$

This law is found to apply not only to static magnetic fields but to time-varying fields as well.

Let us now consider a plane boundary between two media with different values of μ, as shown in Fig. 5.11. We shall wish to apply the divergence theorem; for this we choose a cylindrical surface of integration, as shown in the figure. This surface is a right-circular cylinder (or "pillbox") with its plane faces parallel to the boundary. One plane face is in medium 1, and the other is across the boundary in medium 2. The area of each plane face is A, and the height of the cylinder (distance between the two plane faces) is h. We shall be interested in the limit in which h becomes very small.

We now apply the divergence theorem and (5.19), obtaining

$$\int_S \vec{B} \cdot \vec{dS} = \int_V \nabla \cdot \vec{B}\, dV = 0 \tag{5.20}$$

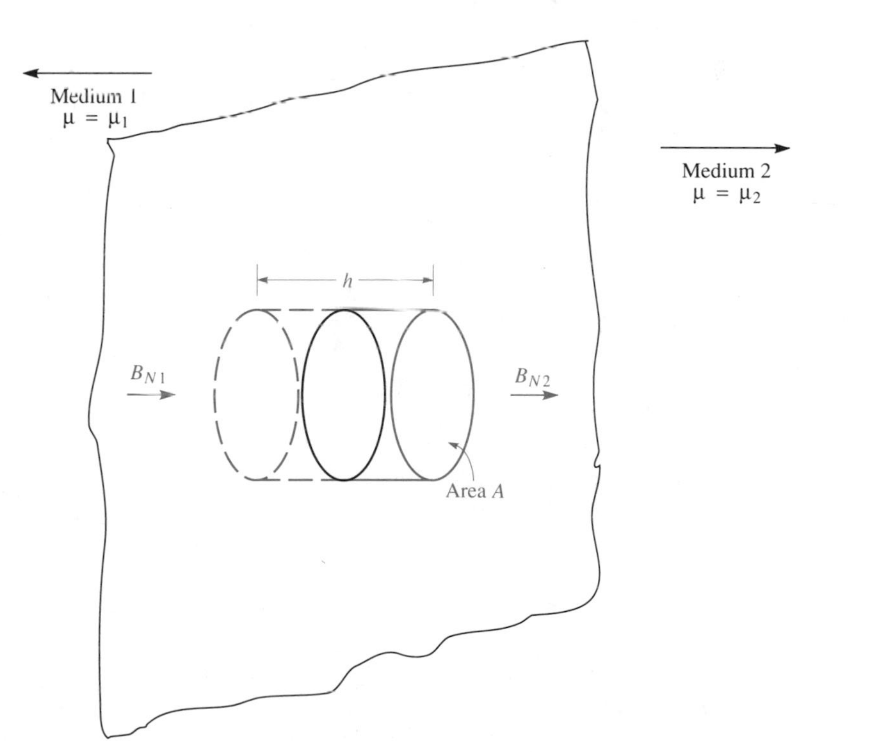

Figure 5.11 Information about fields normal to an interface is obtained using surface integration on a pillbox-shaped surface.

As in the preceding discussion, we separate the fields on the two sides of the boundary into their tangential and normal components. The surface integral consists of three parts: one part for each plane face, and one part for the curved surface of the cylinder. The latter, however, vanishes in the limit at $h \to 0$, because the product of any finite field with an arbitrarily small area must vanish.

Only the field components perpendicular to the boundary can contribute to the surface integrals on the plane faces. In terms of the normal fields shown in Fig. 5.11, (5.20) becomes

$$B_{N2}A - B_{N1}A = 0 \tag{5.21}$$

(The minus sign appears before the second term because $\overrightarrow{dS}$ is the *outward* normal of the integration surface.) Thus we have the desired boundary condition.

$$B_{N1} = B_{N2} \tag{5.22}$$

In words, the normal component of $\vec{B}$ is continuous. Note that (5.22), the condition on the *normal* fields, applies to $\vec{B}$, while (5.18), the condition on the *tangential* fields, applies not to $\vec{B}$, but to $\vec{H}$.

Result (5.22) is an example of the second type of boundary condition. The distinguishing features of this second type are: (a) we are examining the continuity of a *normal* component of the field, and (b) we make use of a *surface* integral of the field in question. The technique for finding the boundary condition involves the use of a thin cylindrical surface of integration, which straddles the boundary. The above proof is based on (5.19), from which, with the help of the divergence theorem, we know that the surface integral of $\vec{B}$ vanishes. A similar proof could not be used for $\vec{H}$ because no equation giving the divergence of $\vec{H}$ is available.

To investigate boundary conditions in other situations, we generally apply either the first of the above methods or the second, depending on whether we wish to know about the tangential component of a field or its normal component. For instance, we can now apply the second method to obtain an important boundary condition for the normal component of $\vec{E}$. Let us consider an electric field at the boundary of two nonconductive materials with electric permittivities ϵ_1 and ϵ_2. There are no real electric charges anywhere in the materials. Thus from (4.13) we have $\nabla \cdot \vec{D} = 0$. Since the divergence of $\vec{D}$ vanishes, the proof based on (5.19) can simply be repeated, and we conclude that

$$D_{N1} = D_{N2} \tag{5.23}$$

Note that (5.23) is derived for the case of two nonconductive dielectrics. If the materials were conductors, some real charge might arrive at the interface. In that case, the volume of integration might contain some charge; the surface integral of D would not vanish and (5.23) would be replaced by $D_{N2} - D_{N1} = \sigma$, where σ is the surface charge per unit area on the interface. In general one must look critically at each different case, and by reviewing the derivation, determine whether or not a boundary condition applies. One should try to develop some skill in doing this, so one can understand unfamiliar situations when they arise. A review of the more common boundary conditions will be found in Appendix B.

EXERCISE 5.8

Consider a certain vector field $\vec{V}(x, y, z)$ that obeys the physical law $\nabla \times \vec{V} = \vec{F}_o$, where $\vec{F}_o$ is a given constant vector. Show that the tangential component of $\vec{V}$ is continuous across any boundary.

5.7 FERROMAGNETIC DEVICES

Most of the important magnetostatic devices make use of ferromagnetic cores to guide the field. Power transformers, motors, and generators all rely on ferromagnetic cores, without which the necessary large magnetic fields would be difficult to attain. In fact, the fields in these devices *do* vary in time, which means that magneto*static* theory cannot rigorously be applied. However, the variations tend to be quite slow. For example, in an American power transformer the fields vary at a frequency of 60 Hz. This is so low that not much accuracy is lost when magnetostatic approximations are applied.

IRON-CORE INDUCTORS

Although for most materials $\mu = \mu_o$, the class of *ferromagnetic materials*—primarily alloys containing iron—is exceptional. In these materials a much larger value of $\vec{B}$ is obtained for a given $\vec{H}$ than in ordinary materials. This characteristic makes ferromagnetic materials very useful in applications such as transformers and electromagnetic machinery, where large, well-confined magnetic fields are required.

Typical relationships between B and H (*magnetization curves*) are shown in Fig. 5.12. We note that B is not linearly proportional to H, so the equation $\vec{B} = \mu\vec{H}$ does not strictly apply. Instead we observe the phenomenon of *saturation,* evidenced by the failure of B to continue to increase as H increases. This effect, present in all ferromagnetic materials, places an upper limit on the value of $\vec{B}$ that can be obtained. Although $\vec{B}$ is not linearly proportional to $\vec{H}$, it is nearly so for small values of the fields. Thus for small fields we can use $\vec{B} = \mu\vec{H}$ as an approximation, assigning to μ a value based on the initial slope of the B-H curve. The approximate values of μ obtained in this way are very large, sometimes as much as $10{,}000\,\mu_o$.

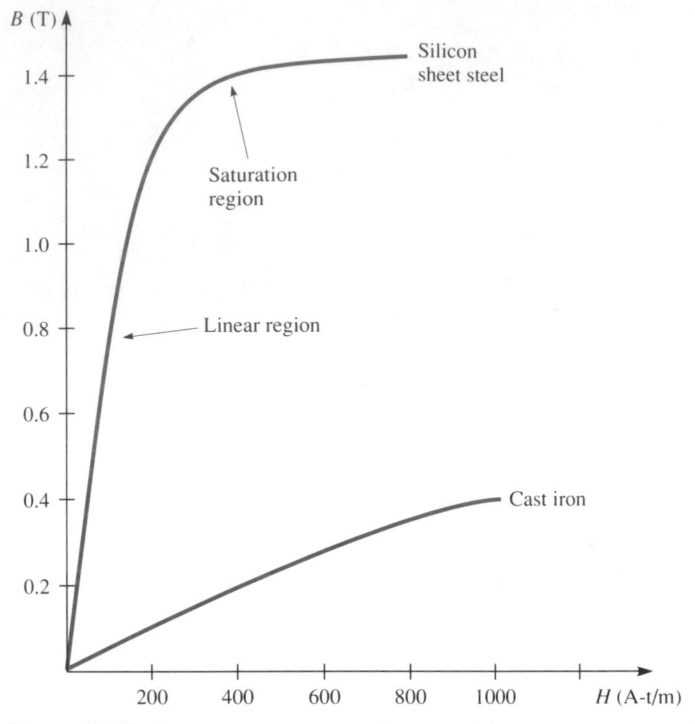

Figure 5.12 Magnetization curves for typical ferromagnetic materials.

EXERCISE 5.9

Estimate μ/μ_o for the silicon sheet steel of Fig. 5.12.

Answer About 6000.

The unusual behavior of ferromagnetic materials results from alignment of microscopic magnets, called *magnetic domains,* inside the material. In unmagnetized iron these domains point randomly in all directions, and the average magnetization $\vec{M}$ is zero. However, when an $\vec{H}$ field is applied, the domains are forced to line up parallel to the applied field, resulting in a large $\vec{M}$. A general relationship, valid for all materials, is that

$$\vec{B} = \mu_o\vec{H} + \mu_o\vec{M} \tag{5.24}$$

(The statement $\vec{B} = \mu\vec{H}$ is thus true only in those special cases in which $\vec{M}$ is linearly proportional to $\vec{H}$. For almost all materials except ferromagnets this assumption is satisfactory.) The large magnetization combines with H to give the large B obtained in ferromagnetic materials. However, $\vec{M}$ cannot increase without limit; it reaches a maximum value when all the domains have been aligned. Thus, at large fields B no longer increases as rapidly as at low fields. This is the saturation effect seen in Fig. 5.12.

Ferromagnetic materials also exhibit the interesting effect known as *hysteresis*. The nature of this effect is that $\vec{B}$ is not a single-valued function of $\vec{H}$; rather, it depends not only on the present $\vec{H}$ but also on values of $\vec{H}$ at earlier times. This effect is illustrated in Fig. 5.13. We begin with a piece of material that has never been magnetized; we are at the origin (point 1) in the diagram. We then apply an $\vec{H}$ field, and $\vec{B}$ increases to the value at point 2. During this step the domains have been aligned, providing a large magnetization $\vec{M}$ that adds to $\vec{H}$ to give a large $\vec{B}$. Now let the applied $\vec{H}$ be reduced to zero. The alignment of the domains partially disappears due to thermal agitation, but some alignment—and hence some magnetization—remains. Thus at point 3 there is still some $\vec{B}$ even though $\vec{H}$ has been reduced to zero. This is the so-called "permanent" magnetization; a piece of iron is made into a permanent magnet in this way. If we next apply a negative $\vec{H}$, the domains are forced to turn around and point the opposite way, giving a negative $\vec{M}$ at point 4. When $\vec{H}$ again returns to zero at point 5, a residual negative magnetization remains. Making $\vec{H}$ positive again will now move us to point 6, and the process continues. In most cases of interest $\vec{H}$ is a sinusoidal function of time. After $\vec{H}$ has cycled from positive to negative a few times, the *B*-*H* curve will settle down into a closed path, as shown in Fig. 5.13(b).[5] Hysteresis can be useful, because it accounts for the existence of permanent magnets. In situations in which the fields are time varying, however, it is generally harmful because it converts electromagnetic energy into heat.

A great advantage of ferromagnetic materials is their ability to "pipe" magnetic flux, almost as water is conducted in a pipe. This characteristic, a consequence of the large value of μ, is responsible for the usefulness of iron cores in transformers and all sorts of electrical machinery. It is difficult to provide a general demonstration of how this piping effect occurs, but a plausibility argument will be given at the end of this chapter. In the meantime, we shall assume that the lines of $\vec{H}$ in a ferromagnetic material do not escape from inside it, and therefore flow parallel to the surfaces, rather like water in a pipe.

Since we know the direction of $\vec{H}$ in the metal, it is easy to find the approximate magnitude of the field by means of Ampère's law. Figure 5.14 shows an N-turn coil wound on a ferromagnetic core. Since $\vec{H}$ inside the metal is parallel to the surfaces, its lines go around the core as shown; roughly speaking, like water through a pipe. Let us choose a path of integration that goes all around the core, as indicated. This path is everywhere approximately

[5]The reader may wonder why hysteresis is not seen in magnetization curves such as Fig. 5.12. This is because the usual *B*-*H* curves are not true graphs of *B*(*H*). They are actually what are known as *normal B*-*H* curves, obtained by plotting the tips of hysteresis loops of increasing magnitude.

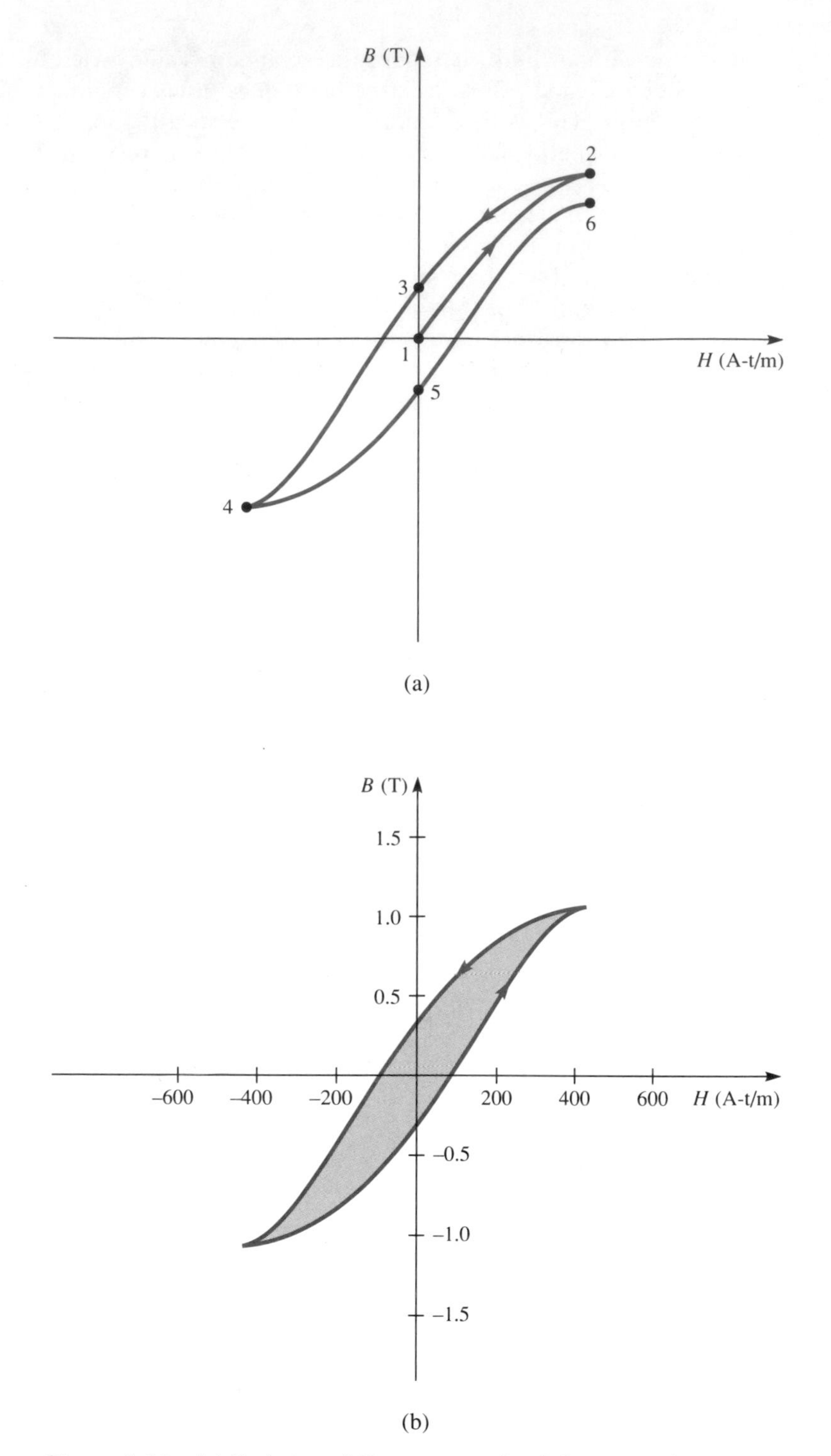

Figure 5.13 (a) Evolution of B in a material with hysteresis, beginning with initially unmagnetized material. (b) Closed path in sinusoidal steady state.

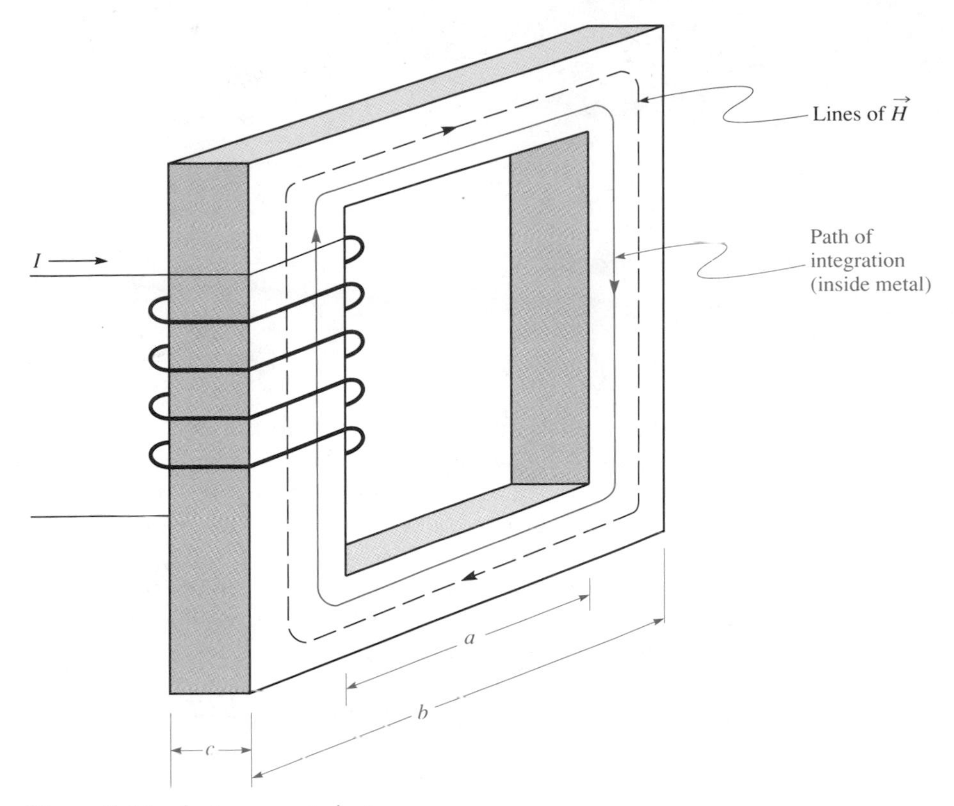

Figure 5.14 An iron-core inductor.

parallel to $\vec{H}$, and the current through its enclosed surface is NI. Thus from Ampère's law,

$$|\vec{H}| \cong \frac{NI}{L_P} \tag{5.25}$$

where L_P is the length of the integration path. (Evidently paths with slightly different lengths can be chosen; L_P would be taken to be an average path length. We are not attempting to find $|\vec{H}|$ exactly, but only to obtain an estimate of its average value.)

The flux obtained in the core is

$$\Phi = \mu|\vec{H}|A \tag{5.26}$$

where A is the cross-sectional area of the core. Flux is of interest because in electrical machinery, the goal of magnet design is usually to obtain maximum flux, subject to limitations of various kinds. There is an analogy between the flux calculated above and the current in a resistive circuit. Let us define the product NI as mmf, the *magnetomotive force* (analogous to *electromotive force,* which is another name for voltage). Then (5.25) and (5.26) can be rewritten as

$$\Phi \cong \frac{\text{mmf}}{R} \tag{5.27}$$

where R is known as the *reluctance* of the *magnetic circuit,* given approximately by

$$R = \frac{L_P}{\mu A} \tag{5.28}$$

Note the resemblance of (5.27) to Ohm's law, with mmf taking the place of voltage, reluctance taking the place of resistance, and flux playing the role of current.

The magnetic circuit analogy is useful because it allows us to estimate flux in cores with more complicated shape. In motors, for example, it is necessary for the flux to pass through air at some point (so it can act on a current-carrying wire and cause it to move). Hence a core with an air gap, schematically like that of Fig. 5.15(a), would be used. If the height of the air gap L_A is not too large, the flux lines go nearly straight across. (A small amount of outward bulging, or "fringing," of the field lines can be neglected.) Using the same path of integration as before, we have

$$|\vec{H}_M|\, L_M + |\vec{H}_A|\, L_A = NI \tag{5.29}$$

where $\vec{H}_M$, $\vec{H}_A$, L_M, and L_A are the fields and path lengths in metal and air respectively. However, from (5.22) this becomes

$$\frac{|\vec{B}|}{\mu} L_M + \frac{|\vec{B}|}{\mu_o} L_A = NI$$

where $\vec{B}$ is the flux density in both air and metal. Since $\Phi = |\vec{B}|A$, we have

$$\Phi = \frac{NI}{R_M + R_A} \tag{5.30}$$

where

$$R_M = \frac{L_M}{\mu A}$$

and

$$R_A = \frac{L_A}{\mu_o A}$$

are the reluctances of the core and the gap, respectively. A diagram of the magnetic circuit is shown in Fig. 5.15(b). Let us generalize that the reluctance of any portion of a magnetic circuit is given approximately by

$$R \cong \frac{L}{\mu A} \tag{5.31}$$

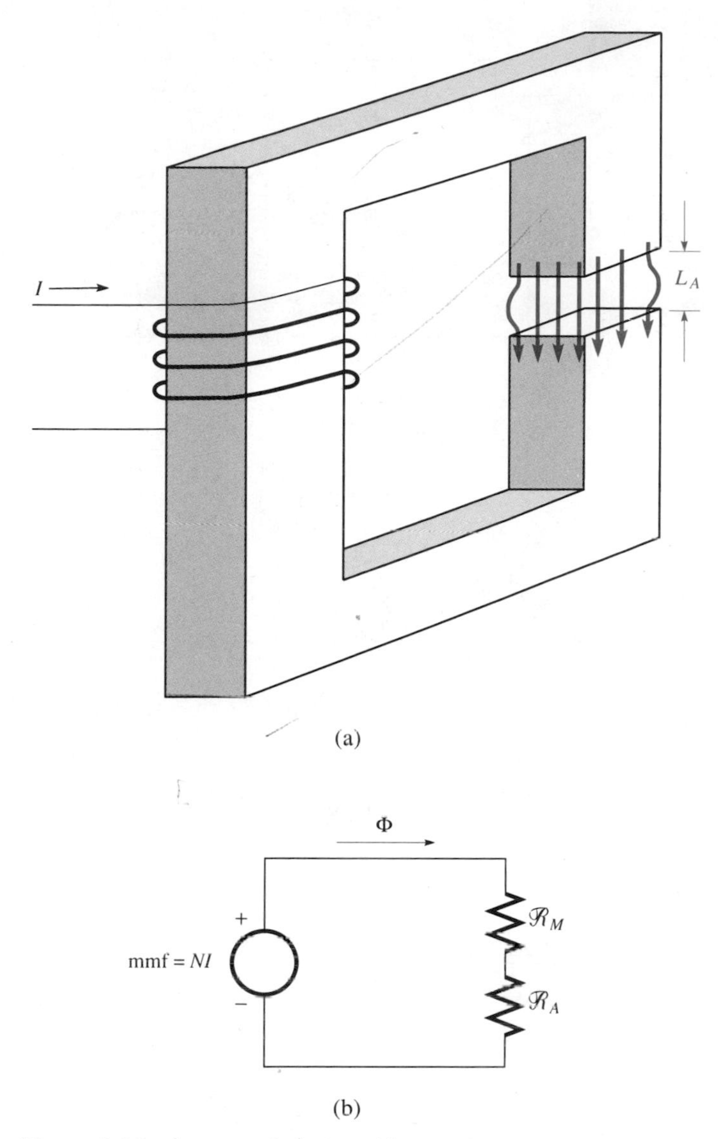

Figure 5.15 Iron-core inductor with gap.

where L, μ, and A are respectively its length, permeability, and cross-sectional area. Estimates of flux in more complicated magnetic circuits can then readily be obtained using the analogy to Ohm's law.

EXAMPLE 5.9

The magnetic circuit shown in Fig. 5.16 is excited by a current $I = 10\text{A}$ passing through a 200-turn coil. The cross-sectional area A of the core is

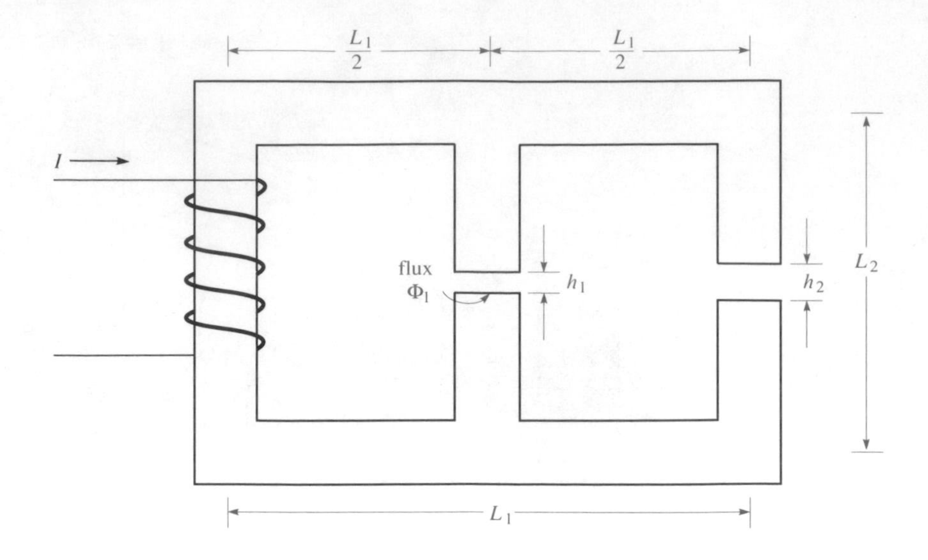

(a)

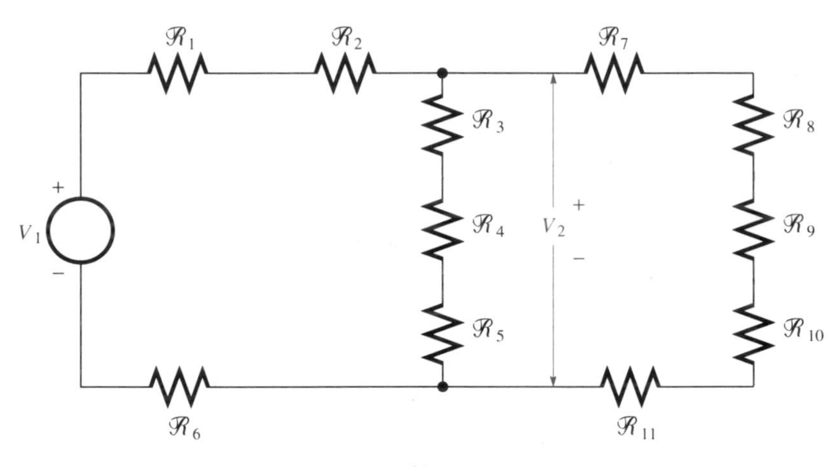

(b)

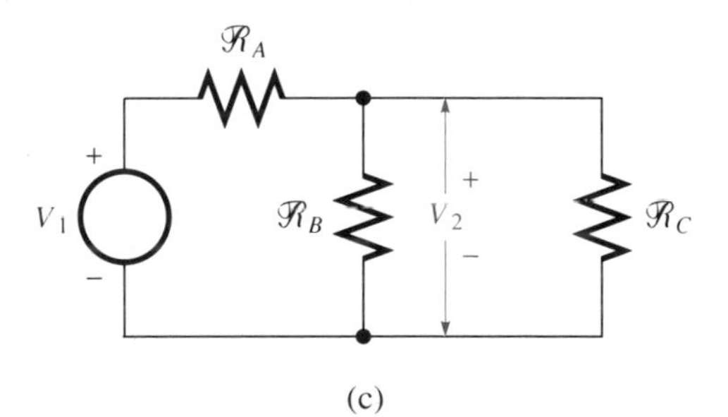

(c)

Figure 5.16 Example of a magnetic circuit. (a) Physical structure; (b) diagram of magnetic circuit; (c) magnetic circuit after simplification.

everywhere 4 cm², and the other dimensions are $L_1 = 10$ cm, $L_2 = 5$ cm, $h_1 = 1$ mm, $h_2 = 2$ mm. The permeability of the core material is $500\,\mu_o$. Find the flux in gap h_1.

Solution A diagram of the magnetic circuit is given in Fig. 5.16(b). Here

$$R_1 = \frac{L_2}{\mu A} \qquad R_2 = R_6 = R_7 = R_{11} = \frac{L_1}{2\mu A}$$

$$R_3 = R_5 = R_8 = R_{10} = \frac{L_2}{2\mu A}$$

$$R_4 = \frac{h_1}{\mu_o A} \qquad R_9 = \frac{h_2}{\mu_o A}$$

Proceeding as with an ordinary dc electric circuit, we can simplify to obtain Fig. 5.16(c), with

$$R_A = R_1 + R_2 + R_6$$

$$R_B = R_3 + R_4 + R_5$$

$$R_C = R_7 + R_8 + R_9 + R_{10} + R_{11}$$

Then

$$V_2 = \frac{(R_B \| R_C)}{(R_B \| R_C) + R_A} NI$$

and

$$\Phi = \frac{V_2}{R_B}$$

Numerically,

$$R_1 = 1.9 \times 10^5 \text{ A/Wb} \qquad R_2 = 2.0 \times 10^5 \text{ A/Wb}$$

$$R_3 = 1.0 \times 10^5 \text{ A/Wb} \qquad R_4 = 2.0 \times 10^6 \text{ A/Wb}$$

$$R_9 = 4.0 \times 10^6 \text{ A/Wb} \qquad \Phi = 6.5 \times 10^{-4} \text{ Wb}$$

The flux density in gap h_1 is $|B| = \Phi/A = 1.6$ tesla.

The idea of magnetic circuits can be used to provide a qualitative justification for our assertion that flux is confined inside magnetic cores. Referring to Fig. 5.14, the flux has different paths available to it. It can take the path through the iron, as assumed up to now, or it can take a path outside the iron. Each of these paths can be characterized by its reluctance, and the magnetic circuit includes the two paths in parallel. Following the electric circuit analogy, the total flux will divide, some following the iron path and some the air path. However, owing to the iron's large μ, the reluctance of the iron path is much lower than that of the air path, and the flux flows almost entirely through the iron. We expect that no lines of flux will pass outside of the iron and then go back in, because of the analogy to electric conduction: electric current takes the path of least resistance, and similarly, we expect the flux to take the path of least reluctance.

5.8 POINTS TO REMEMBER

1. Magnetostatic fields originate in currents. If all currents are known, the magnetic field can be found by vectorially adding the contributions of all current elements (law of Biot and Savart).

2. The magnetic field arising from current in a long straight wire has the form of circles centered on the wire. The direction of the field with respect to the current is given by the right-hand rule.

3. There are two vectors used to describe magnetic fields: the magnetic intensity $\vec{H}$ and the magnetic flux density $\vec{B}$. The two are usually related by $\vec{B} = \mu\vec{H}$, where μ, the magnetic permeability, is a characteristic of the medium. The value of μ has the value $\mu_o = 4\pi \times 10^{-7}$ for vacuum and almost all materials except ferromagnetics, for which μ is much larger.

4. The magnetic flux through a surface is equal to the surface integral of B over that surface.

5. The force on a wire of length $\vec{dl}$ carrying a current I in a magnetic field $\vec{B}$ is given by $\vec{df} = I\,\vec{dl} \times \vec{B}$. Thus the direction of the force is perpendicular to both $\vec{dl}$ and the magnetic field.

6. The curl operator acts on a vector field to produce another vector field. The curl operator can be used to derive the magnetic field from a magnetic vector potential. The curl of the magnetostatic field is related to the current density through $\nabla \times \vec{H} = \vec{J}$.

7. Ampère's circuital law states that the line integral of $\vec{H}$ around a closed path is equal to the current passing through any surface enclosed by that path. This law is a useful tool for calculating magnetic field when the current distribution is known and possesses a high degree of symmetry.

8. The term *boundary conditions* refers to a set of rules describing how electric and magnetic fields behave at various kinds of boundaries, such as between vacuum and metal, or between two dielectrics. For example, the tangential component of H is continuous across a boundary between two materials with different μ (meaning that it has the same value on either side, close to the boundary). The normal component of B is also continuous across this boundary. The normal component of D is continuous across a boundary between two dielectrics. There are quite a few boundary conditions that apply to different situations; they are summarized in Appendix B.

9. Because of their large values of μ, ferromagnetic materials can contain magnetic fields almost as water is contained in a pipe. This makes them useful in devices that require large amounts of magnetic flux, such as transformers, motors, and generators. A ferromagnetic structure used for containing and concentrating magnetic flux is known as a magnetic circuit. Magnetic circuits can be analyzed by means of an analogy to ordinary resistive electric circuits.

REFERENCES

1. S. Ramo, J. R. Whinnery, and T. Van Duzer, *Fields and Waves in Communication Electronics*. New York: Wiley, 1984.
2. N. N. Rao, *Elements of Engineering Electromagnetics*. Englewood Cliffs, N.J.: Prentice-Hall, 1987.
3. D. T. Paris and F. K. Hurd, *Basic Electromagnetic Theory*. New York: McGraw-Hill, 1969.
4. P. Lorrain and D. Corson, *Electromagnetic Fields and Waves*, 2nd ed. New York: Freeman, 1970.

An introductory discussion of ferromagnetism and magnetic circuits will be found in:

5. S. V. Marshall and G. G. Skitek, *Electromagnetic Concepts and Applications*, 2nd ed. Englewood Cliffs, N.J.: Prentice-Hall, 1987.

For more advanced discussions of magnetic circuits and their applications, see:

6. S. A. Nasar and L. E. Unnewehr, *Electromechanics and Electric Machines*. New York: Wiley, 1979.
7. A. E. Fitzgerald, C. Kingsley Jr., and S. J. Umans, *Electric Machinery*, 4th ed. New York: McGraw-Hill, 1983.

On the physics of ferromagnetism, see:

8. R. S. Elliott, *Electromagnetics*. New York: McGraw-Hill, 1966.
9. S. Wang, *Solid State Electronics*. New York: McGraw-Hill, 1966.

PROBLEMS

Section 5.1

5.1 A current element with length 1 mm, located at $x = 2$ cm, $y = 0$, $z = 0$, carries a current of 150 mA in the $+x$ direction. Find the contribution of this element to the magnetic field at $x = 0$, $y = 3$ cm, $z = 0$.

5.2 Repeat the preceding problem assuming that the direction of the current element is changed so that it is in the same direction as the vector $\vec{e}_x + \vec{e}_y$.

5.3 A current element is located at the origin and carries current in the $+z$ direction. Find the direction of its contribution to the magnetic field at:

(a) $(1, 0, 0)$ (d) $(0, 0, 1)$
(b) $(0, 1, 0)$ (e) $(1, 1, 0)$
(c) $(-1, 0, 0)$ (f) $(1, 1, 1)$

5.4 A current I flows clockwise around a square wire loop, the side of which is a. Find $\vec{H}$ at the center of the loop.

*5.5 A current I flows around a square wire loop with side a. The loop lies in the x-y plane with its center at the origin and the current flows clockwise as seen from the $+z$ direction. Find $\vec{H}$ at the point $(0, 0, b)$.

**5.6 A wire lies in the x-y plane and has the form of the parabola $y = x^2$. It carries current I (in the direction from lesser x toward greater x). Find $\vec{H}$ at the point $(0, \frac{1}{4}, 0)$.

**5.7 Current flows in the y direction through a narrow sheet of metal in the $z = 0$ plane. The sheet occupies the space $-\dfrac{h}{2} < y < \dfrac{h}{2}$, $-\infty < x < \infty$. The current in a section of the strip of width dx is $I'\,dx$, where I' is a constant. Find the contribution of this current to $\vec{H}$ at the position $(0, 0, b)$. Assume $b \gg h$.

Section 5.2

5.8 The magnetic intensity $\vec{H}$ is equal to $H_o(\vec{e}_x + \vec{e}_y)$. Find the magnetic flux through a circular surface of radius a lying in the x-z plane.

*5.9 The magnitude of the magnetic intensity is H_1. $\vec{H}$ is perpendicular to the z axis and makes an angle of 30° with the x axis, as shown in Fig. 5.17. Find the flux through a circular surface of radius a lying

(a) in the y-z plane
(b) in a plane perpendicular to the vector $\vec{e}_x + 2\vec{e}_y + \vec{e}_z$

*5.10 The magnetic intensity $\vec{H}$ is given by

$$\vec{H} = H_o|y|\vec{e}_x$$

Find the flux through a circular surface with radius a and center at the origin, oriented perpendicular to the vector $\vec{e}_x + \vec{e}_y$.

*5.11 A uniform magnetic field H_o is in the x direction. A surface is constructed by taking an infinitely long cylinder of radius b, the axis of which coincides with the z axis, and removing the half in the region $x < 0$. Find the flux per unit length through the remaining half-cylinder.

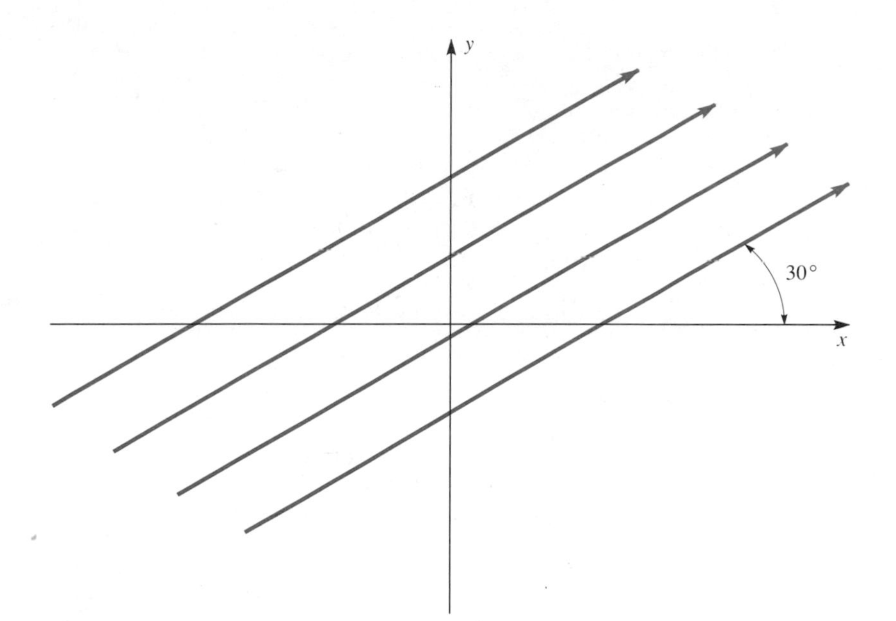

Figure 5.17 For Problem 5.9

**5.12 The magnetic intensity is $\vec{H} = H_o(y\vec{e}_x + x\vec{e}_y)$. Let the polar axis of a spherical coordinate system coincide with the z axis. Find the flux through a hemispherical surface defined by $r = b$, $0 < \theta < 90°$.

Section 5.3

$d\vec{F} = I d\bar{l} \times \bar{B}$
$d\vec{l} = dx\, \hat{e}_x$
$B = \mu H_o(y\hat{e}_x + x\hat{e}_y)$

5.13 Let $\vec{H} = H_o(y\vec{e}_x + x\vec{e}_y)$ and let a current I flow in the $+x$ direction through a wire located on the x axis between $x = 0$ and $x = L$. Find the total force acting on this wire.

*5.14 A charged particle moves through a uniform magnetic field. Show that the total velocity of the particle is constant.

**5.15 A charged particle with mass m and charge q moves in the x-y plane under the influence of the uniform magnetic field $\vec{B} = B_o\vec{e}_z$.
(a) Write Newton's-law equations for v_x and v_y.
(b) Find $v_y(t)$ and $v_x(t)$, assuming that $v_y = v_o$ and $v_x = 0$ at $t = 0$.
(c) Find $y(t)$ and $x(t)$, assuming that $x = 0$ and $y = 0$ at $t = 0$.
(d) Verify that the path of the particle is a circle, and find the frequency with which it returns to its starting point. This is known as the *cyclotron frequency*.

*5.16 The magnetic field in (5.7) is produced by a closed circuit we shall call circuit 1, in accordance with (5.2). The force acting on another closed circuit, circuit 2, is $\vec{F}$. Show that the force acting on circuit 1 because of the field produced by circuit 2 is $-\vec{F}$.

Section 5.4

5.17 Let $\vec{V}(x,y,z) = A(x^2y\vec{e}_x + xz\vec{e}_y - 2yz\vec{e}_z)$. Find $\nabla \times \vec{V}$.

5.18 Determine by inspection, using the "paddlewheel" method, whether or not the following equal zero when evaluated at the origin:
(a) $\nabla \times (y^2\vec{e}_x)$
(b) $\nabla \times (y^2\vec{e}_x + x^2\vec{e}_y)$
(c) $\nabla \times (\tan y\vec{e}_x)$
(d) $\nabla \times (\cos y\vec{e}_x - \sin x\vec{e}_y)$
(e) $\nabla \times [f(y^2)\vec{e}_x + g(x^2)\vec{e}_y + h(z^2)\vec{e}_z]$, where f, g, and h are any differentiable functions.

*5.19 Let $\vec{V}(r,\theta,\phi) = \dfrac{A\vec{e}_r}{r^2}$ in spherical coordinates. Verify that $\nabla \times \vec{V} = 0$
(a) by taking the curl in spherical coordinates, and
(b) by converting to rectangular coordinates before taking the curl.

5.20 Verify that $\nabla \times (\nabla f) = 0$, where ∇f is the gradient of any scalar function.

5.21 Let $\vec{B} = B_o\vec{e}_x$. Find a vector potential which when curled gives $\vec{B}$. Can you find a second, different $\vec{A}$ which also gives this $\vec{B}$? (Do not simply add a constant to the first one.)

5.22 Let $\vec{B} = B_oz\vec{e}_x$. Find an $\vec{A}$ which when curled gives this $\vec{B}$, and also satisfies $\nabla \cdot \vec{A} = 0$.

5.23 Find the magnetic vector potential arising from a current I in the $+z$ direction through a wire lying on the z axis of a system of cylindrical coordinates.

*5.24 A semicircular wire lies in the $z = 0$ plane of a cylindrical coordinate system, at $r = a$, $0 < \phi < 180°$. A current I passes through the wire in the $+\phi$ direction. Find the x component of the vector magnetic potential $\vec{A}$ at $(r = b, \phi = 0, z = 0)$.

**5.25 A wire in the shape of an infinitely long helix is described by the equation (in cylindrical coordinates) $z = a\phi$, $r = b$. A current I passes through it in the $+\phi$ direction. Let the x axis coincide with $\phi = 0$.
(a) Find the z component of $\vec{A}$ at the origin.
(b) Find A_x at the origin.
(c) Construct an integral which, if evaluated, gives A_y at the origin, in terms of I, a, and b. (You are not asked to evaluate the integral.)

5.26 Suppose $\vec{H}(r,\phi,z) = H_o\left(\dfrac{r}{a}\right)\vec{e}_\phi$ for $r < a$.
(a) What is the current density $\vec{J}$ for $r < a$?
(b) What is the vector potential $\vec{A}$ for $r < a$?

Section 5.5

5.27 Let $\vec{V} = \cos\frac{\pi y}{2a}\vec{e}_x$, and consider the line integral of $\vec{V}$ around the closed path in the x-y plane $(0,0) \to (1,0) \to (1,a) \to (0,a) \to (0,0)$.
(a) Evaluate the line integral directly.
(b) Find the surface integral of $\nabla \times \vec{V}$ and verify Stokes' theorem.

*5.28 Let $\vec{H}(r,\phi,z) = Ar^2\vec{e}_\phi$ in cylindrical coordinates.
(a) Let the surface of integration be a hemisphere of radius a centered at the origin in the region $z > 0$. Find the surface integral $\int_S \nabla \times \vec{H} \cdot \vec{dS}$ on this surface.
(b) Let the surface be the portion of the plane $z = 0$ bounded by the circle $r = a$. Find $\int_S \nabla \times \vec{H} \cdot \vec{dS}$. Does the result agree with (a)? Why?
(c) Evaluate $\oint \vec{H} \cdot \vec{dl}$ on the circular path $r = a$, $z = 0$ and verify Stokes' theorem.

5.29 The current density is everywhere in the z direction and has the following dependence on r (in cylindrical coordinates):

$$|J| = 0 \quad \text{for} \quad r < a$$

$$|J| = B \quad \text{for} \quad a < r < b$$

$$|J| = C\frac{r^2}{a^2} \quad \text{for} \quad b < r < c$$

$$|J| = D \quad \text{for} \quad c < r < d$$

$$|J| = 0 \quad \text{for} \quad r > d$$

Find $\vec{H}$ as a function of r for all r.

5.30 In the preceding problem let $a = 1$ cm, $b = 2$ cm, $c = 3$ cm, $d = 4$ cm, $B = C = 1\ \text{A/cm}^2$.
(a) Find the value of D that results in $\vec{H} = 0$ for $r > d$.
(b) Sketch $|\vec{H}(r)|$.

*5.31 The current density (in cylindrical coordinates) is $\vec{J} = Ar^n\vec{e}_z$ where n is an integer. For what values of n is $\vec{H}$ finite at $r = a$?

Section 5.6

5.32 Let $\mu = \mu_o$ for $x < 0$ and $\mu = 2\mu_o$ for $x > 0$. A uniform H field in the region $x < 0$ makes an angle of 30° with the normal to the interface. Find the angle between field and normal in the region $x > 0$.

*5.33 State whether the following field components are continuous, and explain your reasoning.
(a) H_T at the boundary between air and a material with finite conductivity

(b) H_T at the boundary between air and an ideal metal with infinite conductivity

(c) B_N at the boundary of air and an ideal metal with infinite conductivity

(d) D_N at the boundary of two nonconductive dielectrics (Assume there is no charge on the boundary.)

(e) D_N at the boundary of two dielectrics, one of which is nonconductive and one with finite conductivity

*5.34 Let the region $x < 0$ contain conductive material with $\epsilon = \epsilon_1$ and conductivity σ_{E1}. For $x > 0$, $\epsilon = \epsilon_2$ and the conductivity is σ_{E2}. The electric field in the region $x < 0$ is $E_1 \vec{e}_x$.

(a) Find the electric field in the region $x > 0$. (*Hint:* The current densities in the two regions must be equal. Otherwise infinite charge would eventually pile up at the boundary.)

(b) Show that the surface charge density on the boundary is

$$\sigma = \left(\frac{\epsilon_2 \sigma_{E1} - \epsilon_1 \sigma_{E2}}{\sigma_{E2}} \right) E_1$$

*5.35 A certain vector field $\vec{V}$ obeys $\nabla \cdot \vec{V} = K$, where K is a constant. Is the normal component of $\vec{V}$ continuous at boundaries? Prove or disprove.

*5.36 Consider a boundary located in the plane $z = 0$. In the vicinity of the boundary a vector field $\vec{V}$ obeys $\nabla \cdot \vec{V} = \ln|z|$. Is V_N continuous? What if $\nabla \cdot \vec{V} = \dfrac{1}{|z|}$?

Section 5.7

5.37 Estimate the flux in the magnetic circuit of Fig. 5.15 if $N = 500$, $I = 1$ A, the cross-sectional area of the core is 4 cm^2, $\mu = 5000\,\mu_o$, $L_A = 4$ mm, $L_M = 20$ cm. What is $|B|$ in the gap?

*5.38 A portion of a magnetic circuit lying between $z = 0$ and $z = L$ is a tapered cylinder, the radius of which is $r = a + b\left(\dfrac{z}{L}\right)$. Estimate the reluctance of this section.

CHAPTER 6

Time-Varying Fields

When electric and magnetic fields vary with time, important new phenomena come into the picture. These new effects are *interactions* between the electric and magnetic fields. As we have noted earlier, the electric and magnetic fields are not really unrelated; in fact, they are two parts of a single field, the electromagnetic field. Only in a non-time-varying case can electric and magnetic effects be considered separately. Effects arising from the coupling of time-varying electric and magnetic fields are said to be *electrodynamic* in nature.

In the previous two chapters we began our study of electricity and magnetism by considering the case of non-time-varying fields. This was done for the sake of simplicity. However, most fields of practical interest do vary in time. For instance, all the interesting and useful phenomena associated with electromagnetic waves are based on the time-varying nature of the fields.

In this chapter we shall first introduce the two basic electrodynamic effects. This will be done by adding terms containing $\frac{\partial \vec{B}}{\partial t}$ and $\frac{\partial \vec{D}}{\partial t}$ to the static equations studied earlier. Then in section 6.3 all effects, both static and dynamic, will be collected in the four famous equations known as Maxwell's equations. At that point the basic physics will be complete.

6.1 DISPLACEMENT CURRENT

As we have seen, the magnetostatic field is related to the currents that produce it by $\nabla \times \vec{H} = \vec{J}$. In the case of time-varying fields, it is found that an additional term must be added:

$$\nabla \times \vec{H} = \vec{J} + \frac{\partial \vec{D}}{\partial t} \tag{6.1}$$

Physically, this means that a time-varying *electric* field acts as an additional force driving the magnetic field. The term $\frac{\partial \vec{D}}{\partial t}$ acts as though it is an additional current that helps to produce $\vec{H}$. Since it acts like a current in (6.1), the term $\frac{\partial \vec{D}}{\partial t}$ (time derivative of the displacement vector) is known as the *displacement current*.

A second form of (6.1) can be obtained using Stokes' theorem. Let us choose any surface S, and form the surface integrals over it of the vectors on the left and right sides of (6.1). This results in

$$\int_S \nabla \times \vec{H} \cdot \vec{dS} = \int_S \left(\vec{J} + \frac{\partial \vec{D}}{\partial t}\right) \cdot \vec{dS} \tag{6.2}$$

On the left side of this equation we use Stokes' theorem to convert the surface integral into a line integral around the path L that bounds the surface S, just as we did in (5.13)–(5.15) when deriving Ampère's circuital law. On the right side we have two terms. The first, the surface integral of the current density, is just I_S, the current passing through S. The second term on the right is the surface integral of the displacement current. Thus (6.2) becomes

$$\oint_L \vec{H} \cdot \vec{dl} = I_S + \int_S \frac{\partial \vec{D}}{\partial t} \cdot \vec{dS} \tag{6.3}$$

Result (6.3) can be considered the integral form of (6.1), since it contains exactly the same physics, and is obtained from (6.1) by integration. Alternatively, (6.1) can be called the differential form of (6.3). Equation (6.3) is the more general form of Ampère's circuital law that applies when a time-varying $\vec{D}$-field is present.

To illustrate the effect of displacement current in Ampère's law, let us consider the capacitor shown in Fig. 6.1(a). This capacitor is being charged by the constant current I. We apply (6.3), choosing our surface of integration so that it passes through the wire connected to the capacitor. Taking advantage of the cylindrical symmetry to simplify the line integral, (6.3) becomes

$$2\pi r|\vec{H}(r)| = I \tag{6.4}$$

where $\vec{H}(r)$ is the field at a distance r from the wire. Note that no term containing the displacement current appears on the right side of this equation because the current through the wire is constant in time, and thus there are no time-varying fields in the vicinity of S. From (6.4), $|\vec{H}| = I/2\pi r$. Now let us refer to Fig. 6.1(b). Here the surface S has been replaced by a new surface S' which passes between the plates of the capacitor; however, the path L

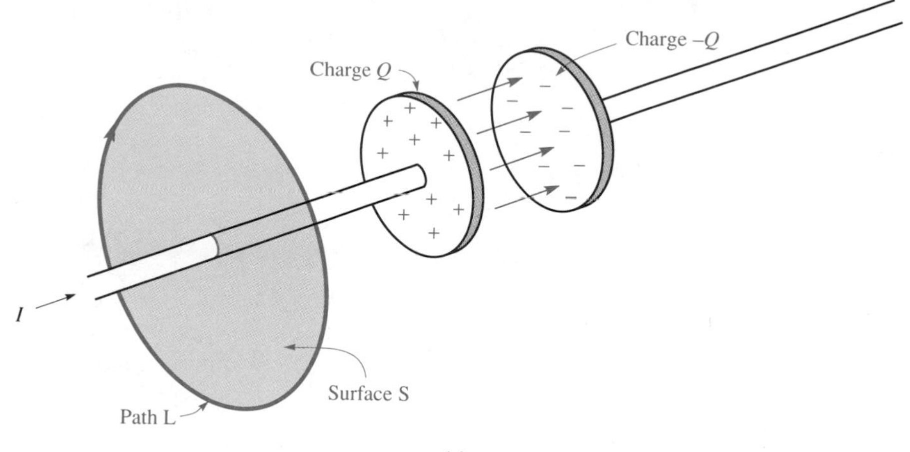

(a)

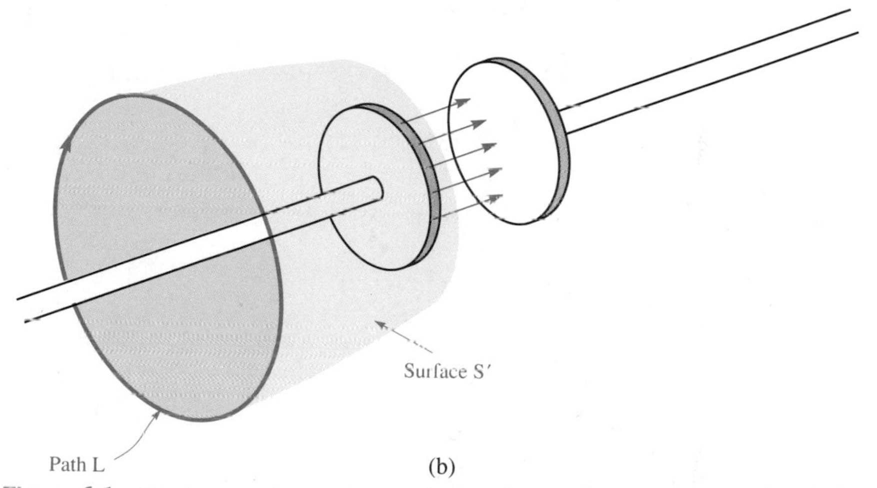

(b)

Figure 6.1 Displacement current can substitute for conduction current in Ampère's circuital law. The surface integral of $\vec{J}$ through surface S in (a) is the same as the surface integral of $\partial\vec{D}/\partial t$ through surface S' in (b).

that bounds S' is exactly the same as path L in Fig. 6.1(a). Now (6.3) becomes

$$2\pi r|\vec{H}(r)| = \int_{S'} \frac{\partial \vec{D}}{\partial t} \cdot \vec{dS} \tag{6.5}$$

Now $\vec{H}(r)$ in (6.5) is exactly the same field as $\vec{H}(r)$ in (6.4); all we have done is changed the arbitrarily chosen surface S. Thus if things are to make sense, the right side of (6.5) must be equal to the right side of (6.4). We will now show that this is the case.

The surface integral of $\vec{D}$ is related to the charge Q on one capacitor plate by Gauss' law, (4.14):

$$\int_{S'} \vec{D} \cdot \vec{dS} = Q$$

Thus the integral on the right side of (6.5) becomes

$$\int_{S'} \frac{\partial \vec{D}}{\partial t} \cdot \vec{dS} = \frac{\partial}{\partial t} \int_{S'} \vec{D} \cdot \vec{dS} = \frac{\partial}{\partial t} Q = I \qquad \textbf{(6.6)}$$

which agrees with (6.4). From this example we see that the displacement current term in (6.1) and (6.3) is necessary for logical consistency. If it were not there, the results from Figs. 6.1 (a) and (b) could not be made to agree.

THE CONTINUITY EQUATION

Although it is not very obvious at first, the displacement current term in (6.1) expresses the fact that a current is a flow of charges. To see this, let us apply the divergence operator to each side of (6.1). This results in

$$\nabla \cdot (\nabla \times \vec{H}) = \nabla \cdot \vec{J} + \nabla \cdot \frac{\partial \vec{D}}{\partial t} = \nabla \cdot \vec{J} + \frac{\partial}{\partial t} (\nabla \cdot \vec{D}) \qquad \textbf{(6.7)}$$

However, it is a general theorem that the divergence of the curl of any vector equals zero. (See the table of vector identities in the endpapers.) Using also (4.13), we obtain

$$\nabla \cdot \vec{J} + \frac{\partial \rho}{\partial t} = 0 \qquad \textbf{(6.8)}$$

This is the differential form of an equation known as the *continuity equation*. By integrating each side over a volume V and applying the divergence theorem, we obtain the integral form of the continuity equation,

$$\int_S \vec{J} \cdot dS = -\frac{\partial}{\partial t} \int_V \rho \, dV = -\frac{\partial}{\partial t} Q_V \qquad \textbf{(6.9)}$$

where S is the closed surface surrounding the volume V, and Q_V is the charge contained in V.

The integral on the left of (6.9) represents the flux of $\vec{J}$ through the surface S. We have already encountered the notion of flux in several ways. In general, the flux of a vector represents the amount of that vector flowing through the surface in question. Thus the flux of $\vec{J}$ through S is simply the total current flowing outward through the surface S. (Remember that $\vec{J}$ is current *density,* with units amperes per square meter.) But a current is a movement of charges. If there is a net outward current, the amount of charge inside S must be decreasing; this is represented by the term $-\frac{\partial Q_V}{\partial t}$ in (6.9).

From (6.9) we can obtain a boundary condition that applies when two media with different conductivities are in contact. Using the usual pillbox-shaped surface of integration, we find that

$$J_{N1} - J_{N2} = -\frac{\partial}{\partial t}\sigma \tag{6.9a}$$

where J_{N1} and J_{N2} are the normal components of the current density and σ is the surface charge density residing on the interface. (It is quite probable that there will be surface charge at the interface, because charges can flow there through the conductive materials.) In the special case of non-time-varying fields, however, we must have $J_{N1} = J_{N2}$. If this were not the case, (6.9a) would imply that σ would increase at a constant rate and would approach infinity; this would make no physical sense.

EXAMPLE 6.1

The current density in a semiconductor is given by $\vec{J} = Ax\vec{e}_x$. Find the rate of change of the charge inside the volume $-a < x < a$, $-b < y < b$, $-c < z < c$.

Solution We shall demonstrate two ways of solving this problem. In the first, we make use of (6.8) to calculate $\partial\rho/\partial t$:

$$\frac{\partial\rho}{\partial t} = -\nabla\cdot\vec{J} = -\frac{\partial}{\partial x}Ax = -A$$

The charge inside the volume of interest is $\int_V \rho\, dV$; hence

$$\frac{\partial Q_V}{\partial t} = \int_V \frac{\partial\rho}{\partial t}\, dV = \int_V (-A)\, dV = -8Aabc$$

We can also obtain this result from (6.9), which states that

$$\frac{\partial Q_V}{\partial t} = -\int_S \vec{J}\cdot dS$$

The surface integral is to be taken over the surface enclosing the volume in question. Since $\vec{J}$ has only an x component, the only contributions to the surface integral come from the surfaces at $x = -a$ and $x = a$. Hence

$$\frac{\partial Q_V}{\partial t} = -4bc[J_x(x = a) - J_x(x = -a)]$$

(Note the minus sign before $J_x(x = -a)$. It appears because the outward normal to the volume at the plane $x = -a$ points in the $-x$ direction; this is easy to overlook!)

$$= -4bc[aA - (-a)A] = -8Aabc$$

in agreement with the answer obtained from (6.8).

6.2 FARADAY'S LAW

We have seen that a time-varying electric field acts as a driving force for the magnetic field. The opposite statement is also found to be true: a time-varying magnetic field produces an electric field. This experimental fact is described by the following equation:

$$\nabla \times \vec{E} = -\frac{\partial \vec{B}}{\partial t} \tag{6.10}$$

Applying Stokes' theorem, using any surface S bounded by a closed path L, we obtain the integral form of (6.10),

$$\oint_L \vec{E} \cdot \vec{dl} = -\frac{\partial}{\partial t}\int_S \vec{B} \cdot \vec{dS} = -\frac{\partial \Phi}{\partial t} \tag{6.11}$$

where Φ is the magnetic flux through the surface S. Equation (6.11) is known as *Faraday's law*.

In Chapter 4 it was stated that the concept of potential difference, or voltage, is in general inapplicable to time-varying systems. From (6.11) we can now see why this is the case. Figure 6.2 illustrates the line integral $\int_{L_1} \vec{E} \cdot \vec{dl}$ from point A to point B, and also the line integral $\int_{L_2} \vec{E} \cdot \vec{dl}$, which connects the same two endpoints by a different path. Suppose that voltage is a well-defined, single-valued function of position. Then there is

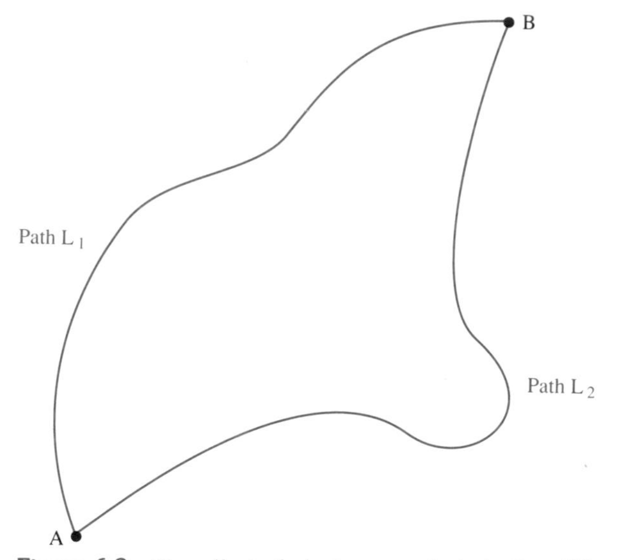

Figure 6.2 The effect of electromagnetic induction. When time-varying magnetic fields are present, the value of the line integral of $\vec{E}$ from A to B may depend on the path one chooses.

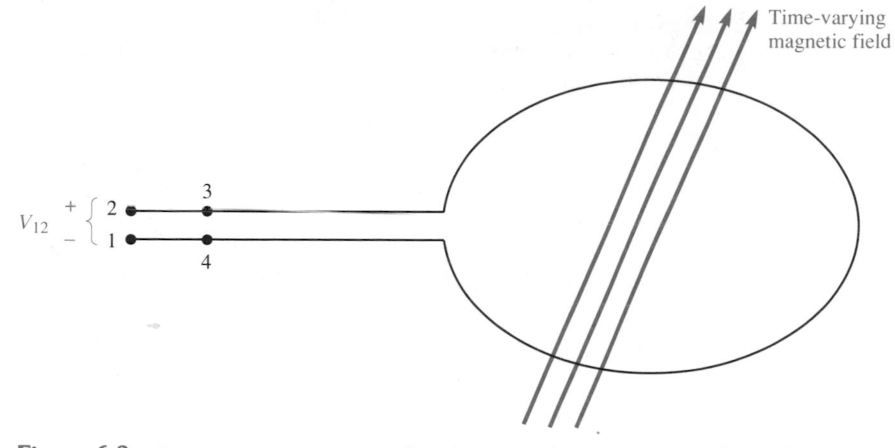

Figure 6.3 A time-varying magnetic flux through a loop of wire results in the appearance of a voltage across its terminals.

only one unique voltage $V_{AB} = V_A - V_B$, and

$$V_{AB} = \int_{L_1} \vec{E} \cdot \vec{dl} = \int_{L_2} \vec{E} \cdot \vec{dl} \tag{6.12}$$

However, we can also consider the complete closed path that goes from A to B, via L_1, and returns to A via L_2. Applying (6.11) to this closed path, we have

$$\oint \vec{E} \cdot \vec{dl} = \int_{L_1} \vec{E} \cdot \vec{dl} - \int_{L_2} \vec{E} \cdot \vec{dl} = -\frac{\partial \Phi}{\partial t} \tag{6.13}$$

(The minus sign before the second integral appears because we are traversing L_2 backwards, from B to A.) However, from (6.12) this means that

$$V_{AB} - V_{AB} = 0 = -\frac{\partial \Phi}{\partial t} \tag{6.14}$$

Thus we conclude that the voltage V_{AB} can be defined only if the time derivative of the magnetic flux is zero; that is, in the magnetostatic case. The student should bear in mind that when time-varying magnetic fields are present, the terms "electric potential" or "voltage" may have no useful meaning.[1] The concept of electric field, however, remains entirely valid.

Faraday's law is the basis of the *electric generator*. The principle of the generator is illustrated in Fig. 6.3. Here a time-varying magnetic flux $\Phi(t)$

[1] "What's this?" the reader may say. "In electronic circuits we frequently speak of voltages, even though the currents in the circuit may be varying at a high frequency." This is true, but in that case one is really using approximations. As a general rule of thumb, it can be shown that a circuit will follow the laws of statics, approximately, if its physical size is much smaller than a wavelength at the frequency of operation. See section 7.5.

passes through a loop. The terminals of the loop are points 1 and 2; these terminals are distant from the time-varying magnetic field. Let us apply (6.11) to a closed path that begins at terminal 1, goes up to terminal 2, and then through the loop, returning to 1. The result is

$$\int_1^2 \vec{E} \cdot \vec{dl} + \int_2^1 \vec{E} \cdot \vec{dl} = -\frac{\partial \Phi}{\partial t} \tag{6.15}$$

However, the path from 2 to 1 is entirely through the wire. We shall assume that the material of the wire is a perfect conductor; therefore the electric field inside it vanishes, and the second line integral must therefore vanish. This leaves us with

$$\int_1^2 \vec{E} \cdot \vec{dl} = -\frac{\partial \Phi}{\partial t} \tag{6.16}$$

It is customary to identify $\int_1^2 \vec{E} \cdot \vec{dl} = V_{12}$ as the output voltage. If instead of a single loop we have an N-turn coil, the voltages of the individual loops add in series, and we have

$$V_{12} = -N \frac{\partial \Phi}{\partial t} \tag{6.17}$$

The reader may ask how an output *voltage* can be defined, in this non-static case. The reason this is possible is that we have assumed that the terminals are far away from the time-varying magnetic field. Suppose we consider the other pair of terminals, 3 and 4, shown in the figure. By assumption, there is no magnetic field in this region. Thus we can apply (6.11) to the closed path $1 \to 2 \to 3 \to 4 \to 1$, and again assuming the wires to be perfect conductors, obtain

$$\int_1^2 \vec{E} \cdot \vec{dl} - \int_4^3 \vec{E} \cdot \vec{dl} = 0 \tag{6.18}$$

Thus the line integral of $\vec{E}$ between the two wires is the same everywhere along the wires, and the definition

$$V_{12} = \int_{\text{wire 1}}^{\text{wire 2}} \vec{E} \cdot \vec{dl}$$

is unambiguous.

EXAMPLE 6.2

An N-turn wire loop with external terminals has area A. It is placed in a uniform magnetic field $\vec{B}$, so that a diameter of the loop is perpendicular to $\vec{B}$. The loop is mounted on a shaft along this diameter, so it can rotate, as shown in Fig. 6.4. The speed of rotation is n revolutions per second. Find the voltage at the terminals.

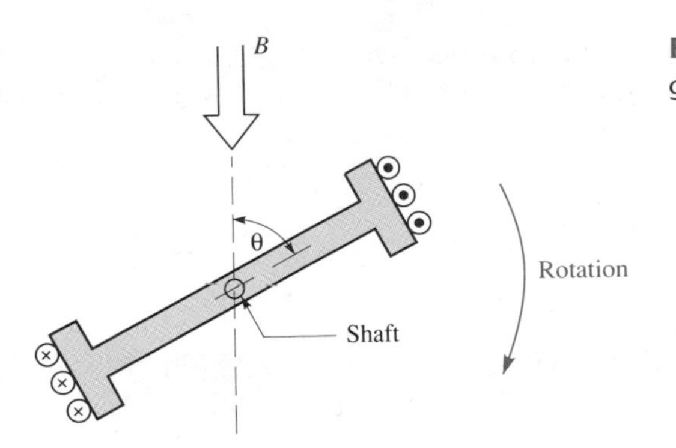

Figure 6.4 Principle of the ac generator.

Solution The flux through the loop is $BA \sin\theta$, where θ is the angle between B and the plane of the loop, as shown in the figure. This angle obeys $\theta = 2\pi nt$. Thus from (6.17)

$$v(t) = -2\pi nNBA \cos(2\pi nt)$$

The voltage is a sinusoid with amplitude $2\pi nNBA$ and angular frequency $2\pi n$.

This example illustrates the principle of the ac electric generator. Of course some way must be found to keep the two wires leading away from the loop from getting wound up as the loop is turned. This can be accomplished by means of a pair of sliding contacts.

EXERCISE 6.1

A rectangular loop of wire lies in the x-y plane with its sides on $x = 0$, $y = 0$, $x = a$, and $y = b$. The loop is located near a transmission line, so that it experiences a traveling-wave magnetic field represented by the phasor

$$\mathbf{B}_z = B_o e^{-jkx}$$

Find the smallest value that can be given to a that will result in no current flowing through the loop.

Answer $2\pi/k$.

BOUNDARY CONDITION ON TANGENTIAL ELECTRIC FIELD

From Faraday's law, (6.11), we can obtain the boundary condition on the tangential component of $\vec{E}$ at a dielectric boundary. The procedure is similar to that used in Chapter 5 for magnetic fields. For the path of integration in

(6.11), we choose an elongated rectangle like that in Fig. 5.10. The line integral of $\vec{E}$ on this path (in the limit as the width of the path approaches zero) is $(E_{T1} - E_{T2})l$. The flux Φ through the area enclosed by the path will approach zero, because the product of the field strength, which is finite, and the area, which approaches zero, will have to vanish. Thus we conclude that

$$E_{T1} = E_{T2} \tag{6.19}$$

The corresponding condition on the normal component, which states that $D_{N1} = D_{N2}$, has already been obtained in Chapter 5.

EXERCISE 6.2

An electric field inside a dielectric material makes an angle of 25° with the normal to an interface between the material and air. The relative permittivity of the dielectric is 1.8. What angle does the field make with the normal, on the air side of the interface?

Answer 14.5°.

6.3 INDUCTANCE

An important application of Faraday's law is the phenomenon of *inductance*. As we have seen, a time-varying magnetic flux can induce a voltage across the terminals of a wire loop. Now suppose a current flows through this loop; then that current gives rise to magnetic fields, and these fields create flux through the loop. Thus a time-varying current through a loop gives rise to a voltage across the same loop. The effect just described is known as *self-inductance*. It is also possible for a current in one loop to produce a time-varying magnetic flux in a second, nearby loop. The resulting effect, which couples the two loops together magnetically, is known as *mutual inductance*.

From (6.17) the voltage induced across an N-turn loop is given by $v = N \dfrac{\partial \Phi}{\partial t}$. Since the flux Φ originates from the current i flowing through the loop, we can assume that Φ is proportional to i with some proportionality constant K, so that $\Phi = Ki$. Let us define the inductance L according to the usual formula of circuit theory $v = L(\partial i/\partial t)$. Then setting the two expressions for voltage equal, we obtain

$$v = N \frac{\partial \Phi}{\partial t} = NK \frac{\partial i}{\partial t} = L \frac{\partial i}{\partial t} \tag{6.20}$$

Thus, for the case of a simple planar coil we find that

$$L = NK = \frac{N\Phi}{i} \tag{6.21}$$

Inductors, consisting of loops and coils of wire in various shapes, are of course often used as elements in electronic circuits.

Calculating the inductance of a coil can be quite difficult, especially if its shape is complicated. However, we can obtain approximate results, at least, for coils of simple shape. Let us consider, for example, a very long rectangular loop, as shown in Fig. 6.5. (This shape is of interest because it might represent a section of parallel-conductor transmission line. The reader will recall that the inductance per unit length plays an important role in the characteristics of such a line.) To calculate the inductance of this loop, let us make the simplifying assumptions that the wires are free of resistance and that $a \ll w$ and $w \ll h$. Because of these assumptions, the following approximate statements can be made:

1. The flux through the loop is produced primarily by the currents flowing in its two long sides. Since we assume that $h \gg w$, the fields at most positions inside the loop are nearly the same as they would be if the loop were infinitely long;
2. The current flows entirely on the surface of the wires and is uniformly distributed around their circumference. Thus,
3. The field produced by each of the long sides is given approximately by (5.3).

We shall also confine our attention to the magnetic fields *outside* the wires, thus obtaining what is known as the *external inductance*. The contribution to total inductance from fields *inside* the wire, known as the *internal inductance,* is usually negligible. (Internal inductance will be considered in section 7.3.)

The flux consists of two terms corresponding to the contributions of the two long sides. Because of symmetry, these two contributions are equal. (Verify, by means of the right-hand rule, that the contributions of the two

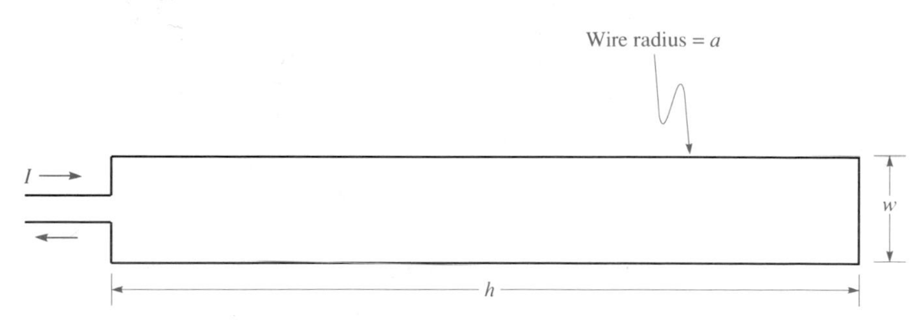

Figure 6.5 Finding the inductance per unit length of a parallel-conductor transmission line.

sides add, rather than cancel.) Thus for the flux per unit length we have

$$\Phi = 2\mu_o \int_a^w \frac{i}{2\pi R}\, dR = \frac{\mu_o i}{\pi} \ln\left(\frac{w}{a}\right) \tag{6.22}$$

Hence the inductance per unit length is given approximately by

$$L = \frac{\mu_o}{\pi} \ln\left(\frac{w}{a}\right) \tag{6.23}$$

For coils of more complex shape, it may not be feasible to apply (6.21) because of difficulty in evaluating Φ. For instance, suppose a coil is made by winding a wire in the form of a loose helix on a pencil. In that case the surface S bounded by the wire has a very complicated shape, and calculating the flux through it would be difficult. In practice, the inductance of coils is usually determined by numerical methods or by measurement.

EXAMPLE 6.3

Find the inductance per unit length of a coaxial transmission line. The radius of the inner conductor is a and the inside radius of the outer conductor is b. Assume the currents flow on the surfaces of the conductors, which have no resistance.

Solution Let us use the path of integration shown in Fig. 6.6. From (6.11) we have that

$$\int_L \vec{E} \cdot \vec{dl} = -\frac{\partial}{\partial t} \int_S \vec{B} \cdot \vec{dS}$$

The segments of the path marked "3" and "4" contribute nothing to the line integral, because there is no tangential electric field at the surface of a perfect conductor. Let us define $\int_1 \vec{E} \cdot \vec{dl} = v(z)$, and $\int_2 \vec{E} \cdot \vec{dl} = -v(z+h)$. (The minus sign occurs because the direction of segment 2 is opposite to

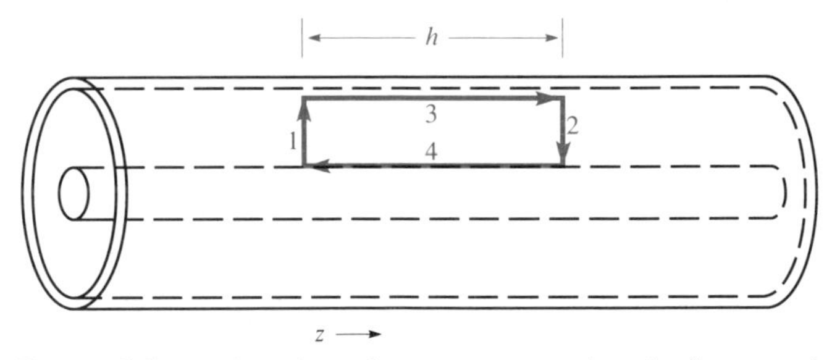

Figure 6.6 Finding the inductance per unit length of a coaxial transmission line.

that of segment 1.)[2] Thus

$$
\begin{aligned}
v(z) - v(z+h) &= -\mu \frac{\partial}{\partial t} \int_S \vec{H} \cdot \vec{dS} \\
&= \mu h \frac{\partial}{\partial t} \int_a^b H_\phi \, dr \\
&= \frac{\mu h}{2\pi} \frac{\partial}{\partial t} i_z \int_a^b \frac{1}{r} \, dr \\
&= \frac{\mu h}{2\pi} \ln\left(\frac{b}{a}\right) \frac{\partial i_z}{\partial t}
\end{aligned}
$$

where (5.3) (or Example 5.8) has been used. Now comparing with (1.4) we recognize the inductance per unit length as

$$\frac{\mu}{2\pi} \ln\left(\frac{b}{a}\right)$$

MUTUAL INDUCTANCE

When a current flowing in one circuit creates a flux through a second circuit, we say there is a *mutual inductance M* between the two circuits. A time-varying current in one circuit will then create additional voltage across the terminals of the other. Let Φ_{12} be the flux through circuit 1 arising from current i_2, and let Φ_{21} be similarly defined. The voltage induced in each winding will now be the sum of two terms, arising from self- and mutual inductance:

$$
\begin{aligned}
v_1 &= L_1 \frac{\partial}{\partial t} i_1 + N_1 \frac{\partial}{\partial t} \Phi_{12} \\
v_2 &= L_2 \frac{\partial}{\partial t} i_2 + N_2 \frac{\partial}{\partial t} \Phi_{21}
\end{aligned}
\tag{6.24}
$$

Let us again assume that each flux is linearly proportional to the current that produces it, so that $\Phi_{12} = K_{12} i_2$ and $\Phi_{21} = K_{21} i_1$, where K_{21} and K_{12} are constants. Then defining additional constants $M_{12} = N_1 K_{12}$ and $M_{21} = N_2 K_{21}$

---

[2]Due care must be taken in introducing the notion of voltage in this situation, which has time-varying fields. We can give the name "voltage" to the line integral of $\vec{E}$ along a radial path. However, we note that the value of this line integral varies with x, since the voltage on a transmission line varies along the line. This is different from the electrostatic case, in which the metals would be equipotentials, and the voltage between them would be the same, no matter where one measured it.

we have

$$v_1 = L_1 \frac{\partial}{\partial t} i_1 + M_{12} \frac{\partial}{\partial t} i_2$$
$$v_2 = L_2 \frac{\partial}{\partial t} i_2 + M_{21} \frac{\partial}{\partial t} i_1 \qquad \textbf{(6.25)}$$

It appears at first that the coefficients M_{12} and M_{21} could be different, but it can be shown that $M_{21} = M_{12}$. To show this we write

$$M_{12} = \frac{1}{i_2} \int_{S_1} \vec{B_2} \cdot \vec{dS_1}$$

where the integral is taken over the surface enclosed by circuit 1 and $\vec{B_2}$ is the field produced by the current in circuit 2. In terms of the vector potential produced by the latter current, this becomes

$$M_{12} = \frac{1}{i_2} \int_{S_1} (\nabla \times A_2) \cdot \vec{dS_1}$$
$$= \frac{1}{i_2} \oint_{L_1} \vec{A_2} \cdot \vec{dl_1}$$
$$= \frac{1}{i_2} \oint_{L1} \left[\oint_{L_2} \frac{\mu i_2 \, \vec{dl_2}}{4\pi |\vec{r_2} - \vec{r_1}|} \right] \cdot \vec{dl_1}$$

where in the last step we have expressed $\vec{A_2}$ explicitly using (5.10). Rearranging slightly we have finally

$$M_{12} = \frac{\mu}{4\pi} \oint_{L_1} \oint_{L_2} \frac{\vec{dl_2} \cdot \vec{dl_1}}{|\vec{r_2} - \vec{r_1}|} \qquad \textbf{(6.25a)}$$

a result which is known as Neumann's formula. Inspecting this expression, we observe that it is completely symmetrical with respect to interchange of subscripts 1 and 2; therefore the expression for M_{21} will be identical, and $M_{21} = M_{12}$. We can dispense with the subscripts by simply setting $M_{21} = M_{12} = M$.

The two self-inductances L_1 and L_2, however, are unequal in general. It can be shown (see texts on circuit theory) that $\sqrt{L_1 L_2} \geq M$. The *coefficient of coupling*, defined by $k_c = M/\sqrt{L_1 L_2}$, is therefore less than or equal to one.

EXAMPLE 6.4

An infinitely long straight wire lies in the same plane as a small circular loop of radius a. The center of the loop is located at a distance b from the wire, and $b \gg a$. Find the mutual inductance between the two circuits, approximately.

Solution Because the loop is small, we can make the approximation that the field of the wire is constant over the loop. Thus

$$M = \frac{\Phi_{21}}{I_1}$$

where Φ_{21} is the flux through circuit 2 produced by the current in circuit 1

$$\cong (\pi a^2)\left(\frac{\mu}{2\pi b}\right) = \frac{a^2\mu}{2b}$$

We note that it may be not be equally easy to calculate M_{12} and M_{21}. In this example, we might, for example, have calculated the voltage induced in an infinitely large loop by the current in the small one, and then have taken the limit as the radius of the larger loop went to infinity. This should give the same answer, since the two M's are equal, but the calculation would be much more difficult.

IRON-CORE INDUCTORS

In Chapter 5 it was mentioned that magnetic field lines tend to be "piped" through ferromagnetic cores. This behavior makes calculation of inductance much easier than is the case with air-core coils. For example, consider the iron-core inductor shown in Fig. 5.15. From (5.25) and (5.26) the magnetic flux is given by $\Phi = \mu ANI/L_P$ and thus for this inductor

$$L = \frac{\mu AN^2}{L_P} \tag{6.26}$$

An iron-core transformer is made by winding two or more separate windings on a single iron core, as illustrated in Fig. 6.7. Assuming that no field lines leak out of the iron core, the flux linking coil 1 (which has N_1 turns) is the same as that linking coil 2 (which has N_2 turns). Thus $v_1 =$

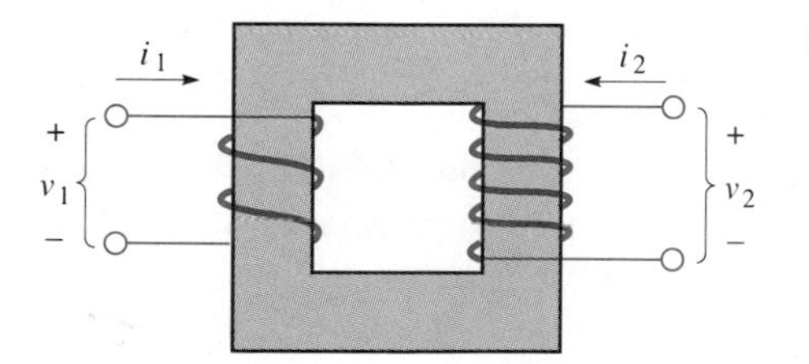

Figure 6.7 An iron-core transformer.

$N_1 \dfrac{\partial \Phi}{\partial t}$ and $v_2 = N_2 \dfrac{\partial \Phi}{\partial t}$, and assuming that $\dfrac{\partial \Phi}{\partial t} \neq 0$,

$$\frac{v_1}{v_2} = \frac{N_1}{N_2} \tag{6.27}$$

Thus iron-core transformers can be used in ac circuits to step voltages up and down, in proportion to the turns ratio.

Let us next write Ampère's circuital law for a path of integration around the core, like that shown in Fig. 5.14. This results in

$$N_1 i_1 + N_2 i_2 = |\vec{\mathbf{H}}| L_P = \frac{\Phi L_P}{\mu A} \tag{6.28}$$

where $L_P/\mu A$ is the reluctance of the core. Assuming that the voltages and currents are sinusoidal, we have $\mathbf{v}_1 = N_1 j\omega\boldsymbol{\Phi}$, where $\boldsymbol{\Phi}$ is the phasor representing $\Phi(t)$. Thus

$$N_1 \mathbf{i}_1 + N_2 \mathbf{i}_2 = \frac{L_P \mathbf{v}_1}{2\pi j f N_1 \mu A} \tag{6.29}$$

Multiplying the complex conjugate of (6.29) by $\mathbf{v}_1$ and using (6.27) we have

$$\mathbf{v}_1 \mathbf{i}_1^* + \mathbf{v}_2 \mathbf{i}_2^* = \frac{j L_P |\mathbf{v}_1|^2}{2\pi f N_1^2 \mu A} \tag{6.30}$$

The right side of (6.30) is purely imaginary. Taking the real part of this equation tells us that the sum of the powers entering the transformer through the two windings is zero. This is not surprising, since our model has no loss mechanisms. The imaginary term on the right side of (6.30) represents a reactive power arising from storage of magnetic energy in the fields in the core. This term corresponds to out-of-phase currents, which do not contribute to power transfer between the windings but do contribute to ohmic loss. In a well-designed transformer the right side of (6.29) may be negligible, in which case we have

$$\frac{\mathbf{i}_2}{\mathbf{i}_1} \cong -\frac{N_1}{N_2}$$

This approximation is often used in ac circuit analysis.

EXERCISE 6.3

Estimate the inductance of the iron-core inductor shown in Fig. 5.15. Let $a = 3$ cm, $b = 6$ cm, $c = 1.5$ cm, $N = 2000$ turns, and $\mu/\mu_o = 5000$. The four sides of the square core are identical.

Answer 31 H.

6.4 MAXWELL'S EQUATIONS

We can now collect the results of the preceding chapters in the form of the four important equations known as Maxwell's equations. The form we present here is the *macroscopic* form of Maxwell's equations. The word "macroscopic" indicates that the equations apply to fields in bulk materials such as gases, liquids, and solids. (The *microscopic* form of Maxwell's equations, with which we shall not be concerned, is only slightly different; it is useful in working with individual atomic particles.) Maxwell's equations are valid for time-varying fields; the equations of electrostatics and magnetostatics are obtained from Maxwell's equations by setting $\partial/\partial t$ equal to zero.

The four equations to be collected are as follows:

1. The source equation for the electric flux density $\vec{D}$;
2. A similar equation for the magnetic induction $\vec{B}$;
3. Faraday's law; and
4. Ampère's circuital law (including the displacement current term).

The four equations are given in Table 6.1.

It is interesting to consider the symmetry (and lack of symmetry) among these four equations. For instance, there is a source equation for $\vec{D}$ and a similar equation for $\vec{B}$. The former states that the divergence of $\vec{D}$ is the electric charge density ρ, but there is no term corresponding to charge in the magnetic equation. This reflects the interesting physical fact that magnetic charges do not exist.[3] It is as though nature intended to set up the equations for electric and magnetic fields identically, but somehow neglected to create the magnetic charges that would be needed to make the equations agree. Since $\nabla \cdot \vec{B} = 0$, there are no sources or sinks of magnetic field, which means that the lines of $\vec{B}$ never start or end. Instead they form closed loops, like rubber bands. Electric fields, on the other hand, can start and end on charges. Note that except in the electrostatic case, electric fields do not *have to* begin on charges; electric fields can also originate from time-varying fields, through Faraday's law, even though no charges are present. The point is that electric fields have the possibility of starting and ending when charges are present, but magnetic field lines are always closed loops. Electric and magnetic field lines are compared in Fig. 6.8. The same similarities and differences appear in the other pair of Maxwell equations, involving the curl. Equation (3) tells us that a time-varying magnetic field creates an electric field; equation (4) tells us that a time-varying electric field creates a magnetic field. So far, the two equations are highly similar. However, equation (4) states that an electric current ($\vec{J}$) also creates a magnetic field. There is no corresponding term in (3), because if there were, it would

[3]Some physicists believe, on theoretical grounds, that a subatomic particle called a *magnetic monopole,* with magnetic charge, ought to exist, but as yet such particles have not been found.

Table 6.1 THE FOUR MACROSCOPIC MAXWELL EQUATIONS IN DIFFERENTIAL FORM

1. $\nabla \cdot \vec{D} = \rho$
2. $\nabla \cdot \vec{B} = 0$
3. $\nabla \times \vec{E} = -\dfrac{\partial \vec{B}}{\partial t}$
4. $\nabla \times \vec{H} = \vec{J} + \dfrac{\partial \vec{D}}{\partial t}$

have to be a magnetic current, consisting of a flow of magnetic charges, and magnetic charges do not exist.

Table 6.1 presents Maxwell's equations in what is known as *differential form*. As we have already seen, the differential operators can be eliminated from these equations by integrating over an arbitrary volume and applying the divergence theorem (for equations (1) and (2)), or by integrating over an arbitrary surface and applying Stokes' theorem (for equations (3) and (4)). Doing this results in the *integral form* of Maxwell's equations, Table 6.2. These equations express exactly the same physics as those in Table 6.1; they are simply presented in a different mathematical way.

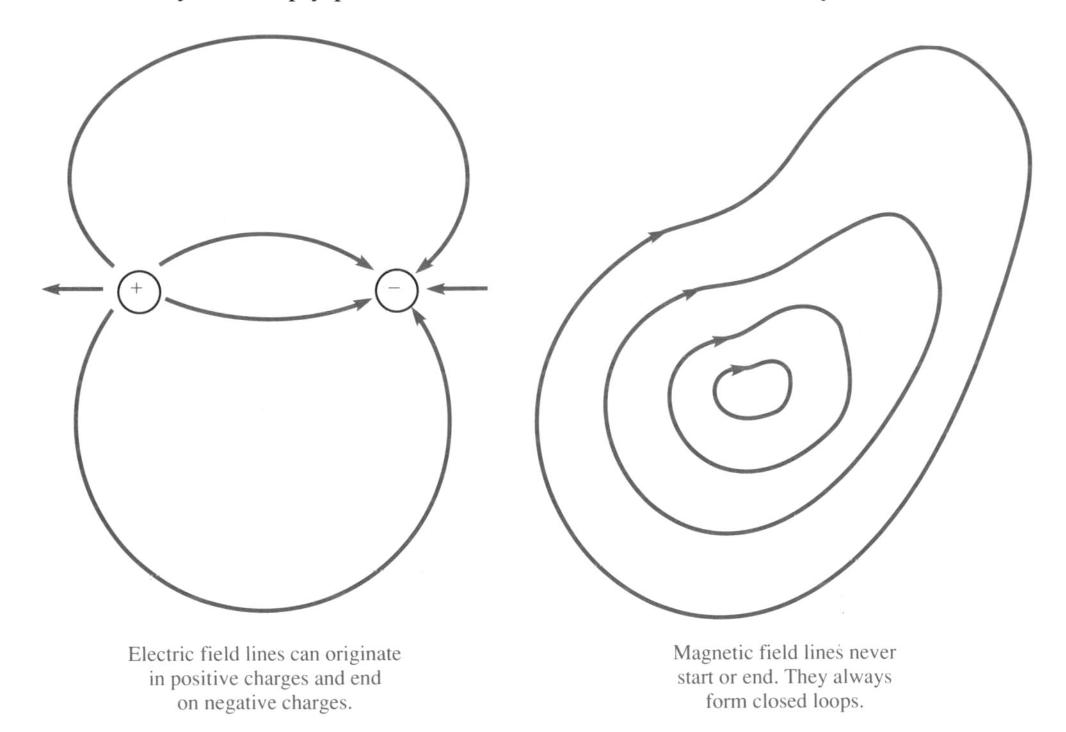

Figure 6.8 Electric fields can originate on positive charges and can end on negative charges. But since nature has neglected to supply us with magnetic charges, magnetic fields cannot begin or end; they can only form closed loops.

Table 6.2 MAXWELL'S EQUATIONS IN INTEGRAL FORM*

1. $\int_S \vec{D} \cdot \vec{dS} = \int_V \rho \, dV$
2. $\int_S \vec{B} \cdot \vec{dS} = 0$
3. $\oint_L \vec{E} \cdot \vec{dl} = -\frac{\partial}{\partial t} \int_S \vec{B} \cdot \vec{dS}$
4. $\oint_L \vec{H} \cdot \vec{dl} = \int_S \vec{J} \cdot \vec{dS} + \frac{\partial}{\partial t} \int_S \vec{D} \cdot \vec{dS}$

*In (1) and (2), S is an arbitrary closed surface enclosing the volume V. In (3) and (4), L is an arbitrary closed path that bounds the surface S.

EXERCISE 6.4

Consider a case in which $\vec{B}$ varies sinusoidally in time with frequency ω. Show, by means of a vector identity, that equation (2) of Table 6.1 can be derived from equation (3).

6.5 POINTS TO REMEMBER

1. When time-varying fields are present, an additional driving term appears in the magnetic curl equation:

$$\nabla \times \vec{H} = \vec{J} + \frac{\partial \vec{D}}{\partial t}$$

The additional term $\frac{\partial \vec{D}}{\partial t}$ acts like a current in that it combines with the real current to act as a driving force for the magnetic field. Hence it is known as *displacement current*.

2. Just as a time-varying electric field can act as a driving force for magnetic fields, a time-varying magnetic field can generate electric fields. This effect, which is the basis of electric generators, is described by Faraday's law:

$$\nabla \times \vec{E} = -\frac{\partial \vec{B}}{\partial t}$$

3. From Faraday's law we conclude that a single-valued voltage may not exist in the presence of time-varying magnetic fields.

4. The tangential component of $\vec{E}$ is continuous at a dielectric interface.

5. Time-varying currents give rise to time-varying magnetic fields, which in turn can induce electric fields in other circuits or in the same circuit. This gives rise to the phenomena of mutual inductance and self-inductance.

REFERENCES

1. S. Ramo, J. R. Whinnery, and T. Van Duzer, *Fields and Waves in Communication Electronics*. New York: Wiley, 1984.
2. N. N. Rao, *Elements of Engineering Electromagnetics*. Englewood Cliffs, N.J.: Prentice-Hall, 1987.
3. S. V. Marshall and G. G. Skitek, *Electromagnetic Concepts and Applications,* 2nd ed. Englewood Cliffs, N.J.: Prentice-Hall, 1987.
4. D. K. Cheng, *Field and Wave Electromagnetics*. Reading, Mass.: Addison-Wesley, 1983.
5. W. H. Hayt, Jr., *Engineering Electromagnetics,* 4th ed. New York: McGraw-Hill, 1981.

For more advanced discussions, see:

6. P. Lorrain and Dale Corson, *Electromagnetic Fields and Waves,* 2nd ed. New York: Freeman, 1970.
7. J. D. Jackson, *Classical Electrodynamics*. New York: Wiley, 1962.

PROBLEMS

Section 6.1

6.1 In a region where there are no currents and $\epsilon = \epsilon_o$,

$$\vec{H} = Ax\vec{e}_y$$

where A is a constant. Find $\vec{E}(x, y, z, t)$. Assume that $\vec{E} = 0$ at time $t = 0$.

6.2 A magnetic field is given, in cylindrical coordinates, by

$$\vec{H} = Ar\vec{e}_\phi$$

where A is a constant. Assume that in the region of interest there are no currents and $\epsilon = \epsilon_o$. Find $\vec{E}(r, \phi, z, t)$. Assume that $\vec{E} = 0$ at time $t = 0$.

6.3 In Fig. 6.1(a), the magnetic field outside the capacitor at a distance a from the axis (where a is greater than the wire radius) is $\vec{H}(a) = \dfrac{k}{a}\vec{e}_\phi$, where k is a given constant. Assume that the capacitor has area A, and that the electric field in the capacitor is uniform (i.e., does not vary with position) and equals zero at $t = 0$. Find the electric field in the capacitor at time t.

6.4 Suppose the current density in a semiconductor is given by

$$\vec{J} = Ax\vec{e}_x$$

Assume that at $t = 0$ the semiconductor was electrically neutral (i.e., $\rho = 0$).
(a) Find $\rho(x, t)$.
(b) Sketch $\vec{J}$ and explain why your answer to part (a) is physically reasonable.

*6.5 (a) Consider a material which obeys Ohm's law, $\vec{J} = \sigma_E \vec{E}$, where σ_E, the electrical conductivity, is a constant. Use (6.8) to show that

$$\frac{\partial \rho}{\partial t} = -\frac{\sigma_E}{\epsilon}\rho$$

(b) Suppose a sphere with radius b is made of the conductive material discussed in part (a). At time $t = 0$, the material contains a uniform charge density ρ_o. Find $\rho(r, t)$ for times $t > 0$. Where is the charge going?
(c) Find the electric field $\vec{E}(r, t)$ inside and outside the sphere, for times $t > 0$.

*6.6 An extremely thin sheet of conductive material lies in the x-y plane. It carries a uniform current, of magnitude $I_W(t)$ amperes per meter, in the y direction. Finite time-varying electric fields are present. Let H_{x1} and H_{y1} be the tangential components of H just above the sheet (that is, where z is a very small positive number) and let H_{x2} and H_{y2} be the tangential components just below the sheet. Find (a) $H_{x1} - H_{x2}$ and (b) $H_{y1} - H_{y2}$.

*6.7 A certain semiconductor contains only one kind of mobile charged particle, with positive charge. (These particles are known as *holes*.) For such a material, the current density is given by

$$\vec{J} = -D\,\nabla\rho + \mu_P \vec{E}\rho$$

Here the first term on the right describes diffusion of the particles, and D is a constant called the *diffusion coefficient*. The second term describes *drift current,* in which particles are pushed by the electric field; the constant μ_P is called the *mobility*. Show that in this material the charge density ρ obeys

$$D\,\nabla^2\rho - \mu_P \vec{E}\cdot\nabla\rho - \mu_P \frac{\rho^2}{\epsilon} = \frac{\partial \rho}{\partial t}$$

Section 6.2

6.8 A square loop of wire has two terminals for external connection, as shown in Fig. 6.9. The loop is in the x-y plane, with sides of length a, and moves in the $+x$ direction with velocity U. At $t = 0$ the center of

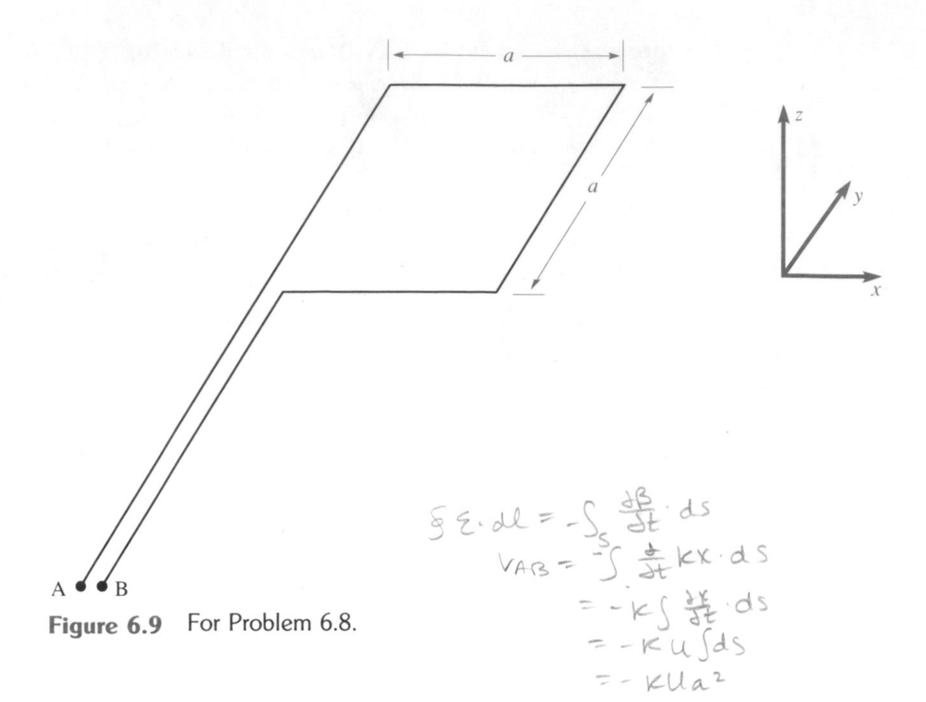

Figure 6.9 For Problem 6.8.

the loop is at the origin. There is a non-time-varying magnetic field in the z direction given by $B_z = Kx$, where K is a given constant. Find the external voltage $V_{AB} = V_A - V_B$. (*Note:* the *sign* of V_{AB} is also to be determined.)

*6.9 A slowly increasing field is given in cylindrical coordinates by

$$\vec{B} = (B_o + \beta t)\vec{e}_z \qquad \text{for } r < r_o$$
$$= 0 \qquad \text{for } r > r_o.$$

where t is the time and B_o and β are given constants. Find the θ component of the electric field, E_θ, as a function of r. (*Suggestion:* exploit the analogy to Ampère's circuital law.)

*6.10 The phasor representing the electric field in a coaxial transmission line is

$$\vec{\mathbf{E}}(r) = E_o \frac{a}{r} e^{-jkz}\, \vec{e}_r \qquad \text{for } r_1 < r < r_2$$

where E_o, a, and the frequency ω are given real constants.

(a) Find the magnetic field in the transmission line, as a function of position and time.

(b) Make a sketch showing both fields at $z = 0$, $t = 0$.

(c) At a given instant of time, are the maxima of E and H at the same value of z, or at different values of z?

6.11 An electric field on the air side of an air–dielectric interface makes an angle θ_A with the normal to the interface. On the dielectric side the field makes an angle θ_D with the normal. The relative permittivity of the dielectric is 2.0. Sketch θ_D as a function of θ_A over the range $0 < \theta_A < 90°$.

*6.12 For the case of the preceding problem, determine the value of θ_D for which $|\theta_D - \theta_A|$ (the angular deviation of the field as it passes through the interface) is maximum.

6.13 Some dielectric media, such as ideal plasmas, can have relative permittivities that are *negative*. Is it possible for an electric field to exist at the interface of air with such a dielectric? If so, draw a sketch indicating the directions of $\vec{E}$ on the two sides.

6.14 Is the electrostatic potential continuous or discontinuous across a dielectric interface? Why?

Section 6.3

6.15 Consider a parallel-conductor transmission line consisting of an infinitely long loop like that shown in Fig. 6.5. The velocity of transmission-line waves is found to be c (the velocity of light). What is the capacitance per unit length between the two conductors?

*6.16 It can be shown that the magnetic field at a large distance r from a small loop of radius a carrying current I is (in spherical coordinates)

$$B_r = \frac{\mu I a^2}{2r^3} \cos\theta$$

$$B_\theta = \frac{\mu I a^2}{4r^3} \sin\theta$$

$$B_\phi = 0$$

where the polar axis ($\theta = 0$) is perpendicular to the plane of the loop. Use this information to estimate the mutual inductance between two small loops of radii c and d, respectively, lying in the same plane with a large distance h between their centers. Show that $M_{12} = M_{21}$.

*6.17 Two loops, one large and one small, lie in parallel planes separated by a distance h, as shown in Fig. 6.10. One loop has radius a and the other radius b, with $a \ll b$ and $a \ll h$. The axes of the two loops coincide. Find the mutual inductance by calculating the flux through the small loop due to a current in the large one.

*6.18 For the case of the preceding problem, find the mutual inductance by calculating the flux through the large loop arising from a current in the small one. Use the fields given in Problem 6.16. (*Suggestion:* integrate over a curved surface that is a portion of a sphere with its center at the small loop.) Your answer should agree with Problem 6.17.

**6.19 Find the mutual inductance between an infinitely long straight wire and a small circular loop of radius a. The wire is in the plane of the loop

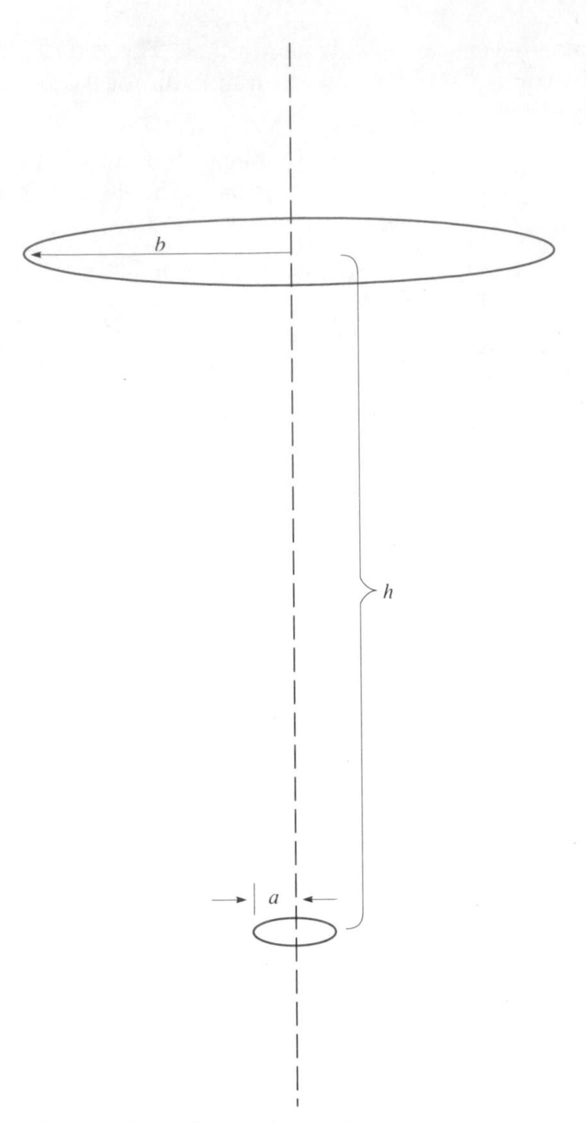

Figure 6.10 For Problem 6.17.

and its distance from the center of the loop is b, with $b \gg a$. Find the mutual inductance by calculating thc flux due to a current in the small loop (see Problem 6.16). Consider the straight wire to be part of an infinitely large loop, as shown in Fig. 6.11. The answer should agree with Example 6.4.

6.20 Find the approximate self-inductance of a toroidal inductor made by winding 1000 turns of wire on a doughnut-shaped ferrite core with a relative permeability of 10,000. The large radius of the toroidal core is

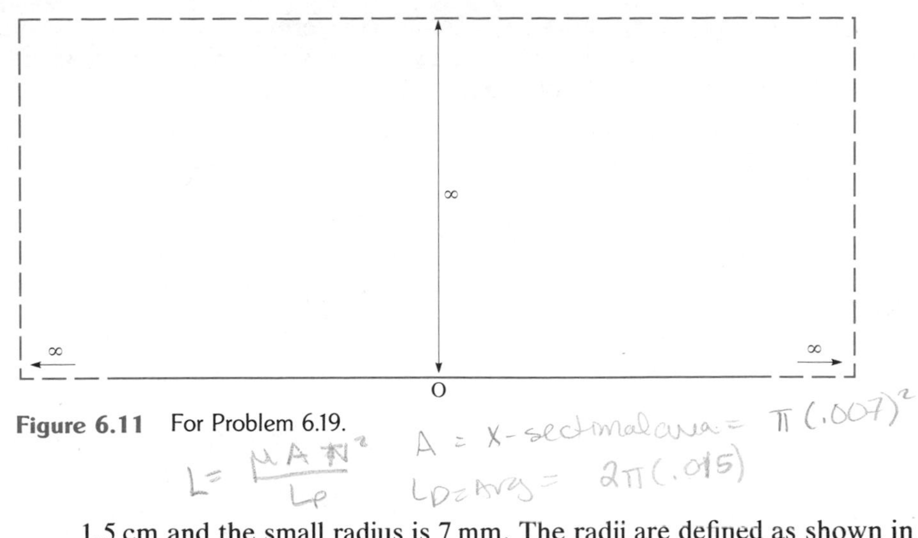

Figure 6.11 For Problem 6.19.

1.5 cm and the small radius is 7 mm. The radii are defined as shown in Fig. 6.12. (*Caution*: this may not be the definition you expect.)

6.21 Iron-core inductors often need to be "trimmed" (turns added or removed) to achieve a precise value of inductance. Show that if the inductance is $P\%$ too high, $\frac{1}{2}P\%$ of the turns should be removed.

Section 6.4

6.22 (a) Suppose that no charged particles existed in the world. (In that case there would also be no currents, as a current is a flow of charged particles.) Write the four differential-form Maxwell equations as they would appear in that case.

(b) Suppose that ordinary charged particles do exist, and magnetic monopoles also exist. Write the four Maxwell's equations in this case.

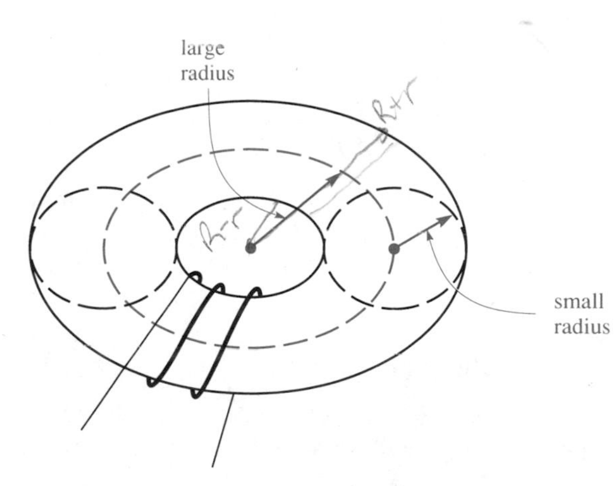

Figure 6.12 A toroidal inductor. Note the definitions of *large radius* and *small radius*. (For Problem 6.20.)

6.23 Beginning with the integral forms of Maxwell's equations, apply the divergence theorem and Stokes' theorem to obtain the differential forms of the four equations.

*6.24 Suppose a surface current of J_s A/m flows in the x direction on a thin sheet of perfectly conducting metal that lies in the x-y plane. Let H_1 be the magnetic induction on one side of the sheet and H_2 the magnetic induction on the other side. Prove that

$$H_{1x} = H_{2x}$$

$$H_{1y} - H_{2y} = J_s$$

$$H_{1z} = H_{2z}$$

*6.25 A certain rectangular resonant cavity is a perfectly conducting box with walls at $x = 0$, $x = a$, $y = 0$, $y = b$, $z = 0$, $z = d$. The magnetic field inside the cavity oscillates at the frequency ω_1 and is represented by the phasors

$$\mathbf{H}_x = -H_o \sin\frac{\pi x}{a} \cos\frac{\pi z}{d}$$

$$\mathbf{H}_z = H_o \left(\frac{d}{a}\right) \cos\frac{\pi x}{a} \sin\frac{\pi z}{d}$$

(a) Sketch the lines of $\vec{H}$ in a plane perpendicular to the y axis.
(b) Find the electric field at all positions inside the resonator.
(c) Is surface charge present on any of the walls? Find the surface charge.

**6.26 Two materials, with the same ϵ but differing resistivities ρ_1 and ρ_2 are in contact. (The resistivity ρ is the reciprocal of the electric conductivity. Thus $\vec{J} = \vec{E}/\rho$.)
(a) Find the relationship between the tangential electric fields on the two sides.
(b) Find the relationship between the normal components of $\vec{E}$ on the two sides. The field does not vary with time. (Use the fact that as much current leaves one material as enters the other.)
(c) Show that a real charge density σ_{12} must be present on the interface. Evaluate σ_{12} in terms of D_{N1}, ρ_1, and ρ_2.

**6.27 (a) Repeat the previous problem for the case of materials that have both different permittivities ϵ_1 and ϵ_2 and different resistivities ρ_1 and ρ_2.
(b) Determine a condition on ϵ_1, ϵ_2, ρ_1, ρ_2 that will make the charge on the interface vanish.
(c) If D_{1N} is directed from material 1 toward material 2, $\epsilon_1/\epsilon_2 = 2$, and $\rho_1/\rho_2 = 3$, is the charge on the interface negative or positive?

*6.28 Material 1 has $\epsilon = \epsilon_1$, $\mu = \mu_o$, and occupies the space $z > 0$. Material

2 has $\epsilon = \epsilon_2$, $\mu = \mu_o$, and occupies the space $z < 0$. Both materials are nonconductive. The magnetic field in material 1 is

$$\vec{H}_1 = Ae^{-\alpha z} \cos(\omega t - kx)\vec{e}_y$$

(a) Find the electric field E_1 in region 1.
(b) Find all components of $\vec{H}$ and $\vec{E}$ just across the boundary in region 2.
(c) Find the phasors representing the fields found in parts (a) and (b).

*6.29 For the case of the previous problem, assume that in region 2 the magnetic field is

$$\vec{H}_2 = Be^{\beta z} \cos(\omega t - k'x)\vec{e}_y$$

where both α and β are positive real numbers.
(a) Show that $A = B$ and $k' = k$.
(b) Use the necessary relationships of $\vec{E}_1$ and $\vec{E}_2$ on the boundary to show that either ϵ_1 or ϵ_2 must be negative.
(c) A wave of this sort, traveling along a dielectric interface, is known as a *dielectric surface wave*. Sketch the fields of the wave, assuming $|\epsilon_1| = 10|\epsilon_2|$. Does the wave extend farther into region 1 or region 2?

**6.30 *The point of this long problem is not so much to find an answer as to illustrate a principle. We have seen that strictly speaking it is incorrect to make use of electrostatic calculations and electrostatic concepts, such as voltage, when time-varying fields are present. Nonetheless, this is done every day: for instance, we usually describe electronic circuits, with time-varying currents in them, in terms of voltages. This must be because when the frequency is sufficiently low, the assumption that the frequency is zero is satisfactory. But in that case, how low must the frequency be for electrostatic calculations to give good results? That is the question that is approached in this problem.*

Suppose the voltage applied to a parallel-plate capacitor with plates at $z = -\frac{1}{2}$ cm and $z = \frac{1}{2}$ cm is $v(t) = 100 \cos \omega t$ volts, where $\omega = 2\pi f = 2\pi(10^6 \text{ Hz})$. One might guess, on the basis of electrostatic theory, that inside the capacitor, $\vec{E}$ would equal $10^4 \cos \omega t\, \vec{e}_z$ V/m. (Do you agree?)

(a) Assuming that $\vec{E}$ is as stated, find $\vec{B}$ inside the capacitor, using

$$\nabla \times \vec{E} = -\frac{\partial B}{\partial t}.$$

(b) Show that the $\vec{B}$ found in (a) does not satisfy the other "curl" equation.
(c) As a consequence, the electric field we guessed cannot be exactly correct. Show that both "curl" equations *can* be satisfied if instead we let

$$\vec{E} = 10^4 \cos \omega t\, f(x,y)\vec{e}_z \quad \text{V/m}$$

provided that the function $f(x, y)$ satisfies

$$\frac{\partial^2 f}{\partial x^2} + \frac{\partial^2 f}{\partial y^2} = -\omega^2 \mu_o \epsilon_o f(x, y)$$

(d) Many different functions $f(x, y)$ satisfy the above equation, and the correct one cannot be chosen without knowing the shape and size of the capacitor plates. However, it is easily seen that one function that *could* satisfy the equation governing f is

$$f(x, y) = \cos kx$$

If this is the correct function, what is the numerical value of k?

(e) Make a sketch showing, to scale, the capacitor and the magnitude of the electric field, as a function of position, using the function $f(x, y)$ given in (d). Assume the center of the capacitor is at the origin and the plates lie between $x = -5$ cm and $x = 5$ cm. Compare the field based on $f(x, y)$ with the original incorrect guess of $\vec{E}$ based on electrostatics. The field based on $f(x, y)$ satisfies Maxwell's equations, and is thus a correct solution for the time-varying case. Is the electrostatic approximation similar to the correct solution in this case?

(f) Repeat part (e), assuming the plates lie between $x = -1.5 \times 10^4$ cm and $x = 1.5 \cdot 10^4$ cm. Now, is the electrostatic guess a good one?

This problem illustrates a general rule: electrostatic solutions are good approximations if the physical structure under consideration is small compared with the wavelength at the frequency in question.

**6.31 The *electromagnetic field tensor* is written as follows:

$$F_{kl} = \begin{bmatrix} 0 & H_z & -H_y & -jcD_x \\ -H_z & 0 & H_x & -jcD_y \\ H_y & -H_x & 0 & -jcD_z \\ jcD_x & jcD_y & jcD_z & 0 \end{bmatrix}$$

(where $j = \sqrt{-1}$). Let us also define a four-dimensional current vector, according to

$$\vec{J} = \begin{bmatrix} J_x \\ J_y \\ J_z \\ jc\rho \end{bmatrix}$$

(where c is the velocity of light). Show that two of the four Maxwell equations are expressed by the single equation

$$\sum_{l=1}^{4} \frac{\partial F_{kl}}{\partial x_l} = J_k$$

where $x_1 = x$, $x_2 = y$, $x_3 = z$, and $x_4 = jct$. (Each of the four values of the subscript k gives one equation.)

This problem demonstrates the "4-vector" (four-dimensional) formulation of electromagnetics used in relativity theory.

COMPUTER EXERCISE 6.1 NUMERICAL CALCULATION OF MUTUAL INDUCTANCE

In this exercise we shall calculate the mutual inductance between two circular loops of radii a and b, as shown in Fig. 6.10. The loops lie in parallel planes and are centered one above the other at a distance h. Their mutual inductance can be found by means of Neumann's formula, (6.25a).

Let $u \equiv a/h$, $v \equiv b/h$. Write a program to perform the necessary integral numerically. Express M in the form

$$M = \mu_0 h f(u, v)$$

Your program should find $f(u, v)$ for any given values of u and v.

Check your result in the limit $u \rightarrow 0$. The mutual inductance in this limit can be estimated by taking the product of the *on-axis* magnetic field of the larger loop (which is easy to find) and the area of the small one.

CHAPTER 7

Electrodynamics

Our study of electromagnetics has now progressed to a point at which our knowledge of basic principles is essentially complete. All the various static and dynamic phenomena we have discussed can now be seen to be special cases included in the four general equations of Maxwell. The beauty of these equations is that they contain so much physics in such brief and elegant form.

Remarkably, these four rather simple equations express the basic physics behind all phenomena commonly thought of as "electromagnetic."[1] Studies of wave propagation, classical optics, antennas, waveguides, and related matters usually begin with Maxwell's equations as a starting point. Therefore, we now change our emphasis and move from fundamental principles to applications. In this chapter we shall study two general kinds of solution for Maxwell's equations, one relevant inside conducting materials, the other applicable in dielectrics or empty space. We shall then be ready to apply Maxwell's equations to cases of practical interest, such as free wave propagation, guided waves, and antennas.

---

[1]Unlike Newton's laws of mechanics, Maxwell's equations are consistent with special relativity. However, they are incapable of describing quantum-mechanical effects. For example, the Einstein photoelectric effect cannot be interpreted in terms of Maxwell's equations. As a rule of thumb, quantum effects tend to become important when $hf \geq kT$ (h = Planck's constant = 6.7×10^{-34} J/Hz, f = frequency in Hz, k = Boltzmann's constant = 1.4×10^{-23} J/°K, and T = absolute temperature in degrees Kelvin.) At room temperature, T = 300° K, this criterion suggests that quantum effects will be important primarily at frequencies above 10^{12} Hz. Thus they are often important in the infrared and visible parts of the spectrum, but seldom at radio or microwave frequencies.

7.1 SINUSOIDAL FIELDS

In most electromagnetic problems, the fields can be considered to vary sinusoidally in time. The reasons are the same as in the field of electric circuits, where sinusoidal analysis is very common. In electromagnetics, however, sinusoidal analysis is even more important than in circuit theory, because there is nothing in electromagnetic technology that corresponds to digital technology in circuits. In electromagnetics, information is usually transmitted by imposing amplitude, frequency, or phase modulation on a sinusoidal carrier. Thus, it is worthwhile to develop phasor techniques for dealing with sinusoidal fields.

To illustrate the phasor technique, let us begin with the "curl H" Maxwell equation.

$$\nabla \times \vec{H} = \vec{J} + \frac{\partial \vec{D}}{\partial t} \tag{7.1}$$

We shall assume that all the fields and currents in this equation are sinusoidal, at a single[2] frequency ω. In that case we can replace (7.1) by its phasor representation, obtaining

$$\nabla \times \vec{\mathbf{H}} = \vec{\mathbf{J}} + j\omega\vec{\mathbf{D}} = \vec{\mathbf{J}} + j\omega\epsilon\vec{\mathbf{E}} \tag{7.2}$$

Note that phasors are distinguished from ordinary vectors by the use of **boldface** symbols.

Now let us pause to notice what sort of quantities appear in this equation. $\vec{\mathbf{H}}$, for instance, is a vector function of position; however it is *not* a function of time. Since it is a phasor, its three numerical components are *complex numbers*. The usual rules relating a sinusoidal function to its phasor continue to apply, but for each component separately. That is, if

$$\vec{H}(x, y, z, t) = f_1(x, y, z) \cos(\omega t + \phi_1)\vec{e}_x + f_2(x, y, z) \cos(\omega t + \phi_2)\vec{e}_y$$

then

$$\vec{\mathbf{H}}(x,y,z) = f_1(x,y,z)e^{j\phi_1}\,\vec{e}_x + f_2(x,y,z)e^{j\phi_2}\,\vec{e}_y$$

We also note that

$$\vec{H}(x,y,z,t) = \text{Re}\,[\vec{\mathbf{H}}(x,y,z)e^{j\omega t}] \tag{7.3}$$

---

[2]This assumption is reasonable because Maxwell's equations are *linear* equations. The situation is similar to that in linear circuits, where all currents and voltages typically have the same frequency.

EXAMPLE 7.1

Find $\vec{\mathbf{E}}_1$ (2, 3, 0), if

$$\vec{E}_1(x, y, z, t) = 3 \sin x \cos\left(\omega t + \frac{\pi}{4}\right)\vec{e}_x + 5 \cos x \sin y \sin\omega t \, \vec{e}_y$$

(Arguments of sin and cos are in radians.)

Solution We first replace $\sin\omega t$ by $\cos\left(\omega t - \frac{\pi}{2}\right)$. We then construct the phasor by finding the phasor for the x component and adding it to the phasor for the y component:

$$\begin{aligned}\vec{\mathbf{E}}_1 &= 3 \sin x \, e^{j\frac{\pi}{4}} \vec{e}_x + 5 \cos x \sin y \, e^{-j\frac{\pi}{2}} \vec{e}_y \\ &= 3 \sin x \left(\frac{1 + j}{\sqrt{2}}\right) \vec{e}_x + 5 \cos x \sin y \, (-j) \, \vec{e}_y\end{aligned}$$

Setting $x = 2$, $y = 3$, $z = 0$, we have the desired result:

$$\vec{\mathbf{E}}_1(2, 3, 0) = (1.93 + 1.93j)\vec{e}_x + 0.293 \, j\vec{e}_y$$

The reader may notice that it is difficult to visualize a vector such as this, with complex coefficients. It is not really useful to think of it as an arrow pointing in some direction in three-dimensional space. However, one could think of it as the sum of *two* vectors, one real and one imaginary, pointing in different directions:

$$\vec{\mathbf{E}}_1(2, 3, 0) = 1.93\vec{e}_x + j(1.93e_x + 0.293\vec{e}_y)$$

Perhaps the best way is to simply fall back on the mathematician's definition of a vector as an "ordered triple" of numbers. That is, complete information about the vector $2\vec{e}_x + 3\vec{e}_y + 0\vec{e}_z$ is conveyed by stating the three numbers 2, 3, 0 in the proper order. A vector such as $\vec{\mathbf{E}}_1$ is then simply an ordered triple of complex numbers; in this example, [(1.93 + j1.93), 0.293j, 0]. In this way we avoid thinking of a vector as an "arrow" at all; it is simply a set of three numbers.

EXAMPLE 7.2

Find the field $\vec{H}_1(r, \theta, \phi, t)$ represented by the phasor $\vec{\mathbf{H}}_1 = (2 + 3j)\vec{e}_r + (4 - 5j)\vec{e}_\phi$.

Solution From (7.3),

$$\begin{aligned}\vec{H}_1(r, \theta, \phi, t) &= Re\,(\vec{\mathbf{H}}_1 e^{j\omega t}) \\ &= (2 \cos\omega t - 3 \sin\omega t)\vec{e}_r + (4 \cos\omega t + 5 \sin\omega t)\vec{e}_\phi\end{aligned}$$

A different form of the answer is obtained by first converting $(2 + 3j)$ to its polar form $3.6e^{j56°}$ and $(4 - 5j)$ to $6.4e^{-j51°}$. Hence

$$\vec{H}_1 = 3.6 \cos(\omega t + 56°)\vec{e}_r + 6.4 \cos(\omega t - 51°)\vec{e}_\phi$$

7.2 THE VECTOR LAPLACIAN OPERATOR

A phasor representation of the "curl E" Maxwell equation is obtained in the same way we obtained (7.2):

$$\nabla \times \vec{\mathbf{E}} = -j\omega\vec{\mathbf{B}} \tag{7.4}$$

Let us now apply the curl operator to each side of (7.4). This results in

$$\nabla \times (\nabla \times \vec{\mathbf{E}}) = -j\omega\nabla \times \vec{\mathbf{B}} \tag{7.5}$$

Now $\nabla \times \vec{\mathbf{B}}$ can be replaced using (7.2):

$$\nabla \times (\nabla \times \vec{\mathbf{E}}) = -j\omega\mu(\vec{\mathbf{J}} + j\omega\epsilon\vec{\mathbf{E}}) \tag{7.6}$$

On the left side of (7.6), the operator $\nabla \times \nabla \times$ acts on the vector field to produce another vector field. It is thus an operator of the same type as the single curl operator $\nabla \times$. Referring to the table of vector identities in the endpapers, we see that the twice-repeated curl operator can be expressed in another way:

$$\nabla \times \nabla \times \vec{\mathbf{E}} = \nabla(\nabla \cdot \vec{\mathbf{E}}) - \nabla^2\vec{\mathbf{E}} \tag{7.7}$$

Each term of this equation is a vector. (Verify for yourself that $\nabla(\nabla \cdot \vec{E})$ is a vector.) The final term contains a new operator that we have not seen before, known as the *vector Laplacian operator.*

The vector Laplacian operator is of the type that acts on a vector to produce a vector. Thus it is fundamentally different from the ordinary Laplacian operator, which acts on a scalar to produce a scalar. Conventionally, however, the same symbol ∇^2 is used for both. To determine which operator is meant, one must look at its argument. If we see, for example, $\nabla^2 V$, where V is the scalar electrostatic potential, then the operator must be the ordinary Laplacian. If we see $\nabla^2\vec{E}$, the vector Laplacian must be meant.

In rectangular coordinates, the vector Laplacian operator is given by

$$\nabla^2\vec{E} = \left(\frac{\partial^2 E_x}{\partial x^2} + \frac{\partial^2 E_x}{\partial y^2} + \frac{\partial^2 E_x}{\partial z^2}\right)\vec{e}_x + \left(\frac{\partial^2 E_y}{\partial x^2} + \frac{\partial^2 E_y}{\partial y^2} + \frac{\partial^2 E_y}{\partial z^2}\right)\vec{e}_y + \left(\frac{\partial^2 E_z}{\partial x^2} + \frac{\partial^2 E_z}{\partial y^2} + \frac{\partial^2 E_z}{\partial z^2}\right)\vec{e}_z$$

This can be written more succinctly in terms of the ordinary Laplacian operator:

$$\nabla^2\vec{E} = \vec{e}_x\, \nabla^2 E_x + \vec{e}_y\, \nabla^2 E_y + \vec{e}_z\, \nabla^2 E_z \tag{7.8}$$

(Note that the operator on the left side is the vector Laplacian, but those on the right have scalar arguments.) Expressions for the vector Laplacian operator in cylindrical and spherical coordinates will be found in the end-paper tables. In those coordinate systems the relationship between the two kinds of Laplacian is much less apparent than in rectangular coordinates, and they are best thought of as quite different operators.

EXERCISE 7.1

Let $\vec{E}(x, y, z, t) = E_o \cos(\omega t - kz)\vec{e}_x$. Find $\vec{\mathbf{E}}$ and $\nabla^2\vec{\mathbf{E}}$.

Answers $\vec{\mathbf{E}} = E_o e^{-jkz}\,\vec{e}_x$. $\nabla^2\vec{\mathbf{E}} = -k^2 E_o e^{-jkz}\,\vec{e}_x$. Note that for this particular $\vec{\mathbf{E}}$, $\nabla^2\vec{\mathbf{E}} = -k^2\vec{\mathbf{E}}$.

In most situations, we shall be dealing with materials that contain no real charge density. This is the case with dielectric materials, and it is also generally the case with conductive materials, because any real charge that may exist will repel itself and thus travel outwards until it resides on the material's outer surfaces. (See Problem 6.5.) Therefore $\nabla \cdot \vec{D} = 0$, and assuming that ϵ is a constant not equal to zero, $\nabla \cdot \vec{E} = 0$. Thus the first term on the right of (7.7) vanishes, and, substituting into (7.6) we obtain

$$\nabla^2\vec{\mathbf{E}} = j\omega\mu(\vec{\mathbf{J}} + j\omega\epsilon\vec{\mathbf{E}}) \tag{7.9}$$

7.3 THE SKIN EFFECT

As our first application of Maxwell's equations, let us investigate the flow of alternating currents in good conductors. We shall see that currents tend to flow on the surface, or "skin." Thus alternating current flow is said to be influenced by the *skin effect*. This effect is of practical importance; it affects resistive losses whenever a high-frequency current flows in an electronic circuit.

Let us assume that the conductive material in question obeys Ohm's law, $\vec{J} = \sigma_E\vec{E}$, where σ_E is the conductivity. Then (7.9) becomes

$$\nabla^2\vec{\mathbf{E}} = j\omega\mu(\sigma_E + j\omega\epsilon)\vec{\mathbf{E}} \tag{7.10}$$

For simplicity, let us now assume that the material in question is a *very good* conductor, so that $\sigma_E \gg |\omega\epsilon|$. (See note[3].) In that case, the displacement-

[3]The absolute-value bars are used because for some materials the value of ϵ can be negative. What matters is that the term containing $\omega\epsilon$ should be negligible compared with the term containing σ_E.

current term, dwarfed by the conduction term, can be neglected, leaving us with

$$\nabla^2\vec{\mathbf{E}} = j\omega\mu\sigma_E\vec{\mathbf{E}} \tag{7.11}$$

Since $\vec{\mathbf{E}} = \vec{\mathbf{J}}/\sigma_E$ we also have

$$\nabla^2\vec{\mathbf{J}} = j\omega\mu\sigma_E\vec{\mathbf{J}} \tag{7.12}$$

while a similar development (substituting (7.4) into (7.2) to eliminate $\vec{\mathbf{E}}$) results in

$$\nabla^2\vec{\mathbf{H}} = j\omega\mu\sigma_E\vec{\mathbf{H}} \tag{7.13}$$

Equations (7.11)–(7.13) are the basic equations of the skin effect.

To see the significance of these equations, let us consider a conductive material filling the half-space $z < 0$, as shown in Fig. 7.1. Let us imagine that current flows through this material in the x direction, with the current density at the surface being J_o A/m^2. The current density is independent of y and x; thus, (7.12) simplifies to

$$\frac{\partial^2\mathbf{J}_x}{\partial z^2} = j\omega\mu\sigma_E\mathbf{J}_x \tag{7.14}$$

which has the solution

$$\mathbf{J}_x = Ae^{-(1+j)z/\delta} + Be^{(1+j)z/\delta} \tag{7.15}$$

where A and B are any constants, and δ, given by

$$\delta = \sqrt{\frac{2}{\omega\mu\sigma_E}} \tag{7.16}$$

is known as the *skin depth*.

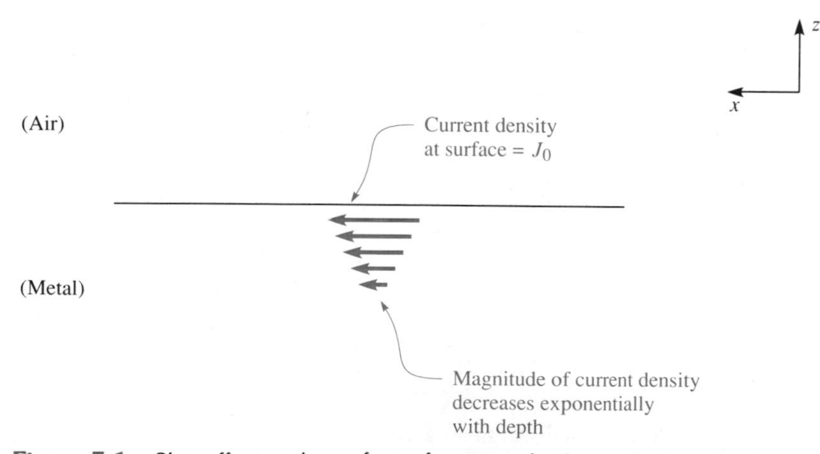

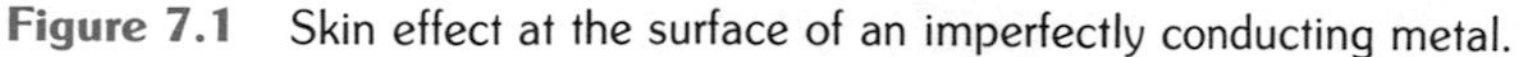
Figure 7.1 Skin effect at the surface of an imperfectly conducting metal.

EXERCISE 7.2

Substitute (7.15) into (7.14) and verify that it is a solution.

Expression (7.15) can be rewritten as

$$\mathbf{J}_x = Ae^{-\frac{z}{\delta}}e^{-j\frac{z}{\delta}} + Be^{\frac{z}{\delta}}e^{j\frac{z}{\delta}} \tag{7.17}$$

Now this result applies only in the region $z < 0$, where the conductive material is present. The first term of (7.17) becomes larger exponentially as $z \to -\infty$, and would eventually give rise to infinitely large currents. This is unreasonable physically; hence, the constant A must vanish. The constant B can be evaluated if it is given that $\mathbf{J}_x = J_o$ when $z = 0$:

$$J_o = Be^{(0)}e^{(0)} = B$$

Hence, (7.17) becomes

$$\mathbf{J}_x(z) = J_o e^{\frac{z}{\delta}}e^{j\frac{z}{\delta}} \tag{7.18}$$

From (7.18) we see that the current density decreases exponentially as one goes deeper into the conductor. (Remember that z is negative.) The significance of the parameter δ, the skin depth, is that at this depth, the current density has decreased to $1/e$ of its value at the surface. For copper, $\sigma_E = 5.9 \times 10^7$ (ohm meter)$^{-1}$ and $\mu = \mu_o$ (as is true for most materials other than ferromagnetics). Thus $\delta = 0.065/\sqrt{f}$ m, where f is in hertz. The dependence of δ on frequency for copper is shown in Fig. 7.2. Values of conductivity and skin depths for some other materials are given in Table 7.1. These values apply at radio and microwave frequencies, typically up to

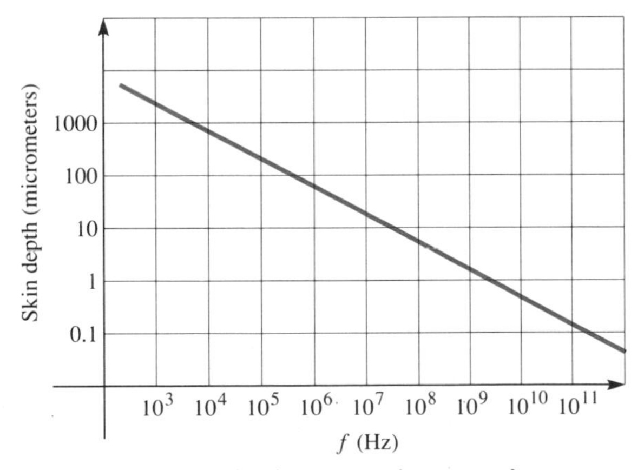

Figure 7.2 Skin depth δ versus frequency for copper.

Table 7.1 SKIN DEPTH AND SURFACE RESISTANCE FOR METALS*

$$\delta = \frac{1}{\sqrt{\pi\mu\sigma_E f}} = \frac{A}{\sqrt{f}} \qquad R_S = \sqrt{\frac{\pi\mu f}{\sigma_E}} = B\sqrt{f}$$

	σ_E $(\Omega m)^{-1}$	A	B
Silver	$6.81 \cdot 10^7$	$6.10 \cdot 10^{-2}$	$2.41 \cdot 10^{-7}$
Copper	$5.91 \cdot 10^7$	$6.55 \cdot 10^{-2}$	$2.59 \cdot 10^{-7}$
Gold	$4.10 \cdot 10^7$	$7.86 \cdot 10^{-2}$	$3.10 \cdot 10^{-7}$
Aluminum	$3.54 \cdot 10^7$	$8.46 \cdot 10^{-2}$	$3.34 \cdot 10^{-7}$
Iron	$1.02 \cdot 10^7$	$1.58 \cdot 10^{-1}$	$6.22 \cdot 10^{-7}$

*All quantities are in mks units (δ in meters, R_S in ohms, f in hertz). The value of μ is assumed equal to μ_o. Values of σ_E, taken from the *American Institute of Physics Handbook*, 2nd ed. (New York: McGraw-Hill, 1963) are for pure single crystals. Values for thin films will be somewhat smaller.

around 10^{11} Hz. At higher frequencies, the values of σ_E and $\omega\epsilon$ are no longer constant; they vary with frequency. Furthermore, as frequency increases, the assumption $\sigma_E \gg \omega\epsilon$ may no longer be justified. Thus, the theory developed here may require modification for use in the infrared or optical regions of the spectrum.

Not only does the amplitude of the current density decrease as depth increases; its *phase* changes as well. From the last factor in (7.18) we see that the phase angle of the current density decreases linearly with depth. In fact, when the depth reaches $\pi\delta$, the phase angle is 180°, and the direction of the current is opposite to that at the surface! (Of course by this depth the magnitude has gotten rather small: 4.3% of its value at the surface.) The total current per unit width is

$$\begin{aligned} \mathbf{I}_w &= \int_{-\infty}^{0} \mathbf{J}_x \, dz \\ &= \int_{-\infty}^{0} J_o e^{\frac{(1+j)z}{\delta}} \, dz \\ &= \frac{J_o \delta}{1 + j} \end{aligned} \tag{7.19}$$

The significance of the factor $(1 + j)$ is that the total current is 45° out of phase with the current density at the surface.

EXERCISE 7.3

Refer to the coordinate system of Fig. 7.1.

(a) The current density at the surface is J_o amperes per unit *area*. In what plane is the *area* referred to?

(b) The current per unit width is $\mathbf{I}_w$ amperes per *meter*. Along which axis are the "*meters*" measured?
(c) At an instant of time when the current at the surface is maximum in the x direction, what is the magnitude and direction of current at a depth $\pi\delta/4$?
(d) At $\pi\delta/2$?

Answers (a) The y-z plane; (b) the y axis; (c) $e^{-\frac{\pi}{4}} J_o/\sqrt{2}$ in the x direction; (d) no direction: it equals zero at that instant of time.

The current density at the surface, J_o, is related to the x-directed electric field at the surface $\mathbf{E}_o$ by $J_o = \sigma_E \mathbf{E}_o$. Thus the current per unit width $\mathbf{I}_w$ is related to the surface field by

$$\mathbf{I}_w = \left(\frac{\sigma_E \delta}{1 + j}\right) \mathbf{E}_o$$

Let us define a *surface impedance* Z_S according to

$$\mathbf{I}_w = \frac{\mathbf{E}_o}{Z_S} \tag{7.20}$$

where evidently

$$Z_S = \frac{(1 + j)}{\sigma_E \delta} = (1 + j)\left(\frac{\mu\omega}{2\sigma_E}\right)^{1/2} \tag{7.21}$$

In words, (7.20) states that *the current per unit width is equal to the voltage per unit length divided by the surface impedance*. The situation is illustrated in Fig. 7.3, where we are looking down on the metal surface from the air side, along the z axis. (That is, the plane of the page is the surface of the metal.) Let the potential difference $\mathbf{E}_o l$ that exists over length l be called $\mathbf{V}$; let the current that flows inside width w be called $\mathbf{I}$. Then

$$\frac{\mathbf{V}}{\mathbf{I}} = Z_S\left(\frac{l}{w}\right) \tag{7.22}$$

From (7.22) it is seen that the dimensions of Z_S are *ohms*. We also observe that if the surface area shown in Fig. 7.3 is a *square* (that is, if $l = w$), then the ratio $\mathbf{V}/\mathbf{I}$ is simply equal to Z_S. For this reason, the surface impedance Z_S is often said to have units of "ohms per square." It is interesting that for any square region, regardless of its size, the ratio of potential difference to current is simply Z_S. (The notion of surface impedance is not confined to the skin effect. For example, a sheet of resistive carbon paper would also be described in terms of its surface resistance, in ohms per square.)

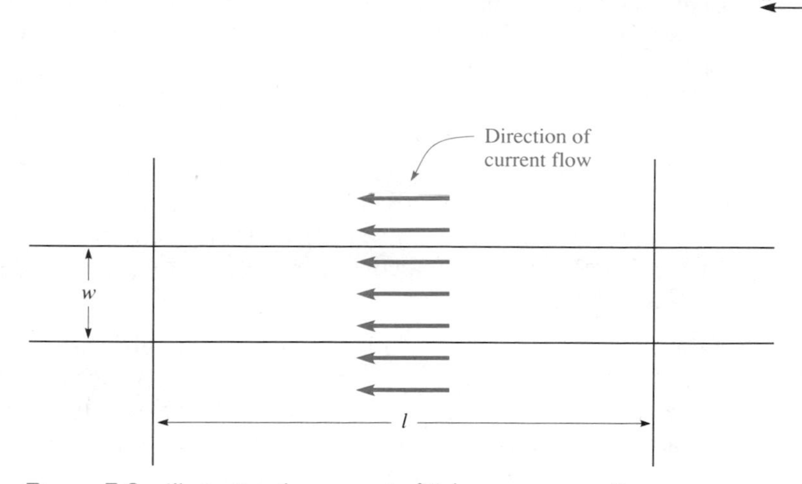

Figure 7.3 Illustrating the concept of "ohms per square."

The fact that Z_S is complex indicates that there is a phase difference between potential difference and current. We can separate Z_S into its real and imaginary parts, just as one would do in circuit analysis:

$$Z_S = R_S + jX_S \tag{7.23}$$

Here R_S is the *surface resistance* and X_S is the *surface reactance,* both in units of ohms per square. From (7.21), however, we see that $X_S = R_S$. Since X_S is positive, we can picture the surface impedance as being composed of a resistance in series with an inductance. Formulas for R_S are given in Table 7.1, and its behavior as a function of frequency, from (7.21) and (7.16), is shown in Fig. 7.4, for the case of copper. As frequency increases, the skin depth decreases in proportion to $1/\sqrt{f}$. The current is thus carried in a layer

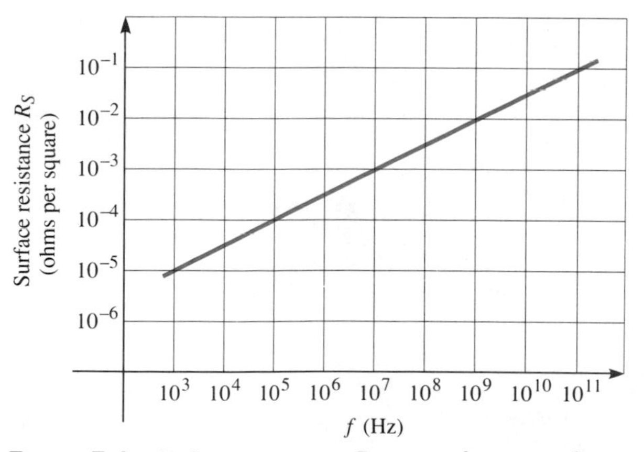

Figure 7.4 Surface resistance R_S versus frequency for copper.

of conductor that becomes thinner; meanwhile the surface resistance increases as $\sqrt{f}$. Although the numerical values of R_S appear small, surface resistance is a major cause of loss in microwave circuits.

EXAMPLE 7.3

A cylindrical copper wire-bond with diameter 0.1 mm and length 1 cm is used to connect two points in a planar circuit operating at 10 GHz, as shown in Fig. 7.5(a). Find the impedance introduced by this wire as a result of its surface impedance.

Solution A fundamental difficulty appears to be present. We are asked to find the impedance of a cylindrical conductor, which should logically be done using cylindrical coordinates. So far we have only obtained results for a planar conductor. However, the planar results suffice to obtain an accurate answer in this case. Because of skin effect, the actual current flows only in a thin outer layer of the wire; the thickness of the current layer is on the order of δ, which is 0.6 μm (Fig. 7.5(b)). Because the current layer is so thin compared with the thickness of the wire, the fact that the surface is slightly curved is rather unimportant. In fact, we can neglect this curvature and treat the conducting layer as though it were planar, as shown in Fig. 7.5(c). The effective width of the planar conductor is equal to the circumference of the wire, which is $\pi \times 100$ μm. The length of the conductor (measured perpendicular to the page in Figs. 7.5(b) and (c)) is 1 cm, the same as the actual wire. From (7.21), $Z_S = (1 + j) \times 0.026$ ohms per square. Then, using (7.22), we find that

$$Z = Z_S \left(\frac{10^4 \ \mu\text{m}}{\pi \times 100 \ \mu\text{m}} \right) = 0.86 + 0.86j \text{ ohms}$$

Note that the success of this method depends on the wire's diameter being much larger than the skin depth. If this were not the case, the fields would extend to the center of the wire, and it would not behave like an infinitely thick slab. In that case, the entire calculation would have to be redone, going back to (7.12) and expressing the vector Laplacian operator in cylindrical coordinates instead of rectangular.[4]

INTERNAL INDUCTANCE AND TOTAL INDUCTANCE

The inductance calculated in Chapter 6, arising from magnetic flux external to the wire, is what is known as *external inductance*. In this chapter, we have seen that the surface impedance also contains an inductive term; this

[4] The solution in cylindrical coordinates is quite interesting, but beyond the scope of this book. See, for example, S. Ramo, J.R. Whinnery, and T. Van Duzer, *Fields and Waves in Communication Electronics*. New York: Wiley, 1984, pp. 180–83.

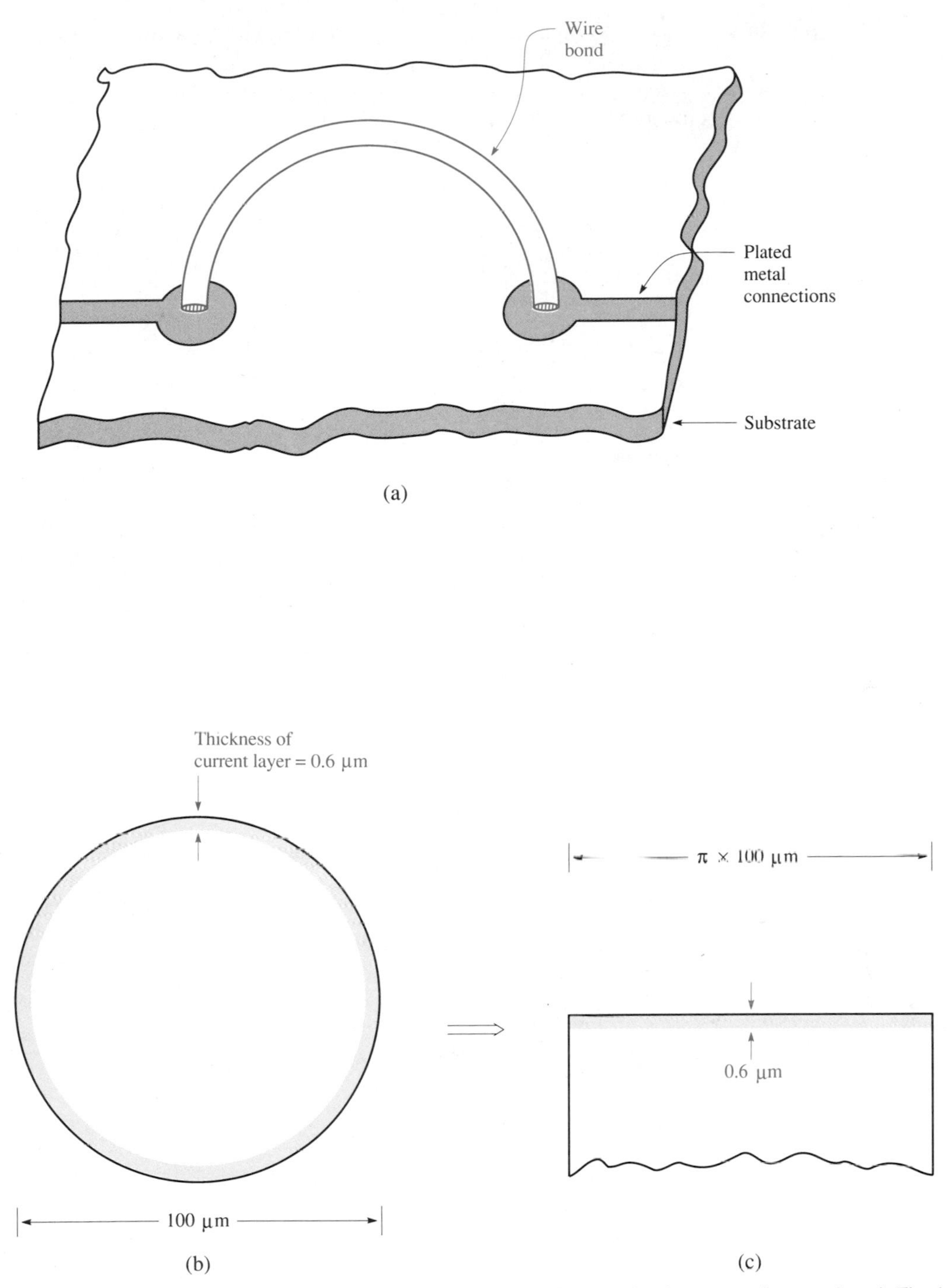

Figure 7.5 Finding the resistance and internal inductance of a wire bond. The bond, a length of wire connecting two pads on an IC, is shown in (a). (b) is a cross-sectional view showing skin depth. In (c), we imagine the conducting layer unfolded into a plane.

contributes what is known as *internal inductance*. The total inductance is the sum of the two, as will now be shown.

Figure 7.6 shows an inductor with two external terminals A and B. Consider a closed path of integration from B to A, then around the inductor just outside the metal and back to B. Writing Faraday's law for this path we have

$$\oint \vec{\mathbf{E}} \cdot \vec{dl} = -j\omega \int_S \vec{\mathbf{B}} \cdot \vec{dS} \tag{7.24}$$

or

$$\mathbf{V}_{BA} + \int_A^B \vec{\mathbf{E}}_T \cdot \vec{dl} = -j\omega\mathbf{\Phi} \tag{7.25}$$

where $\mathbf{V}_{BA} = \int_B^A \vec{\mathbf{E}}_1 \cdot \vec{dl}$ and $\vec{\mathbf{E}}_T$ is the tangential field along the outside of the wire. Note that $\mathbf{\Phi}$, as defined here, involves only the magnetic field that is external to the wire. Thus in analogy to (6.21) we define $\mathbf{\Phi} = L_{\text{ext}}\mathbf{I}$, where $\mathbf{I}$ is the current through the inductor and L_{ext} is the external inductance. From (7.20) and (7.21), $|\vec{\mathbf{E}}_T| = Z_S I_w$. If l is the length of the inductor and w is its circumference, we have

$$\mathbf{V}_{BA} + \frac{Z_S \mathbf{I} l}{w} = -j\omega L_{\text{ext}}\mathbf{I} \tag{7.26}$$

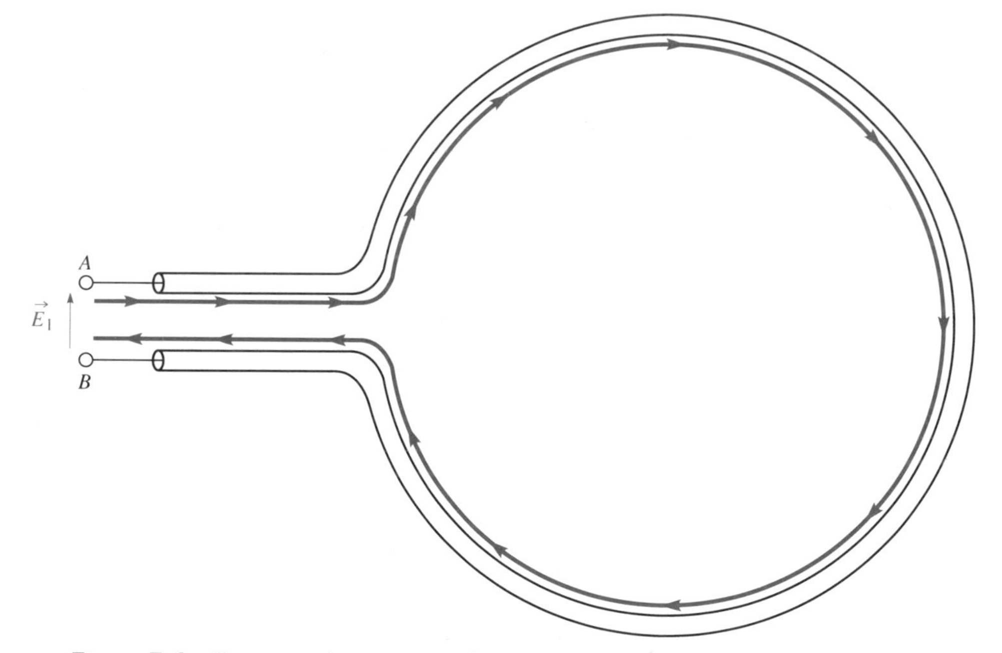

Figure 7.6 Illustrating the meaning of "internal inductance."

Let us define the internal impedance Z_{int} by

$$Z_{int} = Z_S \frac{l}{w} = R_{int} + j\omega L_{int} \tag{7.27}$$

Then we obtain the result

$$\mathbf{V}_{AB} = j\omega(L_{int} + L_{ext})\mathbf{I} + R_{int}\mathbf{I} \tag{7.28}$$

In practice, L_{int} is usually small in comparison with L_{ext}, but R_{int} is important because it causes ohmic loss.

EXERCISE 7.4

An iron-core inductor like that in Fig. 5.15 operates at 100 kHz. It consists of 50 turns of copper wire with a radius of 2 mm. The total length of wire is 15.7 m, the cross-sectional area of the iron core is 0.785 cm^2, its relative permeability is 5000, and the length of the flux path is 11 cm. Estimate the external inductance, internal inductance, and resistance.

Answers $L_{ext} = 11.2$ mH; $L_{int} = 1.64 \times 10^{-7}$ H; $R_{int} = 0.103\ \Omega$.

SKIN-EFFECT LOSSES

If the phasor voltage across a circuit element is $\mathbf{V}$ and the current through it is $\mathbf{I}$, then clearly the time-averaged power that is dissipated is $\frac{1}{2}$ Re $\mathbf{VI}^*$. However, in many electromagnetic problems, the structure in question is comparable in size to wavelength, making voltage difficult to define. Thus it is often convenient to express the loss per unit area in terms of the fields existing outside the conductor at any given point.

Let us again consider the small region of the conductor surface shown in Fig. 7.3. From (7.20) and (7.21), the current per unit width is $\mathbf{E}_o/Z_S$, where $\mathbf{E}_o$ is the tangential electric field. When the field forces the current through a portion of the surface with length l and width w, the time-averaged power that is dissipated is

$$\begin{aligned} P &= \frac{1}{2} \text{Re } (\mathbf{E}_o l)(\mathbf{I}_w w)^* \\ &= \frac{1}{2} |\mathbf{E}_o|^2 (lw) \text{ Re} \left(\frac{1}{Z_S^*}\right) \\ &= \frac{1}{2} |\mathbf{E}_o|^2 A \text{ Re} \left(\frac{1}{Z_S}\right) \end{aligned} \tag{7.29}$$

where $A = lw$ is the area, and we have used the fact that $\text{Re}(1/Z_S^*) = \text{Re}(1/Z_S)$. We now evaluate expression (7.29) by means of some algebra.

Making use of the fact that $X_S = R_S$, we obtain the following expression for the power dissipated per unit area

$$P_A = \frac{|\mathbf{E}_o|^2}{4R_S} \tag{7.30}$$

where $\mathbf{E}_o$ is the tangential electric field at the surface. (Note that the algebra leading to (7.30) is not as obvious as it may seem at first. Where did the 4 in the denominator of (7.30) come from?)

It is often more useful to find the power dissipated in terms of $\mathbf{I}_w$, the current per unit width. Beginning again from (7.29), we have

$$P = \tfrac{1}{2}\,\mathrm{Re}(\mathbf{E}_o l)(\mathbf{I}_w w)^*$$

Substituting (7.20), we find that the power dissipated per unit surface area is

$$P_A = \tfrac{1}{2}\,R_S|\mathbf{I}_w|^2 \tag{7.31}$$

Equation (7.31) can be used to find the power loss in terms of the *magnetic* field at the surface. The tangential component of the magnetic field at the surface is related to $\mathbf{I}_w$ in a simple way, as shown in Fig. 7.7. In this

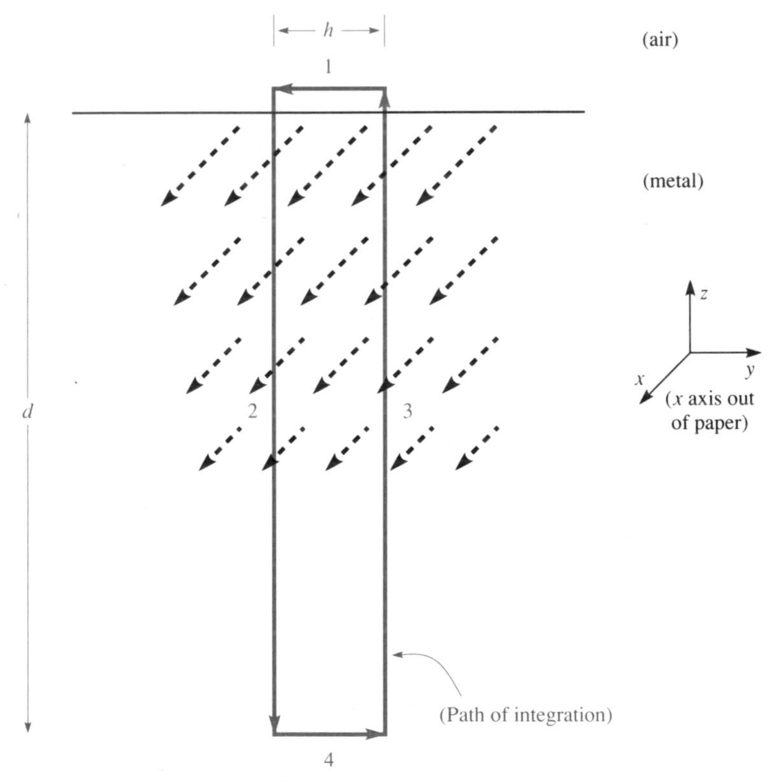

Figure 7.7 The current flowing per unit width is related to the tangential magnetic field at the surface.

figure the current density J_x points out of the paper, and decreases exponentially with depth in the metal. Let us choose a path of integration in the y-z plane as shown, and apply Ampère's law:

$$\oint \vec{\mathbf{H}} \cdot dl = \int_S (\mathbf{J}_x + j\omega\epsilon\mathbf{E}_x)\, dA \tag{7.32}$$

Because of our high-conductivity assumption $\sigma_E \gg \omega\epsilon$, the displacement current $j\omega\epsilon\mathbf{E}_x$ can be neglected in comparison with J_x. Furthermore,

$$\int_S \mathbf{J}_x\, dA = \mathbf{I}_w h$$

where h is the width of the path, as shown in the figure. Now in evaluating the line integral on the left side of (7.32), we let d, the length of the integration path, be much larger than δ; thus the fields have died away to nothing at the depth of segment 4, and that segment contributes nothing to the line integral. We shall assume that the magnetic field has only a y component; hence segments 2 and 3 also do not contribute. Thus, (7.32) becomes $-H_y h = \mathbf{I}_w h$, or

$$-H_y = \mathbf{I}_w \tag{7.33}$$

We can generalize this result to the case of magnetic field in any tangential direction:

$$\vec{\mathbf{I}}_w = \vec{n} \times \vec{\mathbf{H}} \tag{7.34}$$

where $\vec{n}$ is the outward normal unit vector of the metal surface. From (7.31) and (7.34) we then have

$$P_A = \tfrac{1}{2} R_S\, |\vec{\mathbf{H}}_T|^2 \tag{7.35}$$

where $\vec{\mathbf{H}}_T$ is the tangential magnetic field at the surface.

EXAMPLE 7.4

A certain waveguide consists of a hollow cylindrical pipe with copper walls. Its radius a is 2 cm. The magnetic field inside it is (in cylindrical coordinates)

$$\vec{\mathbf{H}} = H_o f(r) e^{-jkz}\, \vec{e}_\phi$$

where $H_o = 100$ mks units. The function of radius $f(r)$ has the property that $f(a) = 0.519$, and the frequency is 7 GHz. Find the power dissipated in the walls of the waveguide, per unit length.

Solution From (7.34), $\mathbf{I}_w$ is in the z direction, and is given by

$$\vec{\mathbf{I}}_w = -H_o f(a) e^{(-jkz)}\, \vec{e}_z$$

(The minus sign appears because the outward normal from the metal is in the *minus-r* direction.) Then from (7.31), the power dissipated per unit length is

$$P_L = \tfrac{1}{2} R_S H_0^2[f(a)]^2(2\pi a)$$
$$= \tfrac{1}{2}(2.18 \times 10^{-2})(10^4)(0.269)(2\pi \times 0.02)$$
$$= 2.97 \text{ watts/m}$$

The same result could also have been obtained directly from (7.35).

BOUNDARY CONDITIONS FOR GOOD CONDUCTORS

We have seen that at the surface of a *perfect* conductor (one with $\sigma_E = \infty$), the tangential component of $\vec{E}$ must vanish. For real conductors with high conductivity, the statement $E_{\text{tan}} = 0$ is not precisely true, but it is usually a good approximation. If E_{tan} had a significant value, the result would be an extremely large current density. Such a current, if set up, would transport charges to new positions, where they would set up new fields that would oppose E_{tan} and tend to cancel it out. Thus $\vec{E}$ is approximately normal to the surface of a good conductor.

From similar arguments we can see that sinusoidal magnetic fields are approximately *tangential* to the surface of a good conductor. From Maxwell's $\nabla \times \vec{E}$ equation we have

$$\oint \vec{\mathbf{E}} \cdot \vec{dl} = -j\mu\omega \int_S \vec{\mathbf{H}} \cdot \vec{dS} \qquad \textbf{(7.36)}$$

Let the path of line integration be a small circle on the surface of a good conductor. Then the right-hand side of (7.36) becomes, approximately, $-j\mu\omega A\,\mathbf{H}_N$, where A is the area of the circle and $\mathbf{H}_N$ is the normal component of magnetic field. However, the line integral of $\vec{\mathbf{E}}$ on the left side of (7.36) involves the tangential component of $\vec{\mathbf{E}}$, which we have shown must be small. Thus $\mathbf{H}_N$ must be nearly zero as well.

7.4 ELECTROMAGNETIC WAVES

The results of the previous section, dealing with skin effect, were based on the assumption that $\sigma_E \gg \omega\epsilon$; that is, we were considering fields inside a good conductor. Now we shall consider the opposite case, that of $\sigma_E \ll \omega\epsilon$. We shall now be dealing with fields inside insulators or in air or vacuum.

Accordingly, let us begin again from (7.9). This time we neglect the conduction current $\vec{J}$, which is now assumed to be much less than the displacement current, and have

$$\nabla^2\vec{\mathbf{E}} + \omega^2\mu\epsilon\vec{\mathbf{E}} = 0 \qquad \textbf{(7.37)}$$

Equation (7.37) is the *vector Helmholtz equation*. The Helmholtz equation appears simple, but it has a rich variety of solutions. We note that *all* sinusoidal electric fields in nonconducting space—light, waves in waveguides, waves transmitted by antennas—of necessity must be solutions of (7.37).

THE UNIFORM PLANE WAVE

To see the nature of these solutions, let us first restrict ourselves to a particular special case in which $\vec{\mathbf{E}}$, by assumption, has only an x component and varies only in the z direction. In that case, (7.37) simplifies to

$$\frac{\partial^2 \mathbf{E}_x}{\partial z^2} + \omega^2 \mu\epsilon \mathbf{E}_x = 0 \tag{7.38}$$

Because of our specific assumptions, the vector Helmholtz equation has now been transformed into a scalar Helmholtz equation. It is identical (assuming sinusoidal time variation) with the wave equation (1.9), which describes the voltage on a lossless transmission line. Thus, its solutions must be similar to transmission-line waves. By analogy with the transmission-line waves (1.20) and (1.21), the Helmholtz equation must have solutions

$$\mathbf{E}_x = E_o e^{\pm j\omega\sqrt{\mu\epsilon}\, z} \tag{7.39}$$

These solutions can be verified by substitution. $E_o = A_o e^{j\phi}$ is an arbitrary complex constant specifying the wave's amplitude and phase; the minus and plus signs in the exponent correspond to waves moving in the $+z$ and $-z$ directions respectively. Expressed as a function of time, the field of the wave going in the $+z$ direction is

$$\begin{aligned} E_x(z, t) &= \text{Re}\,[E_o e^{-j\omega\sqrt{\mu\epsilon}\, z} e^{j\omega t}] \\ &= A_o \cos(\omega t - \omega\sqrt{\mu\epsilon}\, z + \phi_o) \end{aligned} \tag{7.40}$$

The position of a field maximum z_{max} is given by $\omega t - \omega\sqrt{\mu\epsilon}\, z_{\text{max}} + \phi_o = 0$. Thus the phase velocity is given by

$$U_P = \frac{dz_{\text{max}}}{dt} = \frac{1}{\sqrt{\mu\epsilon}} \tag{7.41}$$

When $\mu = \mu_o$ and $\epsilon = \epsilon_o$, as in vacuum, the phase velocity of the waves is c, the *velocity of light*, with the numerical value[5]

$$c = \frac{1}{\sqrt{\mu_o \epsilon_o}} = 2.99793 \times 10^8 \text{ m/sec} \tag{7.42}$$

---

[5]This is an experimental value, based on several different experiments and cited by G. H. Dieke in *The American Institute of Physics Handbook*, 2nd ed. (New York: McGraw-Hill, 1963). The fact that c is so close to being an integer times 10^8 makes the number easy to remember, but of course is only a coincidence.

The wave described by (7.40) is known as a *uniform plane wave*. This solution corresponds to a wave moving in the $+z$ direction. By choosing the plus sign in (7.39) we could have obtained a similar wave

$$E_x(z, t) = A_1 \cos(\omega t + \omega\sqrt{\mu\epsilon}\, z + \phi_1) \tag{7.43}$$

moving in the $-z$ direction; this is also a uniform plane-wave solution of (7.37).

In Chapter 1 we described transmission-line waves in terms of a constant k, known as the propagation constant. In terms of k the positive-going wave (7.40) is

$$E_x(z, t) = A_o \cos(\omega t - kz + \phi_o) \tag{7.44}$$

Comparing with (7.40) we see that for this wave k is equal to k_o, defined by

$$k_o = \omega\sqrt{\mu\epsilon} \tag{7.45}$$

Note that $k = k_o$ is a property of this particular kind of wave, the plane wave. All waves propagating in the $+z$ direction have fields proportional to e^{-jkz}, but only certain waves have $k = k_o$. From (1.53) we see that these are the same waves that have phase velocity $(\mu\epsilon)^{-1/2}$; in other words, these are the waves that travel at the speed of light in the medium. In Chapter 9 waves that have different phase velocity will be encountered.

Uniform plane waves can be visualized in just the same way as the transmission-line waves of Chapters 1 and 2. The positive-going wave, for example, might be represented by a "snapshot" of it, taken at time $t = 0$. This snapshot is shown in Fig. 7.8(a). (The special case shown is for $A_o = 1.5$ V/m, $\phi_o = 45°$.) Seen "frozen" by the snapshot at time $t = 0$, the dependence of E_x on position, as given by (7.44) is as shown in Fig. 7.8(a). Now if we re-graph (7.44) for a slightly later time t_1, such that $\omega t_1 = \pi/2$, we get the result shown as the solid line in Fig. 7.8(b). This corresponds to another snapshot of the wave, taken at the time $t_1 = \pi/2\omega$. The snapshot at $t = 0$ is shown as a dashed line for comparison. Note how the wave appears to move as a rigid structure to the right.

Each field of the wave is proportional to e^{-jk_oz}. For instance, the phasor representing E_x might be $E_oe^{-jk_oz}$, where $E_o = A_oe^{j\phi_o}$ and A_o and ϕ_o are real constants. In that case the phasor representing the field would be $(A_oe^{j\phi_o})\, e^{-jk_oz}$. The amplitude of this phasor is A_o and the phase angle is $\phi_o - k_oz$. Let us now investigate the *equiphase surfaces,* those surfaces on which phase is constant. Evidently these are surfaces of constant z, which are planes. It is for this reason that we call the wave a plane wave. A plane wave is one for which the surfaces of constant phase are planes. A *uniform* plane wave is one in which the amplitude is the same at all points on an equiphase plane.

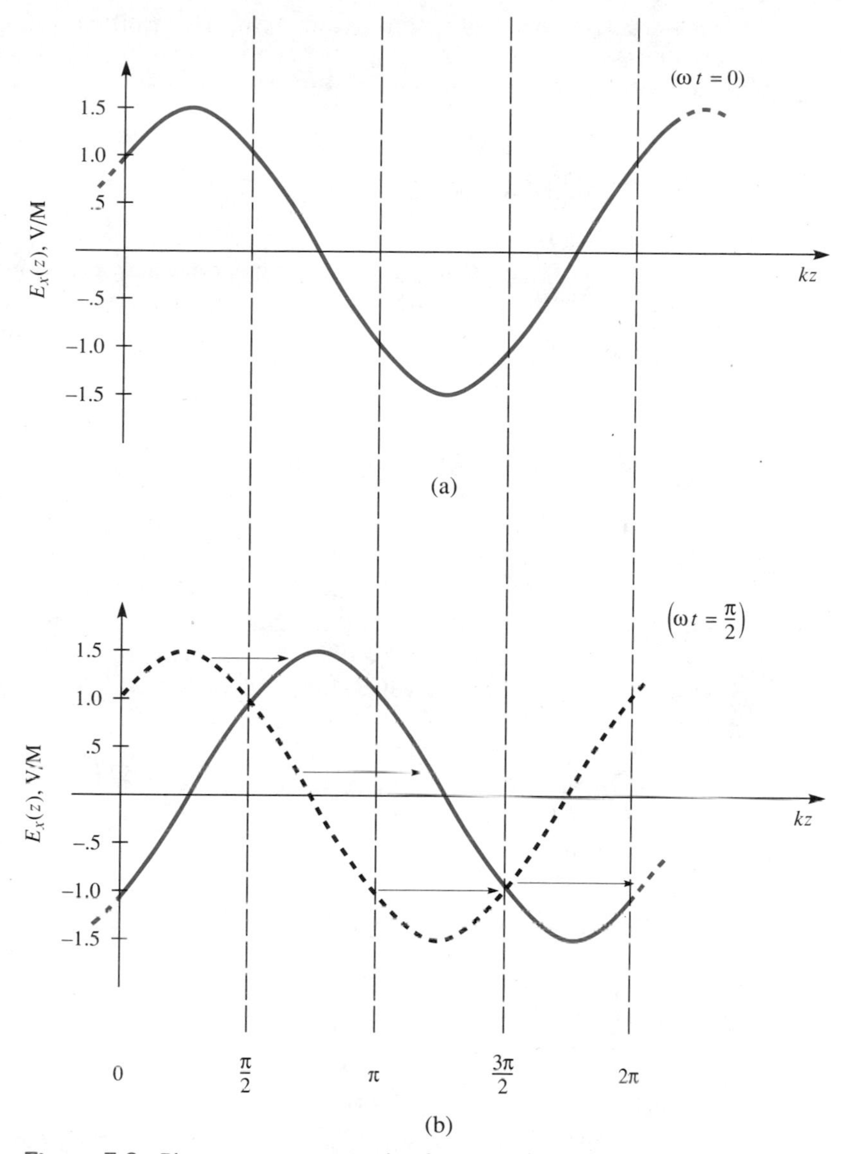

Figure 7.8 Plane waves move in the direction of propagation as a unit, just as transmission-line waves do.

EXERCISE 7.5

A positive-going plane wave in vacuum, described by (7.44), has frequency 1 GHz and phase angle $\phi_o = 60°$. Find (a) k, (b) the wavelength λ, and (c) the position of field maxima at time $t = 0.2$ nsec.

Answers (a) $k = 20.9 \text{ m}^{-1}$; (b) 0.30 m; (c) $0.11 \pm (0.30)N$ meters where N is any integer.

ELECTROMAGNETIC WAVES IN GENERAL

The preceding paragraphs dealt with an important special case: the solution of (7.37) under the assumption that $\vec{E}$ has only an x component and varies only in the z direction. The resulting solution, known as a plane wave, is important, and in fact will be the subject of the entire next chapter. However, we should note that most solutions of (7.37) will not satisfy these assumptions, and therefore are not uniform plane waves. Equation (7.37), the three-dimensional vector Helmholtz equation, has a great many other solutions.

In general, the most common problems in electromagnetics amount to finding solutions of (7.37) *subject to given boundary conditions*. Such problems can be called electrodynamic boundary-value problems. We have already encountered boundary-value problems in electrostatics: those involved finding solutions to Laplace's equation subject to given boundary conditions. In electrodynamics the situation is the same except that instead of Laplace's equation we are finding solutions of (7.37). Plane waves are an especially simple case, in which the boundaries (which must exist, because something produced the wave) are so far away that we do not see them. Suppose, however, that a metal guiding structure is present in the region of interest. In that case the fields will still be electromagnetic waves, but not necessarily *plane* electromagnetic waves. They will still be solutions of the Helmholtz equation, but they must be those solutions that will satisfy the given boundary conditions. For example, it is easy to show by substitution that the wave (in spherical coordinates)

$$\mathbf{E}_r = 0$$

$$\mathbf{E}_\theta = \frac{A}{r \sin\theta} e^{-jk_o r} \tag{7.46}$$

$$\mathbf{E}_\phi = 0$$

satisfies (7.37). (A is an arbitrary amplitude constant.) This example happens to be a wave on the metal biconical transmission line shown in Fig. 8.9 in Chapter 8. It exists only in the region $\theta_o < \theta < (\pi - \theta_o)$, where θ is the vertical angle of the transmission line. But although (7.46) is a solution of the Helmholtz equation, it is clearly not a plane wave; the surfaces of constant phase are not planes. Phase is kept constant by keeping r constant. Thus each equiphase surface is a portion of a sphere.

Other solutions of the Helmholtz equation may not be traveling waves at all. For example, a *cavity resonator* is a completely enclosed, hollow metal structure, inside which fields can be contained. Inside a rectangular resonator with perfectly conductive metal walls at $x = 0$ and a, $y = 0$ and

b, $z = 0$ and d, one solution of (7.37) is

$$\mathbf{E}_y = E_o \sin\frac{\pi x}{a} \sin\frac{\pi z}{d} \qquad \mathbf{E}_x = \mathbf{E}_z = 0 \tag{7.47}$$

provided the frequency is chosen correctly. (See Problem 7.37.) Again, this solution is not a uniform plane wave; on the contrary, it is more like a standing wave on a transmission line. We shall return to cavity resonators in Chapter 9. Here we only mean to illustrate the variety of solutions that the wave equation can have.

The reader may ask what role the magnetic field plays in these solutions of the wave equation. In electrodynamics the magnetic and electric fields are closely linked through (7.1) and (7.4). In fact, when the electric field is known, the magnetic field can immediately be found, using (7.4). Furthermore, our derivation of the wave equation was performed by eliminating $\vec{H}$ from (7.1) and (7.4). We could just as well have eliminated $\vec{E}$ from these two equations; this leads to a similar equation for $\vec{H}$,

$$\nabla^2\vec{H} + \omega^2\mu\epsilon\vec{H} = 0 \tag{7.48}$$

Thus $\vec{H}$ obeys the same equation as $\vec{E}$. Solutions for $\vec{H}$ will also be wavelike in nature.

7.5 THE QUASISTATIC APPROXIMATION

If we consider the limiting case as frequency approaches zero, (7.37) becomes

$$\nabla^2\vec{E} = 0 \tag{7.49}$$

The case of $\omega = 0$ corresponds to electrostatics. Thus electrostatic fields obey (7.49). For a given set of electrostatic boundary conditions, one can find the fields by solving (7.49), or alternatively, one can solve Laplace's equation for the potential, and then take the gradiant of potential to obtain $\vec{E}$. The results of the two methods will be identical.

Often we wish to obtain solutions for the Helmholtz equation in situations in which the frequency is *small* but *not actually zero*. Let us refer to this as the "low-frequency case." We expect that as frequency becomes small, the fields will come to resemble the electrostatic fields for the same boundaries, but this raises the question, how small does the frequency have to be? We shall assert the following: whenever the region under consideration is small compared with $\lambda/2\pi$, where λ is the wavelength of plane waves at the frequency in question, the solution of (7.37) resembles the solution of (7.49). In other words, when the above criterion is satisfied, the correct electrodynamic fields resemble electrostatic fields with the same boundaries.

Such fields are said to be *quasistatic*. In problems involving regions that are small compared to wavelength, one may obtain approximate solutions by solving Laplace's equation, which is simpler than the Helmholtz equation and thus easier to solve. This method is known as the *quasistatic approximation*.[6]

The theory of the quasistatic approximation cannot be discussed here in full detail. However, the reader may wish to know the general line of reasoning. In Chapter 10 it will be shown[7] that if the currents producing an electromagnetic field are specified, the magnetic vector potential is

$$\vec{\mathbf{A}}(\vec{r}) = \int \frac{\vec{\mathbf{J}}(\vec{r}')e^{-\omega\sqrt{\mu\epsilon}\,|\vec{r}-\vec{r}'|}\, dV'}{4\pi\,|\vec{r} - \vec{r}'|} \tag{7.50}$$

From this potential, the magnetic field can be found by taking the curl. We note that if $\omega\sqrt{\mu\epsilon}\,|\vec{r} - \vec{r}'| \ll 1$, which is the same as

$$|\vec{r} - \vec{r}'| \ll \frac{\lambda}{2\pi}$$

then

$$\vec{\mathbf{A}}(\vec{r}') \rightarrow \frac{\mu}{4\pi} \int \frac{\vec{\mathbf{J}}(\vec{r}')\, dV'}{|\vec{r} - \vec{r}'|} \tag{7.52}$$

which is the same as the magnetostatic formula (5.10). We note that since (7.52) represents a magnetostatic field, the corresponding magnetic field satisfies $\nabla \times \vec{\mathbf{H}} = 0$. Meanwhile the electric field satisfies

$$\nabla \cdot \vec{\mathbf{E}} = 0 \qquad \text{and} \qquad \nabla \times \vec{\mathbf{E}} = -j\omega\mu\vec{\mathbf{H}}$$

Hence

$$\begin{aligned} \nabla^2\vec{\mathbf{E}} &= \nabla(\nabla \cdot \vec{\mathbf{E}}) - \nabla \times (\nabla \times \vec{\mathbf{E}}) \\ &= 0 \end{aligned} \tag{7.53}$$

Thus $\vec{\mathbf{E}}$ obeys the same equations as the electrostatic field.

As an example, consider a plane electromagnetic wave incident on a small object, as shown in Fig. 7.9(a). The object is small in comparison with $\lambda/2\pi$. Most of the wave goes right on by, but as it does it induces some charges and currents on the object. These in turn give rise to additional fields, known as *scattered fields*. The total field consists of the incident field plus the scattered field. At distances r from the scatterer such that $r \ll \lambda/2\pi$, the fields resemble electrostatic fields. We could find the fields in the quasistatic approximation by solving Laplace's equation. Figure 7.9(b) shows

[6] L. C. Shen and J. A. Kong, *Applied Electromagnetism*. Belmont, Calif.: Brooks-Cole Publishing Co., 1983.
[7] See also: S. Ramo, J. R. Whinnery, and T. Van Duzer, "Fields and Waves in Communication Electronics," 2nd ed. New York: Wiley, 1984, section 3.21.

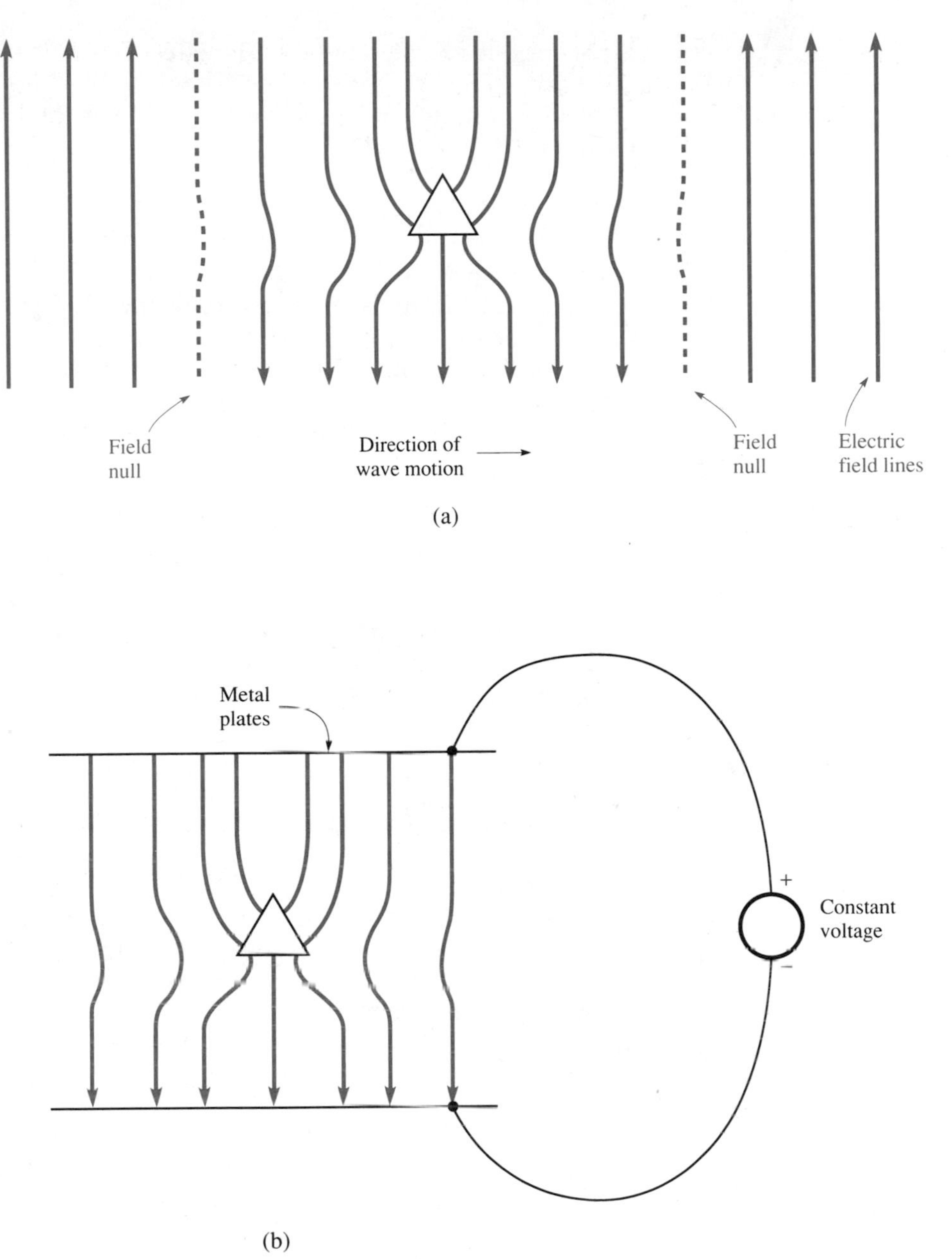

Figure 7.9 Using the quasistatic approximation to study scattering of electromagnetic waves by an object much smaller than the wavelength.

the corresponding electrostatic model. The effect of the incoming wave is modeled by placing the object between the plates of a parallel-plate capacitor, to which a constant voltage is applied. The fields close to the scatterer should have the same shape in (b) as in (a).

LUMPED-ELEMENT CIRCUITS

An important consequence of the quasistatic approximation is that it provides a connection between electromagnetics and ordinary circuit theory. In circuit work, we usually think of circuit elements, such as resistors or inductors, as "lumped." A lumped-circuit element is one whose internal workings can be ignored; lumped-circuit elements are completely described by their terminal voltages and currents, and the relationships between them. For instance, a conventional lumped inductor is completely described by the equation $v = L\frac{di}{dt}$, once the value of L is given. Lumped circuit analysis is used routinely up to frequencies of hundreds of megahertz, and even higher. Yet we have seen that the term "voltage" is really meaningful only in electrostatics. In electrodynamics, on the contrary, the line integral of electric field depends on what path of integration is chosen, and the notion of "potential difference" loses its meaning. Thus we seem to have a paradox. When we describe high-frequency circuits in terms of voltages, what on earth are we really doing?

The answer is provided by the quasistatic approximation. A circuit may be treated as lumped and described in terms of voltages and currents if its size is small compared with a wavelength at the frequency of operation. At 1 GHz, the wavelength is 30 cm; thus even at this high frequency, the internal operation of an integrated circuit can safely be described by Kirchhoff's laws. However, if the integrated circuit is to be connected to another one 5 cm away, the interconnecting wires would not be small compared with $\lambda/2\pi$. The interconnection must be described in terms of electrodynamics, not electrostatics. If it consists of two parallel wires, it can be treated as a transmission line.

7.6 POINTS TO REMEMBER

1. The vector Laplacian operator is represented by the same symbol (∇^2) as the scalar Laplacian operator, but it is a different operator, because it acts on a vector field to give another vector field. One can tell by the argument of the operator which of the two is meant.

2. When time-varying fields are present in a material that has high conductivity, it is found that the fields and the current are confined to a region near the surface of the material. This is known as the "skin

effect." It causes high-frequency currents in a wire to avoid the interior of the wire, flowing mainly in a thin region near the surface. The thickness of this region is called the "skin depth." Skin effect increases the effective resistance of conductors at high frequencies.

3. The ratio of the electric field at the surface to the current that flows in the skin-effect layer, per unit width, is known as the "surface impedance." The surface impedance is useful for calculating the resistance of metal components and for finding resistive power loss.

4. These are the boundary conditions at the surface of a good conductor: the tangential component of $\vec{E}$ is nearly zero, and the normal component of $\vec{H}$ is nearly zero.

5. In nonconductive materials and in vacuum, sinusoidal electric fields obey the Helmholtz equation. The solutions of this equation are electromagnetic waves. One especially important example is the uniform plane wave.

6. At very low frequencies, electromagnetic fields begin to resemble static fields. Such fields are said to be "quasistatic." The quasistatic assumption is important because it underlies ordinary circuit analysis. (For example, one could not speak of "voltage" in an ac circuit unless one had confidence that the fields resemble static fields.) The criterion for making the quasistatic approximation is that the free-space wavelength at the frequency of interest must greatly exceed the size of the system in question.

REFERENCES

1. S. Ramo, J. R. Whinnery, and T. Van Duzer, *Fields and Waves in Communication Electronics*. New York: Wiley, 1984.
2. D. K. Cheng, *Field and Wave Electromagnetics*. Reading, Mass.: Addison-Wesley, 1983.
3. W. H. Hayt, Jr., *Engineering Electromagnetics*, 4th ed. New York: McGraw-Hill, 1981.

The quasistatic approximation is discussed in

4. N. N. Rao, *Elements of Engineering Electromagnetics*, 2nd ed. Englewood Cliffs, N.J.: Prentice-Hall, 1987.

and more extensively in

5. D. T. Paris and F. K. Hurd, *Basic Electromagnetic Theory*. New York: McGraw-Hill, 1969.
6. L. C. Shen and J. A. Kong, *Applied Electromagnetism*. Belmont, Calif.: Brooks-Cole Publishing Co., 1983.

The mathematical and physical fundamentals of electromagnetic theory are treated at a higher level in

7. R. F. Harrington, *Time-Harmonic Electromagnetic Fields*. New York: McGraw-Hill, 1961.

PROBLEMS

Section 7.1

7.1 State whether the following are scalars or vectors, whether they are real or complex, and whether or not they are functions of time:
(a) electrostatic potential V
(b) $\mathbf{v}(z)$, the phasor representing the voltage on a transmission line
(c) $\vec{E}$, the electric field of an electromagnetic wave
(d) $\vec{\mathbf{H}}$, the phasor representing the sinusoidal magnetic field of an electromagnetic wave.

7.2 Find the phasor representing the electric field

$$\vec{E}(x,y,z,t) = [Ax^2 \cos(\omega t + 30°)]\vec{e}_y - [Bz \sin \omega t]\vec{e}_x$$

(A and B are given constants.)

7.3 Find the phasor representing the magnetic field

$$\vec{H}(x,y,z,t) = A \cos(\omega t - kz)\vec{e}_x + B \sin(\omega t + kz)\vec{e}_y$$

7.4 A sinusoidal electric field is described by the phasor $\vec{\mathbf{E}} = A_o e^{jkx}\vec{e}_x + B_o e^{j\phi} e^{-jkx}\vec{e}_y$, where A_o, B_o, k, and ϕ are given real constants. Find $\vec{E}(x,y,z,t)$. = $A_o\cos(\omega t - kx)\vec{e}_y + B_o\cos(\omega t - kx + \phi)\vec{e}_y$

*7.5 An electric field is represented by the phasor $\vec{\mathbf{E}} = E_o(\vec{e}_x + j\vec{e}_y)$, where E_o is a given real constant.
(a) Find the vector $\vec{E}$ as a function of time.
(b) Sketch the vector $\vec{E}$ in the x-y plane at times t_1, t_2, t_3, t_4, and t_5, where $\omega t_1 = 0$, $\omega t_2 = \pi/2$, $\omega t_3 = \pi$, $\omega t_4 = 3\pi/2$, and $\omega t_5 = 2\pi$.

**7.6 The vector $\vec{H}(r,\phi,z,t)$ in cylindrical coordinates is represented by the phasor

$$\vec{\mathbf{H}} = \left(\frac{r}{a}\right)[(2 + 3j)\vec{e}_r + (3 - 2j)\vec{e}_\phi]$$

where a is a constant.
(a) Find $\vec{H}(r,\phi,z,t)$.
(b) Make sketches of $\vec{H}(t)$, in the vicinity of $r = a$ for several different times during a period. (*Suggestion:* Eliminate sine functions and express the field entirely in terms of cosines.)

*7.7 Let the electric field in a certain nonconductive region be $\vec{E}(x,y,z,t) = \vec{f}(x,y,z)\cos[\omega t + \phi(x,y,z)]$, where $\vec{f}$ is a given vector-valued function of position and ϕ is a given scalar function of position. Prove that at any position, the magnetic field $\vec{H}(x,y,z,t)$ varies sinusoidally with frequency ω.

**7.8 Let the phasor representing the electric field in a certain region be $\vec{E}_o e^{-j\vec{k}\cdot\vec{r}}$, where $\vec{k}$ and $\vec{E}_o$ are real constant vectors and $\vec{r}$ is the radius vector.

(a) Find the equation describing a surface on which the phase angle is zero.
(b) Show that this surface is a plane through the origin perpendicular to $\vec{k}$.

Section 7.2

7.9 Find
(a) $\nabla^2(\sin kx)$ (where k is a given constant)
(b) $\nabla^2(\vec{e}_y \sin kx)$
(c) $\nabla^2 [E_o \cos(\omega t - kz)\vec{e}_x]$ (where E_o and k are given constants).

*7.10 Apply the vector Laplacian operator to the field (7.46). Show that this field obeys $\nabla^2\vec{\mathbf{E}} = -k_o^2\vec{\mathbf{E}}$.

7.11 Let $\vec{E}(x,y,z) = E_o[(y^2 - z^2)\vec{e}_x + (z^2 - x^2)\vec{e}_y + (x^2 - y^2)\vec{e}_z]$.
(a) Show that $\nabla^2\vec{E}$ is zero at the origin.
(b) Invent a different $\vec{E}(x,y,z)$ (preferably not too simple) for which $\nabla^2\vec{E}$ also vanishes at the origin.

*7.12 Suppose it is known that in a certain region there are no charges, and also the magnetic field does not vary in time. Prove that $\nabla^2\vec{E} = 0$ in this region.

Section 7.3

In Problems 7.14–7.31, assume $\sigma_E \gg \omega\epsilon$.

7.13 For our skin-effect theory to apply, it is necessary for conductivity to be sufficiently high. Determine whether or not (7.11) can be applied to
(a) a sample of silicon with $\epsilon = 11.7\epsilon_o$, $\sigma_E = 10$ (ohm cm)$^{-1}$ at $f =$ 10 GHz.
(b) a sample of silver with $\sigma_E = 330$ (ohm cm)$^{-1}$, $\epsilon = -8.6\epsilon_o$ at $f =$ 6×10^{14} GHz (visible light).

7.14 If one uses an electric conductor that is much thicker than the skin depth, conductive material is wasted, because no current flows in the interior of the conductor. Let us adopt $t_{MU} \cong 2\delta$ as an approximate value of the maximum useful thickness. Find t_{MU} for copper at
(a) 60 Hz
(b) 10 MHz
(c) 10 GHz.

7.15 The radius of a circular metal wire is a, and $a \gg \delta$, where δ is the skin depth at angular frequency ω. The phasor representing current density at the surface of the wire is J_o A/m^2.
(a) What is $\mathbf{I}_W$, the current per unit width?
(b) What is $\mathbf{I}$, the total current through the wire?

*7.16 Suppose a conductive material fills half of space, and its boundary with air passes through the origin and is perpendicular to the vector $2\vec{e}_x$ +

$\vec{e}_y$. The negative half of the x axis is inside the material. The phasor representing the current density at the surface is $J_o\,\vec{e}_z$. Obtain an expression for $\vec{\mathbf{J}}(x,y,z)$ inside the conductor.

**7.17 A good electric conductor fills the region $z < 0$. At the surface there is a sinusoidal electric field described by the phasor $E_o\vec{e}_x$, where E_o is a real number.
(a) Sketch $|\vec{E}(z,t)|$, the magnitude of the instantaneous electric field, versus z inside the conductor, at three different times, defined by $\omega t = 0$, $\omega t = \pi/2$, and $\omega t = \pi$.
(b) Show, by comparison with (1.50) and (1.52), that $|\vec{E}|$ acts like a propagating wave, with attenuation, in the $-z$ direction. Find the propagation constant β, the attenuation constant α, the phase velocity U_P, and the group velocity U_G. Express your answers in terms of ω and δ only.
(c) Find the numerical values of U_P and U_G for silver at 1 GHz. Express your answers in terms of the speed of light in vacuum (e.g., 0.1 c).

7.18 To study the skin effect in cylindrical wires when the radius is *not* large compared with δ, it is necessary to solve (7.12) in cylindrical coordinates. Assume that the axis of the wire coincides with the z axis. The current density and electric field are in the z direction and are independent of ϕ and z. Obtain the form of (7.12) appropriate for this problem. Your result should be a differential equation describing the dependence of J_x on r, analogous to (7.14), which applies to rectangular coordinates.

(The equation obtained in this problem is a form of *Bessel's equation*. Its solutions belong to the family of functions known as *Bessel functions*. These are transcendental functions, like the trigonometric functions. As with the sine or cosine, the numerical value of a Bessel function is found by looking it up in a table.)

**7.19 A slab of conductive material occupies the region $0 < z < d$. The phasor representing the electric field is $E_o\,\vec{e}_x$ at $z = 0$, and also at $z = d$. The skin depth δ and the conductivity σ_E are given.
(a) Find the electric field phasor everywhere inside the slab.
(b) Find the phasor representing the total current flowing through the slab per unit width. ("Width" is measured in the y direction.)
(c) Find the surface resistance of the slab in ohms per square.

*7.20 When a sheet of material is characterized by a surface impedance (measured in ohms per square), it is sometimes necessary to estimate the impedance (measured in ohms) of an oddly shaped region. To make this estimate, one may use a technique known as "counting squares." This technique is illustrated in Fig. 7.10(a), which shows two perfectly conducting electrodes separated by a thin sheet of resistive material with surface impedance Z_S ohms per square. The sheet has been divided into imaginary squares of equal size.

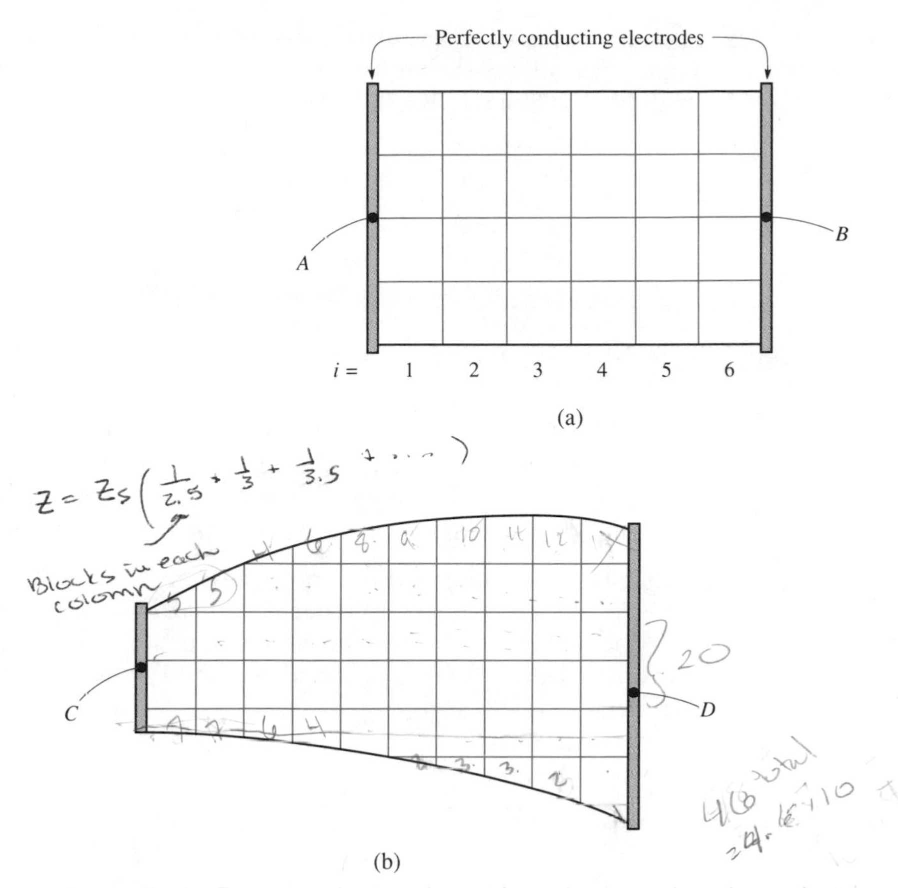

Figure 7.10 Estimating the impedance of two-dimensional conductors by "counting squares." (For Problem 7.20.)

(a) Show that the impedance between terminals A and B is equal to

$$Z_S \sum_i \frac{1}{N_i} = \frac{6}{4} Z_S$$

where N_i is the number of squares in each vertical row. (In Fig. 7.10(a), i varies from 1 to 6, and $N_i = 4$ for all i.)

(b) Find the approximate impedance between terminals C and D in Fig. 7.10(b), by the square-counting method.

*7.21 A conductive material fills the half-space $z < 0$. A sinusoidal current flows in the x direction. The current density at the surface is J_o per unit area. Assuming that the magnetic field is in the y direction, show that

$$\mathbf{H}_y(z) = -\frac{J_o(1-j)\delta}{2} e^{(1+j)z/\delta}$$

7.22 A copper rod has a square cross section, with dimensions 1 mm × 1 mm. Find its resistance (that is, the real part of its impedance) per unit length, at 60 Hz and at 1 GHz.

*7.23 Compare the internal and external inductances per unit length for a coaxial transmission line made of copper, with inner radius 1 mm and outer radius 5 mm, at $f = 100$ MHz. (See Example 6.3.)

7.24 The radius of a cylindrical wire is a, and $a \gg \delta$, where δ is the skin depth. The phasor representing the total current through the wire is I.
(a) Assuming that Ampère's circuital law applies, find the magnetic field at the surface of the wire.
(b) Find the magnetic field at the surface of the wire by means of (7.33). Does your result agree with part (a)?

**7.25 Ampère's circuital law is derived for the case of non-time-varying fields; in general it does not apply to electrodynamics. Explain why the assumption $\sigma_E \gg \omega\epsilon$ justifies its use in the preceding problem.

7.26 A long circular-cylindrical hole is bored parallel to the z axis through a block of metal, and a high-frequency current whose phasor is I flows near the surface of the hole in the $+z$ direction. Find the magnitude and direction of the magnetic field phasor $\vec{\mathbf{H}}$ in the hole just outside the metal. Draw a sketch, and also express $\vec{\mathbf{H}}$ mathematically in cylindrical coordinates.

*7.27 A certain cavity resonator consists of a cube with sides of length a, oriented parallel to the rectangular axes. The walls at $z = 0$ and $z = a$ are made of copper, and the other walls, at $x = 0$, $x = a$, $y = 0$, and $y = a$, are made of an imaginary material whose conductivity is infinitely large. Inside the box there is a magnetic field with frequency 10 GHz, represented by the phasors

$$\mathbf{H}_x = H_o \sin\frac{\pi x}{a} \cos\frac{\pi z}{a}$$

$$\mathbf{H}_y = 0$$

$$\mathbf{H}_z = -H_o \cos\frac{\pi x}{a} \sin\frac{\pi z}{a}$$

where H_o is a given constant. Find the power converted to heat in the walls of the box.

**7.28 Perform the calculation of the preceding problem, assuming that all six walls of the box are made of copper. (You must think carefully about the meaning of $|\vec{\mathbf{H}}_T|^2$, in the case when $\vec{\mathbf{H}}_T = \vec{\mathbf{H}}_x\vec{e}_x + \vec{\mathbf{H}}_z\vec{e}_z$.)

7.29 Show that the magnetic field of Problem 7.27 is consistent with the approximate boundary condition for sinusoidal magnetic fields at a conducting wall (namely, $H_n \cong 0$).

**7.30 Find the electric field that must accompany the magnetic field of Problem 7.27. Show that it obeys the approximate boundary condition for electric fields near a good conductor (namely $E_T \cong 0$). Is your answer consistent with the wall currents required by Eq. (7.34)? Do you think the field in Problem 7.27 can be an exactly correct solution of Maxwell's equations, if the walls of the box are made of perfect conductors? Can it be exactly correct if the walls are made of copper?

7.31 A region of space contains a uniform sinusoidal electric field $\vec{E} = E_o \cos \omega t\, \vec{e}_x$. A small solid copper sphere is introduced into the region, causing the field to be distorted in its vicinity.
(a) Sketch the appearance of the field when the sphere is present.
(b) The electric field is taken away and a magnetic field $\vec{H} = H_o \cos \omega t\, \vec{e}_x$ is applied. Sketch the appearance of the distorted magnetic field when the sphere is introduced.

Section 7.4

7.32 Verify, by substitution, that (7.39) (with either sign) is a solution of (7.38).

*7.33 A uniform plane wave traveling in the $+z$ direction, described by the phasor $\mathbf{E}_x = A_o e^{-jkz+j\phi_o}$, strikes a perfectly conducting wall located at $z = 0$.
(a) What is the total electric field at $z = 0$?
(b) Using the analogy with transmission lines, find the phasor representing the reflected wave.

*7.34 Find the magnetic field associated with the plane wave traveling in the positive direction in (7.39). What is its direction? Is it in phase or out of phase with $\vec{E}$? Repeat for the negative-going wave.

7.35 An *equiphase surface* is a surface on which the phase of a wave is everywhere the same.
(a) For the plane wave $\mathbf{E}_x = A_o e^{-jkz}$ (where A_o is *real*), sketch the surfaces on which the phase angle of the electric field is zero. (Note that there are an infinite number of such surfaces.) Write the expression for E_x as a function of time at any point on these surfaces.
(b) Add to your sketch (in another color) the surfaces at which the phase is $\pi/2$. What is the electric field on these surfaces, as a function of time?
(c) Sketch the zero-phase surfaces for the spherical wave (7.46). (Note that (7.46) applies to the region $\theta_o < \theta < (\pi - \theta_o)$. Let $\theta_o = 30°$.)

**7.36 A plane wave in vacuum with angular frequency ω propagates in the direction of the unit vector $\vec{u} = A\vec{e}_x + B\vec{e}_y$, where $A = 0.65$ and $B = 0.76$. The electric field points in the z direction and has amplitude E_o. The phase angle at the origin is zero. Find the phasor $\mathbf{E}_z$. (*Suggestion:*

Begin with (7.39) and perform a rotation of the coordinate system.) Sketch the equiphase surfaces.

*7.37 Verify that (7.47) is a solution of the Helmholtz equation (7.37), if and only if the frequency ω has the proper value. What is that frequency?

*7.38 Verify that (7.46) is a solution of the Helmholtz equation (7.37) in the region $\theta_o < \theta < (\pi - \theta_o)$.

Section 7.5

7.39 A certain integrated circuit occupies a circular area with diameter 6 mm. Estimate the highest frequency at which this circuit could be safely analyzed by means of conventional circuit theory. Do you think the dielectric constant of the substrate $\left(\frac{\epsilon}{\epsilon_o} = 10\right)$ has an effect on the answer?

**7.40 Consider a section of transmission line with an open circuit at its end. Its length is h, its characteristic impedance is Z_o, and its capacitance per unit length is C.

(a) If the line is sufficiently short, the quasistatic approximation must apply. Under this assumption the line appears to be simply a lumped capacitor. What is its input impedance at frequency ω in that case?

(b) Find the input impedance for arbitrary h, by means of transmission-line theory.

(c) Show that in the limit as $h \to 0$, the result of (b) approaches that of (a).

(d) Sketch the results of parts (a) and (b) as functions of the product $2\pi h/\lambda$ (where $\lambda = 2\pi U_P/\omega$.) For what value of $2\pi h/\lambda$ do the two results differ "substantially" (let us say, by 50%)? Does the quasistatic theorem seem to apply?

CHAPTER 8

Plane Waves

As we have seen in Chapter 7, the plane wave is just one of the many possible solutions of the Helmholtz equation. Because of its widespread importance, however, we must give it special attention. One reason for this importance is that plane waves are especially simple solutions, easily visualized and analyzed. A second reason is that many waves of practical interest *are* very nearly plane waves, or at least tend to approach plane waves as a limit when they have traveled far enough. This behavior is illustrated in Fig. 8.1. Figure 8.1(a) shows a spherical wave being transmitted by an antenna. Figure 8.1(b) shows the antenna viewed from a greater distance. Figure 8.1(c) shows a magnified view of the region inside the dashed box in (b). We see that the curved wavefronts become increasingly planar as they travel. Another, more subtle reason for the importance of plane waves is that in more advanced work, they sometimes play a role similar to that of sinusoids in circuit analysis. Just as it is possible to decompose a non-sinusoidal signal into a sum of sinusoids by means of the Fourier theorem, one can also decompose a non-plane wave into a superposition of plane waves, thus facilitating analysis. For all these reasons, it is useful to study plane waves in some detail.

8.1 CHARACTERISTICS OF PLANE WAVES

As we learned in section 7.4, one solution of the Helmholtz equation (i.e., of the wave equation, with the assumption of sinusoidal time behavior) is

$$\vec{\mathbf{E}} = E_o e^{-jkz}\, \vec{e}_x \tag{8.1}$$

where E_o is a (possibly complex) constant describing amplitude and phase,

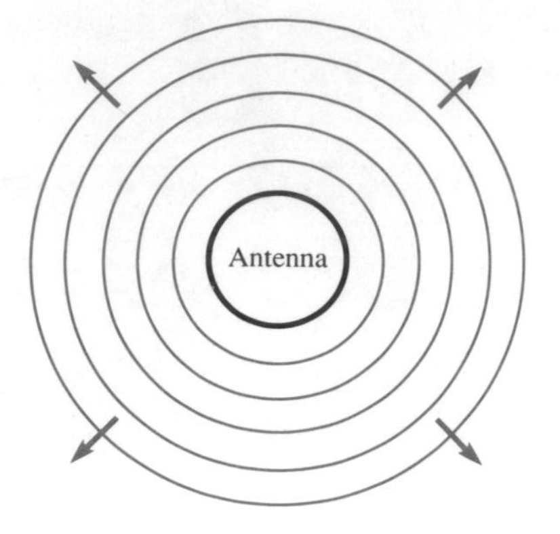

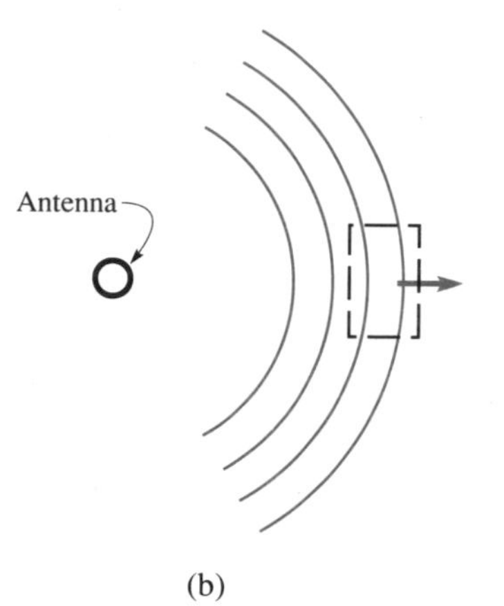

(b)

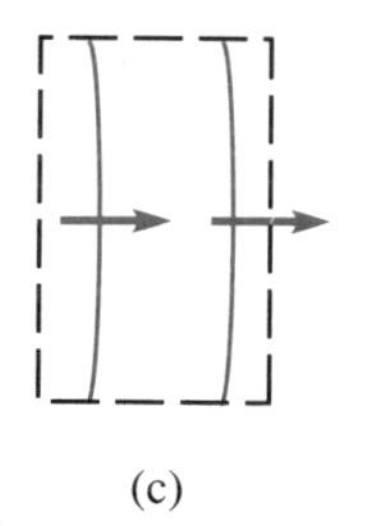

(c)

Figure 8.1 Spherical waves are nearly indistinguishable from plane waves when the observer is far away from the source.

and

$$k = k_o = \omega\sqrt{\mu\epsilon} = 2\pi/\lambda \tag{8.2}$$

This particular wave is one which propagates in the $+z$ direction and has an electric field directed in the x direction.

The time-varying electric field of the wave must, according to Faraday's law, be accompanied by a magnetic field. Thus

$$-j\omega\mu\vec{\mathbf{H}} = \nabla \times \vec{\mathbf{E}} = -jkE_o e^{-jkz}\,\vec{e}_y \tag{8.3}$$

and hence

$$\vec{\mathbf{H}} = \frac{E_o}{\eta}e^{-jkz}\,\vec{e}_y \tag{8.4}$$

where

$$\eta = \sqrt{\mu/\epsilon} \tag{8.5}$$

is called the characteristic impedance of the wave, or, for short, the *wave impedance*. When $\epsilon = \epsilon_o$, $\mu = \mu_o$, η is the characteristic impedance of vacuum, with the value 377 ohms. If we set $E_o = A_o e^{j\phi_o}$, where A_o is a real constant, then the fields of the wave, expressed as functions of time, are

$$\begin{aligned} \vec{E} &= A_o \cos(\omega t - kz + \phi)\,\vec{e}_x \\ \vec{H} &= \frac{A_o}{\eta}\cos(\omega t - kz + \phi)\,\vec{e}_y \end{aligned} \tag{8.6}$$

We now observe the following important characteristics of plane waves:

1. $\vec{E}$, $\vec{H}$, and the direction of propagation are all perpendicular to each other. (In our example, $\vec{E}$ is in the x direction, $\vec{H}$ is in the y direction, and propagation is in the z direction.)
2. E and H are in phase. That is, if at a certain position and time the electric field has its maximum value, then $\vec{H}$ at that position and time also has its maximum value. In fact, at all places $|\vec{H}|$ is simply equal to $|\vec{E}|/\eta$.
3. The wave moves in the $+z$ direction in exactly the same way as the transmission-line waves discussed in Chapters 1 and 2. That is, the wave appears to move as though it were a rigid structure moving toward the right.

It is important that the physical picture of the plane wave be firmly fixed in one's mind. Figure 8.2(a) shows the sinusoidal E and H fields, in the x and y directions respectively, at a certain instant of time. Figure 8.2(b) shows the same wave after a time $\pi/2\omega$, during which the wave has moved a distance $\lambda/4$ to the right. The important point is that the E and H fields move together, as a unit, in the direction of propagation.

Once we have this physical picture, we can think of other waves that are physically the same, but differently oriented with respect to the coordinate system. For example, it would be possible for the wave of Fig. 8.2 to be described in another coordinate system rotated 90°, so $\vec{E}$ is in the y direction (with propagation still in the z direction). The wave still retains the same physical structure, so H is now in the $-x$ direction. The phasors

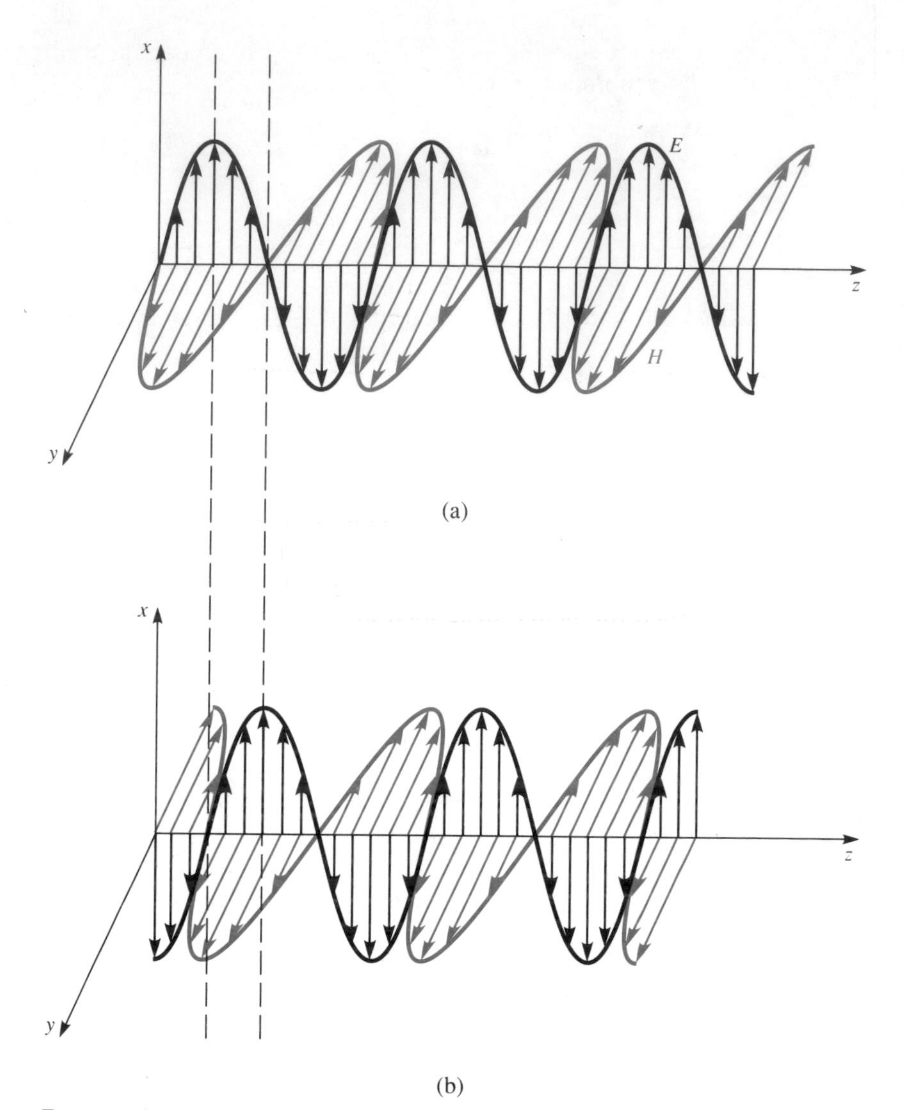

Figure 8.2 Figure (a) shows the electric and magnetic fields of a plane wave moving in the z direction. Figure (b) shows the same wave at a slightly later time. Note that the entire structure appears to move as a unit to the right.

describing this wave are

$$\vec{\mathbf{E}} = E_o e^{-jkz}\,\vec{e}_y$$
$$\vec{H} = -\frac{E_o}{\eta}\,e^{-jkz}\,\vec{e}_x \tag{8.7}$$

Note that physically this is the same wave as that of (8.1) and (8.4); the difference is only in the coordinate system being used to describe it.

When the electric field of a wave is always directed along the same line, the wave is said to be *linearly polarized*. The wave of Fig. 8.2 is said to be linearly polarized in the x direction. The wave of (8.7) is linearly polarized in the y direction.

EXAMPLE 8.1

Find the phasors describing a linearly polarized wave propagating in the $+z$ direction, with electric field in the direction shown in Fig. 8.3.

Solution We use the formulas for coordinate rotation:

$$\vec{e}_{x'} = \vec{e}_x \cos\theta + \vec{e}_y \sin\theta$$

$$\vec{e}_{y'} = -\vec{e}_x \sin\theta + \vec{e}_y \cos\theta$$

Here $\vec{e}_{x'}$ and $\vec{e}_{y'}$ are the unit vectors in one coordinate system, and $\vec{e}_x$ and $\vec{e}_y$ are the unit vectors in another system rotated $\theta°$ clockwise with respect to the first. If the $x'y'$ coordinate system is oriented to place $\vec{E}$ along the x' axis, then the fields, described in that coordinate system, are

$$\vec{\mathbf{E}} = E_o e^{-jkz}\,\vec{e}_{x'}$$

$$\vec{\mathbf{H}} = \frac{E_o}{\eta} e^{-jkz}\,\vec{e}_{y'}$$

Now using the coordinate transformation with $\theta = 60°$, we describe the wave in the x,y system:

$$\vec{\mathbf{E}} = E_o e^{-jkz}\left(\frac{1}{2}\vec{e}_x + \frac{\sqrt{3}}{2}\vec{e}_y\right)$$

$$\vec{\mathbf{H}} = \frac{E_o}{\eta} e^{-jkz}\left(-\frac{\sqrt{3}}{2}\vec{e}_x + \frac{1}{2}\vec{e}_y\right)$$

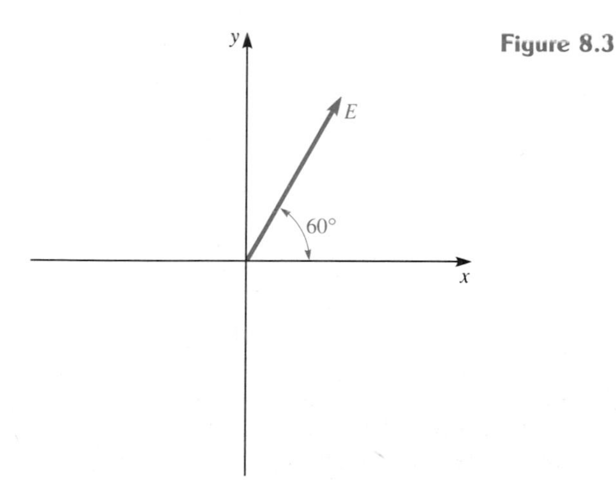

Figure 8.3

CIRCULAR POLARIZATION

Plane waves that are not linearly polarized also exist. These waves are in general said to be *elliptically polarized*. An important special case of elliptical polarization is the *circularly polarized wave*.

We can demonstrate a circularly polarized wave by adding together two linearly polarized waves. Let these waves be polarized respectively in the x and y directions, let them have equal amplitudes, and let one of them be delayed 90° in phase with respect to the other.[1] The sum of these two waves is

$$\vec{E} = E_o e^{-jkz}(\vec{e}_x + j\vec{e}_y) \quad \textbf{(8.8)}$$

(The reader may wish to verify that if the 90° phase shift—the factor j before $\vec{e}_y$—is omitted, the sum is a linearly polarized wave, polarized at 45°.) Let us now express $\vec{E}$ as a function of time, at the position $z = 0$:

$$\vec{E}(t) = \text{Re}\,[E_o(\vec{e}_x + j\vec{e}_y)e^{j\omega t}]$$
$$= E_o\,[\vec{e}_x \cos\omega t - \vec{e}_y \sin\omega t] \quad \textbf{(8.9)}$$

Evidently this wave is not linearly polarized, as the direction of $\vec{E}$ is different at different times, as shown in Fig. 8.4. When $\omega t = 0$, it is in the x direction; when $\omega t = \pi/2$, it points in the $-y$ direction; when $\omega t = \pi$, it points in the $-x$ direction, and so forth. As time passes, the point of the electric field vector seems to go around in a circle. Looking toward the $+z$ direction, the motion is counterclockwise; thus we would describe this wave as having counterclockwise circular polarization.

EXERCISE 8.1

Show that the wave

$$\vec{E} = E_o(\vec{e}_x - j\vec{e}_y)e^{-jkz}$$

is circularly polarized in the clockwise direction.

If we inspect the field at $z = L$ instead of $z = 0$, (8.9) changes into

$$\vec{E}(t) = E_o[\vec{e}_x \cos(\omega t - kL) - \vec{e}_y \sin(\omega t - kL)]$$

Thus the electric field describes the same motion at $z = L$, but with a time delay. The greater L is, the longer the delay. Hence a snapshot of $\vec{E}$ at a fixed instant of time has the corkscrew appearance shown in Fig. 8.5. It is

[1] Since the Helmholtz equation is a linear equation, it is evident that the sum of any two solutions is also a solution.

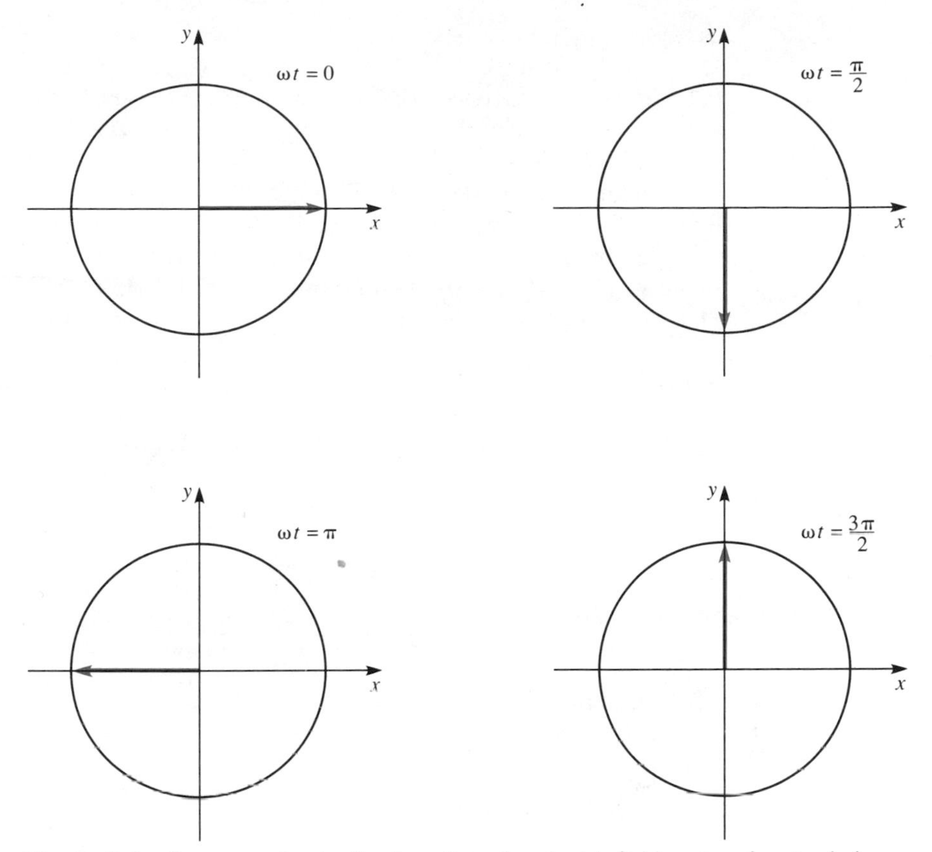

Figure 8.4 If one stands at a fixed position, the electric field vector of a circularly polarized wave appears to go around in a circle.

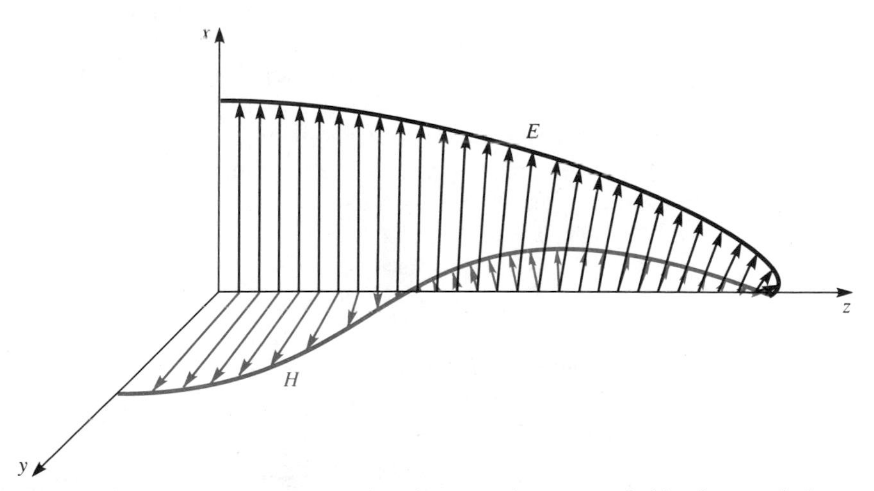

Figure 8.5 "Snapshot" showing the electric and magnetic fields of a circularly polarized wave at a fixed instant of time.

easily shown that $\vec{H}$ is still perpendicular to $\vec{E}$ everywhere. The magnitude of $\vec{E}$ is E_o at all points and times; only the direction of $\vec{E}$ changes. The magnitude of $\vec{H}$ is E_o/η at all points and times. Circularly polarized waves are often used in radio communications because the orientation of the receiving antenna (hard to control in airplanes, for example) with respect to the transmitting antenna is less important with circular polarization than with linear. Circular polarization is also used in optical technology.

The most general case is elliptical polarization. If we add together two plane waves with different amplitudes, or if we add them together with any phase difference except zero or $\pm 90°$, elliptical polarization generally results. In that case the path described by the tip of the $\vec{E}$ vector at any fixed point is an ellipse. Elliptical polarization can be thought of as intermediate between circular and linear polarizations. (If one flattens a circle, one passes through a family of ellipses and finally, when the circle is completely flattened, arrives at a line.)

8.2 POYNTING'S THEOREM

The well-known mathematical statement known as *Poynting's theorem* is an identity based on Maxwell's equations. Basically it is just a statement that one function of the fields must always be equal to another function of the fields. Its importance, however, arises from the fact that some of these functions represent the energies of the electric and magnetic fields. Another term in Poynting's theorem can sometimes be interpreted as representing power flow. Thus the theorem can often be used as an energy-balance equation. Its usual application is to cases when electromagnetic waves are present. In such cases it is customary to identify a certain special vector, known as the *Poynting vector,* as representing the power flow of the wave.

The mathematical derivation of Poynting's theorem is simple and straightforward; it is the physical interpretation of the theorem that is subtle. Thus let us begin with the straightforward mathematics. We begin by writing down the (seemingly arbitrarily chosen) vector $\vec{E} \times \vec{H}$, and taking its divergence. To do this we use the general vector identity, valid for any two vectors $\vec{A}$ and $\vec{B}$, which states that $\nabla \cdot (\vec{A} \times \vec{B}) = \vec{B} \cdot (\nabla \times \vec{A}) - \vec{A} \cdot (\nabla \times \vec{B})$. Thus we have

$$\nabla \cdot (\vec{E} \times \vec{H}) = \vec{H} \cdot (\nabla \times \vec{E}) - \vec{E} \cdot (\nabla \times \vec{H}) \qquad \textbf{(8.10)}$$

Now making use of the two Maxwell's equations expressing $\nabla \times \vec{H}$ and $\nabla \times \vec{E}$, we obtain

$$\nabla \cdot (\vec{E} \times \vec{H}) = -\vec{H} \cdot \frac{\partial \vec{B}}{\partial t} - \vec{E} \cdot \frac{\partial \vec{D}}{\partial t} - \vec{E} \cdot \vec{J} \qquad \textbf{(8.11)}$$

We can proceed a further step by assuming that $\vec{B} = \mu\vec{H}$ and $\vec{D} = \epsilon\vec{E}$, and noting that

$$\frac{\partial}{\partial t}(\vec{E}\cdot\vec{E}) = \frac{\partial}{\partial t}|\vec{E}|^2 = 2\vec{E}\cdot\frac{\partial\vec{E}}{\partial t}$$

hence

$$\nabla\cdot(\vec{E}\times\vec{H}) = -\frac{\partial}{\partial t}\left(\frac{\epsilon|\vec{E}|^2}{2}\right) - \frac{\partial}{\partial t}\left(\frac{\mu|\vec{H}|^2}{2}\right) - \vec{E}\cdot\vec{J} \tag{8.12}$$

Finally, we can integrate each side of (8.12) over any arbitrary volume V. Applying the divergence theorem to the integral on the left, we have

$$\int_S (\vec{E}\times\vec{H})\cdot\vec{dS} = -\int_V \left[\frac{\partial}{\partial t}\left(\frac{\epsilon}{2}|\vec{E}|^2\right) + \frac{\partial}{\partial t}\left(\frac{\mu}{2}|\vec{H}|^2\right) + \vec{E}\cdot\vec{J}\right] dV \tag{8.13}$$

Equation (8.13) is Poynting's theorem. From its derivation we know that it is a generally true statement about the fields, valid whenever the rather general assumptions used in the derivation (such as $\vec{D} = \epsilon\vec{E}$, $\vec{B} = \mu\vec{H}$) are correct. However, the significance of (8.13) is not at all obvious. Why should we *care* that this particular relationship between the fields is true? This is what we must next discuss.

ELECTRIC AND MAGNETIC ENERGY

In general, whenever electric charges are moved around in each other's presence, mechanical work must be done. This is because the charges exert forces on each other, so that forces must be applied to move them. As the charges are moved, the electric field distribution changes. It can be shown that the amount of work required to assemble a particular set of charges can be expressed as an integral over the electric field of the charge assembly. This energy is

$$W_E = \int \frac{\epsilon}{2}|\vec{E}|^2\, dV \tag{8.14}$$

where the integral is taken over all regions of space. The derivation of (8.14) is rather long, and would interrupt our discussion if presented here. However, the interested reader will find a derivation of (8.14) in Appendix C. Since according to (8.14), the energy is equal to the volume integral of $\frac{1}{2}\epsilon|\vec{E}|^2$ over all space, it is customary to regard $\frac{1}{2}\epsilon|\vec{E}|^2$ as the *electric energy density* U_E. In this point of view, the energy stored in an elementary volume

ΔV is $U_E \Delta V$. (See note[2].) In similar fashion it can be shown that the energy required to assemble a system of currents, which interact through magnetic fields, is

$$W_H = \int \frac{\mu}{2} |\vec{H}|^2 \, dV \tag{8.15}$$

One may consider this to be the volume integral of a *magnetic energy density* $U_H = \frac{1}{2}\mu|\vec{H}|^2$.

EXAMPLE 8.2

A capacitor consisting of two long coaxial cylinders with radii a and b is charged to a voltage V. Calculate the energy required to charge a length L of this capacitor by means of (8.14).

Solution Using Gauss' law, it is easily shown that the electric field of the charged capacitor is

$$\vec{E} = \left(\frac{V}{\log \dfrac{b}{a}} \right) \frac{1}{r} \vec{e}_r \tag{8.16}$$

Integrating over the volume between the conductors, we find that the energy of a length L is

$$W_E = \frac{\epsilon L}{2} \left(\frac{V}{\log \dfrac{b}{a}} \right)^2 \int_a^b \frac{2\pi r \, dr}{r^2}$$

(where we have used $dV = 2\pi r \, dr \, dz$.) Thus

$$W_E = \frac{\pi \epsilon L V^2}{\log \dfrac{b}{a}} \tag{8.17}$$

From circuit theory we know that the energy required to charge a capacitor C to a voltage V is $\frac{1}{2}CV^2$. From this and (8.17) we can find the

[2]The designation of U_E as the energy density, though customary, is not really justifiable. Although Appendix C shows that the integral of U_E over *all space* is equal to energy needed to establish the charges in their positions, this does not prove that the energy stored in any *partial* volume V' is equal to the integral of U_E over V'. However, in the usual applications of Poynting's theorem, contradictions are not found to arise.

We also note that not all the energy W_E can be recovered by separating the charges. An isolated electron will still have fields and hence W_E will not vanish. The remaining W_E of an isolated electron would have to be regarded as the energy required to assemble the electron. This energy is permanently attached to the electron, since in fact the electron cannot be taken apart. (See Problem 8.13.)

capacitance of the cylindrical capacitor

$$\frac{1}{2}CV^2 = W_E = \frac{\pi\epsilon L V^2}{\log\dfrac{b}{a}}$$

$$C = \frac{2\pi\epsilon L}{\log\dfrac{b}{a}}$$

in agreement with Exercise 4.5.

INTERPRETATION OF POYNTING'S THEOREM

The final term of (8.13),

$$\int_V \vec{E}\cdot\vec{J}\,dV \tag{8.18}$$

has a clear interpretation, as illustrated in Fig. 8.6. Here a rectangular volume element has length L and is oriented so that L is parallel to the current density $\vec{J}$. The faces perpendicular to $\vec{J}$ have area A. Applying the integral (8.18) to this volume, we have

$$\int_V \vec{E}\cdot\vec{J}\,dV = |\vec{E}||\vec{J}|\,LA\cos\theta$$

where θ is the angle between $\vec{E}$ and $\vec{J}$. This can be rewritten as

$$\begin{aligned}\int_V \vec{E}\cdot\vec{J}\,dV &= (|\vec{E}|\,L\cos\theta)(|\vec{J}|A)\\ &= VI\end{aligned} \tag{8.19}$$

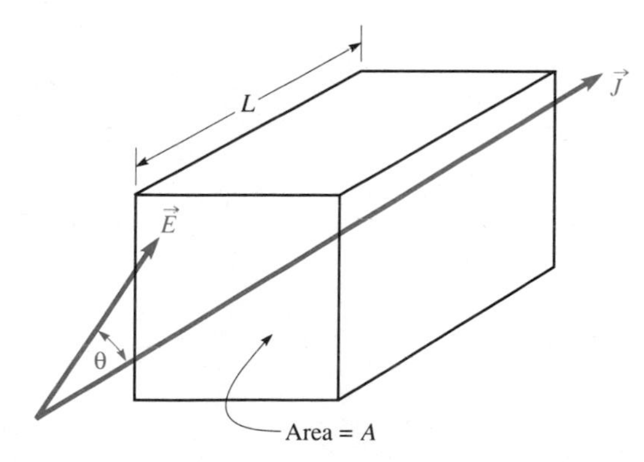

Figure 8.6 Significance of the $\vec{E}\cdot\vec{J}$ term in Poynting's theorem.

where V is the potential difference across the volume and I is the current flowing through it. Thus, (8.18) represents the power being dissipated (turned into heat) in the volume V.

We can now interpret Poynting's theorem as a power-balance equation. Repeating (8.13) for convenience, we have

$$-\int_S (\vec{E} \times \vec{H}) \cdot \vec{dS}$$

$$= \frac{\partial}{\partial t} \int_V \frac{\epsilon|\vec{E}|^2}{2}\, dV + \frac{\partial}{\partial t} \int_V \frac{\mu|\vec{H}|^2}{2}\, dV + \int_V \vec{E} \cdot \vec{J}\, dV \quad \textbf{(8.20)}$$

The first term on the right represents the rate of increase of electric energy inside the volume V. The second term on the right represents the rate of increase of magnetic energy. The last term represents power converted into heat inside the volume. Energy conservation requires that these three terms be balanced by a flow of energy into the volume, and this is the meaning of the term on the left. We designate the vector $\vec{E} \times \vec{H}$ as the *Poynting vector* $\vec{S}$:

$$\vec{S} = \vec{E} \times \vec{H} \quad \textbf{(8.21)}$$

We then interpret $\vec{S}$ as representing power density, just as the vector $\vec{J}$ represents current density. In other words, when $\vec{S}$ makes an angle θ with the normal to an element of surface with area A, as shown in Fig. 8.7, the power flowing through that surface is

$$P = (\vec{S} \cdot \vec{n})\, A = |\vec{S}|\, A \cos\theta \quad \textbf{(8.22)}$$

The term on the left side of (8.20) is then seen to represent the total power flowing through the surface S, which encloses V. The minus sign enters because $\vec{dS}$ is the *outward* normal, and hence $(\vec{E} \times \vec{H}) \cdot \vec{dS}$ represents power flowing outward, rather than inward as is needed for energy balance. Thus our interpretation of Poynting's theorem is

$$\begin{pmatrix}\text{Power}\\ \text{entering } V\end{pmatrix} = \begin{pmatrix}\text{Rate of increase}\\ \text{of electrical energy}\\ \text{stored in } V\end{pmatrix} + \begin{pmatrix}\text{Rate of increase}\\ \text{of magnetic energy}\\ \text{stored in } V\end{pmatrix} + \begin{pmatrix}\text{Power converted}\\ \text{to heat inside } V\end{pmatrix} \quad \textbf{(8.23)}$$

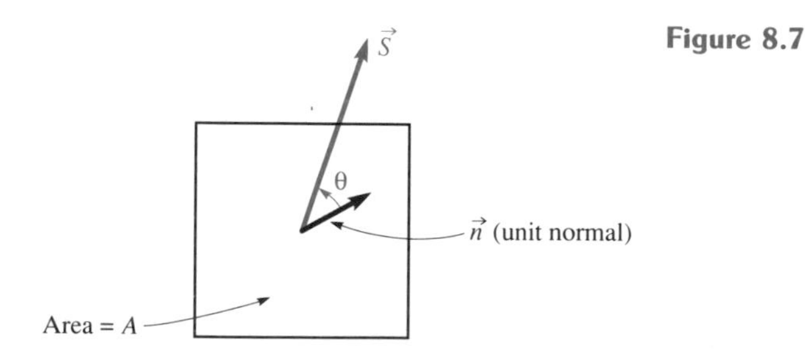

Figure 8.7

EXAMPLE 8.3

A wire with radius a, made of material having conductivity σ_E, conducts a constant, uniformly distributed current I in the z-direction. Find the electrostatic and magnetostatic fields at the surface of the wire. Then calculate, by means of Poynting's theorem, the power dissipated in a length L of the wire.

Solution The current density in the wire is $(I/\pi a^2)\vec{e}_z$. Hence the electric field everywhere in the wire (also at the surface) is $(I/\pi a^2\sigma_E)\vec{e}_z$. The magnetic field H at the surface, from Ampère's circuital law, is $(I/2\pi a)\vec{e}_\phi$. The Poynting vector at the surface is then

$$\vec{S} = \left(\frac{I}{\pi a^2\sigma_E}\vec{e}_z\right) \times \left(\frac{I}{2\pi a}\vec{e}_\phi\right) = -\frac{I^2}{2\pi^2 a^3\sigma_E}\vec{e}_r \tag{8.24}$$

In this case everything is constant in time; thus the terms in Poynting's theorem containing $\frac{\partial}{\partial t}$ both vanish. The power dissipated per unit length of wire can now be found from (8.20):

$$P_D = \int_V \vec{E}\cdot\vec{J}\,dV = -\int_S (\vec{E}\times\vec{H})\cdot\overrightarrow{dS} = \frac{I^2}{2\pi^2\sigma_E a^3}(2\pi a L)$$

$$= \frac{I^2 L}{\pi\sigma_E a^2} \tag{8.25}$$

This result can be checked by means of ordinary circuit theory. The resistance of a length L of the wire is $R = L/\pi a^2\sigma_E$, and thus the dissipated power I^2R agrees with (8.25).

The preceding example illustrates both a strength and a weakness of Poynting's theorem. The strength is that Poynting's theorem allows us to calculate power transfer as a function of the fields. On the other hand, the identification of $\vec{S}$ as the power density vector, in this example, goes against one's intuition. In fact, it is easy to find examples in which $\vec{S}$ does not represent power flow at all. (See Problem 8.16.) One must thus use care in interpreting the various terms of Poynting's theorem. On the other hand, one can always[3] feel confident that equation (8.20) will be mathematically correct, as it is in Example 8.3.

It is in connection with electromagnetic waves that the Poynting vector is found to be most useful. No one will question the fact that the electro-

[3] presuming that the few assumptions of our derivation are satisfied.

magnetic waves transport energy; after all, this is how the earth receives energy from the sun. Let us imagine a plane wave entering a box, as shown in Fig. 8.8. The box is absorptive, so all power that enters, through an entry hole with area A, is dissipated (turned into heat) inside the box. As we have seen, $\vec{E}$ and $\vec{H}$ are perpendicular, and their cross product $\vec{S}$ is in the direction of propagation. (Note that $\vec{S}$ is never *opposite* to the direction of propagation. When $\vec{E}$ becomes negative, $\vec{H}$ also becomes negative, so the product $\vec{E} \times \vec{H}$ remains positive.) The power entering the box is therefore

$$p = (\vec{E} \times \vec{H}) \cdot \vec{n} A \tag{8.26}$$

where $\vec{n}$ is the unit normal of the entrance aperture. When the direction of propagation (and hence $\vec{S}$) is normal to the aperture, the power entering the box is simply $|\vec{E}||\vec{H}|A$. Of course, both $\vec{E}$ and $\vec{H}$ are sinusoidal functions of time, and the power p we have found is the instantaneous power. Suppose that at the entrance to the box, $\vec{E}$ is given by

$$\vec{E} = E_o \cos \omega t \, \vec{e}_x$$

Then at that point

$$\vec{H} = \frac{E_o}{\eta} \cos \omega t \, \vec{e}_y$$

and

$$p = \frac{E_0^2 A}{\eta} \cos^2 \omega t$$

In most cases one is interested in the time-averaged power P. Since the time average of $\cos^2 \omega t$ is $\frac{1}{2}$, we have

$$P = \frac{E_0^2 A}{2\eta} \tag{8.27}$$

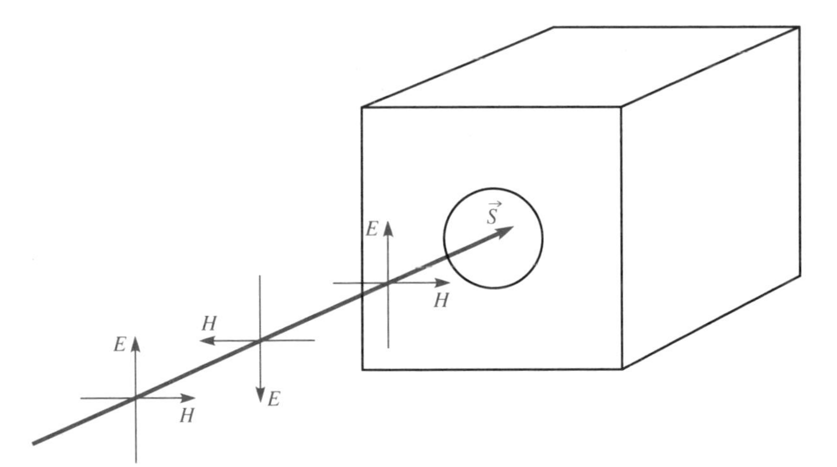

Figure 8.8 The power entering the box is equal to the surface integral of the Poynting vector over the entrance hole.

The time-averaged power transported by a plane wave, per unit area, is thus $E_0^2/2\eta$. This quantity is known as the *intensity* of the wave.

EXAMPLE 8.4

A powerful carbon dioxide laser, with output power 1000 watts, is used for cutting metal in a factory. After focusing, the radius of the beam is 0.1 mm. Find the amplitude of the electric field in the beam.

Solution The power per unit area is (in mks units)

$$\frac{E_0^2}{2(377)} = \frac{1000}{\pi(10^{-4})^2}$$

Hence

$$E_o = 4.9 \times 10^6 \text{ V/m}$$

This electric field exceeds the breakdown strength of air (nominally about 10^6 V/m, but a function of frequency), so the laser beam may have to be used in vacuum.

If the fields are described by their phasors, $\vec{\mathbf{E}}$ and $\vec{\mathbf{H}}$, time-averaged power density can be found in terms of these phasors. By a fairly straightforward computation (Problem 8.20) it is found that

$$\text{time average of } \vec{S} = \tfrac{1}{2}\,\text{Re}(\vec{\mathbf{E}} \times \vec{\mathbf{H}}^*) \qquad \textbf{(8.28)}$$

This is easily remembered by its similarity to $P = \frac{1}{2}\,\text{Re}(\mathbf{VI}^*)$ in conventional circuit theory.

Waves other than plane waves can also be described in terms of the Poynting vector. In non-plane waves, the ratio $|\vec{E}|/|\vec{H}|$ will not in general be equal to η, and $\vec{E}$ may not be perpendicular to $\vec{H}$; however the vector $\vec{E} \times \vec{H}$ still represents the power density.

EXAMPLE 8.5

The electric field of a wave propagating on a biconical transmission line (Fig. 8.9) is (in spherical coordinates)

$$\vec{E} = \frac{A}{r\sin\theta}\, e^{-jkr}\, \vec{e}_\theta \qquad \text{for } \theta_o < \theta < (\pi - \theta_o)$$

$$= 0 \qquad \text{otherwise}$$

Find

(a) the magnetic field
(b) the time-averaged Poynting vector
(c) the total time-averaged transmitted power of the wave.

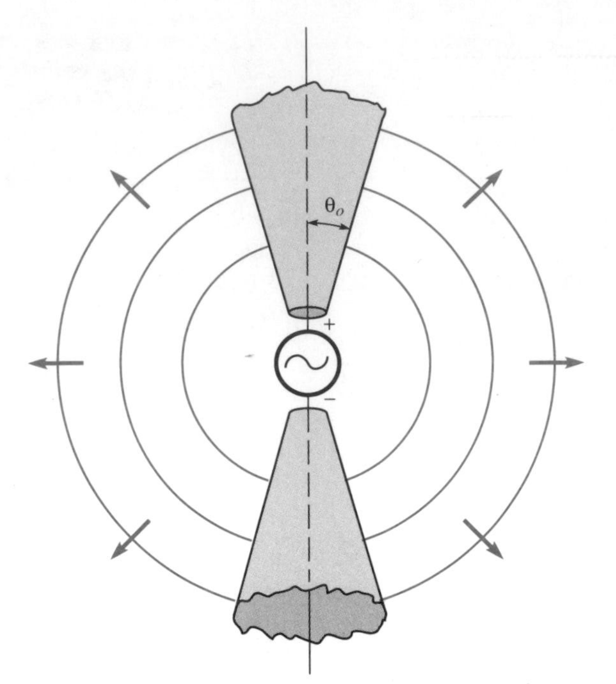

Figure 8.9 Waves on a biconical transmission line.

Solution

(a) From $\nabla \times \vec{\mathbf{E}} = -j\omega \vec{\mathbf{B}}$, we have

$$\vec{\mathbf{H}} = \frac{A}{\eta r \sin\theta} e^{-jkr}\, \vec{e}_\phi$$

(We note that in this wave $|E|/|H| = \eta$, as in a plane wave. However, the reader should not think that this is a general result for non-plane waves. We have chosen an especially simple spherical wave as an example.)

(b)

$$\text{time average of } \vec{S} = \frac{1}{2} \text{Re}\,(\vec{\mathbf{E}} \times \vec{\mathbf{H}}^*) = \frac{1}{2}\left(\frac{A}{\eta r \sin\theta}\right)^2 \vec{e}_r$$

(c) To find the total power we must integrate the Poynting vector over the portion of the spherical surface between θ_o and $(\pi - \theta_o)$:

$$P = \int_{\theta_o}^{\pi-\theta_o} \frac{1}{2} \text{Re}\,(\vec{\mathbf{E}} \times \vec{\mathbf{H}}^*) \cdot \vec{dS}$$

$$= \int_{\theta_o}^{\pi-\theta_o} \frac{1}{2\eta}\left(\frac{A}{r \sin\theta}\right)^2 2\pi r^2 \sin\theta\, d\theta$$

$$= -\frac{2A^2\pi}{\eta} \ln\left(\tan\frac{\theta_o}{2}\right)$$

(The minus sign appears because $\tan\frac{\theta_o}{2}$ is less than unity; hence its logarithm is negative.) We observe that the result is independent of r. This is as one would expect, since although the fields become weaker as the wave propagates outward, their total power should remain the same.

EXERCISE 8.2

At a certain position in a complicated wave,

$$\vec{E} = (2 + 3j)\vec{e}_x - 2\vec{e}_y$$

$$\vec{H} = 4\vec{e}_z + 2j\vec{e}_y$$

Find the time-averaged Poynting vector.

Answer $-4\vec{e}_x - 4\vec{e}_y + 3\vec{e}_z$.

8.3 REFLECTION AND TRANSMISSION AT NORMAL INCIDENCE

In general, plane waves have a great deal in common with waves on a transmission line. When plane waves strike discontinuities of various kinds, they are transmitted and reflected in much the same way as transmission-line waves. The similarity is particularly great when the plane wave strikes a barrier that is planar and perpendicular to the wave's direction of propagation. In that case the wave is said to be *normally incident* on the barrier in question.

An especially simple case occurs when a plane wave strikes a perfectly conducting wall at normal incidence. In this case the wall imposes a boundary condition: the electric field tangential to the wall must vanish. Let the wave be linearly polarized in the x direction and propagating in the $+z$ direction, and let the conducting wall be located at $z = 0$. The incident wave is then

$$\vec{\mathbf{E}}_i = E_o e^{-jkz}\,\vec{e}_x$$

We expect that there will be a reflected wave

$$\vec{\mathbf{E}}_r = E_1 e^{jkz}\,\vec{e}_x$$

Since the electric field in the x direction must vanish at $z = 0$, we have

$$(E_o + E_1)\,\vec{e}_x = 0$$

Thus, $E_1 = -E_o$. The reflected wave has the same amplitude as the incident wave; this is as we expect, since a perfectly conducting wall can absorb no power, and must reflect with 100% efficiency. The total electric field is

$$\vec{\mathbf{E}} = \vec{\mathbf{E}}_i + \vec{\mathbf{E}}_R = E_o(e^{-jkz} - e^{jkz})\,\vec{e}_x$$
$$= -2j\,E_o \sin kz\,\vec{e}_x \tag{8.29}$$

This result is identical with the expression for the voltage on a short-circuited transmission line.

The magnetic field of the incident wave is

$$\vec{\mathbf{H}}_i = \frac{E_o}{\eta}\,e^{-jkz}\,\vec{e}_y \tag{8.30}$$

However, the magnetic field of the reflected wave is

$$\vec{\mathbf{H}}_r = -\frac{E_1}{\eta}\,e^{jkz}\,\vec{e}_y \tag{8.31}$$

The reader should convince himself of this important minus sign. It arises because $\vec{E} \times \vec{H}$ must be in the direction of propagation. If the wave moves in the $-z$ direction, and $\vec{E}$ is in the $+x$ direction, the direction of $\vec{H}$ has to be $-y$. (To visualize, hold the thumb, index, and middle fingers of your right hand at right angles to each other. Let $\vec{E}$ be in the direction of your thumb, and $\vec{H}$ in the direction of your index finger; then the direction of propagation is in the direction of your middle finger.) Equation (8.31) is similar to the equation

$$\mathbf{i}_r = -\frac{\mathbf{v}_r}{Z_o}$$

for a wave moving in the $-z$ direction on a transmission line. By analogy with the transmission line, we would say that the conducting barrier has a reflection coefficient ρ_o equal to -1, resulting in a standing-wave ratio that is infinitely large. The total magnetic field is

$$\vec{\mathbf{H}} = \frac{E_o}{\eta}(e^{-jkz} + e^{jkz})\,\vec{e}_y = \frac{2E_o}{\eta}\cos kz\,\vec{e}_y \tag{8.32}$$

In transmission-line problems, the concept of impedance is very useful; thus we expect it will be useful in plane-wave problems as well. Let us define the *wave impedance* for an x-polarized plane wave by

$$Z(z) = \frac{\mathbf{E}_x(z)}{\mathbf{H}_y(z)} \tag{8.33}$$

This impedance is analogous to $Z(z) = \mathbf{v}(z)/\mathbf{i}(z)$ on a transmission line. For a single x-polarized wave propagating in the $+z$ direction, it becomes the characteristic wave impedance, which is simply η.

EXERCISE 8.3

Find the wave impedance $Z(z)$ for the short-circuited plane wave of (8.29).

Answer $Z(z) = -j\eta \tan kz.$

Let us next consider the case of a plane wave normally incident on a dielectric discontinuity, as shown in Fig. 8.10(a). Here the incident wave E_i, travelling in the z direction, strikes a boundary between two different dielectric materials. In the region $z < 0$, $\epsilon = \epsilon_1$, and for $z > 0$, $\epsilon = \epsilon_2$. The situation is analogous to that of an abrupt change of characteristic impedance in a transmission line, as shown in Fig. 8.10(b). In the case of transmission lines, the boundary conditions at the discontinuity were (a) voltage continuous across the discontinuity, and (b) current continuous across the discontinuity. By now it is evident that in the case of normally incident plane waves, $\vec{E}$ plays the role of voltage on the transmission line, and $\vec{H}$ plays

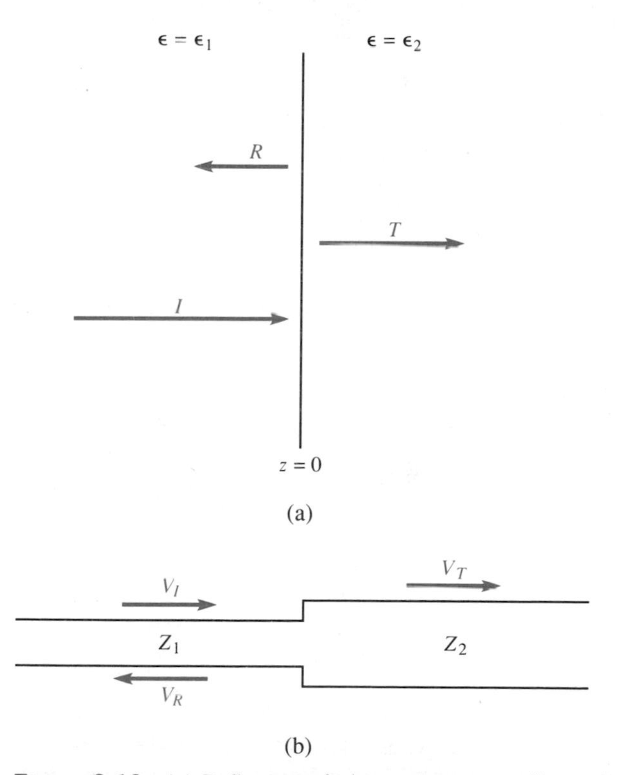

Figure 8.10 (a) Reflection of plane waves normally incident on the interface between two dielectrics. (b) The transmission-line analogy.

the role of current. Moreover, the boundary condition on $\vec{E}$ is that its tangential component (in this case its only component) is continuous; the boundary condition on $\vec{H}$ is that its tangential component (also its only component) is continuous. Thus we have

$$\vec{E}_i(z = 0) + \vec{E}_r(z = 0) = \vec{E}_t(z = 0)$$
$$\vec{H}_i(z = 0) + H_r(z = 0) = H_t(z = 0) \tag{8.34}$$

Letting $\vec{\mathbf{E}}_i = E_o e^{-jkz}\vec{e}_x$, $\vec{\mathbf{E}}_r = E_1 e^{jkz}\vec{e}_x$, $\vec{\mathbf{E}}_t = E_2 e^{-jkz}\vec{e}_x$, using (8.31), and substituting into (8.34), we obtain

$$E_o + E_1 = E_2$$
$$\frac{E_o}{\eta_1} - \frac{E_1}{\eta_1} = \frac{E_2}{\eta_2} \tag{8.35}$$

where $\eta_1 = \sqrt{\mu_o/\epsilon_1}$ and $\eta_2 = \sqrt{\mu_o/\epsilon_2}$. Solving, we have

$$\frac{E_1}{E_o} = \rho = \frac{\eta_2 - \eta_1}{\eta_2 + \eta_1}$$
$$\frac{E_2}{E_o} = \tau = \frac{2\eta_2}{\eta_2 + \eta_1} \tag{8.36}$$

The reader will observe that these calculations are identical with the corresponding transmission-line calculations in Chapter 2. In fact, once we have noted the relationships between transmission-line and wave quantities, we can simply take over all the techniques of transmission-line analysis, including the Smith chart, and make use of them in problems involving plane waves at normal incidence. The analogous quantities for the two cases are summarized in Table 8.1.

EXERCISE 8.4

For the case of Fig. 8.10, find the fraction of incident power that is reflected if $\epsilon_1 = \epsilon_o$ and $\epsilon_2 = 2\epsilon_o$.

Answer 2.94%.

EXAMPLE 8.6

A plane wave in air, with frequency $f = 10$ GHz, is normally incident on a dielectric window with thickness 1 cm and $\epsilon = 3\epsilon_o$, as shown in Fig. 8.11(a). Find the reflection coefficient $\rho_o = \mathbf{E}_r/\mathbf{E}_i$ at $z = 0$.

Table 8.1 ANALOGY BETWEEN TRANSMISSION-LINE WAVES AND PLANE WAVES AT NORMAL INCIDENCE

Transmission-line Quantity	Plane-wave Quantity
$v(z, t)$	$E_x(z, t)$
$i(z, t)$	$H_y(z, t)$
$Z(z) = \dfrac{v}{i}$	$Z(z) = \dfrac{E_x}{H_y}$
$Z_o = \sqrt{\dfrac{L}{C}}$	$Z_o = \sqrt{\dfrac{\mu}{\epsilon}}$
$U_P = \dfrac{1}{\sqrt{LC}}$	$U_P = \dfrac{1}{\sqrt{\mu\epsilon}}$
$\lambda = U_P/f$	$\lambda = U_P/f$
$k = \dfrac{2\pi}{\lambda} = \dfrac{\omega}{U_P} = \omega\sqrt{LC}$	$k = \dfrac{2\pi}{\lambda} = \dfrac{\omega}{U_P} = \omega\sqrt{\mu\epsilon}$
$P = \dfrac{1}{2}\dfrac{\lvert\mathbf{v}\rvert^2}{Z_o}$	$P = \dfrac{1}{2}\dfrac{\lvert\mathbf{E}_x\rvert^2}{\eta}$

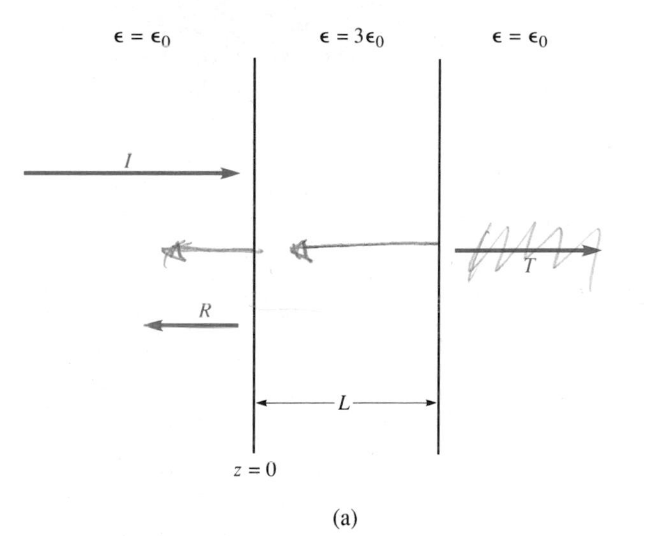

(a)

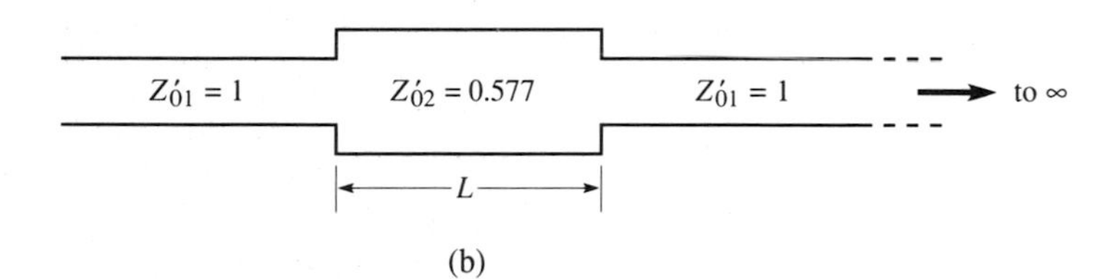

(b)

Figure 8.11 (a) Transmission and reflection at a dielectric window. (b) Transmission-line analogy.

Solution Let us normalize the characteristic impedances to that of air. Then the normalized characteristic impedance of the window material is 0.577. The phase change of the wave in passing through the window is given by

$$kL = 2\pi f\sqrt{\mu_o\epsilon}\, L = \frac{2\pi f}{c}\sqrt{3}\, L = 3.63 \text{ radians}$$

The analogous transmission-line circuit is shown in Fig. 8.11(b). A wave moving to the right inside the window sees a real load impedance larger than its own characteristic impedance by the factor $(1/0.577) = 1.732$. Locating the point $R = 1.732$, $X = 0$ on the Smith chart and translating it 3.63 rad toward the generator (remember that this involves moving around the Smith chart clockwise through an angle that is twice as large, or 7.26 rad), we find $Z'(z = 0) = (1.2 - 0.58j)(0.577) = 0.69 - 0.335j$. From the Smith chart, the corresponding reflection coefficient is $\rho_o = 0.27\, e^{j238°}$.

8.4 REFLECTION AND REFRACTION AT OBLIQUE INCIDENCE

When a plane wave impinges on a plane discontinuity, and incidence is not normal, it is said to be *oblique*. This situation is more complicated than that of normal incidence, but is worthy of study because of its interesting optical applications. These include Snell's law, and the phenomena of total internal reflection and Brewster's angle.

OBLIQUE INCIDENCE

The plane containing both the normal to the surface and the direction of propagation is known as the *plane of incidence*. There are two different cases to consider: either (1) the incident wave is linearly polarized with its

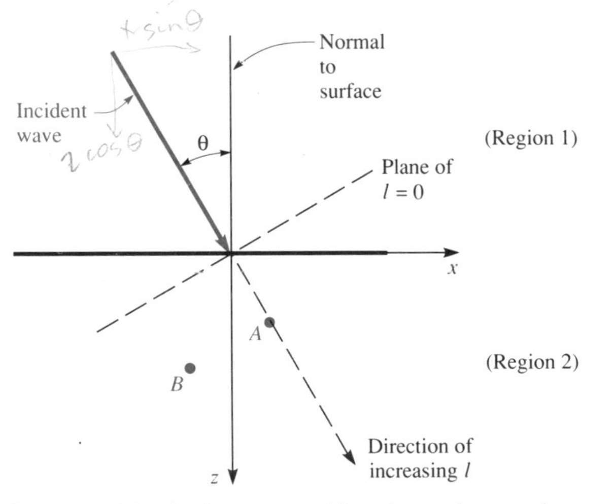

Figure 8.12 A plane wave obliquely incident on the x-y plane.

electric field *in* the plane of incidence, or else (2) the incident wave is polarized with *E perpendicular* to the plane of incidence. Any other incident waves can be decomposed into linear combinations of these two.

Figure 8.12 shows a wave of either polarization incident on some surface. The angle θ, between the normal to the surface and the propagation direction, is called the *angle of incidence*. We choose the plane of incidence to be the x-z plane, with z directed downward into the surface, as shown. We also pass a plane perpendicular to the propagation direction through the origin, in order to measure distance in the propagation direction. The distance from this plane, measured in the direction of propagation, is called l. (Note that points A and B have the same value of l.) If the coordinates of a certain point are x, z, it can be shown that the corresponding value of l is given by

INCIDENT WAVE
$$l = x \sin\theta + z \cos\theta \tag{8.37}$$

If the amplitude of the incident wave is E_o, and its electric field is perpendicular to the plane of incidence, then its fields are described by

$$\begin{pmatrix} E \text{ perpendicular to} \\ \text{plane of incidence} \end{pmatrix} \quad \begin{cases} \vec{\mathbf{E}} = E_o e^{-jk_{01}l}\,\vec{e}_y \\ \vec{\mathbf{H}} = \dfrac{E_o}{\eta_1} e^{-jk_{01}l}\,(-\cos\theta\,\vec{e}_x + \sin\theta\,\vec{e}_z) \end{cases} \tag{8.38}$$

If its electric field is in the plane of incidence,

$$(E \text{ in plane of incidence}) \quad \begin{cases} \vec{\mathbf{E}} = E_o e^{-jk_{01}l}\,(\cos\theta\,\vec{e}_x - \sin\theta\,\vec{e}_z) \\ \vec{\mathbf{H}} = \dfrac{E_o}{\eta_1} e^{-jk_{01}l}\,\vec{e}_y \end{cases} \tag{8.39}$$

Here $k_{01} = \omega\sqrt{\mu\epsilon_1}$ and $\eta_1 = \sqrt{\mu/\epsilon_1}$ are the values of the propagation constant and characteristic impedance in region 1.

There may also be reflected and transmitted waves E_R and E_T, as shown in Fig. 8.13. These have angles of incidence θ' and θ'', as shown. Distances along their propagation directions are l' and l'', as shown in the figure. For the reflected wave

REFLECTED WAVE
$$l' = x \sin\theta' - z \cos\theta' \tag{8.40}$$

and

$$\begin{pmatrix} E \text{ perpendicular to} \\ \text{plane of incidence} \end{pmatrix} \quad \begin{cases} \vec{\mathbf{E}}_r = E_1 e^{-jk_{01}l'}\,\vec{e}_y \\ \vec{\mathbf{H}}_r = \dfrac{E_1}{\eta_1} e^{-jk_{01}l'}\,(\cos\theta'\,\vec{e}_x + \sin\theta'\,\vec{e}_z) \end{cases} \tag{8.41}$$

$$(E \text{ in plane of incidence}) \quad \begin{cases} \vec{\mathbf{E}}_r = E_1 e^{-jk_{01}l'}\,(\cos\theta'\,\vec{e}_x + \sin\theta'\,\vec{e}_z) \\ \vec{\mathbf{H}}_r = -\dfrac{E_1}{\eta_1} e^{-jk_{01}l'}\,\vec{e}_y \end{cases} \tag{8.42}$$

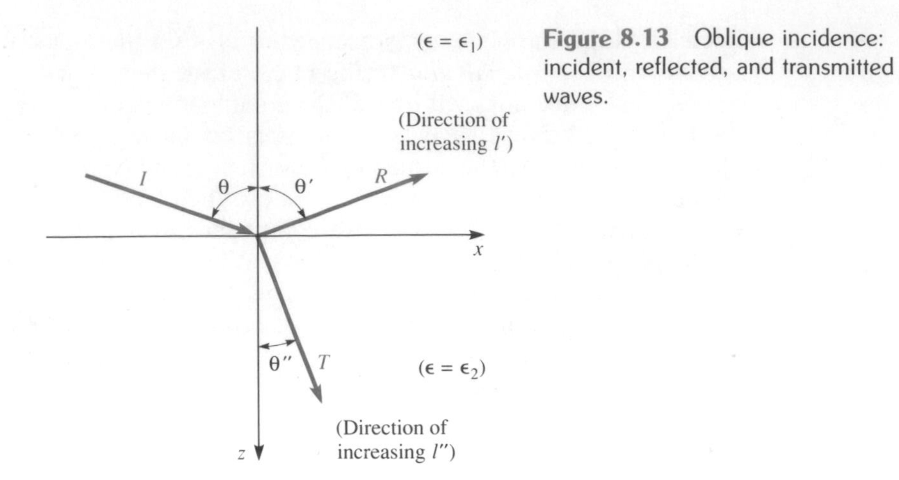

Figure 8.13 Oblique incidence: incident, reflected, and transmitted waves.

and for the transmitted wave the relationships (which are similar to those for the incident wave) are

TRANSMITTED WAVE

$$l'' = x \sin\theta'' + z \cos\theta'' \tag{8.43}$$

$$\begin{pmatrix}E \text{ perpendicular to} \\ \text{plane of incidence}\end{pmatrix} \quad \begin{cases} \vec{\mathbf{E}}_t = E_2 e^{-jk_{02}l''}\,\vec{e}_y \\ \vec{\mathbf{H}}_t = \dfrac{E_2}{\eta_2}\, e^{-jk_{02}l''}\,(-\cos\theta''\,\vec{e}_x + \sin\theta''\,\vec{e}_z) \end{cases} \tag{8.44}$$

$$(E \text{ in plane of incidence}) \quad \begin{cases} \vec{\mathbf{E}}_t = E_2 e^{-jk_{02}l''}\,(\cos\theta''\,\vec{e}_x - \sin\theta''\,\vec{e}_z) \\ \vec{\mathbf{H}}_t = \dfrac{E_2}{\eta_2}\, e^{-jk_{02}l''}\,\vec{e}_y \end{cases} \tag{8.45}$$

All these relationships can readily be checked using vector analysis.

EXAMPLE 8.7

Suppose only the incident wave is present. Find the speed at which field maxima appear to move along the surface in the x direction. (This speed is known as the *phase velocity in the x direction.*)

Solution For a wave with E perpendicular to the plane of incidence

$$\vec{\mathbf{E}} = E_o e^{-jk_{01}l}\,\vec{e}_y = E_o e^{-jk_{01}x\sin\theta}\, e^{-jk_{01}z\cos\theta}\,\vec{e}_y$$

On the surface, $z = 0$, leaving

$$\vec{\mathbf{E}} = E_o e^{-jk_{01}x\sin\theta}\,\vec{e}_y$$

As a function of time,

$$\vec{E}(x, y, z, t) = \text{Re}\,(\vec{\mathbf{E}}\,e^{j\omega t}) = E_o \cos(\omega t - k_{01}x\sin\theta)\,\vec{e}_y$$

The field is maximum when the argument of the cosine is zero; that is, when

$$x = x_{\max} = \frac{\omega t}{k_{01} \sin\theta}$$

Thus

$$U_{Px} = \frac{dx_{\max}}{dt} = \frac{\omega}{k_{01} \sin\theta} = \frac{U_{P1}}{\sin\theta}$$

where U_{P1} is the phase velocity of waves in region 1. For the other polarization, the result is the same.

Interestingly, the phase velocity in the x direction is *larger* than U_{P1}, not smaller, as one might guess at first.[4]

As a first application, let us consider a wave with E in the plane of incidence, striking a perfectly conducting surface. In this case the incident and reflected waves must add together in such a way as to give $E_x = 0$ everywhere in the plane $z = 0$. The total electric field in the x direction, at $z = 0$, is obtained from (8.37), (8.39), (8.40), and (8.42), and set equal to zero:

$$\mathbf{E}_x(z = 0) = E_o \cos\theta \, e^{-jkx\sin\theta} + E_1 \cos\theta' \, e^{-jkx\sin\theta'} = 0 \tag{8.46}$$

This equation must hold for all values of x, which is possible only if $\sin\theta = \sin\theta'$. Thus we have the familiar rule of mirrors, that angle of incidence equals angle of reflection. Once θ' has been set equal to θ, (8.46) immediately states that $E_1 = -E_o$. The total field, incident plus reflected, is then

$$\begin{aligned}\mathbf{E}_x(z) &= E_o \cos\theta \, e^{-jkx\sin\theta} \, [e^{-jkz\cos\theta} - e^{jkz\cos\theta}] \\ &= -2jE_o \cos\theta \, e^{-jkx\sin\theta} \sin[kz\cos\theta]\end{aligned}$$

From this result we observe that there are planes in which $\mathbf{E}_x$ always vanishes. To be specific, $\mathbf{E}_x$ vanishes in the planes determined by $kz\cos\theta = n\pi$, where n is any integer. The situation resembles that of a short-circuited transmission line, where, it will be recalled, the voltage vanishes at distances L from the load given by $kL = n\pi$. In fact, in the case of a normally incident wave the analogy is exact. However, when incidence is oblique, k is replaced by $k\cos\theta$ in the expression for the field nulls. Thus as the angle of incidence increases, the planes of vanishing $\mathbf{E}_x$ remain parallel to the conducting surface but become farther apart.

Next let us consider the case of a wave, with E in the plane of incidence, incident on the discontinuity between two dielectrics with ϵ_1 and ϵ_2. In this

---

[4]If region 1 contains air or vacuum, U_{Px} is larger than the speed of light! However, it is not possible to send messages faster than the speed of light. Do you see why?

case there will be both reflected and transmitted waves. The boundary conditions that apply at $z = 0$ require continuity of tangential E and tangential H, that is, of E_x and H_y. Setting $E_x(z = 0)$ in region 1 equal to $E_x(z = 0)$ in region 2, we have

$$E_o \cos\theta \, e^{-jk_{01}x \sin\theta} + E_1 \cos\theta' \, e^{-jk_{01}x \sin\theta'} = E_2 \cos\theta'' \, e^{-jk_{02}x \sin\theta''} \tag{8.47}$$

Again this equation must hold for all values of x, and this is possible only if

$$k_{01} \sin\theta = k_{01} \sin\theta' = k_{02} \sin\theta'' \tag{8.48}$$

We see that again $\theta' = \theta$. Furthermore

$$\frac{\sin\theta''}{\sin\theta} = \frac{k_{01}}{k_{02}} = \frac{\sqrt{\epsilon_1}}{\sqrt{\epsilon_2}} = \frac{n_1}{n_2} \tag{8.49}$$

The quantity $n = \sqrt{\epsilon/\epsilon_o}$ is known in optics as a material's *index of refraction*. Equation (8.49) is the familiar *Snell's law*. It holds for the other polarization also.

Canceling the exponential terms of (8.47) by means of (8.48), we have

$$E_o \cos\theta + E_1 \cos\theta = E_2 \cos\theta'' \tag{8.50}$$

Using (8.48), the boundary condition on H_y becomes

$$\frac{E_o}{\eta_1} - \frac{E_1}{\eta_1} = \frac{E_2}{\eta_2} \tag{8.51}$$

We can now solve (8.50) and (8.51) simultaneously for E_1 and E_2, with the result

$$\begin{aligned} \frac{E_1}{E_o} &= \frac{\eta_2 \cos\theta'' - \eta_1 \cos\theta}{\eta_2 \cos\theta'' + \eta_1 \cos\theta} \\ \frac{E_2}{E_o} &= \frac{2\eta_2 \cos\theta}{\eta_2 \cos\theta'' + \eta_1 \cos\theta} \end{aligned} \tag{8.52}$$

TOTAL INTERNAL REFLECTION

If, in Fig. 8.13, $\epsilon_1 > \epsilon_2$, the value of $\sin\theta''$ indicated by Snell's law (8.49) can be greater than one. In such a case, no power goes into the transmitted wave, and all of the incident power goes into the reflected wave. This effect is used in optics, for example in the totally reflecting glass prism shown in Fig. 8.14(a). If the index of refraction of the glass is larger than $\sqrt{2}$, no power is transmitted through the diagonal face of the prism.

Although no power goes into the transmitted wave in this case, it would not be correct to conclude that there are no fields in region 2. On the contrary, if one assumes that only the incident and reflected waves exist, it proves impossible to satisfy the two boundary conditions on E and H. For the boundary conditions to be satisfied, some fields have to be present on the

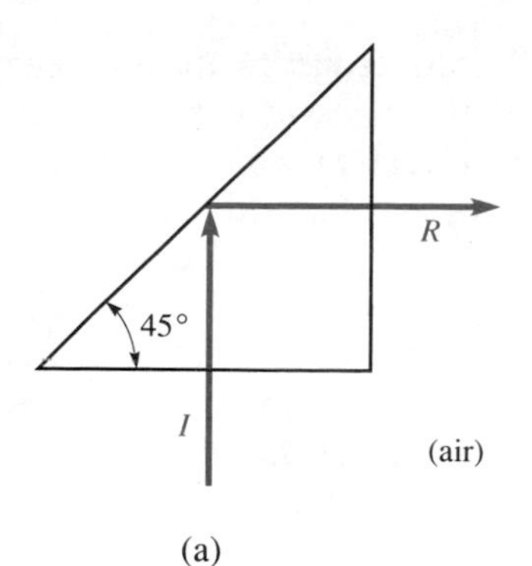

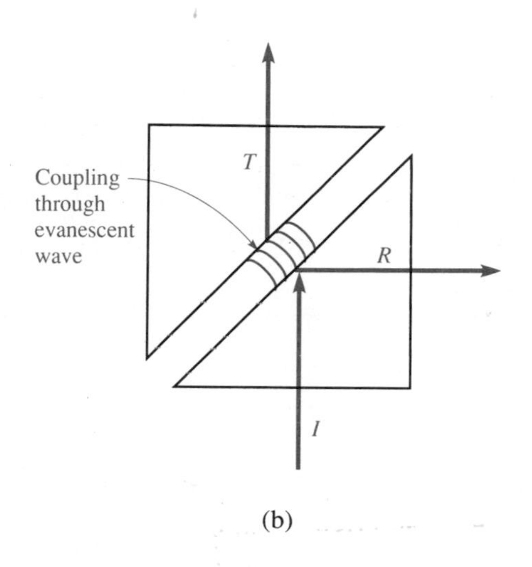

Figure 8.14 (a) Total internal reflection in a glass prism. (b) Energy can be coupled into a second prism through the evanescent fields.

"transmitted" side of the boundary. However, they are not the usual fields of a plane wave, and they do not carry away any power.

Actually, the solutions we have already obtained do satisfy the boundary conditions, and can still be used, provided that we interpret them correctly. Suppose (8.49) indicates that $\sin\theta''$ is equal to A, a real number greater than one. Then $\cos\theta'' = \sqrt{1 - A^2}$ is an imaginary number which we may call $-jB$. (Either the positive or negative square root could be chosen. We choose the negative one because the positive root leads to unphysical results.) Now from (8.45) the electric field in region 2 is

$$\vec{\mathbf{E}}_t = E_2\,(-jB\,\vec{e}_x - A\,\vec{e}_z)\,e^{-jk_{02}l''} \tag{8.53}$$

Inserting the expression for l'' given in (8.43), we have

$$\vec{\mathbf{E}}_t = E_2\,(-jB\,\vec{e}_x - A\,\vec{e}_z)\,e^{-jk_{02}Ax}\,e^{-k_{02}Bz} \tag{8.54}$$

We observe that because of the last factor, the strength of the field decreases exponentially as one moves away from the boundary into region 2. It can also be shown that the time average of the z component of the Poynting vector is zero, indicating that no power is carried away by the "transmitted wave." A field of this type is known as an *evanescent* field.

Experimentally, one can demonstrate the presence of the evanescent field by placing a second prism close to the first, as shown in Fig. 8.14(b). If the spacing between the prisms is small enough, the evanescent field can penetrate the second prism, where it once again becomes a "real" wave. In that case the transmitted wave reappears. Another way to think of this is that when an air gap is sufficiently small, the wave can jump across it, as though the gap weren't there. This interesting effect is mathematically similar to the phenomenon of "tunneling" in quantum mechanics.

EXERCISE 8.5

The index of refraction of the prism in Fig. 8.14(a) is 2.3. The incident wave, coming from a visible red helium-neon laser has wavelength (in air) 633 nm. (One nm $= 10^{-9}$ meter.) Find the distance from the diagonal face of the prism at which the field has decreased to $1/e$ of its value at the face.

Answer 79 nm.

BREWSTER'S ANGLE

Inspection of (8.52) reveals that the reflected wave vanishes if

$$\eta_2 \cos \theta'' - \eta_1 \cos \theta = 0 \tag{8.55}$$

By means of (8.49) and a bit of algebra, we find that (8.55) is satisfied if $\theta = \theta_B$, where

$$\tan \theta_B = \sqrt{\frac{\epsilon_2}{\epsilon_1}} = \frac{n_2}{n_1} \tag{8.56}$$

The angle θ_B is known as *Brewster's angle*. At this angle of incidence no power is lost to a reflected wave; all incident power is transmitted. It should be noted that since we used (8.52), our conclusion applies to the case of *electric field in the plane of incidence*. If the calculations are repeated for E perpendicular to the plane of incidence, we find that there is *no* angle at which reflections disappear.

The phenomenon of Brewster's angle can be useful when reflections from dielectric surfaces, such as glass windows, would be harmful. For example, Fig. 8.15 shows a typical gas laser. The ionized gas in the laser tube has an absorption coefficient that is negative. As a result, waves passing through it tend to grow, instead of decreasing in amplitude. When the laser is working correctly, these waves bounce back and forth between the mir-

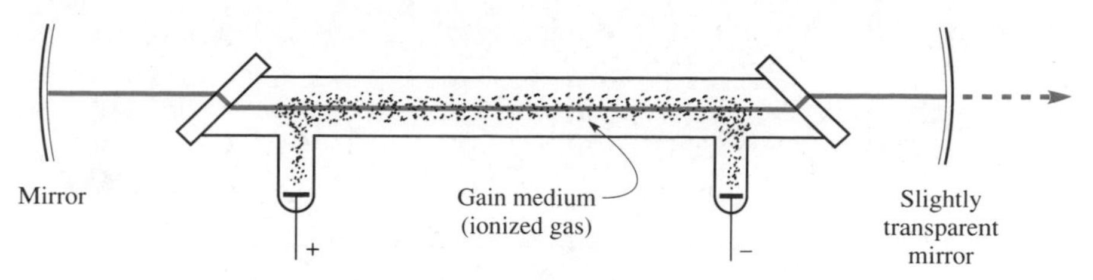

Figure 8.15 Gas laser with Brewster-angle windows.

rors, being amplified at each pass through the gas, so that they eventually become large. However, the actual amplification that occurs on each trip through the gas is sometimes quite small, just a few percent. In that case, the glass windows at the ends of the tube might reflect away more power on each pass than is gained from the gain medium. To avoid these reflection losses, the windows are mounted at Brewster's angle. The laser's output (obtained by a small amount of transmission through one of the mirrors) is then found to be linearly polarized with E in the plane of incidence on the Brewster-angle windows (that is, in the plane of the paper, in Fig. 8.15.) The other polarization has large reflection losses, and thus fails to grow to observable size.

IMPEDANCE METHOD

In section 8.3 we saw how transmission-line techniques can be used in problems involving plane waves. These techniques can also be used when incidence is oblique, provided that certain modifications are made. Let all dielectric surfaces be parallel to each other and to the x-y plane, and let the plane of incidence be the x-z plane, as shown in Fig. 8.16. Then the quantities that are continuous across the various boundaries are the tangential components of E and H; hence, we define the *wave impedance in the z direction* in terms of the components perpendicular to z (and hence parallel to the surfaces). For the case of E in the plane of incidence, let

$$Z_{\|}(z) = \frac{E_x}{H_y} \tag{8.57}$$

For a single wave propagating at an angle θ to the normal, $E_x = E_o \cos\theta e^{-jkl}$, $H_y = (E_o/\eta)e^{-jkl}$, and the wave's characteristic impedance is

$$(E \text{ in plane of incidence}) \qquad Z_{\|} = \eta \cos\theta \tag{8.58}$$

For a wave having E perpendicular to the plane of incidence, we let $Z_{PERP}(z) = -E_y/H_x$. For a single wave $E_y = E_o e^{-jkl}$, $H_x = -(E_o/\eta)\cos\theta e^{-jkl}$, and the wave's characteristic impedance is

$$(E \text{ perpendicular to plane of incidence}) \qquad Z_{PERP} = \frac{\eta}{\cos\theta} \tag{8.59}$$

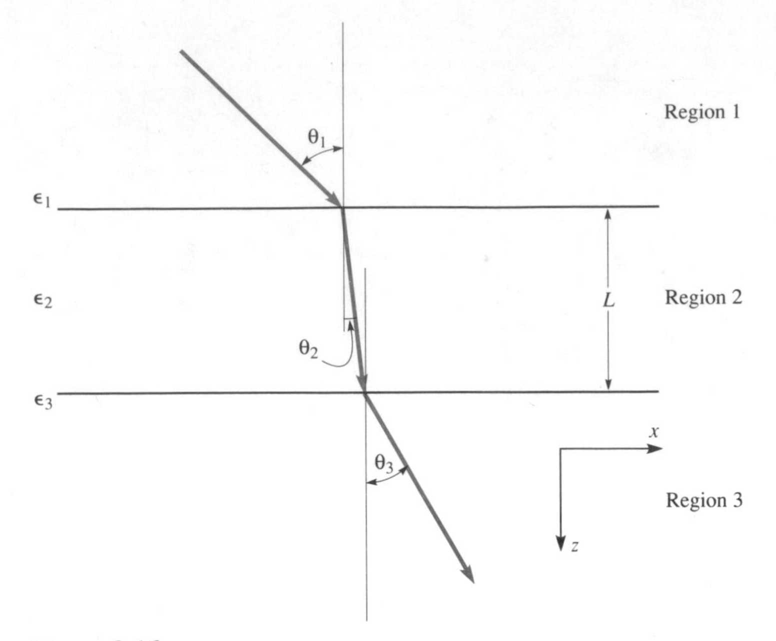

Figure 8.16

We also note that the surfaces perpendicular to the z direction only produce reflections in that direction.[5] Thus the relevant phase delay is the phase delay in the z direction, which, from (8.38) or (8.39) and (8.37) is

$$k_z = k \cos \theta \tag{8.60}$$

for either polarization. In these equations η and k are respectively $\sqrt{\mu/\epsilon}$ and $\omega\sqrt{\mu\epsilon}$ for the medium in which the wave is propagating.

With these modifications, one can solve complex problems involving oblique incidence on multilayered structures, using transmission-line analogies and the Smith chart. A summary of the analogous quantities will be found in Table 8.2.

EXAMPLE 8.8

A sheet of perfectly conducting metal is covered by a layer of dielectric of thickness L, as shown in Fig. 8.17. An incident wave with E in the plane of incidence arrives at angle θ_1, as shown; this wave is described by

$$\vec{\mathbf{E}}_i = E_o \left(\cos\theta_1 \, \vec{e}_x - \sin\theta_1 \, \vec{e}_z\right) e^{-jk_{01}(x \sin\theta_1 + z\cos\theta_1)}$$

[5] Propagation parallel to a surface is unaffected by the surface. From (8.47) and (8.48) we see that the propagation coefficient in the x direction of all the waves is the same as that of the incident wave.

Table 8.2 ANALOGY BETWEEN TRANSMISSION LINES AND PLANE WAVES AT OBLIQUE INCIDENCE

	Plane Waves	
Transmission Line	***E* in Plane of Incidence**	***E* ⊥ to Plane of Incidence**
$\mathbf{v}$	$\mathbf{E}_x$	$\mathbf{E}_y$
$\mathbf{i}$	$\mathbf{H}_y$	$\mathbf{H}_x$
k	$k_z = k\cos\theta$	$k_z = k\cos\theta$
$Z(z) = \dfrac{\mathbf{v}}{\mathbf{i}}$	$Z_{\parallel} = \dfrac{E_x}{H_y}$	$Z_{\perp} = -\dfrac{E_y}{H_x}$
Z_o	$\eta\cos\theta$	$\dfrac{\eta}{\cos\theta}$

Note: $k = \omega\sqrt{\mu\epsilon}$, $\eta = \sqrt{\mu/\epsilon}$. The values to be used for μ and ϵ are those for the particular layer in question.

Let $\theta_1 = 45°$, $\epsilon_1 = \epsilon_o$, $\epsilon_2 = 4\epsilon_o$, $f = 10$ GHz, $L = 0.175$ cm, and E_o is a given constant. Find the phasor describing the reflected wave in region 1.

Solution We need to find the reflection coefficient at $z = 0$. In transmission-line theory this is given by

$$\rho = \frac{Z_L - Z_o}{Z_L + Z_o} \tag{8.61}$$

where Z_o is the characteristic impedance of region 1 and Z_L is the "load" impedance seen looking from $z = 0$ toward the load. From Table 8.2 the

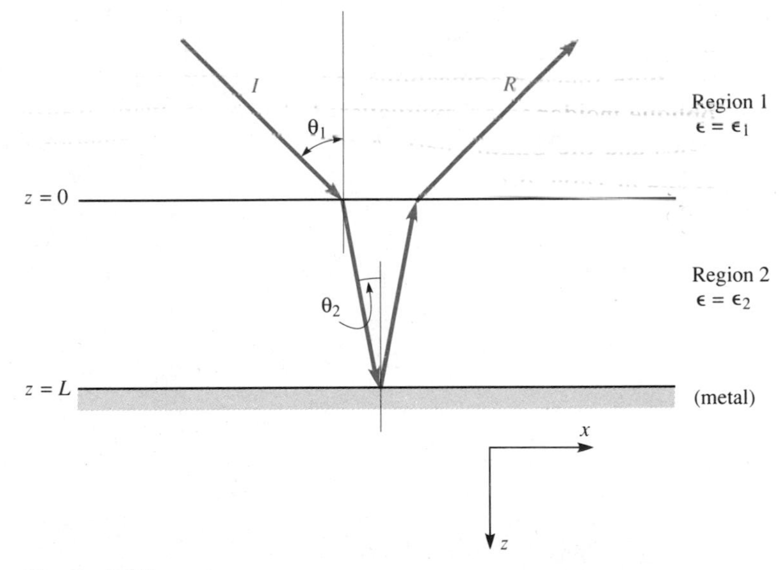

Figure 8.17

corresponding formula for the present case is

$$\rho = \frac{Z_L - \eta_1 \cos\theta_1}{Z_L + \eta_1 \cos\theta_1} \tag{8.62}$$

We still must find $Z_L = Z_{\parallel}(z = 0)$. We note that because of the continuity of the tangential fields, Z_L is the same on both sides of the boundary.

To find $Z_{\parallel}(z = 0)$ we must work back from the known load impedance at the metal. Here the boundary condition is $E_x = 0$, and therefore $Z_{\parallel}(z = L) = 0$. The transmission-line formula which gives the impedance at a distance L before a load Z_L is, from (2.12)

$$Z = Z_o \left[\frac{Z_L \cos kL + jZ_o \sin kL}{Z_o \cos kL + jZ_L \sin kL}\right] \tag{8.63}$$

In the present case we have $Z_L = 0$. Replacing Z_o and k by the analogous quantities from Table 8.2 results in

$$Z_{\parallel}\,(z = 0) = j\eta_2 \cos\theta_2 \tan\left[k_{02}(\cos\theta_2)L\right] \tag{8.64}$$

From Snell's law we find that $\theta_2 = 20.7°$. Inserting this and the given numerical data we have

$$Z_{\parallel}\,(z = 0) = j\left(\frac{377}{2}\right) \cos(20.7°) \tan\left[4.19(\cos 20.7°)(0.175)\right]$$

(note that the argument of the tangent is in radians)

$$= 144j \text{ ohms}$$

Inserting this into (8.62) we have

$$\rho = \frac{144j - 377(.707)}{144j + 377(.707)} = \frac{302e^{j\,152°}}{302e^{j\,28°}} = (1)e^{j\,124°}$$

That the magnitude of ρ should be unity is expected. Since neither the perfect dielectric nor the perfect metal can absorb power, all power must be reflected. The effect of the dielectric is only to alter the phase of the reflected wave.

The significance of the reflection coefficient is that at $z = 0$, E_x of the reflected wave is equal to ρ times E_x of the incident wave. The general expression for the reflected wave in region 1, from (8.42) and (8.40) is

$$\vec{\mathbf{E}}_r = E_1\,(\cos\theta_1\,\vec{e}_x + \sin\theta_1\,\vec{e}_z)\,e^{-jk_{01}(x\sin\theta_1 - z\cos\theta_1)}$$

Setting the x component of $\vec{E}_r$ at $z = 0$ equal to ρ times E_x of the given incident wave, we have

$$E_1 \cos\theta_1\, e^{-jk_{01}x\sin\theta_1} = \rho E_o \cos\theta_1\, e^{-jk_{01}x\sin\theta_1}$$

Thus, $E_1 = \rho E_o$, and the reflected wave is

$$\vec{\mathbf{E}}_r = E_o\,(0.707\vec{e}_x + 0.707\vec{e}_z)\,e^{-j(1.48x - 1.48z - 2.16\,\text{rad})}$$

(where x and z are in centimeters).

8.5 PLANE WAVES IN LOSSY MEDIA

Until now in this chapter we have been discussing plane waves in ideal lossless media. That is, we have been assuming that the medium is nonconductive and also that the value of ϵ is positive and real. Most real media, however, are somewhat absorptive, or, as we say, "lossy." We shall now see how the phenomenon of absorption arises from the theory.

The process of current conduction is usually thought about in connection with *free electrons,* which are electrons that are not tied down, but are free to wander about. The systematic motion of such electrons is what we normally think of as a current. In many materials these currents obey Ohm's law, $\vec{J} = \sigma_E \vec{E}$, where σ_E is the conductivity of the material. As we are about to see, conductivity contributes to absorption of electromagnetic waves.

If a material has no free electrons, however, one cannot assume that it is nonabsorptive. The reason for this is that even in nonconductive materials, the value of ϵ is not real; it contains an imaginary part. Let us imagine that we could somehow hold the free electrons in a crystal (if there are any) motionless. In that case the permittivity is that of the crystal lattice itself; we shall refer to this permittivity as ϵ_C. It can be separated into its real and imaginary parts according to

$$\epsilon_C = \epsilon_C' - j\epsilon_C'' \tag{8.65}$$

Both ϵ_C and σ_E enter Maxwell's equations through the "curl H" equation:

$$\begin{aligned}
\nabla \times \vec{\mathbf{H}} &= j\omega\epsilon_C \vec{\mathbf{E}} + \sigma_E \vec{\mathbf{E}} \\
&= j\omega\left[(\epsilon_C' - j\epsilon_C'') - j\frac{\sigma_E}{\omega}\right]\vec{\mathbf{E}} \\
&= j\omega\left[\epsilon_C' - j\left(\epsilon_C'' + \frac{\sigma_E}{\omega}\right)\right]\vec{\mathbf{E}}
\end{aligned} \tag{8.66}$$

We observe that the imaginary part of ϵ_C adds to the conductivity in this equation, which means that these two terms contribute to absorption in similar fashion. Thus it is reasonable to combine them, and regard the total ϵ of the material as being the sum of their two contributions. Defining

$$\epsilon = \epsilon' - j\epsilon'' \tag{8.67}$$

where

$$\begin{aligned}
\epsilon' &= \epsilon_C' \\
\epsilon'' &= \epsilon_C'' + \frac{\sigma_E}{\omega}
\end{aligned} \tag{8.68}$$

we recover Maxwell's equation in its usual form

$$\nabla \times \vec{\mathbf{H}} = j\omega\epsilon\vec{\mathbf{E}}$$

where, however, we remember that ϵ is now a complex number.

The derivation of (7.44) now goes through exactly as before. However,

$$k_o \equiv \omega\sqrt{\mu\epsilon} = \omega\sqrt{\mu\epsilon'\left(1 - j\frac{\epsilon''}{\epsilon'}\right)} \tag{8.69}$$

is now complex. Defining, in analogy to (1.49),

$$k_o = \beta - j\alpha \tag{8.70}$$

we find, upon equating (8.70) and (8.69)

$$\begin{aligned} \alpha &= \omega\left\{\frac{\mu\epsilon'}{2}[\sqrt{1 + (\epsilon''/\epsilon')^2} - 1]\right\}^{1/2} \\ \beta &= \omega\left\{\frac{\mu\epsilon'}{2}[\sqrt{1 + (\epsilon''/\epsilon')^2} + 1]\right\}^{1/2} \end{aligned} \tag{8.71}$$

To interpret these results we note that all fields of the wave are proportional to e^{-jk_oz}. For example, for a linearly polarized wave with $\vec{E}$ in the x direction,

$$\vec{\mathbf{E}}_x \propto e^{-jk_oz} = e^{-\alpha z}\, e^{-j\beta z} \tag{8.72}$$

Thus we see that the amplitude of E decreases exponentially as the wave travels, in proportion to $e^{-\alpha z}$. From (8.27), the power per unit area, or intensity, of the wave decreases as $e^{-2\alpha z}$. The phase velocity of the wave is given by

$$U_P = \frac{\omega}{\beta}$$

and thus is also affected by ϵ''.

Expressions (8.71) are rather daunting, but they can be simplified. If a material is even slightly transparent the ratio ϵ''/ϵ', known as the *loss tangent,* is usually much less than one. Therefore we can expand our expressions for α and β by means of the binomial theorem. The result is

$$\begin{aligned} \alpha &\cong \frac{k_{00}\epsilon''}{2\epsilon'} \\ \beta &\cong k_{00}\left[1 + \frac{1}{8}\left(\frac{\epsilon''}{\epsilon'}\right)^2\right] \end{aligned} \tag{8.73}$$

where $k_{00} \equiv \omega\sqrt{\mu\epsilon'}$.

In the opposite limit, that of $\sigma_E \to \infty$, we find that $\epsilon \to -j\epsilon''$, and $\epsilon'' \to \frac{\sigma_E}{\omega}$. This case is identical to what we have considered earlier under the heading of the *skin effect.*

EXERCISE 8.6

Let the intensity of a plane wave be I. Verify *Beer's law*, which states that

$$\frac{dI}{dz} = -2\alpha I \tag{8.74}$$

8.6 POINTS TO REMEMBER

1. In a plane electromagnetic wave, $\vec{E}$, $\vec{H}$, and the direction of propagation are all mutually perpendicular. The electric and magnetic fields are in phase. The wave is analogous to a transmission-line wave, with electric field playing the role of voltage and magnetic field playing the role of current.

2. When the electric field of a plane wave always points in the same direction, the wave is said to be "linearly polarized." Waves that are not linearly polarized are said to be "elliptically polarized." An important special case of the latter is the circularly polarized wave.

3. Poynting's theorem is an identity based on Maxwell's equations. In connection with electromagnetic waves it can often be interpreted as a statement about power flow and energy conservation. In such cases the Poynting vector $\vec{E} \times \vec{H}$ represents power flow.

4. The ratio of the electric field phasor to the magnetic field phasor in a plane wave is known as the "wave impedance." It can be used to calculate the reflection coefficient in the case of normal incidence on a dielectric, much as characteristic impedance is used in transmission-line problems.

5. When a plane wave strikes a dielectric interface that is not perpendicular to the direction of propagation, the incidence is said to be "oblique." The plane containing the normal to the interface and the propagation direction is called the "plane of incidence." Reflection and transmission behavior depends on whether the electric field of the incident wave is in the plane of incidence or perpendicular to it.

6. Snell's law,

$$\frac{\sin\theta''}{\sin\theta} = \frac{n_1}{n_2}$$

relates the angle of refraction to the ratio of the refractive indices. If the angle of refraction predicted by Snell's law is greater than 90°, no transmitted wave exists at all; this effect is known as "total internal reflection." An "evanescent field" exists in the region into which no transmission occurs.

7. For waves with electric field in the plane of incidence, there is an angle of incidence, known as "Brewster's angle," at which no reflection occurs.

8. Transmission-line analogies, including the use of the Smith chart, can be used to analyze the reflection and transmission of plane waves at dielectric interfaces. These methods can be extended to the case of oblique incidence as well.

REFERENCES

1. S. Ramo, J. R. Whinnery, and T. Van Duzer, *Fields and Waves in Communication Electronics*. New York: Wiley, 1984.
2. N. N. Rao, *Elements of Engineering Electromagnetics*. Englewood Cliffs, N.J.: Prentice-Hall, 1987.
3. S. V. Marshall and G. G. Skitek, *Electromagnetic Concepts and Applications*, 2nd ed. Englewood Cliffs, N.J.: Prentice-Hall, 1987.
4. P. Lorrain and D. Corson, *Electromagnetic Fields and Waves*, 2nd ed. New York: W. H. Freeman, 1970.
5. D. K. Cheng, *Field and Wave Electromagnetics*. Reading, Mass.: Addison-Wesley, 1983.
6. D. T. Paris and F. K. Hurd, *Basic Electromagnetic Theory*. New York: McGraw-Hill, 1969.

Readers interested in the optical applications of plane waves may wish to consult:

7. F. A. Jenkins and H. E. White, *Fundamentals of Optics*, 4th ed. New York: McGraw-Hill, 1976.
8. H. A. Haus, *Waves and Fields in Optoelectronics*. Englewood Cliffs, N.J.: Prentice-Hall, 1984.
9. A. E. Siegman, *Lasers*. Mill Valley, Calif.: University Science Books, 1986.

PROBLEMS

Section 8.1

8.1 Find the phasors representing $\vec{E}$ and $\vec{H}$ for a linearly polarized plane wave with $\vec{E}$ in the x direction, propagating in the $-z$ direction. What are the corresponding expressions for the fields as functions of time?

*8.2 Find the phasors representing $\vec{E}$ and $\vec{H}$ for a linearly polarized plane wave with $\vec{E}$ in the direction of the vector $\vec{e}_x - \vec{e}_y$, propagating in the $-z$ direction. What are the corresponding expressions for the fields as functions of time?

8.3 Consider a plane wave in vacuum described by $\vec{E} = E_o e^{-jkz} \vec{e}_x$, with $k = 1\ \text{cm}^{-1}$. You are an observer stationed at $z = 0$.
(a) Sketch $E(t)$ and $H(t)$ at your position as functions of time.
(b) On the same scales, sketch E and H as seen by an observer located at $z = 1/k$.

*8.4 Show that for a linearly polarized plane wave with any direction of polarization, propagating in any direction,

$$\vec{k} \times \vec{\mathbf{E}} = \omega\mu\vec{\mathbf{H}}$$
$$\vec{k} \times \vec{\mathbf{H}} = -\omega\epsilon\vec{\mathbf{E}} \tag{8.75}$$

where $\vec{k}$ is a vector with magnitude k, in the direction of propagation.

8.5 Use the result of the preceding problem to find the unit vector pointing in the direction of $\vec{E}$ for a wave propagating in the direction $-2\vec{e}_x + \vec{e}_y$, if $\vec{H}$ is in the z direction.

*8.6 Derive equations (8.75) (Problem 8.4) for the case of circularly polarized waves propagating in any arbitrary direction.

*8.7 Show that a plane wave propagating in an arbitrary direction is described by the phasor

$$\vec{\mathbf{E}} = \vec{E}_o e^{-j\vec{k}\cdot\vec{r}}$$

where $\vec{r}$ is the radius vector $x\vec{e}_x + y\vec{e}_y + z\vec{e}_z$ and $\vec{k}$ is a vector with magnitude k, in the direction of propagation.

8.8 (a) Find the phasor representing magnetic field for the circularly polarized wave (8.8).
(b) Express the magnetic field as a function of time.
(c) Show that the magnetic field is perpendicular to the electric field at all positions and times.

8.9 For the circularly polarized wave (8.8):
(a) Show that the strength of the electric field is the same at all positions and times.
(b) Let θ be the angle the electric field makes with the x axis. Show that $\theta = kz - \omega t$.

**8.10 In a certain material, waves propagating in the z direction propagate with different phase velocities, depending on their polarizations. The linearly polarized wave with E in the x direction behaves as though $\epsilon = \epsilon_1$, whereas that with E in the y direction behaves as though $\epsilon = \epsilon_2$. (Such a material is said to be *birefringent*.) A plane wave with angular frequency ω enters the material at $z = 0$. When it enters, it is linearly polarized with $\vec{E}$ in the direction $(\vec{e}_x + \vec{e}_y)$.
(a) Show that the incident wave can be represented as the sum of two linearly polarized waves respectively polarized in the x and y directions.

(b) Show that after propagating a certain distance L through the material, the wave has become circularly polarized. What is the value of L? (This device, often used in optics to create circularly polarized waves, is known as a *quarter-wave retardation plate*. Do you see why it has that name?)

**8.11 An expression for the electric field of a general elliptically polarized wave is

$$\vec{\mathbf{E}} = E_o(\vec{e}_x + Ae^{j\psi}\,\vec{e}_y)e^{-jkz} \tag{8.76}$$

where A and ψ are real constants.

(a) Show that the orbit traced by the vector $\vec{E}$ at $z = 0$ is an ellipse. (*Suggestion:* Express E_x and E_y as functions of time; then eliminate time to obtain an equation relating E_x and E_y.)

(b) Show that the axes of the ellipse traced by $\vec{E}$ are rotated with respect to the x-y axes by an angle θ given by

$$\tan 2\theta = \frac{2A\cos\psi}{A^2 - 1}$$

(c) Find the maximum and minimum values of the electric field.

Section 8.2

*8.12 A spherical capacitor is made of two concentric spheres with radii a and b ($a < b$).

(a) Find the energy stored in the electric field when the capacitor is charged to a voltage V.

(b) Use the result of part (a) to find the capacitance between the spheres. Does your result agree with Example 4.10?

*8.13 Let us imagine that an electron is a tiny, perfectly conducting sphere, on the surface of which resides the charge $q = 1.6 \times 10^{-19}$ coulombs. According to Einstein's famous postulate, the total energy available if the electron could be entirely turned into energy is mc^2, where $m = 9.1 \times 10^{-31}$ kg is the electron's mass. On the other hand, we have seen that the electrical energy required to assemble the electron is given by (8.14). By setting this energy equal to mc^2, find the radius of the electron.

(The radius thus found is one-half of the number known as the *classical electron radius*. Although its order of magnitude is useful, the derivation is faulty, because electrons also contain energy that is not electrostatic in nature.)

**8.14 Consider a parallel-plate capacitor with area A and spacing L. It is connected to a voltage source so that the potential difference V between the plates is constant. Suppose one of the plates now moves closer to the other, so that the spacing decreases by ΔL, where $\Delta L \ll L$. This causes W_E to change. Also during the motion, mechanical work $F\Delta L$

is done (where F is the attractive force on the plates), and the voltage source does work $V\Delta Q$ (where Q is the charge on the plates). Write an energy-balance equation relating the change in the energy of the system to the mechanical and electrical work done. (Be careful of algebraic signs.) From your equation, determine the force F.

*8.15 A toroidal inductor (similar to Fig. 5.15) has an iron core with permeability μ. The coil has N turns and carries a current I. The cross-sectional area of the core is A, and the length of the flux path is L_p.
(a) Find the stored magnetic energy W_H.
(b) From circuit theory, the stored energy must also be equal to $\frac{1}{2}LI^2$. Use this and the result of part (a) to find the inductance.

**8.16 A point charge Q is stationary in a constant, uniform magnetic field $\vec{H}$.
(a) Sketch the direction of the Poynting vector at various positions surrounding Q.
(b) Does the Poynting vector represent power flow in this case? Discuss.
(c) Does (8.20) hold in this case?
(d) Prove that $\nabla \cdot \vec{S} = 0$ everywhere except at the point charge.

8.17 The time-averaged power per unit area of an optical beam is correctly known as its *intensity*, with symbol I. Assuming that the beam is a plane wave, show that $I = KE_o^2$, where E_o is the instantaneous maximum electric field. Find the numerical value of the constant K in mks units.

*8.18 A plane wave is described by $\vec{\mathbf{E}} = E_o e^{-jkz}\,\vec{e}_x$.
(a) Find the time-averaged electric energy density $\langle U_E \rangle$. *Answer:* $\epsilon E_o^2/4$.
(b) Find the time-averaged magnetic energy density $\langle U_H \rangle$, and show that $\langle U_E \rangle = \langle U_H \rangle$.

*8.19 In a moving fluid, the mass flow per unit area is equal to the product of density and velocity. (Does this seem reasonable?) Similarly, the ratio of time-averaged power per unit area to total time-averaged energy density ($\langle U_E \rangle + \langle U_H \rangle$) in a wave is known as the *energy velocity*. Using the result of Problem 8.18, show that the energy velocity of a plane wave is c.

*8.20 In a general case we may have

$$\vec{\mathbf{E}} = E_1 e^{j\phi_1}\,\vec{e}_x + E_2 e^{j\phi_2}\,\vec{e}_y + E_3 e^{j\phi_3}\,\vec{e}_z$$

$$\vec{\mathbf{H}} = H_1 e^{j\psi_1}\,\vec{e}_x + H_2 e^{j\psi_2}\,\vec{e}_y + H_3 e^{j\psi_3}\,\vec{e}_z$$

where the E_i, H_i, ϕ_i, and ψ_i are arbitrary real numbers. Show that the time average of the Poynting vector is equal to $\frac{1}{2}\,\mathrm{Re}\,(\vec{\mathbf{E}} \times \vec{\mathbf{H}}^*)$.

*8.21 Find the time-averaged power per unit area transmitted by the circularly polarized wave (8.8).

8.22 Equations (8.29) and (8.32) describe the total fields that arise when a plane wave is normally incident on a perfectly conductive surface. Find the time-averaged Poynting vector as a function of z. Does your answer make sense physically?

Section 8.3

8.23 Use Equation (8.75) (Problem 8.4) to verify (8.31).

*8.24 A *magnetic wall* is an imaginary surface that behaves toward magnetic fields the way a perfectly conducting surface behaves toward electric fields. That is, at a magnetic wall, the tangential component of $\vec{H}$ must vanish. (Although true magnetic walls do not exist, the concept is useful in certain electromagnetic field problems—much in the way ideal current sources, which do not exist, are useful in circuit theory.)

The plane wave $\vec{\mathbf{E}} = E_o e^{-jkz} \vec{e}_x$ is normally incident on a magnetic wall located at $z = 0$.

(a) Find the phasors representing the total electric and magnetic fields as functions of z.
(b) Find the wave impedance as a function of z.
(c) Find the time-averaged Poynting vector as a function of z.
(d) Find the electric and magnetic fields as functions of time and z.
(e) To what transmission-line load is the magnetic wall analogous?

8.25 Verify using (8.36) that the sum of the incident and reflected powers is equal to the incident power.

*8.26 For a certain dielectric interface, $\eta_2 \gg \eta_1$. A plane wave is normally incident from region 1.

(a) Show that the approximate value of the transmission coefficient τ is 2. Does this seem reasonable physically?
(b) Find the ratios of transmitted and reflected intensities to the incident intensity.
(c) Verify that the values of E_1 and E_2 predicted by (8.36) are consistent with tangential E being continuous at the interface.

*8.27 A "quarter-wave optical coating" is often used in optics to eliminate reflections from dielectric interfaces. Suppose a plane wave is normally incident from air onto an infinitely thick glass slab with relative permittivity ϵ_G. Deposited on the glass is a thin film of dielectric material with relative permittivity ϵ_F and one-quarter-wave thickness (that is, the thickness L of the film satisfies

$$\frac{L}{\left(\dfrac{c}{\sqrt{\epsilon_F}\, f}\right)} = \frac{1}{4}$$

where f is the frequency).

(a) Show, by means of the analogy to the quarter-wave transformer described in Chapter 2, that reflections are eliminated when ϵ_F has the proper value. What is this value?

(b) Suppose the glass in question is a camera lens with $\epsilon_G = 2.5$, and the coating is designed to work at the wavelength (in air) $\lambda = 550$ nm. However, pictures are in fact made with white light, so reflections at other visible wavelengths are of concern. Sketch a graph of the percentage *power* reflected versus wavelength over the visible range (400–700 nm). Assume ϵ_G and ϵ_F are independent of frequency.

*8.28 A plane wave with free-space wavelength λ is normally incident on a window with index of refraction n and thickness L. Find the smallest value of L (other than zero) that allows 100% of the incident power to be transmitted through the window.

*8.29 A plane wave with frequency 5 GHz is normally incident on a dielectric window of thickness 0.6 cm and relative permittivity 12.9. Find the fraction of the incident power that is reflected.

*8.30 A plane wave with frequency 3 GHz is normally incident from the $-z$ direction on a perfectly conducting plane located at $z = 0$. The region -1 cm $< z < 0$ contains a dielectric material with $\epsilon = 3\epsilon_o$. Find the location of the electric field maximum that is closest to $z = 0$ but outside the dielectric.

*8.31 Although in practice most electromagnetic waves are nearly sinusoidal, it is possible to launch waves of other forms, just as on transmission lines. For example, suppose the electric field in the plane $z = 0$ is 0 for $t < 0$, and $E_o\vec{e}_x$ for $t > 0$.

(a) Sketch E_x as a function of z for several times after the field is turned on.

(b) Sketch E_x as a function of time at several positions.

(c) Suppose a perfectly conducting metal plane is placed at $z = L$. Sketch E_x as a function of z at times $t = 2L/3c$, $4L/3c$, and $7L/3c$.

**8.32 A plane wave is normally incident from air onto a plane-parallel slab of dielectric. The permittivity of the slab is large, so that $\eta_1 \ll \eta_o$ (where the subscripts 1 and o refer to dielectric and air respectively). The slab is one-half wave thick at the frequency ω_o (i.e., $\omega_o L/U_{P1} = \pi$, where L is the thickness of the slab and U_{P1} is the velocity of light in the dielectric). Find an expression for the fraction of the incident power reflected by the slab at the frequency $\omega_o + \Delta\omega$, where $\Delta\omega \ll \omega_o$. Show that 50% reflection is obtained when

$$\Delta\omega = \pm \frac{2\omega_o}{\pi(\eta_o/\eta_1)}$$

approximately.

Since transmission exceeding 50% occurs only within a small frequency range of width $2\Delta\omega$, this device can be used as a narrow bandpass filter. It is a variation of a well-known instrument called the *Fabry-Perot interferometer*.

Section 8.4

8.33 Verify (8.37).

*8.34 Verify (8.38) and (8.39).

*8.35 Verify equations (8.40)–(8.45).

8.36 The direction of propagation of the wave $\vec{\mathbf{E}} = \vec{E}_o e^{-jkl}$ is in the x-z plane, as shown in Fig. 8.12.

(a) Show that the wave can be rewritten in the form

$$\vec{\mathbf{E}} = \vec{E}_o\, e^{-j(k_x x + k_y y)} = \vec{E}_o e^{-j\vec{k}\cdot\vec{r}}$$

and find k_x and k_y.

(b) Show that the phase velocities of the wave in the x and z directions are ω/k_x and ω/k_z respectively.

8.37 Sketch the equiphase surfaces of the evanescent wave (8.54). Locate the surfaces on which E_x and E_z are maximum at $t = 0$. What is the phase velocity of the wave in the x direction? Is this velocity less or more than the velocity of light in the medium?

*8.38 (a) Find the magnetic field associated with evanescent field (8.54).

(b) Construct the Poynting vector. What is the time average of its z component? Of its x component?

*8.39 The 45° prism shown in Fig. 8.18 has index of refraction $n = 2.0$. It is being used to couple an incident optical beam to a slow wave propagating on a metal surface. Coupling occurs when the phase velocity in the x direction of the evanescent wave below the prism is equal to U_{Po}, the velocity of the slow wave on the surface.

Let $U_{Po} = 0.58\,c$, where c is the velocity of light in vacuum. Find the angle θ that the incident wave must make with the normal to the prism, in order for coupling to occur.

This device is known as a *prism coupler*.

**8.40 Derive equation (8.56) from (8.55).

*8.41 Derive results corresponding to (8.52) for the case of E perpendicular

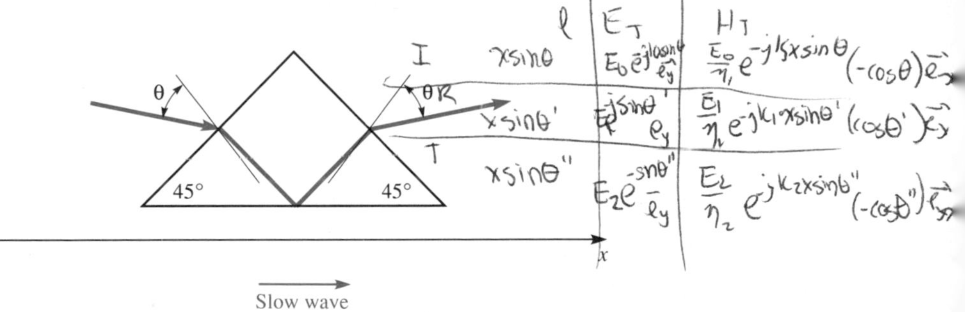

Figure 8.18 Coupling to a slow surface wave with a prism coupler. (For Problem 8.39).

B.C.'s: $E_{T_1} = E_{T_2}$; $E_i + E_r = E_{Trans}$
Snell's law (for same exponents) → $E_0 + E_1 = E_2$

to the plane of incidence. Show that for this polarization, there is no angle of incidence for which the reflected wave vanishes. (Assume $\mu = \mu_o$ for both media.)

*8.42 The gas laser shown in Fig. 8.15 is designed so that the beam meets the glass windows ($n = 1.8$) at Brewster's angle. However, because of a fabrication error, the angle of incidence turns out to be 5° too small. Find the percentage of incident power that is reflected from a single interface, when the beam in the air strikes the glass.

**8.43 A plane wave with E in the plane of incidence strikes a dielectric at an angle θ, as shown in Fig. 8.13. Verify that the sum of the time-averaged powers transmitted through and reflected from an area A of the surface is equal to the time-averaged power incident on that area.

**8.44 A plane wave having amplitude E_o, with E in the plane of incidence, strikes a dielectric slab of thickness L, as shown in Fig. 8.16.

(a) Find the power striking an area A of the surface.

(b) Find the power emerging from an equal area A into region 3.

(c) Let the ratio of the power found in (b) to that in (a) be called the power transmission ratio. Show that the power transmission ratio is the same, if the beam is incident from region 3, with angle of incidence θ_3.

*8.45 Demonstrate the transmission-line analogy by deriving equations (8.52) from the transmission-line formula (1.33), using Table 8.2. (*Note:* This is not as obvious as it seems at first. Watch out for the factor $\cos\theta$ in the numerator of the second equation of (8.52)).

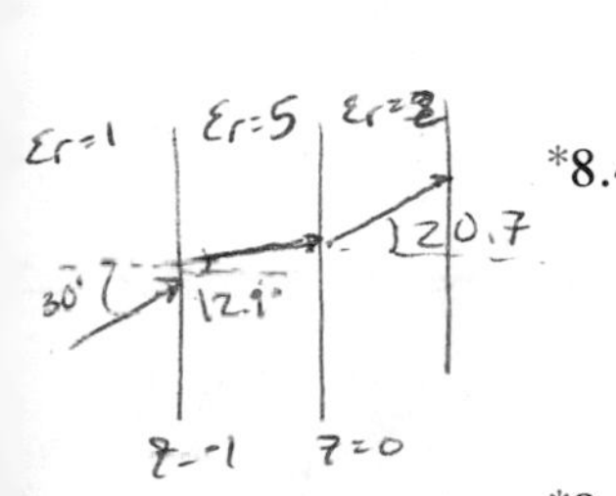

*8.46 The region $z > 0$ is filled with dielectric having $\epsilon = 2$. The region -1 cm $< z < 0$ contains material with $\epsilon = 5$, and the region $z < -1$ cm contains air. A wave with frequency 10 GHz and E in the plane of incidence is incident on the materials from the $-z$ direction, with angle of incidence 30°. Find the fraction of the incident power that is reflected.

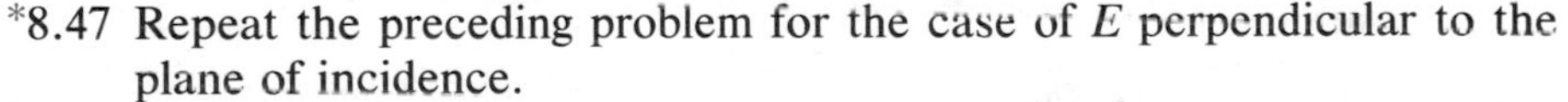

*8.47 Repeat the preceding problem for the case of E perpendicular to the plane of incidence.

**8.48 A plane wave with frequency 30 GHz strikes a plane-parallel slab of dielectric with $\epsilon = 6\epsilon_o$ and thickness 0.236 cm. Construct a graph of fractional reflected power versus angle of incidence:

(a) for E perpendicular to the plane of incidence

(b) for E in the plane of incidence.

Section 8.5

*8.49 Consider the wave (8.72) propagating in lossy material.

(a) Is the wave a *plane* wave?

(b) Are $\vec{E}$ and $\vec{H}$ perpendicular to each other? To the direction of propagation?

(c) Are E and H in phase with each other? (That is, at a given position, are E and H both maximum at the same instant of time?)

*8.50 Let the impedance of a wave in a lossy medium be described by the usual formula

$$\eta = \frac{\mathbf{E}_x}{\mathbf{H}_y}$$

Show that in the limit of small loss tangent,

$$\eta \cong \eta_{oo} \left\{ \left[1 - \frac{3}{8} \left(\frac{\epsilon''}{\epsilon'} \right)^2 \right] + j \frac{\epsilon''}{2\epsilon'} \right\}$$

where $\eta_{oo} = \sqrt{\mu/\epsilon'}$.

*8.51 A plane wave from a carbon dioxide laser ($f = 3 \cdot 10^{13}$ Hz) is traveling through undoped silicon ($\epsilon/\epsilon_o = 11.7$, assumed to be real). The undoped semiconductor contains no free charge carriers. The wave then enters a doped region of the semiconductor, with conductivity $\sigma_E = 10^3$ (ohm meter)$^{-1}$. Using the result of Problem 8.50, find the fraction of the power reflected at the interface between doped and undoped material.

CHAPTER 9

Guided Waves

As we have seen, electromagnetic waves can propagate freely through vacuum. However, in many cases one wishes to conduct signals from a certain starting point to a certain end point. Plane waves, which by their nature are large and unconfined, are unsuitable for conducting signals from point to point. What is needed is something analogous to the wires used in low-frequency circuits. But when wires become comparable in size to the wavelength at the frequency of interest, they cease to act as lumped circuit elements. In fact, they are all too likely to act as antennas, radiating energy away from the desired end point and out of the system. Thus we are led to consider waveguiding structures, which are, in general, elongated structures that act as pipes for electromagnetic waves, carrying electromagnetic energy from place to place. Examples include microstrip, a planar guide used for interconnections in microwave circuits, and hollow metal waveguide, the "classical" microwave guide. A recent development of great importance is the optical fiber guide, which is the basis of the new field of optical communications technology. Conventional transmission lines, such as were discussed in Chapters 1 and 2, are in fact waveguiding structures of an especially simple type. Our previous analysis was in terms of a lumped-circuit model. In this chapter we shall obtain an improved analysis based on their electromagnetic fields.

In order to describe guided waves, we must find solutions of Maxwell's equations that satisfy the boundary conditions for the elongated boundaries. These boundary conditions of course vary from one guide to another, depending on the shape of the guide, whether it is made of metal or dielectric, and so forth. The general sort of solutions we seek are ones that propagate in the z direction in the same way as transmission-line waves and plane waves; that is, the fields are proportional to e^{-jkz}, where k is a real constant that determines the phase velocity. In the case of transmission-line waves and plane waves, we found that $k = k_o$, where $k_o \equiv \sqrt{\mu\epsilon}$. However, for other kinds of waves this may not be true, and we should be prepared to encounter waves for which $k \neq k_o$. In such cases, k is an unknown for which we must solve. The value of k will then determine the phase velocity through $U_P = \omega/k$.

As functions of the transverse coordinates (x and y), the fields may have a complicated form—how complicated depends on the shape of the guide—but the transverse form does not depend on z. In other words, the general form of the guided wave is

$$\vec{\mathbf{E}}(x, y, z) = \vec{E}_o(x, y)e^{-jkz} \tag{9.1}$$

Note that the vector $\vec{E}_o$ can have x, y, and z components, but that it depends only on x and y. The z variation of the fields is all contained in the "propagator" e^{-jkz}. The plane wave is an especially simple special case of (9.1), in which $\vec{E}_o$ is a constant, independent of x and y. (Compare the plane waves (8.1) and (8.8).) The magnetic field of the wave has similar properties, and is described by

$$\vec{\mathbf{H}}(x, y, z) = \vec{H}_o(x, y)e^{-jkz} \tag{9.2}$$

EXERCISE 9.1

A guided wave with angular frequency ω is described by (9.1). Find (1) the electric field as a function of time; and (2) the phase velocity.

Most guiding structures are, in fact, capable of supporting several different waves. By this we mean that there are several different solutions of Maxwell's equations that satisfy the boundary conditions; the different solutions have different propagation constants k, and also different field distributions $\vec{E}_o(x, y)$, $\vec{H}_o(x, y)$. These different solutions are known as *modes* of the guiding structure. In practice it is undesirable for more than one mode of the structure to be present. Therefore, it is best to design the guiding structure in such a way that only one solution to Maxwell's equation exists at the frequency of interest.

9.1 TEM WAVES ON TRANSMISSION LINES

As our first class of guided waves, we shall consider two-conductor metallic guiding structures. The two conductors are not in contact, and are separated by a uniform dielectric material. The conductors must have the same cross section for all values of z. For example, the coaxial transmission line (Fig. 9.1(a)) is such a structure; a transmission line consisting of two parallel cylindrical wires (Fig. 9.1(b)) also belongs to this class. Many other structures

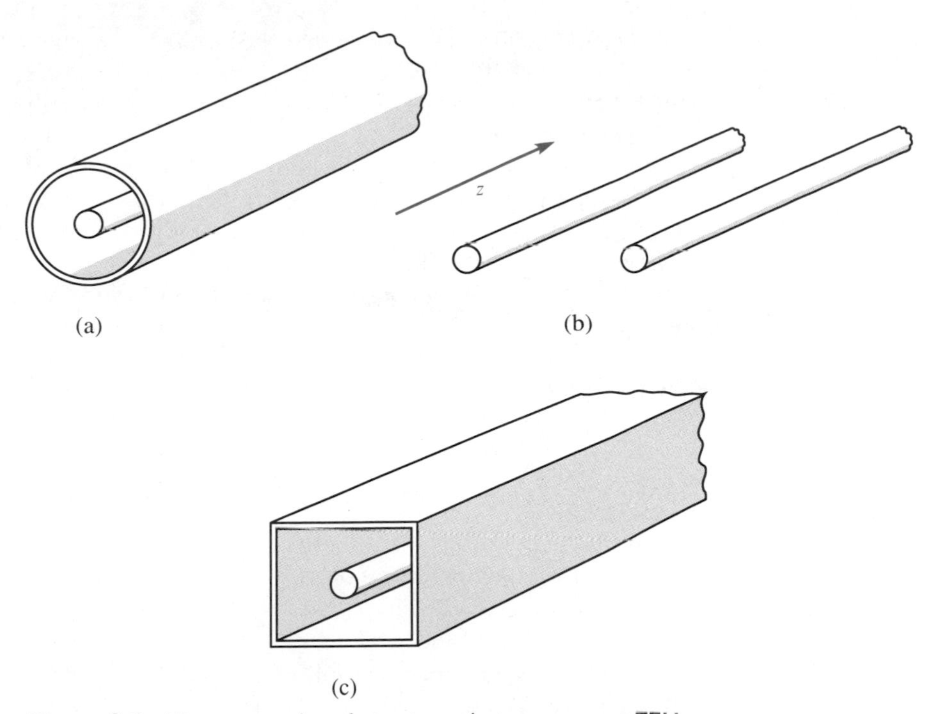

Figure 9.1 Three examples of structures that can support TEM waves.

of this type can be imagined, such as Fig. 9.1(c). This unusual structure has a rectangular outer conductor and an off-center, circular inner conductor. It has little symmetry and thus would be difficult to analyze mathematically, but it does belong to the family of guiding structures under discussion.

Although two-conductor metallic guides can support numerous modes, most of these modes can exist only at high frequencies. In fact there is only one mode of the two-conductor guide, known as the *principal mode,* that exists at low frequencies. This is a very useful property of the two-conductor guide, because so long as the frequency is not too high, single-mode operation is automatically obtained. The principal mode of the two-conductor line has a simplifying mathematical property. As we shall see, the electric and magnetic fields of this wave lie entirely in the x-y plane; they have no z components. Such fields are said to be *transverse*. When a wave's electric field is transverse, it is called a transverse electric (TE) wave; if the magnetic field of a wave is transverse, the wave is called a transverse magnetic (TM) wave. For the principal mode of the parallel-conductor guide, both E and H are transverse, and the wave is called a transverse electromagnetic, or TEM, wave. Mathematically, TEM waves turn out to be the simplest sort of guided waves.[1]

[1]In the literature, TE waves are sometimes referred to as "H waves," meaning that a z component of H exists, and TM waves as "E waves." When the boundaries are sufficiently complicated, E_z and H_z can both be present. Such waves are sometimes called *hybrid* modes.

The reader will have noticed that the guiding structures being discussed in this section are the same as the transmission lines discussed in Chapters 1 and 2. What is the purpose of discussing these waves all over again? The reason is that the point of view taken in Chapters 1 and 2 was based on a circuit model. Like any model, the lumped-element model of the transmission line gives some useful results, but fails to tell the whole story. Now, by applying Maxwell's equations, we shall be able to describe transmission-line waves in a more realistic way. We shall be able to verify the results of the model, prove statements that were stated before without proof, and obtain new information, such as the field configurations of the waves.

To derive the properties of TEM waves, let us begin with the two "curl" Maxwell equations,

$$\nabla \times \vec{\mathbf{E}} = -j\omega\mu\vec{\mathbf{H}} \tag{9.3a}$$

$$\nabla \times \vec{\mathbf{H}} = j\omega\epsilon\vec{\mathbf{E}} \tag{9.3b}$$

Each of these equations in fact provides us with three separate equalities, one for each rectangular component of the curl. In the present case, some of these equalities can be simplified because we are seeking solutions of the form (9.1) and (9.2). Since the dependence of the fields on z is assumed to be only through the factor e^{-jkz}, differentiation with respect to z is equivalent to multiplication by $-jk$. We are also seeking TEM solutions; thus we set H_z and E_z equal to zero. In this way we obtain from (9.3) the following six equations:

$$\frac{\partial \mathbf{E}_x}{\partial y} - \frac{\partial \mathbf{E}_y}{\partial x} = 0 \tag{9.4a}$$

$$-jk\mathbf{E}_x = -j\omega\mu\mathbf{H}_y \tag{9.4b}$$

$$-jk\mathbf{E}_y = j\omega\mu\mathbf{H}_x \tag{9.4c}$$

$$\frac{\partial \mathbf{H}_x}{\partial y} - \frac{\partial \mathbf{H}_y}{\partial x} = 0 \tag{9.5a}$$

$$-jk\mathbf{H}_x = j\omega\epsilon\mathbf{E}_y \tag{9.5b}$$

$$-jk\mathbf{H}_y = -j\omega\epsilon\mathbf{E}_x \tag{9.5c}$$

Combining (9.4b) with (9.5c), we obtain

$$\mathbf{E}_x(k^2 - \omega^2\mu\epsilon) = 0 \tag{9.6}$$

This interesting result shows that for E_x to be nonzero, we must have

$$k^2 - \omega^2\mu\epsilon = 0 \tag{9.7}$$

Thus for TEM waves, $k = k_o = \omega\sqrt{\mu\epsilon}$. Furthermore, by combining other pairs of equations we can show that *all* the components of E and H must vanish, unless (9.7) is correct. Thus we have found the phase velocity of the TEM wave; for, from (9.7) and (1.15),

$$U_P = \frac{\omega}{k} = \frac{1}{\sqrt{\mu\epsilon}} \tag{9.8}$$

The phase velocity of the TEM wave is equal to the velocity of light in the medium. This is the same result obtained earlier for the case of plane waves. Plane waves, though unguided, are also TEM waves, and hence belong to the class of waves now being discussed. The special case of uniform plane waves is that in which $\vec{E}_o$ is a constant.

From (9.4b) and (9.4c) we have

$$\begin{aligned} E_x(t) &= \eta H_y(t) \\ E_y(t) &= -\eta H_x(t) \end{aligned} \tag{9.9}$$

where η, as before, is $\sqrt{\mu/\epsilon}$. From (9.9) one can show that: (a) $\vec{E}$ and $\vec{H}$ are everywhere perpendicular to each other; and (b) $|\vec{E}(t)| = \eta|\vec{H}(t)|$ (Problems 9.3 and 9.4). We recall that plane waves also have these properties.

Applying the vector identity $\nabla \cdot (f\vec{V}) = \vec{V} \cdot \nabla f + f\nabla \cdot \vec{V}$ to (9.1), and observing that $\nabla \cdot \vec{\mathbf{E}} = 0$ in the region of interest and that $\vec{E}_o \cdot \nabla(e^{-jkz}) = 0$ for the TEM wave, we find that $\nabla \cdot \vec{E}_o = 0$. Furthermore,

$$\begin{aligned} \nabla \times \vec{E}_o &= \left(\frac{\partial E_{oy}}{\partial x} - \frac{\partial E_{ox}}{\partial y}\right)\vec{e}_z \\ &= e^{+jkz}\left(\frac{\partial \mathbf{E}_y}{\partial x} - \frac{\partial \mathbf{E}_x}{\partial y}\right)\vec{e}_z \\ &= 0 \end{aligned} \tag{9.10}$$

(The last step from (9.4a).) Since both the divergence and curl of $\vec{E}_o$ vanish, this field obeys precisely the same laws as the electrostatic field. $\vec{E}_o$ has no z component; it lies entirely in the transverse plane. Thus a map of $\vec{E}_o$ in any transverse plane must be identical with the electrostatic field bounded by the same guiding structure. This is illustrated for the case of coaxial line in Fig. 9.2. The electric field in the transverse plane is the same as the electrostatic field for a cylindrical capacitor, and therefore is

$$\vec{E}_o = \frac{A}{r}\vec{e}_r \tag{9.11}$$

where A is an arbitrary constant that specifies the amplitude and phase of the wave. Since $\vec{H}$ is everywhere perpendicular to $\vec{E}$, the lines of $\vec{H}$ can be drawn in as well. It can be shown that $\vec{H}$ is identical with the two-dimensional magnetostatic field for the same boundaries.

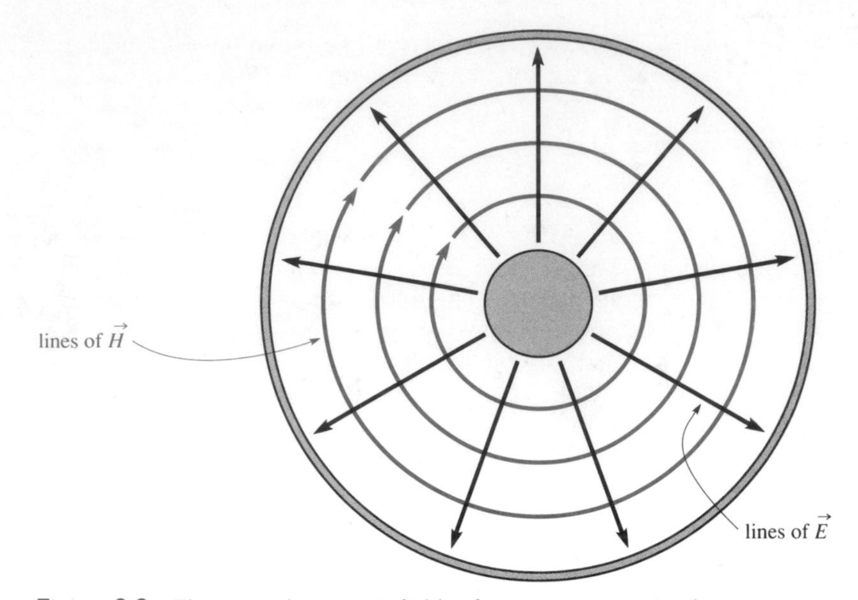

Figure 9.2 Electric and magnetic fields of a coax transmission line.

EXERCISE 9.2

Find the magnetic field associated with a wave on a coaxial line whose electric field is

$$\vec{\mathbf{E}} = \frac{A}{r} e^{-jkz} \vec{e}_r$$

Answer $\vec{\mathbf{H}} = (A/\eta r)e^{-jkz} \vec{e}_\phi$.

It is interesting to compare the electrodynamic picture of the coaxial line with the lumped-circuit model of Chapter 1. Let the inner and outer radii of the line be a and b, and let the fields in the line be

$$\vec{\mathbf{E}} = \frac{A}{r} e^{-jkz} \vec{e}_r$$
$$\vec{\mathbf{H}} = \frac{A}{\eta r} e^{-jkz} \vec{e}_\phi \tag{9.12}$$

Then the voltage on the line is

$$\mathbf{v}(z) = \int_a^b \vec{\mathbf{E}}(r, z) \cdot \vec{dr} = A \ln\left(\frac{b}{a}\right) e^{-jkz} \tag{9.13}$$

and the current, from Ampère's circuital law, is

$$\mathbf{i}(z) = \oint \vec{H} \cdot \vec{dl} = \frac{2\pi A}{\eta} e^{-jkz} \tag{9.14}$$

The characteristic impedance of the line is

$$Z_o = \frac{\mathbf{v}}{\mathbf{i}} = \frac{\eta \ln\left(\frac{b}{a}\right)}{2\pi} \tag{9.15}$$

The power being transmitted is, in the circuit point of view,

$$P = \frac{1}{2}\frac{|\mathbf{v}|^2}{Z_o} = \frac{\pi |A|^2}{\eta} \ln\left(\frac{b}{a}\right) \tag{9.16}$$

In the electrodynamic picture, the time-averaged power can be found by integrating the Poynting vector over the cross-sectional area:

$$\begin{aligned} P &= \int \frac{1}{2} \operatorname{Re} (\vec{\mathbf{E}} \times \vec{\mathbf{H}}^*) \cdot \vec{dS} \\ &= \frac{1}{2} \int_a^b \frac{|A|^2}{\eta r^2} 2\pi r \, dr = \frac{\pi |A|^2}{\eta} \ln\left(\frac{b}{a}\right) \end{aligned} \tag{9.17}$$

in agreement with (9.16). In the circuit picture, the line is characterized by C and L, its capacitance and inductance per unit length. These quantities can be found by equating the expressions for U_P and Z_o from the two points of view:

$$\begin{aligned} U_P &= \frac{1}{\sqrt{LC}} = \frac{1}{\sqrt{\mu\epsilon}} \\ Z_o &= \sqrt{\frac{L}{C}} = \frac{\eta}{2\pi} \ln\left(\frac{b}{a}\right) \end{aligned} \tag{9.18}$$

Solving simultaneously for L and C, we have

$$\begin{aligned} C &= \frac{2\pi\epsilon}{\ln(b/a)} \\ L &= \frac{\mu}{2\pi} \ln\left(\frac{b}{a}\right) \end{aligned} \tag{9.19}$$

These results agree with electrostatic and magnetostatic calculations. (Exercise 4.5 and Example 6.3.) Thus we find that for transmission-line waves, the two points of view are consistent with each other. The lumped-circuit point of view is simple and convenient for transmission-line problems. However, the electromagnetic point of view is far more powerful, and as we shall see, can be used to analyze other kinds of guided waves, which, not being TEM, bear less resemblance to waves on transmission lines.

To summarize, TEM waves have the following properties:

1. Electric and magnetic fields lie entirely in the transverse plane.
2. Phase velocity is equal to the velocity of light in the medium.
3. Electric and magnetic fields are perpendicular.
4. $|\vec{E}(x, y, z, t)| = \eta|\vec{H}(x, y, z, t)|$
5. The forms of the electric and magnetic fields are identical with two-dimensional electrostatic and magnetostatic solutions for the same boundaries.
6. TEM waves are identical with the waves predicted by the lumped-element transmission-line model.

9.2 HOLLOW METAL WAVEGUIDES

Coaxial line, the most common TEM guiding system, is useful at frequencies up to somewhere around 30 GHz. However, as frequency increases, the radius of the line must decrease (to avoid the appearance of additional non-TEM propagation modes), which makes it difficult to fabricate. Furthermore, attenuation, resulting from ohmic loss, increases with frequency and ultimately becomes prohibitive. The traditional alternative, in the frequency range 1–100 GHz, is the hollow metal waveguide. Unlike transmission lines, the hollow metal waveguide consists of only a single conductor. (Therefore it cannot support a TEM wave.) This conductor is in the form of a pipe, with the inside surface smoothly polished to reduce loss. Although pipes of circular cross section are sometimes used, the most common shape is rectangular, as shown in Fig. 9.3. Because of their convenience and low loss, guides of this type have in the past been very widely used in microwave technology. Indeed, when one hears the word *waveguide,* with no further description, it is usually hollow metal waveguide that is meant.

The waves supported by hollow metal guide fall into two classes: TE and TM. Since z-directed fields are present in these waves, (9.4) and (9.5) no longer apply. However, the general form of the guided waves, (9.1) and (9.2), will still be assumed. Rewriting the curl equations (9.3) to include

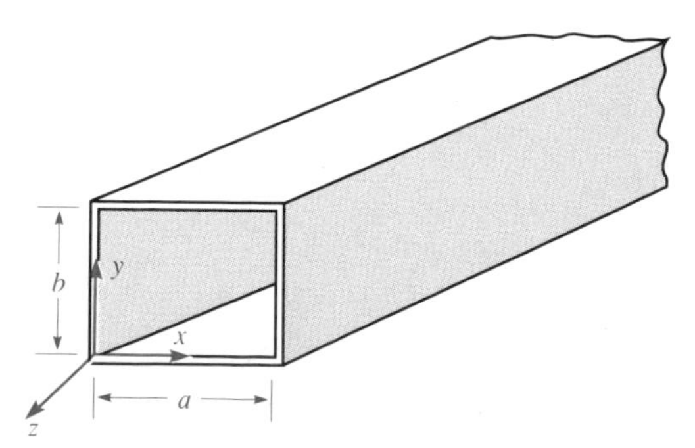

Figure 9.3 Rectangular hollow metal waveguide.

z-directed fields, we obtain

$$\frac{\partial \mathbf{E}_z}{\partial y} + jk\mathbf{E}_y = -j\omega\mu\mathbf{H}_x \tag{9.20a}$$

$$-jk\mathbf{E}_x - \frac{\partial \mathbf{E}_z}{\partial x} = -j\omega\mu\mathbf{H}_y \tag{9.20b}$$

$$\frac{\partial \mathbf{E}_y}{\partial x} - \frac{\partial \mathbf{E}_x}{\partial y} = -j\omega\mu\mathbf{H}_z \tag{9.20c}$$

$$\frac{\partial \mathbf{H}_z}{\partial y} + jk\mathbf{H}_y = j\omega\epsilon\mathbf{E}_x \tag{9.21a}$$

$$-jk\mathbf{H}_x - \frac{\partial \mathbf{H}_z}{\partial x} = j\omega\epsilon\mathbf{E}_y \tag{9.21b}$$

$$\frac{\partial \mathbf{H}_y}{\partial x} - \frac{\partial \mathbf{H}_x}{\partial y} = j\omega\epsilon\mathbf{E}_z \tag{9.21c}$$

These equations can be manipulated algebraically to express the x and y components in terms of the z components. Suppressing the algebraic details, (9.20) and (9.21) become

$$\mathbf{E}_x = -\frac{j}{\omega^2\mu\epsilon - k^2}\left(k\frac{\partial \mathbf{E}_z}{\partial x} + \omega\mu\frac{\partial \mathbf{H}_z}{\partial y}\right) \tag{9.22a}$$

$$\mathbf{E}_y = \frac{j}{\omega^2\mu\epsilon - k^2}\left(-k\frac{\partial \mathbf{E}_z}{\partial y} + \omega\mu\frac{\partial \mathbf{H}_z}{\partial x}\right) \tag{9.22b}$$

$$\mathbf{H}_x = \frac{j}{\omega^2\mu\epsilon - k^2}\left(\omega\epsilon\frac{\partial \mathbf{E}_z}{\partial y} - k\frac{\partial \mathbf{H}_z}{\partial x}\right) \tag{9.22c}$$

$$\mathbf{H}_y = -\frac{j}{\omega^2\mu\epsilon - k^2}\left(\omega\epsilon\frac{\partial \mathbf{E}_z}{\partial x} + k\frac{\partial \mathbf{H}_z}{\partial y}\right) \tag{9.22d}$$

The electric and magnetic fields of course obey the Helmholtz equations,

$$\nabla^2\vec{\mathbf{E}} = -\omega^2\mu\epsilon\vec{\mathbf{E}} \tag{9.23a}$$

$$\nabla^2\vec{\mathbf{H}} = -\omega^2\mu\epsilon\vec{\mathbf{H}} \tag{9.23b}$$

Writing out the vector Laplacian operator in (9.23a), we have[2]

$$\vec{e}_x\left(\frac{\partial^2}{\partial x^2} + \frac{\partial^2}{\partial y^2} + \frac{\partial^2}{\partial z^2}\right)\mathbf{E}_x + \vec{e}_y\left(\frac{\partial^2}{\partial x^2} + \frac{\partial^2}{\partial y^2} + \frac{\partial^2}{\partial z^2}\right)\mathbf{E}_y$$
$$+ \vec{e}_z\left(\frac{\partial^2}{\partial x^2} + \frac{\partial^2}{\partial y^2} + \frac{\partial^2}{\partial z^2}\right)\mathbf{E}_z = -\omega^2\mu\epsilon\,(\mathbf{E}_x\vec{e}_x + \mathbf{E}_y\vec{e}_y + \mathbf{E}_z\vec{e}_z) \tag{9.24}$$

[2]Note: $\left(\frac{\partial^2}{\partial x^2} + \frac{\partial^2}{\partial y^2} + \frac{\partial^2}{\partial z^2}\right)\mathbf{E}_x$ is a shorthand notation for $\frac{\partial^2\mathbf{E}_x}{\partial x^2} + \frac{\partial^2\mathbf{E}_x}{\partial y^2} + \frac{\partial^2\mathbf{E}_x}{\partial z^2}$.

Using the assumption (9.1), which makes $\frac{\partial}{\partial z}$ equivalent to multiplication by $-jk$, the z component of (9.24) becomes

$$\left(\frac{\partial^2}{\partial x^2} + \frac{\partial^2}{\partial y^2}\right) \mathbf{E}_z = (k^2 - \omega^2 \mu \epsilon)\, \mathbf{E}_z \tag{9.25}$$

and similarly

$$\left(\frac{\partial^2}{\partial x^2} + \frac{\partial^2}{\partial y^2}\right) \mathbf{H}_z = (k^2 - \omega^2 \mu \epsilon)\, \mathbf{H}_z \tag{9.26}$$

With this machinery we are now ready to describe TE and TM waves on the hollow rectangular guide.

TM WAVES

To describe TM waves, we assume that $\mathbf{H}_z$ vanishes. The phasor $\mathbf{E}_z$ must obey (9.25). A solution of (9.25) is easily guessed. Let

$$\mathbf{E}_z = (A \cos k_x x + B \sin k_x x)\,(C \cos k_y y + D \sin k_y y)\, e^{-jkz} \tag{9.27}$$

where A, B, C, D are arbitrary constants and k_x and k_y are constants to be determined. Substituting (9.27) into (9.25), we find that (9.25) is satisfied, provided that

$$(k_x^2 + k_y^2) = \omega^2 \mu \epsilon - k^2 \tag{9.28}$$

The form of (9.27) is further constrained by the boundary conditions. Since the walls of the guide are assumed to be perfect conductors, E_z, which is tangential to the walls, must vanish at $x = 0$, $x = a$, $y = 0$, and $y = b$. (See coordinate system in Fig. 9.3.) From (9.27), E_z will not vanish at $x = 0$ unless $A = 0$; similarly to satisfy the boundary condition at $y = 0$, we must set $C = 0$. The boundary conditions at $x = a$ and $y = b$ can be satisfied by setting[3]

$$k_x = \frac{m\pi}{a}$$
$$k_y = \frac{n\pi}{b} \tag{9.29}$$

where m and n are any integers. Thus we have shown that the following $\mathbf{E}_z$ is a solution of the Helmholtz equation that satisfies the boundary conditions:

$$\mathbf{E}_z = K \sin \frac{m\pi x}{a} \sin \frac{n\pi y}{b} e^{-jkz} \tag{9.30}$$

---

[3]The boundary conditions at $x = a$ and $y = b$ could also be satisfied by setting $B = 0$, $D = 0$, but then the entire wave would vanish.

where K (the product of the arbitrary constants B and D) is another arbitrary constant and m and n are any integers. The other field components can now be found immediately from $\mathbf{E}_z$ by means of (9.22).

EXERCISE 9.3

Find E_x and E_y for the TM wave (9.30) and show that they satisfy the boundary conditions.

Answer

$$\mathbf{E}_x = -\frac{jKkm\pi}{(\omega^2\mu\epsilon - k^2)a}\cos\frac{m\pi x}{a}\sin\frac{n\pi y}{b}e^{-jkz}$$

$$\mathbf{E}_y = -\frac{jKkn\pi}{(\omega^2\mu\epsilon - k^2)b}\sin\frac{m\pi x}{a}\cos\frac{n\pi y}{b}e^{-jkz}$$

A very important point is that our solution (9.30) represents not just one wave, but an infinitely large family of waves, characterized by different values of the integers m and n. These waves differ from one another in field configuration, and also (as we shall see momentarily) in velocity. They are known as the *TM modes* of the rectangular waveguide. They are distinguished by their values of m and n. For example, the wave solution with $m = 1$, $n = 2$ is called the TM_{12} mode of the waveguide.

From (9.28), the propagation constant of the TM_{mn} mode is given by

$$k^2 = \frac{\omega^2}{c^2} - \left(\frac{m\pi}{a}\right)^2 - \left(\frac{n\pi}{b}\right)^2 \tag{9.31}$$

The $\omega - k$ plot for the case $m = 1$, $n = 1$, $a/b = 1.7$ is shown in Fig. 9.4. This plot has several interesting features:

1. At high frequencies, the curve is asymptotic to the line $\omega = kc$. Thus at high frequencies, both the phase and group velocities approach c.
2. However, the $\omega - k$ plot is always *above* the line $\omega = kc$. This means that the phase velocity $U_P = \omega/k$ is always *larger* than c.
3. The group velocity U_G is equal to $d\omega/dk$, and hence is determined by the slope of the curve. This slope is always *less* than the slope of the line $\omega = kc$, and hence U_G is always *less* than c.
4. We observe that there is a minimum frequency, known as the *cutoff frequency,* below which k becomes imaginary. Below the cutoff frequency, normal wave propagation ceases and the wave is evanescent. Different modes have different cutoff frequencies. From (9.31), the

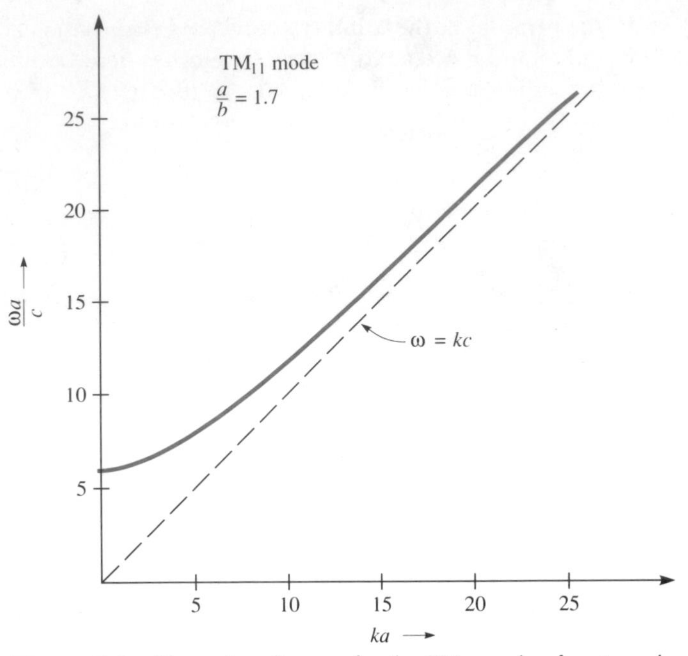

Figure 9.4 Dispersion diagram for the TM_{11} mode of rectangular waveguide. In this example $a/b = 1.7$.

cutoff frequency of the TM_{mn} mode is

$$\omega_{mn} = \pi c \left[\left(\frac{m}{a}\right)^2 + \left(\frac{n}{b}\right)^2 \right]^{1/2} \tag{9.32}$$

As frequency decreases and approaches the cutoff frequency, the phase velocity approaches infinity and the group velocity approaches zero.

EXERCISE 9.4

Find the width of the smallest square waveguide capable of propagating the TM_{11} mode, if the frequency corresponds to a free-space wavelength of 10 cm.

Answer 7.07 cm.

TE WAVES

Much of the preceding derivation applies also to TE waves. Equations (9.22)–(9.26) remain valid. The difference between the TE and TM cases arises from their different boundary conditions at the waveguide walls. In the TM case, we were able to set E_z equal to zero at the walls, because the tangential E field vanishes at a perfect conductor. However, for TE waves there is no E_z. Instead, we must find the boundary condition on H_z that causes tangential E to vanish.

Consider first the upper and lower walls of the guide. On these walls the tangential component of $\vec{E}$ is E_x; thus $E_x = 0$ at $y = 0$ and at $y = b$. From (9.22a), this requires that $\frac{\partial \mathbf{H}_z}{\partial y} = 0$ at those positions. (Remember that $E_z = 0$ for TE waves.) Similarly, from (9.22b) we find that $\frac{\partial \mathbf{H}_z}{\partial x}$ must equal zero at $x = 0$ and at $x = a$. The general solution of (9.26), analogous to (9.27), is

$$\mathbf{H}_z = (A \cos k_x x + B \sin k_x x)(C \cos k_y y + D \sin k_y y)\, e^{-jkz} \tag{9.33}$$

Thus

$$\frac{\partial \mathbf{H}_z}{\partial y} = (A \cos k_x x + B \sin k_x x)(-Ck_y \sin k_y y + Dk_y \cos k_y y)\, e^{-jkz}$$

For $\frac{\partial \mathbf{H}_z}{\partial y}$ to vanish at $y = 0$, as required by the boundary conditions, we must have $D = 0$. Furthermore, $\frac{\partial \mathbf{H}_z}{\partial y}$ can be made to vanish at $y = b$ by setting

$$k_y = \frac{n\pi}{b} \tag{9.34a}$$

where n is any integer. By similar reasoning with respect to the other two walls at $x = 0$ and $x = a$, we find that $B = 0$ and

$$k_x = \frac{m\pi}{a} \tag{9.34b}$$

Thus

$$\mathbf{H}_z = K \cos\frac{m\pi x}{a} \cos\frac{n\pi y}{b}\, e^{-jkz} \tag{9.35}$$

where K is an arbitrary constant. Substituting (9.33) into (9.26), we find that

$$k_x^2 + k_y^2 = \omega^2\mu\epsilon - k^2 \tag{9.36}$$

which is identical with (9.28) for TM waves.

Since (9.36) is the same for TE as for TM waves, the $\omega - k$ plot of the TE_{mn} mode is the same as that of the TM_{mn} mode, and their phase and group velocities and cutoff frequencies, given by (9.32), are the same. There is, however, one important difference, as a result of which one particular TE mode is, in practice, the most important mode.

THE TE_{10} MODE

A significant difference between TM and TE modes is that when either $m = 0$ or $n = 0$, the TM mode fails to exist. This can be seen from (9.30): if either m or n is zero, E_z vanishes, and from (9.22) all the other field components vanish as well. However, for TE waves the equation corresponding to (9.30) is (9.35). Thus $\mathbf{H}_z$ does not necessarily vanish if either m or n is zero.[4]

The importance of the TE_{10} mode results from the fact that of all the modes, it is the one with the lowest cutoff frequency. (We are assuming that $a > b$.) From (9.32) this cutoff frequency is

$$\omega_{10} = \frac{\pi c}{a} \tag{9.37}$$

As an aid to memory, we note that the free-space wavelength corresponding to the TE_{10} cutoff frequency is $2a$. The smaller dimension of the waveguide (b) has no influence on the cutoff frequency of this mode. Figure 9.5 shows the $\omega - k$ plots for the lowest modes of a guide with $a/b = 1.7$. Since TE_{10}

[4]A TE solution with *both* $m = 0$ and $n = 0$, however, does not exist. If such a solution did exist, k would be equal to ω/c. Then the denominators in (9.22) would vanish. Thus one could not conclude from (9.22) that tangential E vanishes at the walls. Note also that H_z would be independent of x and y. Thus the line integral of E, around a closed path in the transverse plane along the walls, would be nonzero, and the boundary condition on tangential E would not be satisfied.

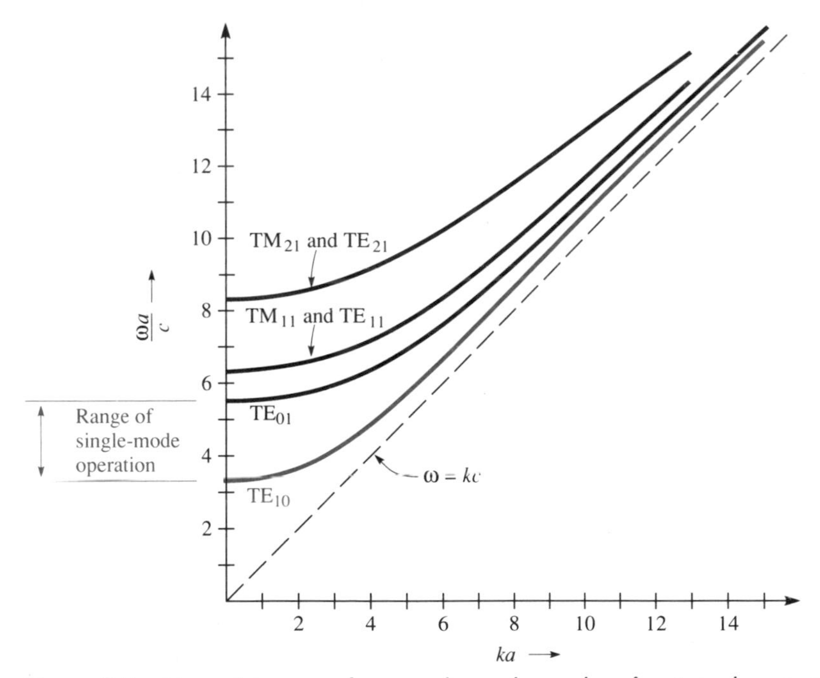

Figure 9.5 Dispersion curves for some low-order modes of rectangular waveguide with $a/b = 1.7$. Single-mode operation is possible when $3.2 < \omega a/c < 5.3$.

has the lowest cutoff frequency, there is a range of frequencies [in this case $(\pi c/a) < \omega < (5.34\, c/a)$] inside which only one mode, the TE_{10} mode, can propagate. This allows us to obtain single-mode operation over that frequency range.

In practice, standard sizes of rectangular waveguides have evolved. Each standard size is recommended for single-mode operation over a certain range of frequencies: larger waveguides for lower frequencies, smaller waveguides for higher frequencies. The recommended frequency range for a given standard waveguide is smaller than the theoretical single-mode range, in order to avoid working too close to cutoff.[5] Some standard waveguides and their characteristics are listed in Table 9.1.

[5]If the walls have finite resistivity, as of course real walls do, the waves suffer attenuation. It can be shown that this attenuation becomes excessively large as frequency approaches cutoff. See (9.45) and (9.46).

Table 9.1 STANDARD RECTANGULAR WAVEGUIDES

Electronics Industry Association Designation	Outer Dimensions and Wall Thickness (inches)	Recommended Frequency Range for TE_{10} Mode (GHz)	Cutoff Frequency for TE_{10} Mode (GHz)	Theoretical Attenuation Lowest to Highest Frequency of Recommended Range (dB per 100 ft.)
WR975	10.000 × 5.125 × 0.125	0.75 – 1.12	0.605	
WR770	7.950 × 4.100 × 0.125	0.96 – 1.45	0.767	
WR650	6.660 × 3.410 × 0.080	1.12 – 1.70	0.908	0.317 – 0.212
WR510	5.260 × 2.710 × 0.080	1.45 – 2.20	1.16	
WR430	4.460 × 2.310 × 0.080	1.70 – 2.60	1.375	0.588 – 0.385
WR340	3.560 × 1.860 × 0.080	2.20 – 3.30	1.735	
WR284	3.000 × 1.500 × 0.080	2.60 – 3.95	2.08	1.102 – 0.752
WR229	2.418 × 1.273 × 0.064	3.30 – 4.90	2.59	
WR187	2.000 × 1.000 × 0.064	3.95 – 5.85	3.16	2.08 – 1.44
WR159	1.718 × 0.923 × 0.064	4.90 – 7.05	3.71	
WR137	1.500 × 0.750 × 0.064	5.85 – 8.20	4.29	2.87 – 2.30
WR112	1.250 × 0.625 × 0.064	7.05 – 10.00	5.26	4.12 – 3.21
WR90	1.000 × 0.500 × 0.050	8.20 – 12.40	6.56	6.45 – 4.48
WR75	0.850 × 0.475 × 0.050	10.00 – 15.00	7.88	
WR62	0.702 × 0.391 × 0.040	12.4 – 18.00	9.49	9.51 – 8.31
WR51	0.590 × 0.335 × 0.040	15.00 – 22.00	11.6	
WR42	0.500 × 0.250 × 0.040	18.00 – 26.50	14.1	20.7 – 14.8
WR34	0.420 × 0.250 × 0.040	22.00 – 33.00	17.3	
WR28	0.360 × 0.220 × 0.040	26.50 – 40.00	21.1	21.9 – 15.0
WR22	0.304 × 0.192 × 0.040	33.00 – 50.00	26.35	31.0 – 20.9
WR19	0.268 × 0.174 × 0.040	40.00 – 60.00	31.4	
WR15	0.228 × 0.154 × 0.040	50.00 – 75.00	39.9	52.9 – 39.1
WR12	0.202 × 0.141 × 0.040	60.00 – 90.00	48.4	93.3 – 52.2
WR10	0.180 × 0.130 × 0.040	75.00–110.00	59.0	

Because the mode index n is zero, the fields of the TE_{10} mode are somewhat simpler than those of other modes. From (9.35), (9.22), (9.36), and (9.34),

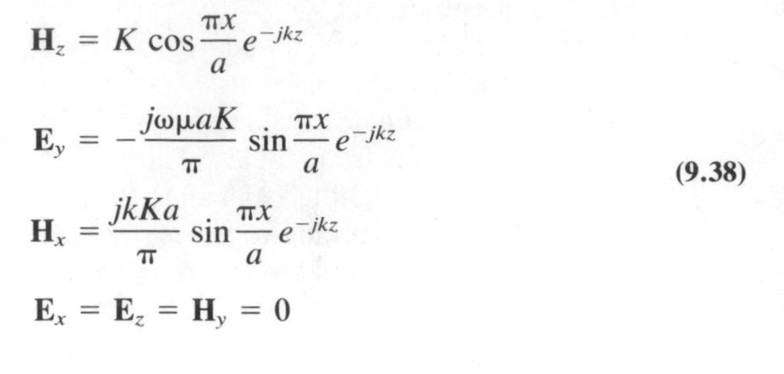

$$
\begin{aligned}
\mathbf{H}_z &= K \cos\frac{\pi x}{a} e^{-jkz} \\
\mathbf{E}_y &= -\frac{j\omega\mu a K}{\pi} \sin\frac{\pi x}{a} e^{-jkz} \\
\mathbf{H}_x &= \frac{jkKa}{\pi} \sin\frac{\pi x}{a} e^{-jkz} \\
\mathbf{E}_x &= \mathbf{E}_z = \mathbf{H}_y = 0
\end{aligned}
\tag{9.38}
$$

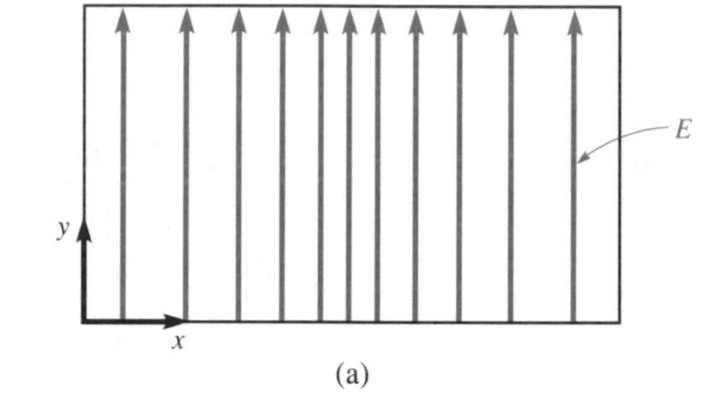

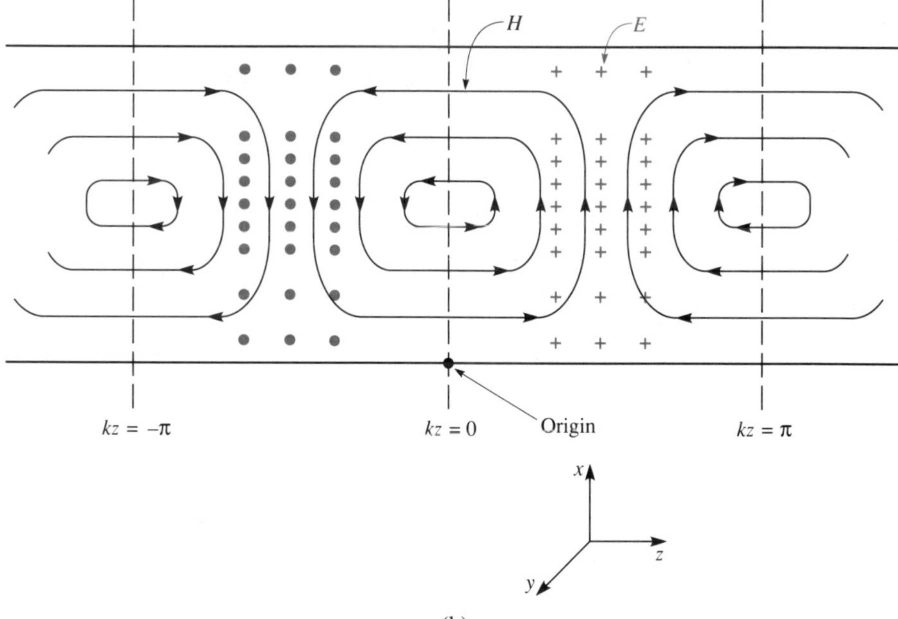

Figure 9.6 Fields of the TE_{10} mode of rectangular waveguide. (a) seen looking down the guide; (b) top view.

We observe that $\vec{E}$ has only a y component. $\vec{H}$ has z and x components, but those are 90° out of phase. Thus at a point on the guide where H_z is maximum, H_x vanishes, and vice versa. The fields of the TE_{10} mode are illustrated in Fig. 9.6, for time $t = 0$, assuming that K is a positive real number. Figure 9.6(a) shows the electric field looking at the waveguide's cross section, at a position where E_y is maximum. Figure 9.6(b) is a view from the top of the guide; the wave is moving to the right. The lines of H form closed loops in the x-z plane. The electric field is also shown (crosses for E directed into the page, dots for E out of the page). E_y and H_x are strongest at positions where H_z is zero.

EXERCISE 9.5

Sketch the fields of Fig. 9.6(b) as they appear at a time one-quarter period later (i.e., at time $t = \pi/2\omega$).

EXAMPLE 9.1

Find the time-averaged power carried by the TE_{10} wave (9.38).

Solution We integrate the z component of the Poynting vector over the cross section. The position (value of z) at which we do this does not matter, since presumably the time-averaged power flow is the same at all values of z. Let us arbitrarily choose $z = 0$. We have

$$P_{AV} = -\int_0^b \int_0^a \frac{1}{2} \text{Re} \, (\mathbf{E}_y \mathbf{H}_x^*) \, dx \, dy$$

$$= \frac{|K|^2}{2} \frac{\omega \mu k a^2}{\pi^2} b \int_0^a \sin^2 \left(\frac{\pi x}{a} \right) dx$$

$$= |K|^2 \frac{\omega \mu k a^3 b}{4\pi^2}$$

The minus sign in the first step arises from taking the component of the cross product in the $+z$ direction. We have written $|K|^2$ instead of K^2 to include the possibility that K is complex. If $K = K_o e^{j\phi}$, the only effect of ϕ is to change the phase of the entire wave. The reader may wish to verify that the units of this result are correct.

WAVEGUIDE LOSSES

The walls of real waveguides are of course not perfect conductors; on the contrary they possess a finite surface resistance R_S. As a result, some of the field energy is converted to heat in the walls. The power lost per unit wall area is given by (7.35), repeated here for convenience:

$$P_A = \tfrac{1}{2} R_S |\vec{\mathbf{H}}_T|^2 \tag{9.39}$$

where $\vec{\mathbf{H}}_T$, the tangential magnetic field at the wall, is a function of position. Usually one is interested in the total power loss per wavelength of guide, and this is obtained by integrating (9.39) over the waveguide surface.

One may object that the fields we have calculated were found under the assumption of perfectly conductive walls. The fields that exist with imperfectly conducting boundaries should be somewhat different, and it is the latter, rather than (9.38), that should be substituted into (9.39). This objection is fundamentally correct. However, the walls, though not perfect conductors, are nonetheless very good conductors; hence the fields we have found are a good approximation to those that exist with slightly lossy walls. From this reasoning, we expect that a good approximation to the loss of a TE_{10} wave (for example) can be obtained by substituting (9.38) into (9.39).

When there is attenuation, the propagation constant k becomes complex. Just as in the case of transmission lines (Section 1.6), we replace k by γ, and express γ in terms of its real and imaginary parts:

$$\gamma = \beta - j\alpha \tag{9.40}$$

Thus all field components are proportional to $e^{-j\beta z}e^{-\alpha z}$. The time-averaged power in the wave is proportional to the square of the field amplitude (Example 9.1); thus

$$P(z) = P_o e^{-2\alpha z} \tag{9.41}$$

where P_o is the time-averaged power at $z = 0$. It follows that

$$\frac{dP}{dz} = P_o(-2\alpha)e^{-2\alpha z} = -2\alpha P \tag{9.42}$$

Equation (9.42) can be used to find the attenuation constant α. Let ΔP be the time-averaged power converted to heat in a length of waveguide Δz. Then from (9.42),

$$\alpha \cong \frac{\Delta P}{2P\,\Delta z} \tag{9.43}$$

This approximation is based on the assumption that $\Delta P/P$ is small, so that P is nearly constant throughout the length Δz.

To evaluate ΔP for the TE_{10} mode, we integrate (9.39) over all surfaces at which a tangential magnetic field is present. There are three contributions: that of H_z on the side walls, that of H_z on the top and bottom walls, and that of H_x on the top and bottom walls. Thus

$$\Delta P = 2\,\frac{R_S}{2}\,|K|^2 b\,\Delta z + 2\,\frac{R_S}{2}\,|K|^2\,\Delta z \int_0^a \cos^2\left(\frac{\pi x}{a}\right) dx + 2\,\frac{R_S}{2}\,|K|^2\,\Delta z \int_0^a \sin^2\left(\frac{\pi x}{a}\right) dx \tag{9.44}$$

Performing the integrations and using (9.43) and the result of Example 9.1, we obtain

$$\alpha = \frac{2\pi^2 R_S}{\omega\mu k a^3 b}\left(b + \frac{\omega^2 a^3}{2\pi^2 c^2}\right) \tag{9.45}$$

where $c = (\mu\epsilon)^{-1/2}$ and

$$k = \frac{\omega}{c}\sqrt{1 - (\pi c/\omega a)^2} \tag{9.46}$$

9.3 CAVITY RESONATORS

When a waveguiding structure is terminated in an ideal short circuit or open circuit, waves incident on the termination must be entirely reflected. If the guide is short-circuited at *both* ends, and if it has small losses, then one might expect that waves could bounce back and forth along its length repeatedly. This can indeed happen, as we shall see in this section. We shall find that wave energy can be stored in such a structure, but only if the frequency of the waves has a suitable value. Because of their preference for specific frequencies, such structures are called *resonators*. They are analogous to LC resonant circuits used in low-frequency electronics. Any type of waveguiding structure can be provided with reflectors at each end to make a resonator. Here, in order to be specific, we shall discuss the particular case of a rectangular waveguide with short circuits at both ends. The resulting structure, a rectangular volume entirely surrounded by metal walls, is known as a rectangular *cavity resonator*.[6]

Let us adopt the waveguide shown in Fig. 9.3; across it we place metal walls at $z = 0$ and $z = d$. These walls impose additional boundary conditions: we must now have $E_x(z = 0) = E_y(z = 0) = E_x(z = d) = E_y(z = d) = 0$. Let us assume that the wave present in the guide is the TE_{mn} mode. From

[6]We shall not deal with the question of how energy can be put into the totally enclosed guide. This may be done with a small wire probe or through a small hole in the cavity wall.

(9.35) and (9.22), the wave propagating in the $+z$ direction has fields

$$\mathbf{H}_z = K \cos k_x x \ \cos k_y y \ e^{-jkz}$$

$$\mathbf{E}_x = K \frac{j\omega\mu k_y}{k_x^2 + k_y^2} \cos k_x x \ \sin k_y y \ e^{-jkz}$$

$$\mathbf{E}_y = K \frac{j\omega\mu k_x}{k_x^2 + k_y^2} \sin k_x x \ \cos k_y y \ e^{-jkz} \tag{9.47}$$

$$\mathbf{H}_x = K \frac{jk}{k_x^2 + k_y^2} \sin k_x x \ \cos k_y y \ e^{-jkz}$$

$$\mathbf{H}_y = K \frac{jk}{k_x^2 + k_y^2} \cos k_x x \ \sin k_y y \ e^{-jkz}$$

where $k_x = m\pi/a$, $k_y = n\pi/b$, and $k^2 = \omega^2\mu\epsilon - k_x^2 - k_y^2$. The wave moving in the $-z$ direction has the same fields, except that k is replaced by $-k$ wherever it appears. The arbitrary constant K can take on complex values (signifying the amplitude and phase of the wave). Let there be waves moving in the $+z$ and $-z$ directions with coefficients K^+ and K^- respectively. Then at $z = 0$,

$$\mathbf{E}_x(z = 0) = \frac{j\omega\mu k_y}{k_x^2 + k_y^2} \cos k_x x \sin k_y y \ (K^+ + K^-) \tag{9.48}$$

and at $z = d$

$$\mathbf{E}_x(z = d) = \frac{j\omega\mu k_y}{k_x^2 + k_y^2} \cos k_x x \sin k_y y \ (K^+ e^{-jkd} + K^- e^{jkd}) \tag{9.49}$$

To satisfy the new boundary conditions imposed by the end walls, these two fields must vanish. This can be accomplished by setting

$$K^+ + K^- = 0$$

$$K^+ e^{-jkd} + K^- e^{jkd} = 0$$

which in turn require $K^- = -K^+$ and $e^{-jkd} - e^{jkd} = 0$; the latter condition is equivalent to

$$\sin kd = 0 \tag{9.50}$$

To satisfy (9.50), we set

$$k = \frac{q\pi}{d} \tag{9.51}$$

where q is an integer. The fields of the resonator are now specified by the three integers m, n, and q; they are the fields of the TE_{mnq} mode of the resonator. The integers m, n, q are also related by $k^2 = \omega^2\mu\epsilon - k_x^2 - k_y^2$. Thus, the frequency of the TE_{mnq} mode is determined:

$$\omega_{mnq}^2 = \pi^2 c^2 \left[\left(\frac{m}{a}\right)^2 + \left(\frac{n}{b}\right)^2 + \left(\frac{q}{d}\right)^2 \right] \tag{9.52}$$

where $c = (\mu\epsilon)^{-1/2}$. These are the resonant frequencies of the cavity resonator. There are an infinite number of such frequencies, corresponding to different choices of the integers m, n, q. Any one of these field distributions can "ring," when properly excited, just as acoustic vibrations in a bell ring, when it is struck with a hammer. In addition to the modes we have found, TM modes exist as well. Their fields are found in similar fashion, and their frequencies are given by exactly the same formula, (9.52).

It is possible for different modes of a resonator to have the same resonant frequency, in which case the modes are said to be *degenerate*.[7] Degeneracy is often undesirable; to prevent degeneracies it is helpful if a, b, and d are given different values. However, pairs of TE and TM modes with the same mode indices will still be degenerate. If nondegenerate modes are desired, one can use the TE_{10q} modes, since, as we recall, the TM_{10} waveguide modes do not exist.

EXERCISE 9.6

Find the lowest resonant frequency of a cubic rectangular cavity, with $a = b = d = L$. Express this frequency in terms of its corresponding free-space wavelength.

Answer $\lambda_{fs} = \sqrt{2}L$.

9.4 PLANAR WAVEGUIDES AND MICROSTRIP

Two important trends in modern electronics are those (a) toward higher frequencies, and (b) toward planar fabrication technology. As frequency increases, circuit dimensions become larger in comparison with wavelength, and consequently transmission-line techniques become important even *inside* the circuits. Meanwhile, circuits and small systems are increasingly being built in planar form. What is needed, then, is a form of transmission line or waveguide that leads itself to planar technology. Neither coaxial line nor hollow metal waveguide would be suitable, because they are inherently three-dimensional.

Figure 9.7 illustrates three waveguides that can be used. All of these are constructed on a dielectric substrate, which is a thin slab of dielectric material that physically supports the circuit. When the substrate is a semiconductor, such as silicon or gallium arsenide, transistors and other devices can be built

[7]The word *degenerate* is also used in quantum mechanics, to describe different states that happen to have the same energy. There is, in fact, quite a strong similarity between the two subjects. Quantum-mechanical energy levels are eigenvalues of the Schroedinger equation, just as the resonant frequencies of a cavity resonator are eigenvalues of the wave equation.

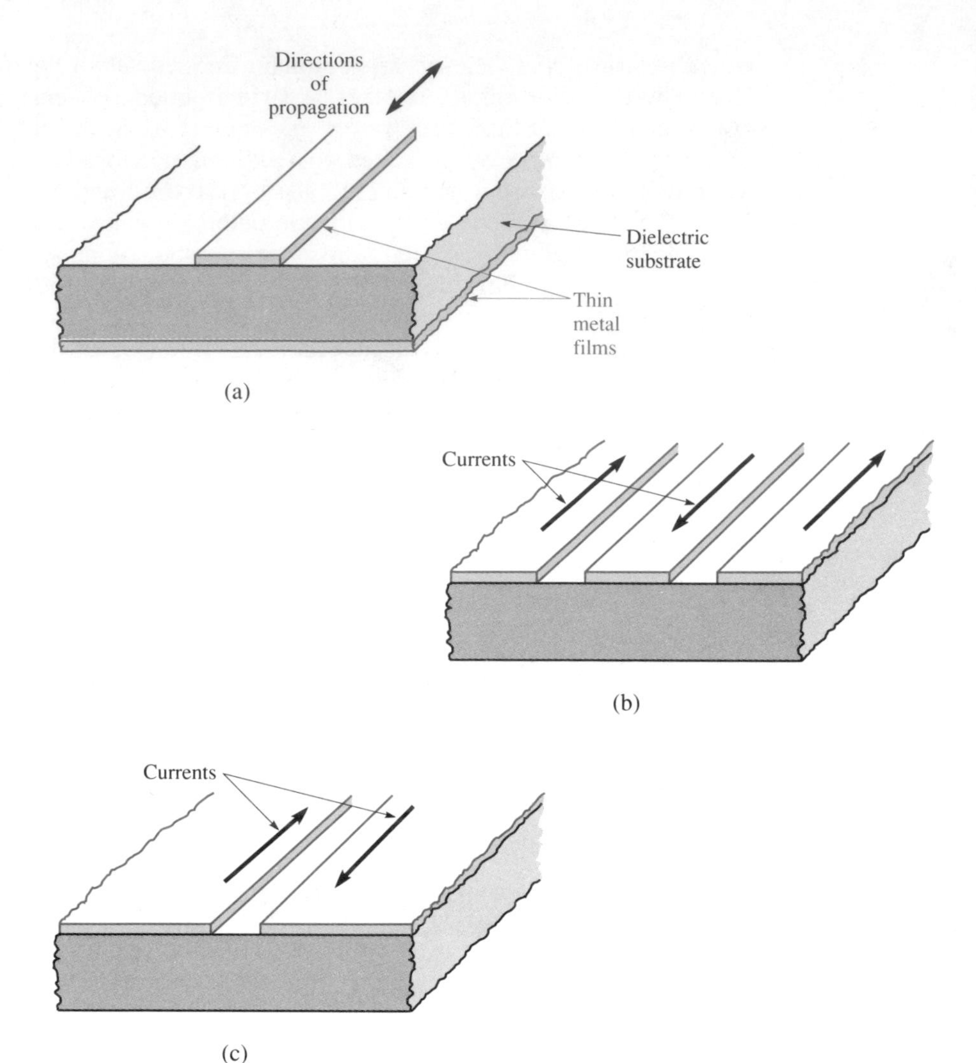

Figure 9.7 Three guiding structures suitable for use in planar circuits. (a) microstrip; (b) coplanar waveguide; (c) slotline.

directly in the substrate material; the result is known as a *monolithic* integrated circuit. In microwave technology, however, the substrate is often a plastic. In that case transistors are soldered onto the circuit, resulting in a *hybrid* integrated circuit. In either case, metallic films on the surfaces of the substrate are used to form the waveguides that act as interconnections in the circuit.

The most widely used planar guide is *microstrip,* shown in Fig. 9.7(a). In microstrip technology, the entire bottom surface of the substrate is covered with a uniform thin metal film, sometimes called the *ground plane*. On the upper surface is a metal strip which, with the ground plane, forms a kind of transmission line. Microstrip is popular because it is compact, comparatively well characterized, and has fairly low ohmic loss, and also because the ground plane is convenient for physical support and heat sinking. A disadvantage of microstrip is the difficulty of making shunt connections between the two conductors; to do this one must bore a hole in the substrate, known as a *via hole*. Via holes add cost and complexity to the circuit, and undesirable inductance as well. To avoid via holes one can substitute *coplanar waveguide,* shown in Fig. 9.7(b). Here the center conductor is similar to that of a microstrip, but instead of the ground plane there are two semi-infinite planar conductors on the upper surface. The guide functions in a way reminiscent of coaxial cable: current flows one way in the center conductor and returns in the two outer conductors.[8] A ground plane is not needed, although one is sometimes used anyway for purposes of heat sinking. Coplanar waveguide permits both shunt and series connections on the easily accessible top surface; however it has larger ohmic loss than microstrip and takes up more space. (An example of a planar circuit based on coplanar waveguide will be found in the frontispiece of this book.) Figure 9.7(c) shows a third guide known as *slotline*. This consists of two semi-infinite metal planes separated by a slot; the two conductors act something like the two conductors of a transmission line. Although simple, slotline has a pronounced tendency toward loss of energy, because it tends to *radiate*—that is, energy tends to be converted from the guided mode into unguided waves that go off into the space above or below the guide. All planar guides in fact exhibit some radiation loss, unless they are surrounded by a metal enclosure.

MICROSTRIP

Let us now discuss microstrip, the most common planar guide, in greater detail. A cross-sectional view is shown in Fig. 9.8. (The direction of propagation is perpendicular to the plane of the page.) The relative permittivity of the dielectric varies from 2, for some plastics, to the order of 13 for

[8]This is the principal, or "odd" mode of the guide. It also supports another mode, the "even" mode, which has higher radiation loss.

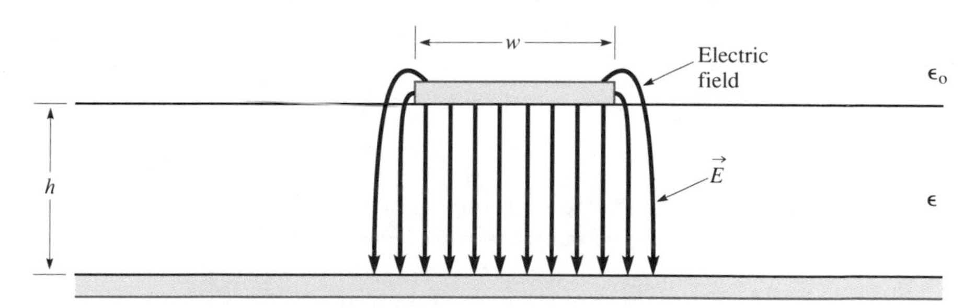

Figure 9.8 Electric field of microstrip.

semiconductors. The substrate thickness h must be less than about 5% of the free-space wavelength at the highest frequency of operation; otherwise unguided modes may become excited. The metal films are not as thick as they appear in the figure. When made by evaporation they are only about 1 μm thick, whereas the substrate thickness will be 100 μm or more.

Microstrip consists of two parallel conductors, and thus resembles the two-conductor transmission lines, such as coax. However, there is an important difference: in microstrip the dielectric is not uniform. Most of the fields are inside the dielectric slab, but some are in the air above it, as shown in Fig. 9.8. Because of this, the structure does not support a true TEM wave, as ordinary transmission lines do. The reason for this is that TEM waves travel at the speed of light in the medium. If there were a TEM wave, the part of it in the air would have to travel at a different speed than the part in the dielectric.

The wave that does propagate on microstrip is, therefore, not a TEM wave; in fact both E_z and H_z are present. A rough picture of this complicated wave can be obtained through the quasistatic approximation. If the dimensions of the guide (w and h) are small compared with the wavelength (which they always are), we expect that the electric and magnetic fields will become similar to the electrostatic and magnetostatic fields of the same structure. This is the reason for guessing that the electric field is as indicated in Fig. 9.8. If we then neglect the small part of the field that extends into the air, we expect that the wave will behave much like a TEM transmission-line wave in the dielectric. This indicates that its phase velocity is given approximately by

$$U_P \cong \frac{1}{\sqrt{\mu_o \epsilon}} \tag{9.53}$$

If w/h is large, we may also neglect the fringing fields and estimate the capacitance per unit length as being that of a parallel-plate capacitor of width w:

$$C \cong \frac{\epsilon w}{h} \tag{9.54}$$

The inductance per unit length can be estimated from the expression for the phase velocity of a transmission line, $U_P = 1/\sqrt{LC}$. Thus

$$L \cong \frac{\mu_o h}{w} \tag{9.55}$$

and the characteristic impedance, from the transmission-line formula $Z_o = \sqrt{L/C}$, is given approximately by

$$Z_o \cong \sqrt{\mu_o/\epsilon}\,\frac{h}{w} \tag{9.56}$$

Because of their neglect of fringing fields, (9.53)–(9.56) are only crude, order-of-magnitude estimates. However, they do give a useful physical feeling for the way impedance varies with guide dimensions.

Much better estimates can be made by retaining the quasistatic approximation (that is, the assumption of a TEM, or transmission-line wave, which is expected to be valid at low frequencies) while improving the crude estimate of C to account for fringing fields and the nonuniformity of the dielectric. More advanced calculations, details of which are beyond us here,[9] provide more accurate values of C. Similar calculations are also applied to another structure identical with Fig. 9.8, except that the dielectric is replaced by air; call the capacitance per unit length of the air-filled structure C_A. Since the air-filled structure supports a true TEM wave, its inductance per unit length, L, can be found from $\sqrt{LC_A} = \sqrt{\mu_o \epsilon_o}$. We then assume that the inductances per unit length of the actual and air-filled structures are the same.[10] Characteristic impedance and phase velocity are then found from the transmission-line expressions $\sqrt{L/C}$ and $(LC)^{-1/2}$, respectively. Separate approximations are required for the case of narrow (small w/h) and wide (large w/h) microstrips. For $w/h > 2$, Wheeler's results give

$$Z_o \cong \frac{377}{\sqrt{\epsilon_R}} \left\{ \frac{w}{h} + 0.883 + \frac{\epsilon_R + 1}{\pi \epsilon_R} \left[\ln \left(\frac{w}{2h} + 0.94 \right) + 1.451 \right] + 0.165 \frac{\epsilon_R - 1}{\epsilon_R^2} \right\}^{-1} \quad \textbf{(9.57)}$$

where $\epsilon_R \equiv \epsilon/\epsilon_o$. Note that for large values of w/h this approaches our simple estimate (9.56). For $w/h < 2$ Wheeler's approximation is

$$Z_o \cong \frac{377}{2\pi \left(\frac{\epsilon_R + 1}{2} \right)^{1/2}} \left\{ \ln \frac{8h}{w} + \frac{1}{8} \left(\frac{w}{2h} \right)^2 - \frac{1}{2} \frac{\epsilon_R - 1}{\epsilon_R + 1} \left[\ln \frac{\pi}{2} + \frac{1}{\epsilon_R} \ln \frac{4}{\pi} \right] \right\} \quad \textbf{(9.58)}$$

The phase velocity for $w/h < 2$ is found to be

$$U_P \cong c \left[\left(\frac{\epsilon_R + 1}{2} \right) + \left(\frac{\epsilon_R - 1}{2} \right) \frac{\ln \frac{\pi}{2} + \frac{1}{\epsilon_R} \ln \frac{4}{\pi}}{\ln \frac{8h}{w}} \right]^{-1/2} \quad \textbf{(9.59)}$$

---

[9]Wheeler, H.A., "Transmission Line Properties of Parallel Wide Strips by Conformal Mapping Approximation." *IEEE Trans. Microwave Theory Tech. MTT-12,* pp. 280–89 (1964); Wheeler, H.A., "Transmission Line Properties of Parallel Strips Separated by a Dielectric Sheet," *IEEE Trans. Microwave Theory Tech. MTT-13,* pp. 172–85 (1965); Owens, R.P., "Accurate Analytical Determination of Quasi-static Microstrip Line Parameters," *The Radio and Electronic Engineer 46,* pp. 360–64 (July, 1976). See also the following general references: K.C. Gupta, R. Garg, and I.J. Bahl, *Microstrip Lines and Slotlines,* Dedham, Mass.: Artech House, 1979; T.C. Edwards, *Foundations for Microstrip Circuit Design,* New York: Wiley, 1981.

[10]This is also an approximation, because the current distributions on the two structures will differ slightly.

Wheeler's estimate of U_p for wide strips cannot be expressed in a single formula. However, Owens provides the following estimate, said to be highly accurate for $8 < \epsilon_R < 12$, $w/h > 1.3$:

$$U_P \cong c\left[\frac{\epsilon_R + 1}{2} + \frac{\epsilon_R - 1}{2}\left(1 + \frac{10h}{2}\right)^{-0.555}\right]^{-1/2} \tag{9.60}$$

The estimates (9.59) and (9.60) are graphed in Fig. 9.9 for values of ϵ_R corresponding to various substrate materials. We note that the two estimates of phase velocity agree well but not precisely in the range $1.3 < w < 2$, where they overlap. Estimates of Z_o, from (9.57) and (9.58), are graphed in Fig. 9.10.

There is a considerable body of literature dealing with various estimates of U_P and Z_o and with the accuracy of these estimates. The subject is still active, especially in connection with computer-aided design of planar microwave circuits. It is difficult to make adjustments in such circuits once they are built; the design must, therefore, be highly accurate. Computer-aided design is used, but of course the resulting design cannot be accurate if the formulas for U_P and Z_o contain errors. As we have seen, the foregoing estimates are based on the resemblance of the wave to a TEM wave. However, as frequency increases the dimensions of the structure become more

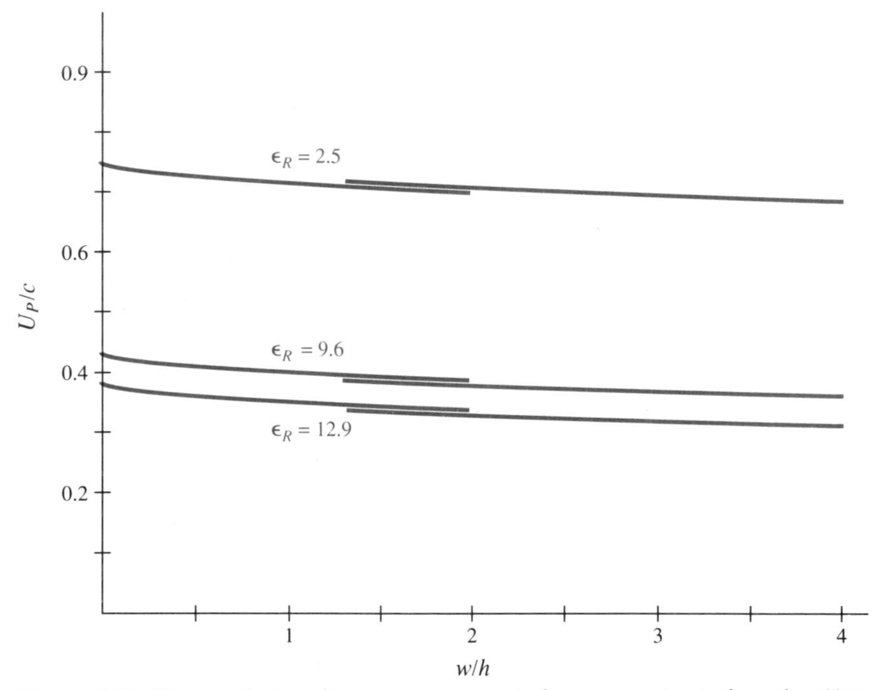

Figure 9.9 Phase velocity of waves on microstrip from approximate formulas. Note slight disagreement where regions of validity overlap.

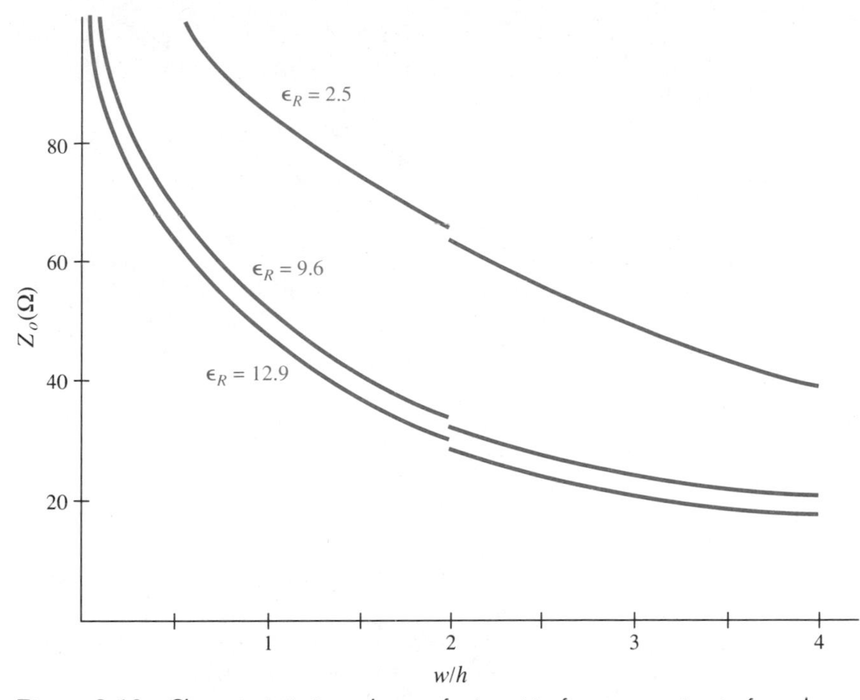

Figure 9.10 Characteristic impedance of microstrip from approximate formulas.

significant compared with wavelength, and the quasistatic approximation becomes less accurate. Phase velocity and characteristic impedance are found to vary with frequency (which would not be the case if the wave were TEM), and for accurate designs, these variations must be taken into account. A number of theories, known as "fullwave analyses," have been developed. These are numerical solutions of the Helmholtz equations using the rather complicated boundary conditions appropriate to microstrip. Some recent results, illustrating the dispersion at high frequencies, are shown in Fig. 9.11. According to a rule of thumb, dielectric waveguide modes (see the following section) are likely to be excited when $h \gtrsim 0.05\lambda_{FS}$; this will be the case in Fig. 9.11 for $f > 150$ GHz. The microstrip would have to be used with some care above that frequency.

9.5 DIELECTRIC WAVEGUIDES AND OPTICAL FIBERS

The waveguides discussed up to this point have all been based on metallic conductors. These have a serious disadvantage for use at high frequencies: the surface resistance of the metal, R_S, increases as the square root of the

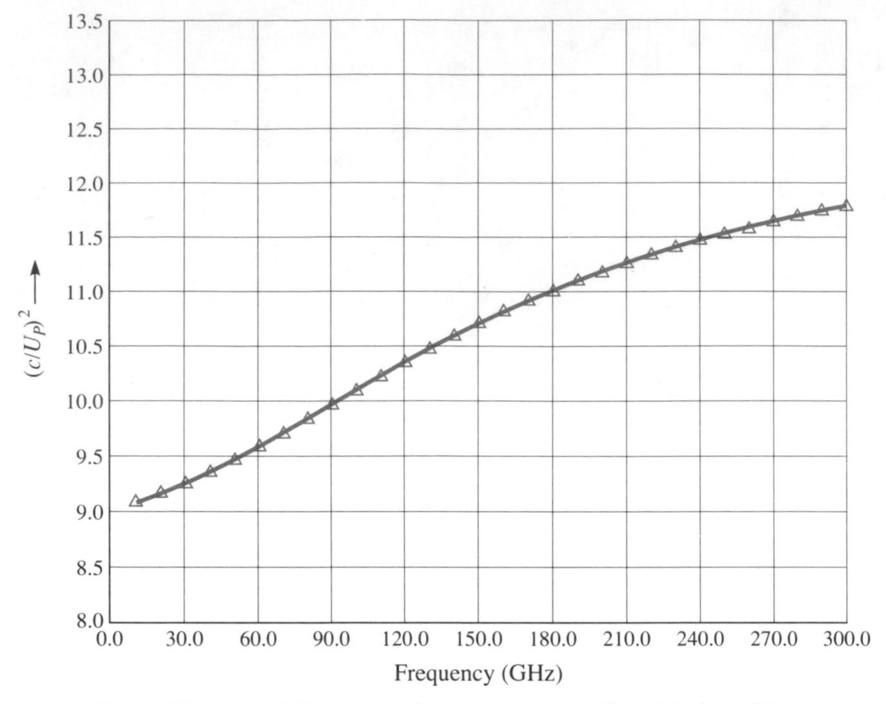

Figure 9.11 Theoretical dispersion of microstrip with $w/h = 1.5$, $h = 0.1$ mm, $\epsilon = 12.9\epsilon_o$ (corresponding to gallium arsenide substrate material.) From X. Zhang, J. Fang, K. K. Mei, and Y. Liu, "Calculations of the Dispersive Characteristics of Microstrips by the Time-domain Finite Difference Method." *IEEE Transactions on Microwave Theory and Techniques*, Dec. 1988.

frequency. This in turn causes the attenuation of the guide to increase. For example, consider the coaxial line shown in Fig. 2.13. As frequency increases, it is necessary to decrease the dimensions of the line, in order to prevent the appearance of higher-order modes. Let us assume that dimensions a and b are both made proportional to $1/\omega$; this keeps the characteristic impedance constant. The series resistance of the line, per unit length, is then, using (7.21),

$$R = R_S\left(\frac{1}{2\pi a} + \frac{1}{2\pi b}\right) = \frac{1}{2\pi}\left(\frac{2\mu\omega}{\sigma_E}\right)^{1/2}\left(\frac{1}{a} + \frac{1}{b}\right) \tag{9.61}$$

which, taking the dependence of a and b into account, is proportional to $\omega^{3/2}$. From (1.60), the attenuation coefficient α is proportional to $R/2Z_{oo}$ and thus α is also proportional to $\omega^{3/2}$. One might argue that what is important is not the attenuation per unit length (α), but rather the attenuation per wavelength, since it can be expected that structures will become smaller at higher frequencies. However even the attenuation per wavelength increases as $\omega^{1/2}$. For this reason coaxial line is seldom used above 20–30 GHz. Similar arguments apply to other metallic guiding structures. Hollow metal waveguide is somewhat less lossy, because of its larger ratio of volume to surface area, and thus can be used at somewhat higher frequencies, on the order of 100 GHz. However, at optical frequencies (around 10^6 GHz), the ohmic loss of any metal waveguide is so large as to be completely impractical.

As an alternative, structures made entirely out of dielectric material can be used. For example, let us consider a plane-parallel slab of material with permittivity ϵ, as shown in Fig. 9.12(a). We can convince ourselves that such a slab can act as a waveguide through the argument illustrated in Fig. 9.12(b). Imagine that a wave bounces back and forth inside the slab as shown, always striking the wall at a large enough angle of incidence to produce total internal reflection. In that case no energy can escape from the guide, and the net effect will be to carry energy in the z direction. It is possible to analyze the wave in terms of this bouncing-ray picture, but instead we shall use the wave formulation developed earlier in this chapter.

Both TE and TM waves are propagated by the dielectric slab. (TE waves correspond to waves with E perpendicular to the plane of incidence in Fig. 9.12(b). TM waves correspond to waves with E in the plane of incidence.) As an example, let us consider TE waves. Since the slab is assumed to be infinitely wide in the x direction, we shall look for solutions that are independent of x. According to (9.26), in region 1 H_z obeys

$$\frac{\partial^2}{\partial y^2}\mathbf{H}_z = (k^2 - \omega^2\mu\epsilon_o)\mathbf{H}_z \qquad \textbf{(9.62)}$$

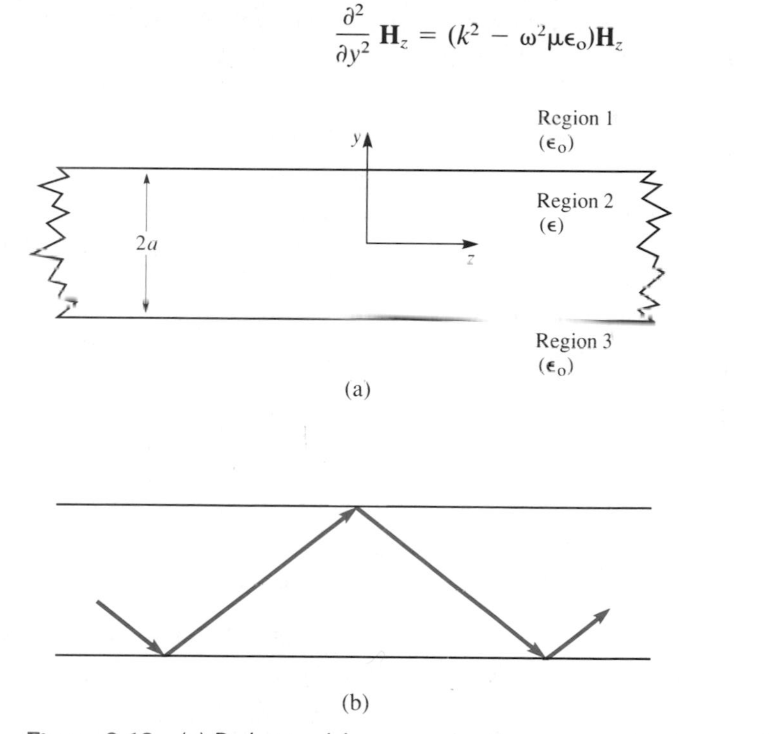

Figure 9.12 (a) Dielectric slab waveguide. The wave may be thought of as a plane wave repeatedly striking the dielectric–air interfaces and experiencing total internal reflection, as shown in (b).

Since physically we expect solutions that decay exponentially as $y \rightarrow \infty$, we expect $k^2 - \omega^2\mu\epsilon_o > 0$, and define

$$k_1^2 = k^2 - \omega^2\mu\epsilon_o \tag{9.63}$$

The physically allowed solution of (9.62) is then

$$\mathbf{H}_z = Ae^{-k_1 y}e^{-jkz} \tag{9.64}$$

where A is an arbitrary constant, and from (9.22)

$$\mathbf{E}_x = -\frac{j\omega\mu A}{k_1}e^{-k_1 y}e^{-jkz} \tag{9.65}$$

$$\mathbf{H}_y = -\frac{jkA}{k_1}e^{-k_1 y}e^{-jkz} \tag{9.66}$$

$$\mathbf{E}_y = \mathbf{H}_x = 0 \tag{9.67}$$

In region 2, H_z satisfies

$$\frac{\partial^2}{\partial y^2}\mathbf{H}_z = (k^2 - \omega^2\mu\epsilon)\mathbf{H}_z \tag{9.68}$$

The fields of the guided wave are partially inside the dielectric and partially outside. Thus we expect that its phase velocity will lie between $(\mu\epsilon_o)^{-1}$ and $(\mu\epsilon)^{-1}$, which implies $\omega\sqrt{\mu\epsilon_o} < k < \omega\sqrt{\mu\epsilon}$. This means that the quantity k_2, defined by

$$k_2^2 = \omega^2\mu\epsilon - k^2 \tag{9.69}$$

will be real and positive.

Because of the symmetry of the structure with respect to the plane $y = 0$, it is possible to find solutions of (9.68) that have either even or odd symmetry in y. These solutions are $\cos k_2 y$ and $\sin k_2 y$. Let us choose the latter. The fields in region 2 are then

$$\mathbf{H}_z = B \sin k_2 y \; e^{-jkz} \tag{9.70}$$

$$\mathbf{E}_x = -\frac{j\omega\mu B}{k_2}\cos k_2 y \; e^{-jkz} \tag{9.71}$$

$$\mathbf{H}_y = -\frac{jkB}{k_2}\cos k_2 y \; e^{-jkz} \tag{9.72}$$

$$\mathbf{E}_y = \mathbf{H}_x - \mathbf{E}_z = 0 \tag{9.73}$$

where B is another arbitrary constant. We now must satisfy the boundary conditions at $y = a$, where region 1 meets region 2.[11] It is necessary that H_z

[11] The boundary conditions at $y = -a$ will be satisfied automatically because of the symmetry of the problem.

and E_x be continuous across the boundary. The former condition requires, from (9.64) and (9.70)

$$Ae^{-k_1 a} = B \sin k_2 a \tag{9.74}$$

while the continuity of E_x requires, from (9.65) and (9.71)

$$\frac{Ae^{-k_1 a}}{k_1} = \frac{B \cos k_2 a}{k_2} \tag{9.75}$$

Dividing (9.75) by (9.74) and eliminating k_1 by (9.63) and k by (9.69), we obtain the characteristic equation for k_2:

$$\tan k_2 a = \left(\frac{\omega^2 \mu(\epsilon - \epsilon_o) a^2}{(k_2 a)^2} - 1 \right)^{1/2} \tag{9.76}$$

Once k_2 is found from this equation, the propagation constant k can be found from (9.69).

The transcendental equation (9.76) cannot be solved explicitly, but lends itself to graphical solution. Let us define

$$\omega_a^2 = \frac{1}{\mu(\epsilon - \epsilon_o) a^2} \tag{9.77}$$

$$q = k_2 a$$

in terms of which (9.76) becomes

$$\tan q = \left(\frac{\omega^2/\omega_a^2}{q^2} - 1 \right)^{1/2} \tag{9.78}$$

We now plot the left and right sides of (9.78) separately as functions of q; the crossing points give the values of q that are solutions. This is done in Fig. 9.13 for several values of the parameter ω/ω_a. For $\omega/\omega_a = 2$, the function on the right side of (9.78) is real only for values of q less than 2. Thus the $\omega/\omega_a = 2$ curve intersects only the first branch of the $\tan q$ curve; this occurs at point A. Hence for $\omega/\omega_a = 2$ there is only one solution of the sort we are considering (TE waves with the sine function, not the cosine, chosen in (9.70)). This wave is called a *symmetric* wave because the electric field is an even function of y. Choosing the cosine function in (9.70) gives *asymmetric* modes.

If frequency increases to $\omega/\omega_a = 4$, the solution moves to point B in Fig. 9.13. Furthermore, the function on the right side of (9.78) is now real out to 1.27π, and hence a second solution appears at C. This means that at this higher frequency, two different modes are propagated. When frequency increases to $\omega/\omega_a = 6$, the lowest order mode is at point D, and the next lowest mode is at E. A third mode, with q larger than 2π, will soon appear when ω becomes slightly larger.

We observe that unlike the higher modes, the lowest order mode has no cutoff. As $\omega \to 0$, the solution point simply moves down the tangent

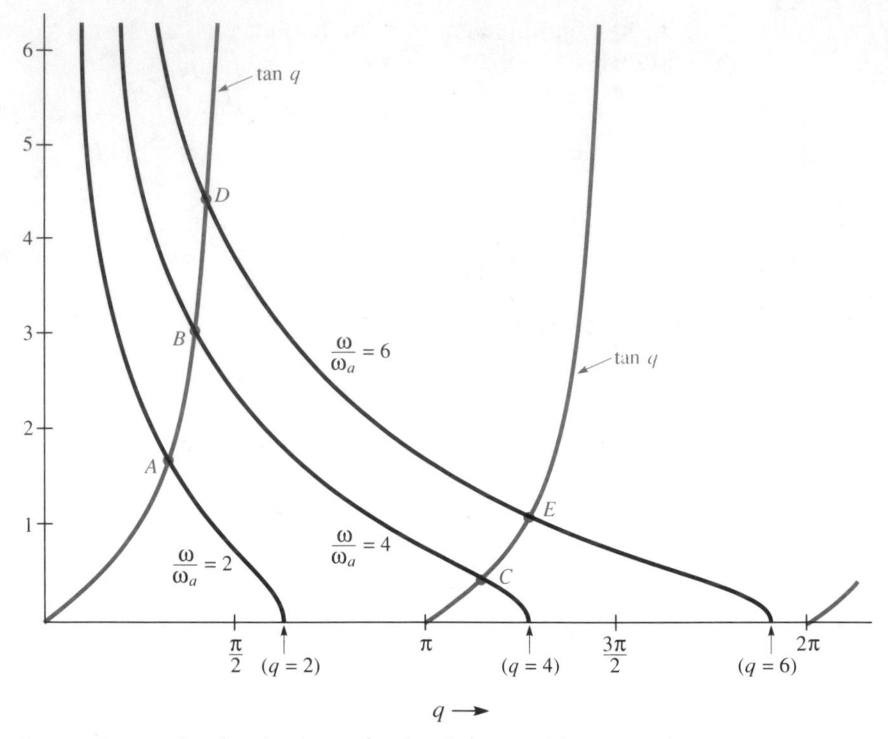

Figure 9.13 Graphical solution for the dielectric slab waveguide.

curve and approaches the origin. In this low-frequency limit $q \ll 1$, and $\tan q \cong q$. Using this and the binomial approximation, (9.78) yields

$$\lim_{\omega \to 0} q = \omega[\mu(\epsilon - \epsilon_o)]^{1/2} a \tag{9.79}$$

From (9.77) and (9.69) this leads to

$$\lim_{\omega \to 0} k = \omega\sqrt{\mu\epsilon_o} \tag{9.80}$$

and since $U_P = \omega/k$,

$$\lim_{\omega \to 0} U_P = \frac{1}{\sqrt{\mu\epsilon_o}} \tag{9.81}$$

On the other hand, as $\omega \to \infty$, the solution for the lowest mode obeys $q \to \pi/2$. Hence

$$\lim_{\omega \to \infty} U_P = \frac{1}{\sqrt{\mu\epsilon}} \tag{9.82}$$

By solving 9.78 for different frequencies and then using (9.77) and (9.69), we can obtain a graph of U_P as a function of frequency. Such a graph is shown in Fig. 9.14 for the case of $\epsilon/\epsilon_o = 9$.

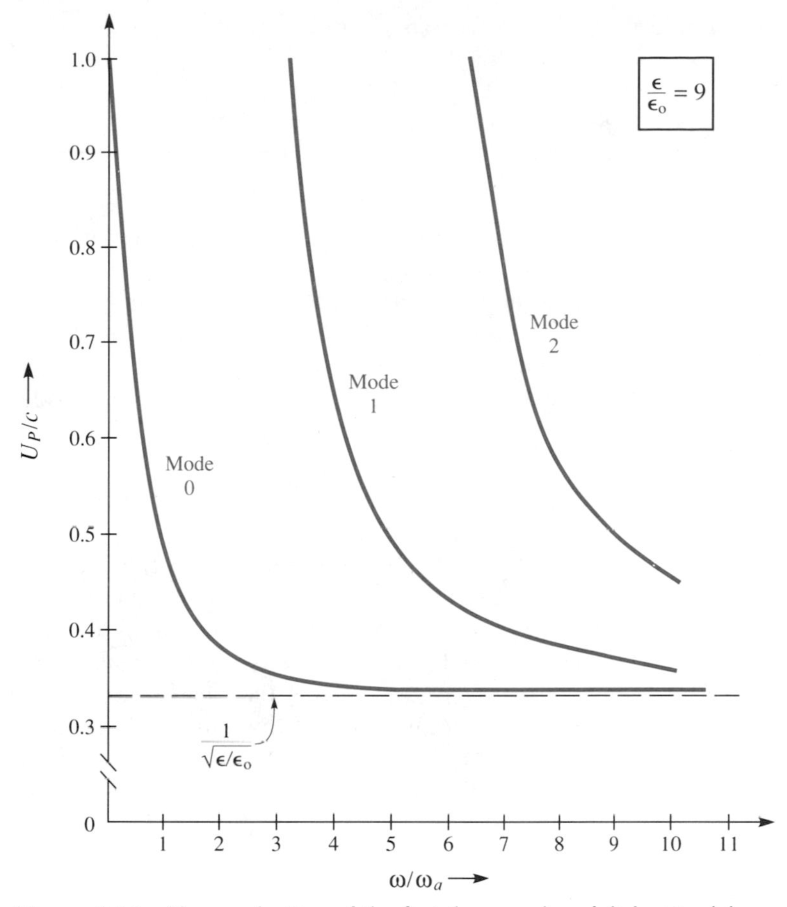

Figure 9.14 Phase velocities of the first three modes of dielectric slab waveguide.

EXAMPLE 9.2

Find the frequency at which the next-to-lowest symmetric TE mode cuts off, for $a = 1$ cm, $\epsilon/\epsilon_o = 3$.

Solution Inspecting Fig. 9.13, we see that the next-lowest mode comes into existence when the curve representing the right side of (9.78) moves to the right far enough to meet the tangent curve at $q = \pi$. The intersection of the former curve with the q axis occurs at $q = \omega/\omega_a$. Thus the next-lowest mode appears at a cutoff frequency given by

$$\omega_{co} = \pi\omega_a$$

or, using (9.77)

$$f_{co} = \frac{c}{2a}\left(\frac{\epsilon}{\epsilon_o} - 1\right)^{-1/2}$$

corresponding to a free-space wavelength

$$\lambda_{co} = 2a\sqrt{(\epsilon/\epsilon_o) - 1}$$

For the given parameters, $f_{co} = 10.6\,\text{GHz}$, $\lambda_{co} = 2.83\,\text{cm}$.

As we have seen, the fields of these waves are largest inside the dielectric, but also exist outside the dielectric, where they decay exponentially with distance from the slab. The fields outside the slab, which resemble the evanescent waves discussed in section 8.4, carry energy in the direction of propagation, but do not carry energy in the direction perpendicular to the slab. As frequency increases, $k_2 a$ for the lowest order mode approaches $\pi/2$, and from (9.69) and (9.63),

$$\lim_{\omega \to \infty} k_1 = \frac{\omega}{c}\sqrt{(\epsilon/\epsilon_o) - 1)} \quad \textbf{(9.83)}$$

Using this result in (9.64)–(9.66), we see that the fields are confined more and more closely to the dielectric as frequency increases. Thus it does not seem surprising that in the high-frequency limit the phase velocity approaches the speed of light in the dielectric material. On the other hand, as $\omega \to 0$ we find that $k_1 \to 0$, implying that the extension of the fields outside the slab becomes infinitely large. In this limit H_y becomes much larger than H_z, and the wave resembles a TEM plane wave; it is now only weakly attached to the slab. In this low-frequency limit most of the field energy is actually outside the dielectric, and the phase velocity approaches the speed of light in vacuum.

The reader may be curious about the other guided waves that can be propagated. The asymmetric TE waves, obtained by choosing cosine instead of sine in (9.70), lead to a characteristic equation like (9.78), with the tangent function replaced by the cotangent. As a result the lowest order solution, unlike the symmetric wave, has a cutoff frequency. The TM solutions are only slightly different from the TE; their phase velocities are slightly larger than the corresponding TE waves'. The lowest order asymmetric TM wave possesses a cutoff frequency, but the lowest order symmetric TM wave does not. Hence it is impossible to choose an operating frequency at which only a single mode can propagate. The reader may wish to consult Haus[12] or Kogelnik[13] for more extensive discussions of the subject.

OPTICAL FIBERS

Although free of metallic conductors, with their attendant ohmic loss, dielectric waveguides are not absolutely lossless. The loss tangents of dielectric

---

[12]H. A. Haus, *Waves and Fields in Optoelectronics*. Englewood Cliffs, N.J.: Prentice-Hall, 1984.

[13]H. Kogelnik, "Theory of Dielectric Waveguides." In *Integrated Optics*, T. Tamir, ed. 2nd ed. New York: Springer-Verlag, 1979.

materials, though small, are still greater than zero, and thus give rise to what is known as *dielectric loss*. Recently, however, it has been discovered that in certain carefully purified glasses, dielectric loss at optical frequencies can be amazingly low. This discovery has led to the use of optical-guiding fibers in communication systems. Optical fibers are now replacing copper wire for local telephone communications, and may also replace coaxial cables, microwave relays, and even satellite links for long-distance transmission. Optical fibers are cheaper than copper wires, take up much less space, are relatively free of crosstalk, and potentially offer large communication bandwidths.

One type of optical fiber, known as the *step-index fiber,* is made as shown in Fig. 9.15. A central glass core of radius r_o and permittivity ϵ is enclosed in a coating with a slightly smaller permittivity ϵ', known as a *cladding*. Waves are confined within the high-ϵ center region by total internal reflection, much as shown in Fig. 9.12(b). Evanescent waves extend outward from the center section, just as they do in the case of the dielectric slab; these fields decay exponentially with increasing distance from the guide. The thickness of the cladding is not important, except that it must be thick enough to contain the evanescent fields.[14] In this way the fields of the guided wave are protected from contact with materials outside the optical glass, which would cause reflections and absorption loss.

The modes of the circular fiber are physically similar to those of the rectangular slab, but the circular shape complicates the mathematics. Descriptions of the modes, which involve Bessel functions, will be found in the references.[15] Most of the modes are neither TE nor TM, but contain all six field components; such modes are called *hybrid modes*. Physically they can be thought of as rays following spiral paths, like threads on a screw, along the guide. The modes are classified as EH modes and HE modes; in

[14] Of course the evanescent fields cannot be *completely* contained in the cladding, since they extend to infinity. However, since they decrease very rapidly with radius, they can be extremely small by the time they reach the end of the cladding.

[15] J. Gower, *Optical Communication Systems*. Englewood Cliffs, N.J.: Prentice-Hall International, 1984. D. Marcuse, *Theory of Dielectric Optical Waveguides*. New York: Academic Press, 1974.

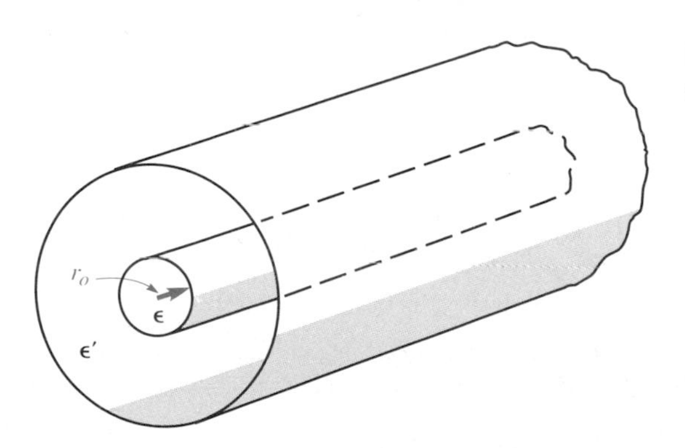

Figure 9.15 Optical fiber waveguide.

the former, the most important transverse fields are those arising (through eqs. (9.22)) from E_z, while for the latter, those from H_z are most important. All the modes cut off at low frequencies except the HE_{11} mode. Hence it is possible to construct single-mode fibers, in which only the HE_{11} mode can propagate. The condition for single-mode propagation is

$$\frac{2\pi r_o f}{c}(\epsilon - \epsilon')^{1/2} < 2.405 \tag{9.84}$$

where f is the frequency. This condition is difficult to satisfy at optical frequencies because r_o becomes inconveniently small. To increase the allowed value of r_o, it is helpful to reduce $(\epsilon - \epsilon')$. This, however, creates a new difficulty: small values of $(\epsilon - \epsilon')$ make it easy for rays to escape from the guide. Referring to Fig. 9.12(b), we observe that as $(\epsilon - \epsilon')$ decreases, the angle of incidence required for total internal reflection approaches 90°; in other words, any rays that are not nearly parallel to the axis are lost out the sides. Because of this difficulty, single-mode fibers are usually used with laser sources, which produce a nearly parallel beam.

Alternatively, it is possible to use fibers with larger cores and tolerate the existence of multiple modes. The number of modes goes up quadratically with core size: an approximate expression for M, the number of propagating modes, is

$$M \cong \frac{2\pi^2 r_o^2 f^2}{c^2}(\epsilon - \epsilon') \tag{9.85}$$

Multimode fibers are easier to make than single-mode fibers and accept rays that are more off-axis, so they can be used with non-laser light sources such as light-emitting diodes. However, the presence of multiple modes causes distortion of signals. Energy propagating in different modes will move with different phase velocities. Thus if a short pulse of light is put into the fiber, it will emerge at the far end as a pulse of longer duration, with a distorted shape. This kind of distortion, known as *intermode dispersion,* tends to restrict the use of multimode fibers to low-speed, short-distance applications.

Even in the case of single-mode fibers, limitations on bandwidth exist. One such limitation is imposed by *material dispersion,* the tendency of ϵ to vary with frequency. Even small material dispersion has a cumulative effect that becomes important over long distances, with the result that different frequency components of the signal arrive at the receiver at different times. Material dispersion can be reduced by operating at frequencies where $\frac{d\epsilon}{df}$ is smallest; infrared wavelengths around 1.5 μm presently seem best for this purpose. Finally, there is also dispersion arising from the inherent variation of phase velocity with frequency for the propagating mode. As with the dispersion of slab modes (Fig. 9.14), this effect can be thought of as resulting from the changing energy distribution: at low frequencies more of the energy is in the cladding, with its lower ϵ, and the wave is "fast." At higher frequencies the energy concentrates in the core and the wave is "slow."

It is possible to reduce intermode dispersion by using more sophisticated refractive-index profiles than that of Fig. 9.15. In *graded-index fibers* the sharply defined central core is replaced by material having a refractive index that varies with radius, often according to

$$n(r) = n_1 \left[1 - \Delta' \left(\frac{r}{r_o} \right)^2 \right] \tag{9.86}$$

where

$$\Delta' = \frac{n_1 - n_2}{n_1} \tag{9.87}$$

and $n_i = \sqrt{\epsilon_i/\epsilon_o}$. Here n_1 is the refractive index on the axis, and n_2 is the refractive index (not a function of position) of the cladding, which as before begins at $r = r_o$. Graded-index fibers are capable of propagating multiple modes with much less intermode dispersion than is obtained with a step-index fiber.

9.6 POINTS TO REMEMBER

1. Structures consisting of parallel, non-contacting conductors can propagate TEM (transverse electromagnetic) waves. The fields of TEM waves all lie in the transverse plane (i.e., the plane perpendicular to the direction of propagation), and $\vec{E}$ and $\vec{H}$ are perpendicular to each other. The fields in any transverse plane have the same configuration as static fields for the same boundaries. The phase velocity is equal to the velocity of light in the medium, and $|\vec{E}|/|\vec{H}| = \sqrt{\mu/\epsilon}$. These waves are identical with the transmission-line waves described in terms of circuit theory in Chapter 1. Plane electromagnetic waves can also be considered to belong to this class.

2. Waves in hollow metal waveguides can be TE (transverse electric) or TM (transverse magnetic). For each type there are an infinite number of modes characterized by integers m and n, such as for example the TE_{mn} mode. All of these modes have phase velocity greater than c and group velocity less than c, and each mode is characterized by a cutoff frequency ω_{mn} below which it does not propagate.

3. The most important mode of the hollow rectangular waveguide is the TE_{10} mode, which has the lowest cutoff frequency of any of the modes. Because of this there is, for any set of waveguide dimensions, a range of frequencies over which single-mode operation in the TE_{10} mode can be obtained.

4. When a waveguide is terminated at both ends by perfectly conducting walls, a type of resonant structure known as a cavity resonator is obtained. Inside an ideal cavity resonator, sinusoidal steady-state solutions of Maxwell's equations exist only at certain frequencies, known as the resonant frequencies. Each resonant frequency corresponds to a certain distribution of the fields referred to as a mode of the resonator. These modes derive from the TE_{mn} and TM_{mn} waveguide modes; for example the qth resonator mode derived from the TE_{mn} waveguide mode is called the TE_{mnq} mode of the resonator.

5. Planar circuits require waveguiding structures that can be fabricated by means of planar-fabrication technology. The most common of these is microstrip, which consists of a strip of metal on one face of a dielectric slab, with a uniform metal groundplane on the other face. Waves on microstrip resemble TEM waves, particularly at low frequencies, but they are not true TEM waves, because of the nonuniform nature of the dielectric environment. In fact, both E_z and H_z are present in these waves. At low frequencies approximate descriptions of these waves can be quite accurate, but at higher frequencies non-TEM behavior such as dispersion may need to be considered.

6. At very high microwave frequencies, the ohmic loss in metallic conductors becomes quite high, due to skin effect, and at optical frequencies these losses are quite prohibitive. Thus in optical technology metallic guides are replaced by all-dielectric guiding structures, such as optical fibers. Similar modes are propagated by plane dielectric slabs, such as, for example, wafers of highly resistive semiconductor. Modes of dielectric waveguides can be TE or TM or they may possess both E_z and H_z ("hybrid modes"). Some have lower cutoff frequencies, but others exist at all frequencies.

7. In optical fibers it is easiest to excite several modes simultaneously, but since different modes have different phase velocity, this results in distortion due to intermode dispersion. Intermode dispersion can be reduced through the use of graded-index fibers (with spatially varying dielectric constant). Alternatively, by using very small fibers one can confine propagation to the single HE_{11} mode, which lacks a cutoff. Material dispersion, arising from the variation of ϵ with frequency, can be reduced by operating at a frequency at which $d\epsilon/d\omega$ nearly vanishes.

REFERENCES

1. S. Ramo, J. R. Whinnery, and T. Van Duzer, *Fields and Waves in Communication Electronics*. New York: Wiley, 1984.
2. D. K. Cheng, *Field and Wave Electromagnetics*. Reading, Mass.: Addison-Wesley, 1983.
3. S. Y. Liao, *Microwave Devices and Circuits*. Englewood Cliffs, N.J.: Prentice-Hall, 1980.

4. K. F. Sander, *Microwave Components and Systems*. Reading, Mass.: Addison-Wesley, 1987.
5. T. C. Edwards, *Foundations for Microstrip Circuit Design*. New York: Wiley, 1981.
6. K. C. Gupta, R. Garg, and I. J. Bahl, *Microstrip Lines and Slotlines*. Dedham, Mass.: Artech, 1979.

A reference with emphasis on numerical solution is:

7. V. F. Fusco, *Microwave Circuits: Analysis and Computer-Aided Design*. Englewood Cliffs, N.J.: Prentice-Hall, 1987.

With regard to optical fiber guides, see:

8. J. Gowar, *Optical Communication Systems*. Englewood Cliffs, N.J.: Prentice-Hall, 1984.
9. A. Yariv, *Introduction to Optical Electronics*, 2nd ed. New York: Holt, Rinehart & Winston, 1976.
10. T. Tamir, ed., *Integrated Optics*, 2nd ed. New York: Springer, 1979.
11. D. Marcuse, *Theory of Dielectric Optical Waveguides*. New York: Academic Press, 1974.

A classic reference at a higher level is:

12. R. E. Collin, *Field Theory of Guided Waves*. New York: McGraw-Hill, 1960.

PROBLEMS

Section 9.1

9.1 Suppose that a guided wave is described by (9.1) and (9.2), and let $\vec{E}_o(x,y) = E_{ox}(x,y)\vec{e}_x + E_{oy}(x,y)\vec{e}_y + E_{oz}(x,y)\vec{e}_z$, and $\vec{H}_o(x,y) = H_{ox}\vec{e}_x + H_{oy}(x,y)\vec{e}_y + H_{oz}(x,y)\vec{e}_z$. Find H_{ox}, H_{oy}, and H_{oz} in terms of E_{ox}, E_{oy}, E_{oz}, and k.

9.2 A certain guided wave has the form (9.1), with

$$\vec{E}_o(x,y) = A(1+j)\sin kx\,\vec{e}_y \qquad \text{for } -\pi < kx < \pi$$
$$= 0 \qquad \text{otherwise}$$

where A is a positive real constant. Sketch $\vec{E}$ as a function of x for several values of z in the range $0 < kz < 2\pi$, at time $t = 0$.

9.3 Show from (9.9) that $|\vec{E}(t)| = \eta|H(t)|$ at all times.

*9.4 Prove that for equations (9.9) to hold, $\vec{E}$ and $\vec{H}$ must be perpendicular.

9.5 Show, by applying the $\nabla \cdot (f\vec{V})$ vector identity to (9.1), that $\nabla \cdot \vec{E}_o = 0$.

9.6 Find the magnetic field associated with a negative-going wave on a coaxial line, if the electric field is $\vec{\mathbf{E}} = (A/r)e^{jkz}\,\vec{e}_r$.

9.7 Sketch the general appearance of the electric and magnetic fields at $z = 0$ for TEM waves propagating on the structures shown in Figs. 9.1(b) and 9.1(c).

9.8 An electrostatic measurement on the transmission line shown in Fig. 9.1(c) shows that when the space between the conductor is filled with air, the capacitance of a section 10 cm long is 100 pF. Assuming that end effects are negligible, find the inductance per unit length of this line.

9.9 For a certain transmission line, the characteristic impedance is 50 ohms and the phase velocity is $0.7c$. Find the capacitance per unit length of this line.

*9.10 Show that for a TEM wave on a two-conductor line, the line integral $\int \vec{E} \cdot \vec{dl}$, taken from one conductor to the other in a plane of constant z, is independent of the path of integration. Why is this result important?

**9.11 The material separating the two conductors of a transmission line has permittivity ϵ and also a conductivity σ.

(a) Obtain a result corresponding to (9.6).

(b) Find the propagation constant β and attenuation constant α as defined in (1.49), if $\epsilon = \epsilon_o$ and $\sigma = \omega\epsilon_o/10$.

(c) For a general TEM wave in this material, is $H(t)$ always perpendicular to $E(t)$? Prove or disprove.

Section 9.2

9.12 Verify that (9.27) and (9.28) are a solution of (9.25).

*9.13 Sketch the electric field of the TM_{11} waveguide mode

(a) in the transverse plane $z = 0$; and

(b) as seen from the side. (These sketches should be of the same sort as Fig. 9.6.)

(c) Add the magnetic field to your sketches.

(d) How do the fields appear at a time π/ω later?

*9.14 Find the time-averaged power carried by the TM_{mn} wave (9.30), for arbitrary integer values of m and n.

*9.15 Find the attenuation coefficient for a TEM wave in a coaxial line with inner and outer copper conductors having radii of 1 mm and 2.77 mm, at 1 GHz. The dielectric between the conductors has $\epsilon = 1.5\,\epsilon_o$.

**9.16 Show that the attenuation coefficient α for the TM_{mn} waveguide mode is

$$\alpha_{mn} = \frac{2R_S}{b\eta\sqrt{1 - \omega_{mn}^2/\omega^2}}\left[\frac{m^2(b/a)^3 + n^2}{m^2(b/a)^2 + n^2}\right]$$

where ω_{mn} is the cutoff frequency of the mode and $\eta = \sqrt{\mu/\epsilon}$.

*9.17 Sketch the attenuation coefficient α for the TE_{10} mode of WR-90 waveguide with copper walls as a function of frequency, for frequencies above cutoff. Estimate the frequency at which α is minimum. (Note that R_S depends on frequency.)

*9.18 Find and sketch the group velocity of the TE_{10} mode, as a function of the normalized frequency $\omega' = \omega/\omega_{10}$, where ω_{10} is the cutoff frequency of the mode.

**9.19 A waveguide operating at a frequency at which many modes can propagate is said to be *overmoded*. Show that at high frequencies the number of modes (both TE and TM) that can propagate in a square waveguide with side a at frequency ω is approximately $a^2\omega^2/2\pi c^2$.

*9.20 Consider the TE_{10} mode (9.38) at a frequency below cutoff. For simplicity assume K is a real constant.
(a) Obtain expressions for the three fields as functions of time. Include the fields' dependence on z. Is this a propagating wave?
(b) Show that the time average of the z component of the Poynting vector vanishes.
(c) Let L be the "penetration length" of the evanescent wave; this is the distance in the z direction over which the field amplitude decreases by a factor of e. Find L.

Section 9.3

9.21 Write expressions for the electric and magnetic fields of the TE_{101} mode of the rectangular cavity resonator. Sketch the fields.

*9.22 A resonator is made from a length d of coaxial transmission line, with inner and outer radii a and b, by placing short-circuit planes at each end. Obtain expressions for the electric and magnetic fields of the lowest-order mode. What is the resonant frequency?

**9.23 The transmission line in the preceding problem is made of copper, and the dielectric is air. Find the Q of the resonator. Note:

$$Q = 2\pi \frac{\text{(Energy stored in the resonator)}}{\text{(Energy lost per cycle)}}$$

*9.24 Let us regard a resonator mode as a transmission system in which a wave travels to one end, is reflected, travels to the other end, is reflected again, and returns to the starting point. In that case the phase of the returned wave must be the same as the phase of the original wave (since the phase at any point has a single unique value). Use this reasoning to find the resonant frequencies of the TE_{mnq} resonator mode. The result should agree with (9.52).

**9.25 Let T be the time required for the fields in a resonator to decrease by a factor of e. Find T for the TE_{101} mode of a copper resonator at 10 GHz. The resonator consists of a section of WR-90 waveguide with short circuits at each end.

Section 9.4

9.26 Find the characteristic impedance and phase velocity of a microstrip with $w/h = 1.5$, $\epsilon = 12.9\epsilon_o$, by means of (9.58) and (9.59). By what percentage do the rough approximations (9.53) and (9.56) differ? Do you think agreement between the two levels of approximation will become better or worse if w/h is increased? If ϵ is reduced?

*9.27 Show that if it is assumed (contrary to fact) that the microstrip wave is a TE wave, the four boundary conditions at the air–dielectric interface cannot be satisfied.

9.28 A resonator is constructed by placing short circuits at each end of a section of microstrip with $w/h = 3$, $\epsilon = 2.5\epsilon_o$, and length 3 cm. Assume the quasistatic approximation is accurate. Find the lowest and next-lowest resonant frequencies.

*9.29 Repeat the previous problem assuming that one end of the microstrip is terminated in an ideal short circuit and the other in an ideal open circuit.

9.30 A 50-ohm microstrip delivers power to a load with impedance Z_L. Show how impedance matching can be accomplished
(a) by a quarter-wave transformer, when Z_L is real
(b) with a double-stub tuner composed entirely of microstrip, when Z_L is complex. Your circuits should be entirely planar.

*9.31 The input port of a certain microwave transistor can be modeled as a 0.31 pF capacitor and an 11-ohm resistor connected in series. Design an impedance-matching circuit to eliminate reflected waves at 10 GHz, when the transistor is driven by 50-ohm microstrip. Assume the permittivity of the substrate is $12.9\epsilon_o$ and its thickness is 200 μm.

9.32 Estimate the phase and group velocities of the microstrip of Fig. 9.11,
(a) at 120 GHz, and
(b) in the limit of low frequencies.

Section 9.5

9.33 For the symmetric TE mode of the dielectric slab guide (Eq. (9.70)), find the fields in region 3. Show that the boundary conditions are satisfied at the interface between regions 2 and 3.

*9.34 Sketch the fields of the lowest-order TE slab mode.

*9.35 Make sketches of $|E_y|$ as a function of y, for symmetric TE slab modes in the following cases:
(a) lowest-order mode, $\omega/\omega_a = 0.5$
(b) lowest-order mode, $\omega/\omega_a = 2.5$
(c) next-to-lowest mode, $\omega/\omega_a = 5$.

9.36 Estimate the group velocity of the zero-order mode of Fig. 9.14 at $\omega = \omega_a$.

*9.37 Show that the number of symmetric TE modes propagated by the slab waveguide at frequency ω is

$$N(\omega) = \frac{a\omega}{\pi c}\left(\frac{\epsilon}{\epsilon_o} - 1\right)^{1/2} + 1$$

**9.38 Study the asymmetric TE slab modes obtained by choosing the cosine function instead of the sine in (9.70). Show that (9.78) is replaced by

$$\cot q = -\left(\frac{\omega^2/\omega_a^2}{q^2} - 1\right)^{1/2}$$

find the cutoff frequency of the lowest mode in terms of a and ϵ.

**9.39 Surface waves can also propagate on a dielectric slab with a metal ground plane on one surface. Solutions for this structure can be found from the waves on a dielectric slab of twice the thickness. Take $y = 0$ at the metal. Are the boundary conditions satisfied by
(a) the symmetric TE modes
(b) the asymmetric TE modes
(c) the symmetric TM modes
(d) the asymmetric TM modes?
Is there a mode that lacks a cutoff frequency?

**9.40 Consider an interface between medium 1, which has $\epsilon = \epsilon_o$ and occupies the half-space $y > 0$, and medium 2, which has a negative permittivity $\epsilon_R\epsilon_o$ ($\epsilon_R < 0$) and occupies $y < 0$. Show that a TM surface wave can propagate on the interface, provided that $|\epsilon_R| > 1$. Find the phase velocity as a function of ϵ_R. (*Note:* Some metals have negative permittivity at optical frequencies. The permittivity of a plasma can also be negative.)

*9.41 Figure 9.16 shows a ray being injected into a cylindrical optical fiber by an external light source. Assuming that Snell's law applies in the same way as for planar interfaces, find $\sin\alpha$, where α is the largest angle (between the ray and the axis) allowed by the requirement of total reflection at the core–cladding interface. (The quantity $\sin\alpha$ is known as the *numerical aperture* (NA) of the fiber.) What is the numerical value of α, in degrees, for $\epsilon = 2.16\epsilon_o$, $\epsilon' = 1.99\epsilon_o$?

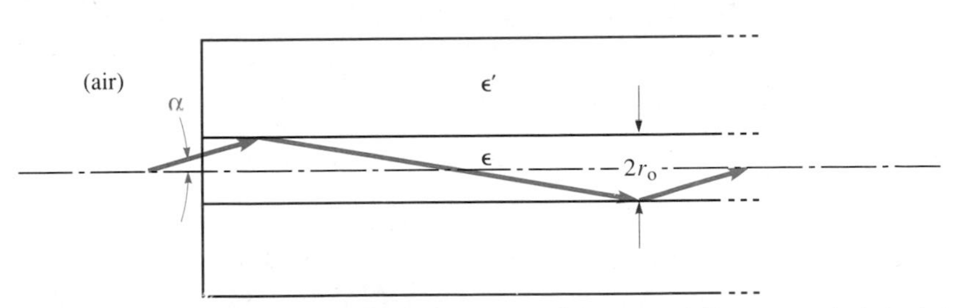

Figure 9.16 Waves are guided by the core of the optical fiber, provided the injection angle α is sufficiently small. (For Problem 9.41.)

*9.42 Suppose the end of the fiber in Fig. 9.16 is illuminated by an isotropic point radiator with total power P_o on the axis at a distance ar_o outside the glass. Show that in terms of the numerical aperture (NA) (see previous problem) the power guided by the fiber will be approximately $(\mathrm{NA})^2 P_o/4$, provided that $\mathrm{NA} \ll 1$ and $a \sin\alpha < 1$.

*9.43 Answer the following questions about the step-index optical fiber guide by using the analogy with the dielectric slab guide:
(a) Sketch the general dependence of field strength on radius for the lowest-order mode, for small $(\epsilon - \epsilon')$ and for larger $(\epsilon - \epsilon')$.
(b) Estimate the phase velocity of the lowest mode in the limits of low and high frequency.

**COMPUTER EXERCISE 9.1 NUMERICAL SOLUTION OF AN EIGENVALUE PROBLEM

In this exercise we shall find the cutoff frequencies and propagation constants for some modes of a *circular* hollow metal waveguide. Actually circular waveguides are very well known, and their modes are described in many textbooks in terms of tabulated special functions known as *Bessel functions.* Here we shall use a numerical method that can be extended to waveguides of more complicated shape.

Let us consider a circular metal pipe of radius a, and for simplicity let us restrict our search to TM modes with circular symmetry. In that case our unknown field is $\mathbf{E}_z(r)$, which is a function of r but independent of ϕ. Helmholtz' equation in cylindrical coordinates becomes

$$r\frac{d^2\mathbf{E}_z}{dr^2} + \frac{d\mathbf{E}_z}{dr} + k_c^2 r\mathbf{E}_z = 0 \qquad \textbf{(9.88)}$$

where $k_c^2 \equiv \omega^2\epsilon - k^2$. One boundary condition is $\mathbf{E}_z(r = a) = 0$. The other can be obtained by inspecting the behavior of (9.88) at $r = 0$. Since $\mathbf{E}_z$ and its derivatives are assumed to be finite, we see that the value of $\frac{d\mathbf{E}_z}{dr}$ at $r = 0$ must be zero.

Let us now choose $N + 1$ equally spaced points along a radius as shown in Fig. 9.17(a), with point 0 at the center. In the figure we have assumed $N = 7$, a number that can be handled on an Hp-15C (if every single register, including the index register, is used!). As usual, better accuracy will be obtained if N is larger, but since we shall have to solve N simultaneous equations, the computational effort will be greater.

As usual in finite-difference methods, we now replace $\frac{d^2\mathbf{E}_z}{dr^2}$ at point i (where $r = hi$) by $[\mathbf{E}_z(i + 1) + \mathbf{E}_z(i - 1) - 2\mathbf{E}_z(i)]/h^2$ and $\frac{d\mathbf{E}_z}{dr}$ by

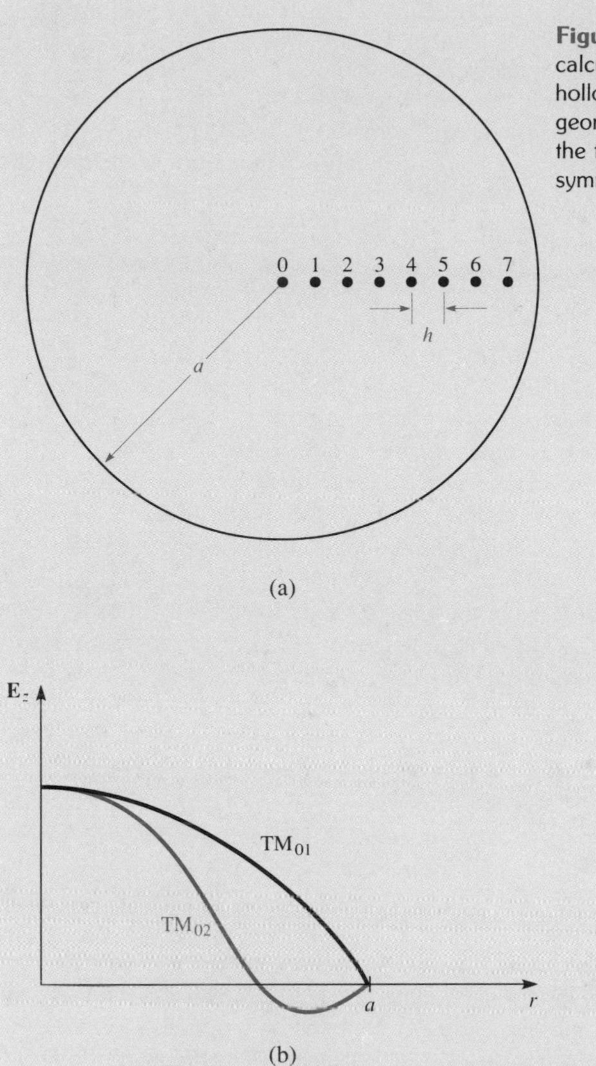

Figure 9.17 Numerical calculation of fields of circular hollow-metal waveguide. (a) geometry; (b) electric fields of the two lowest order circularly symmetric modes.

$[\mathbf{E}_z(i+1) + \mathbf{E}_z(i-1)]/2h$. For points 2 through N, the Helmholtz equation becomes

$$(i - \tfrac{1}{2})\mathbf{E}_z\,(i-1) + i(K^2 - 2)\mathbf{E}_z(i) + (i + \tfrac{1}{2})\mathbf{E}_z\,(i+1) = 0 \qquad \textbf{(9.89)}$$

where $K^2 \equiv k_c^2 h^2$. Point 1 requires special treatment because of the boundary condition $\dfrac{d\mathbf{E}_z}{dr}(r = 0) = 0$. In the spirit of the finite-difference approach we write

$$\mathbf{E}_z(i = 1) \cong \mathbf{E}_z(i = 0) + \left.\frac{d\mathbf{E}_z}{dr}\right|_{z=0} h \qquad \textbf{(9.90)}$$

Then since $\left.\frac{d\mathbf{E}_z}{dr}\right|_{z=0}$ vanishes, we can simply set $\mathbf{E}_z(i = 0) = \mathbf{E}_z(i = 1)$. This conveniently eliminates $\mathbf{E}_z(i = 0)$ as an unknown and we have only N remaining unknowns. From (9.89) we obtain $(N - 1)$ equations. The remaining equation, the Helmholtz equation at point 1, becomes (using $\mathbf{E}_z(0) = \mathbf{E}_z(1)$ in (9.89)),

$$(K^2 - \tfrac{3}{2})\mathbf{E}_z(1) + \tfrac{3}{2}\mathbf{E}_z(2) = 0 \qquad \textbf{(9.91)}$$

We now have N simultaneous equations in the N unknowns $\mathbf{E}_z(1) \cdots \mathbf{E}_z(N)$. Since the right side of all these equations is zero, all the unknown voltages would have to vanish, *unless the determinant of the equations' matrix vanishes.* It is this condition that allows us to determine the constant K^2, which in turn determines the cutoff frequency and phase velocity for a given frequency. The number K^2 is known as an *eigenvalue* of the equation, and this exercise demonstrates numerical solution of an *eigenvalue problem.*

We proceed by setting up the matrix of the N simultaneous equations for some arbitrary starting value of K^2. We then calculate the determinant, which we wish to vanish, but which does not, because we have the wrong value of K^2. We then use a "solve" routine to substitute new values of K^2 and recalculate the determinant until the latter vanishes and the correct K^2 has been found.

In fact, there are many different values of K^2 that cause the determinant to vanish. These correspond to different modes of the waveguide. The axial electric fields for the lowest two of these modes, TM_{01} and TM_{02}, are sketched in Fig. 9.17(b). (The first mode subscript, zero, indicates that the modes are circularly symmetrical.) Not all of these modes will propagate at a given frequency, of course; the higher ones, for which $k^2 = \omega^2\epsilon - K^2/h^2$ is negative, will be cut off. Which mode is found by your "solve" routine depends on your initial guess for K^2. From data on Bessel functions, the correct value of the smallest K^2 is 0.09797, corresponding to the TM_{01} mode; the value corresponding to the TM_{02} mode is 0.47069. You should be able to compute K^2 for these two modes by beginning somewhere near these values. You can then find the cutoff frequencies and construct the ω–k plots for any assumed radius; for convenience take $a = 10$ cm.

(If you are ambitious, you can go on to find $\mathbf{E}_z(i)$, for an arbitrarily assumed axial amplitude, such as $\mathbf{E}_z(0) = 1$. You can then sketch out the field distribution of the mode.)

**COMPUTER EXERCISE 9.2 MICROSTRIP: CHARACTERISTIC IMPEDANCE AND PHASE VELOCITY

In this rather advanced exercise we shall calculate Z_o and U_P for microstrip, in the low-frequency TEM approximation. Our approach will be to calculate the electrostatic capacitance per unit length C, and also C_A, the electrostatic capacitance of a transmission line that is identical, except that the dielectric material is replaced by air. Following the approach outlined in section 9.4, we can then find Z_o and U_P from

$$Z_o \cong \frac{1}{c\sqrt{CC_A}} \quad \textbf{(9.92)}$$

$$U_P \cong c\sqrt{C_A/C} \quad \textbf{(9.93)}$$

where c is the velocity of light.

To find the electrostatic capacitance, we shall use the method of Computer Exercise 4.4, which should be reviewed at this time. To obtain good accuracy, a large number of iterations (say, 250) and a large number of voltage points (say, 2500) will be required. Thus a computer, rather than a programmable calculator, should be used.

Figure 9.18(a) illustrates the problem to be solved. Note that artificial walls, at potential zero, have been placed at the sides and top, so that the microstrip is enclosed in a grounded box. These imaginary walls are needed because without them we would have an unbounded region, requiring an infinite number of voltage points. However, we can ensure that the effect of the imaginary walls is negligible by placing them far away from the microstrip, where the fields have died off to nearly zero. It will only be necessary to find the potential in the left half of the structure, shown in Fig. 9.18(b). From symmetry we know that the potentials in the right-hand half must be the same as those on the left; to solve for them would simply double the computing time without giving any more information. Symmetry requires that $E_x = 0$ at the center of the microstrip. (Do you see why?) The boundary condition to be applied at this plane is thus $E_{normal} = 0$, corresponding to what is known as a "magnetic wall."

Figure 9.18(b) shows a suitable set of dimensions for the case $w/h = 2$. The voltage grid points are identified by index numbers i and j. All points with $i = 1$ lie on the top wall, and thus $V_{1,j} = 0$ for all j. Similarly, $V_{47,j} = 0$ and $V_{i,1} = 0$. The upper conductor of the microstrip, which is assumed to be infinitely thin, lies in the plane $i = 39$. If we arbitrarily set its potential at 1 V we have $V_{39,j} = 1$ for $45 \leq j \leq 53$; however, at the other points in the plane $i = 39$ the voltages are unknown.

To find the unknown voltages we use the iterative method of Computer Exercise 4.4. At interior points we wish the potentials to converge to the average of those at the four nearest-neighbor points; thus we use

$$V_{ij}^{(new)} = V_{ij}^{(old)} + \beta R \quad \textbf{(9.94)}$$

where the "residual" R is given by

$$R_{ij} = \tfrac{1}{4}[V_{i+1,j}^{(old)} + V_{i-1,j}^{(old)} + V_{i,j+1}^{(old)} + V_{i,j-1}^{(old)}] - V_{ij}^{(old)} \quad \textbf{(9.95)}$$

For points in the column $j = 53$ we make use of symmetry. In column $j = 54$ (which physically exists, but is outside our problem space) the potentials

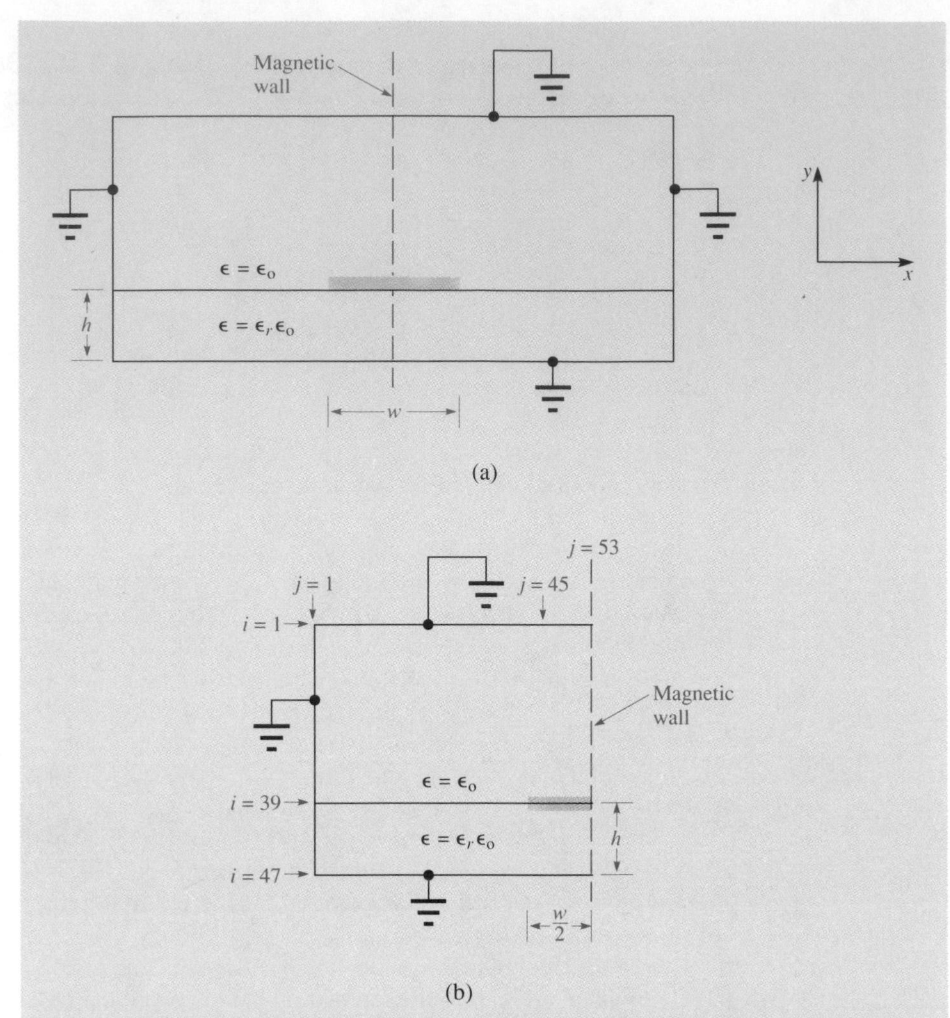

Figure 9.18 Numerical calculation of the electric fields in microstrip.

are identical with those in column 52. Thus, for points in column 53 (9.95) is replaced by

$$R_{i,53} = \tfrac{1}{4}[V^{(\text{old})}_{i+1,53} + V^{(\text{old})}_{i-1,53} + 2V^{(\text{old})}_{i,52}] - V^{(\text{old})}_{i,53}$$

for $2 \leq i \leq 38$ and $40 \leq i \leq 46$.

The points in row 39 (with $2 \leq j \leq 44$) lie on the dielectric boundary and thus require special treatment. The relation between the potential at a boundary point and those at its neighbors can be found by means of Gauss' law. Referring to Fig. 9.19(a), the field E_1 is given approximately by

$$E_1 \cong \frac{V_{39,j} - V_{38,j}}{\delta l}$$

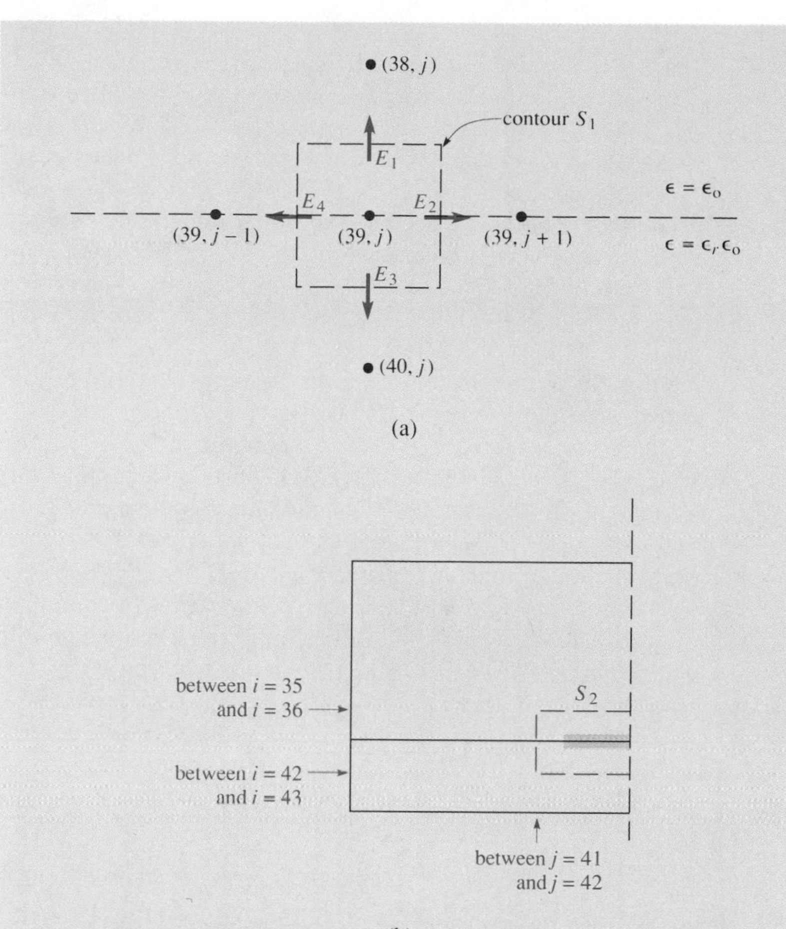

Figure 9.19

where δl is the distance between nearest-neighbor points. Fields E_2, E_3, and E_4 are found in similar fashion. The surface integral of $\vec{D}$ on the contour S_1, which according to Gauss' law must vanish, thus becomes

$$\left(\frac{V_{39,j} - V_{38,j}}{\delta l}\right) \delta l + \left(\frac{V_{39,j} - V_{40,j}}{\delta l}\right) \epsilon_r \, \delta l$$

$$+ \left(\frac{V_{39,j} - V_{39,j-1}}{\delta l}\right)\left(\epsilon_r \frac{\delta l}{2} + \frac{\delta l}{2}\right) + \left(\frac{V_{39,j} - V_{39,j+1}}{\delta l}\right)\left(\epsilon_r \frac{\delta l}{2} + \frac{\delta l}{2}\right) = 0$$

or, collecting terms

$$V_{39,j} = \frac{1}{2(1 + \epsilon_r)} [V_{38,j} + \epsilon_r V_{40,j}] + \tfrac{1}{4}[V_{39,j-1} + V_{39,j+1}]$$

The residual is then obtained by subtracting from this the value of $V_{39,j}^{(old)}$. Note that the residual reduces correctly to (9.95) in the case $\epsilon_r = 1$.

The quantity we actually wish to calculate is the charge on the microstrip's upper conductor. To find this we can construct the surface integral $\int \vec{D} \cdot dS$ on contour S_2 in Fig. 9.19(b). In principle it should not matter where this contour is placed, but in practice the finite-difference estimates of $\vec{E}$ are less accurate close to the microstrip, while convergence is slow and roundoff error large close to the outside walls. A possible choice is the one shown in Fig. 9.19(b). The construction of the field integral proceeds as in Fig. 9.19(a); the details are left to the reader. Note that fields at the magnetic wall contribute nothing to the integral, and that the total charge is *twice* that obtained by integrating over S_2.

You may now proceed to write a program in FORTRAN or another higher language with which to solve for the unknown voltages and the total charge. You may wish to give some attention to the computational efficiency of your program; for example, constructing $\Sigma\ 2\ V_i$ requires approximately twice as many multiplications as constructing $2\ \Sigma\ V_i$. In addition to printing out the total charge for a given value of ϵ_r, you may wish to also print out the matrix V_{ij}. This will allow you to verify that the fields are small near the imaginary walls. Check to see that your potential distribution seems physically reasonable. This will help you to spot any programming errors.

Find the characteristic impedance and phase velocity for a microstrip with $\epsilon_r = 9.6$, $w/h = 2$, and compare your results with Figs. 9.9 and 9.10.

CHAPTER 10

Radiation and Antennas

In earlier chapters we considered the sources of electrostatic and magnetostatic fields. We found that electrostatic fields originate with stationary charges and that magnetostatic fields are produced by non-time-varying currents. However, we have until now neglected the question of how electromagnetic waves originate. Regardless of whether we deal with waves in free space or waves guided by some structure, we eventually must address the question of how free-space waves or guided waves can be produced. That process, in which energy in some other form is converted into electromagnetic waves, is known as *radiation*. The most important and familiar device associated with radiation is the *antenna*. The inverse process, in which electromagnetic waves are captured and converted to another form of energy, is also of interest; an antenna used for this purpose is known as a *receiving antenna*.[1]

We shall find that electromagnetic waves originate with time-varying currents. Expressions will be found relating the radiated fields to the currents that produce them. The problem of antenna design, in this point of view, boils down to producing the right currents: currents that will give rise to the radiated fields one wants. A second, quite different point of view on antennas results from thinking in terms of boundary-value problems. In this picture, the radiated fields are the unique solution of the Helmholtz equation, subject to whatever boundary conditions are imposed by the antenna. Still another point of view is that an antenna is a transition between guided waves and unguided waves; a waveguide horn antenna might be thought of in that way. So many different points of view are needed because, unfortunately, there

[1]In this chapter we shall deal with what are known as *classical* sources of radiation, those described by Maxwell's equations. Quantum-mechanical mechanisms, as in lasers, will not be considered. The classical theory of radiation is generally adequate at radio and microwave frequencies. However, at infrared and optical frequencies (above, say, 10^{12} Hz), quantum effects become important.

are no antenna problems that can be solved exactly, in closed form. Antenna design involves the use of various approximate methods, with different methods and points of view being useful with different kinds of antennas. As a result, the subject of antennas is inherently more subtle and sophisticated than other subjects we have seen.

10.1 SOURCES OF RADIATION

REVIEW OF ELECTROSTATICS AND MAGNETOSTATICS

In electrostatics we find the electrostatic potential by first calculating the potential of a single infinitesimal point charge, and then adding up the contributions of all charges to find the total potential. This is what is known as a *Green's function* approach to solving Poisson's equation. The equation being solved is

$$\nabla^2 V = -\frac{\rho}{\epsilon} \tag{10.1}$$

and the *Green's function,* which is the potential at point $\vec{r}$ arising from a point charge q at $\vec{r}'$, is

$$V(\vec{r}, \vec{r}') = \frac{q}{4\pi\epsilon|\vec{r} - \vec{r}'|} \tag{10.2}$$

The total potential is found by adding up the contributions of all charges in a charge distribution $\rho(\vec{r}')$:

$$V(\vec{r}) = \int \frac{\rho(\vec{r}')\, dv'}{4\pi\epsilon|\vec{r} - \vec{r}'|} \tag{10.3}$$

Similarly the magnetostatic field $\vec{H}$ obeys

$$\nabla \times \vec{H} = \vec{J} \tag{10.4}$$

and if, as in Chapter 5, we set $\nabla \times \vec{B} = \vec{A}$, where $\vec{A}$ is the vector magnetic potential, we have

$$\nabla \times \nabla \times \vec{A} = \nabla(\nabla \cdot \vec{A}) - \nabla^2\vec{A} = \mu\vec{J} \tag{10.5}$$

It must be noted at this point that the vector potential $\vec{A}$ is not unique. There are any number of possible functions which, when operated upon with the curl operator, give the same $\vec{B}$. In magnetostatics it is convenient to choose that particular $\vec{A}$ that not only gives the proper $\vec{B}$ when curled, but also satisfies $\nabla \cdot \vec{A} = 0$. The selection of one particular function for $\vec{A}$ is known as a choice of *gauge,* and this particular choice is called the *Coulomb gauge*. Thus from (10.5), $\vec{A}$ in the Coulomb gauge obeys

$$\nabla^2\vec{A} = -\mu\vec{J} \tag{10.6}$$

which is very similar to (10.1), except that it is a vector equation. This is really an insignificant difference: equation (10.6) simply consists of three scalar equations, $\nabla^2 A_x = -\mu J_x$ and two others for A_y and A_z, each of which is identical in form with (10.1). Thus for each rectangular component of (10.6) there is a Green's function solution of the same form as (10.2). The potential at $\vec{r}$ arising from an infinitesimal current element $i_x\, dx$, located at $\vec{r}'$, is hence

$$A_x(\vec{r}, \vec{r}') = \frac{\mu i_x\, dx}{4\pi|\vec{r} - \vec{r}'|} \tag{10.7}$$

If there is a general current $i(\vec{r}')\,\vec{dl}$ located at $\vec{r}'$, the resulting vector potential is obtained by adding up the three rectangular components:

$$\vec{A}(\vec{r}, \vec{r}') = \frac{\mu i(\vec{r}')\,\vec{dl}}{4\pi|\vec{r} - \vec{r}'|} \tag{10.8}$$

THE RETARDED POTENTIALS

We now wish to proceed in a similar way for the electrodynamic case. However, we must be cautious about the electric potential, since we have seen that the usual electrostatic potential V is inappropriate when time-varying magnetic fields are present. Let us define a new electric potential function $\Phi(\vec{r}, t)$, such that

$$\vec{E} = -\nabla\Phi - \frac{\partial\vec{A}}{\partial t} \tag{10.9}$$

The point of redefining the scalar potential in this way is that (10.9) is consistent with Faraday's law, $\nabla \times \vec{E} = -(\partial\vec{B}/\partial t)$, as is easily seen by substitution. The electrostatic definition $\vec{E} = -\nabla V$ is inconsistent with Faraday's law, and thus cannot be used in dynamics. Note that $\vec{B} = \nabla \times \vec{A}$ continues to be correct.

EXERCISE 10.1

Verify that (10.9) is consistent with Faraday's law and that $\vec{E} = -\nabla V$ is not.

Substituting (10.9) into the source equation $\nabla \cdot \vec{E} = \rho/\epsilon$, we obtain

$$\nabla^2\Phi + \frac{\partial}{\partial t}(\nabla \cdot \vec{A}) = -\frac{\rho}{\epsilon} \tag{10.10}$$

Substituting $\vec{B} = \nabla \times \vec{A}$ into the magnetic curl equation $\nabla \times \vec{H} = J + \partial\vec{D}/\partial t$ results in

$$\nabla^2\vec{A} - \nabla(\nabla \cdot \vec{A}) = -\mu\vec{J} - \mu\epsilon\frac{\partial\vec{E}}{\partial t} \tag{10.11}$$

We now recall that different vector potentials $\vec{A}$ are possible, and that we can select a particular one by choosing $\nabla \cdot \vec{A}$ to suit our convenience. Instead of the Coulomb gauge used in statics, it is convenient to choose a *Lorentz gauge,* characterized by

$$\nabla \cdot \vec{A} = -\mu\epsilon \frac{\partial \Phi}{\partial t} \tag{10.12}$$

The choice (10.12), known as the *Lorentz condition,* causes (10.10) and (10.11) to simplify, and they become

$$\nabla^2 \Phi - \mu\epsilon \frac{\partial^2 \Phi}{\partial t^2} = -\frac{\rho}{\epsilon} \tag{10.13}$$

and

$$\nabla^2 \vec{A} - \mu\epsilon \frac{\partial^2 \vec{A}}{\partial t^2} = -\mu \vec{J} \tag{10.14}$$

Equation (10.14) is the equation to be solved to obtain the fields arising from a time-varying current $\vec{J}$. It is usually unnecessary to solve (10.13) in addition, because if $\vec{A}$ can be found, Φ can be obtained from (10.12).

What is required, then, is a Green's function for (10.14). That is, we need a solution of (10.14) when $\vec{J}$ is an infinitesimal current element, just as (10.8) is the corresponding solution of the static equation (10.6). We shall find that solutions of (10.14) look very much like solutions of (10.6), with the interesting difference that they are *delayed in time*. Intuitively, we expect that if a current is suddenly turned on at the origin, the effects of the current will not be felt far away at the same instant. Information cannot travel faster than the speed of light, so no change in the fields at a distance r can occur until at least a time r/c has passed. The fields and potentials are said to be *retarded,* and the solution we seek for (10.14) is known as a *retarded potential.*

Let us consider a single rectangular component of (10.14). This gives us a scalar equation,

$$\nabla^2 A_x - \mu\epsilon \frac{\partial^2 A_x}{\partial t^2} = -\mu J_x \tag{10.15}$$

We now find the A_x resulting from a current $i_x(t)$ located at the origin. Expecting that this function will have spherical symmetry, we re-express the scalar Laplacian operator in spherical coordinates:

$$\frac{1}{r^2} \frac{\partial}{\partial r} \left(r^2 \frac{\partial A_x}{\partial r} \right) - \mu\epsilon \frac{\partial^2 A_x}{\partial t^2} = -\mu J_x \tag{10.16}$$

We then make the substitution $A_x = f(r, t)/r$. Furthermore, we shall seek a solution valid everywhere except at the origin. The source current is located exclusively at the origin. Hence away from the origin (10.16) becomes

$$\frac{\partial^2 f(r, t)}{\partial r^2} - \mu\epsilon \frac{\partial^2 f(r, t)}{\partial t^2} = 0 \tag{10.17}$$

This equation is identical with the transmission-line equation (1.9). Its solutions consist of waves propagating in the $+r$ or $-r$ directions, $f^+\left(t - \frac{r}{U}\right)$ and $f^-\left(t + \frac{r}{U}\right)$, where U, the phase velocity, is equal to $(\mu\epsilon)^{-1/2}$. The functions f^+ and f^- can be of any form.[2] However, the solution moving in the $-r$ direction makes no physical sense in this case, and we shall discard it.

To determine the form of f^+, we make use of the quasistatic approximation. In the limit as $r \to 0$, the electrodynamic solution must approach the static solution (10.7). Thus

$$\lim_{r\to 0} A_x = \lim_{r\to 0} \frac{f^+\left(t - \frac{r}{U}\right)}{r} = \frac{f^+(t)}{r} = \frac{\mu i_x(t)\, dx}{4\pi r} \tag{10.18}$$

Thus the function $f^+(t)$ is given by

$$f^+(t) = \frac{\mu i_x(t)\, dx}{4\pi} \tag{10.19}$$

and the required solution is

$$A_x(r, t) = \frac{f^+\left(t - \frac{r}{U}\right)}{r} = \frac{\mu i_x\left(t - \frac{r}{U}\right) dx}{4\pi r} \tag{10.20}$$

The meaning of (10.20) is that A_x, at time t and at a distance r from the source, is determined by the current that flowed at a time r/U earlier. This is because it requires a time r/U for the information to travel from the source to the observation point. If the current element at the origin is directed in the direction $\vec{dl}$, (10.20) clearly becomes

$$\vec{A}(r, t) = \frac{\mu i\left(t - \frac{r}{U}\right) \vec{dl}}{4\pi r} \tag{10.21}$$

For example, if the observation point is located as in Fig. 10.1(a) and the current at the origin is a rectangular pulse as shown in (b), then the vector potential is retarded as shown in (c). Note that the direction of $\vec{A}$ is still the same as that of the current which produces it.

Finally, if currents are located at positions $\vec{r}'$ other than the origin, we add together the contributions of all the currents. To allow for the appropriate travel time, each current must be evaluated at a different *retarded time* t_R,

[2] It can easily be shown, by direct substitution, that $f^+[t - (r/U)]$ and $f^-[t + (r/U)]$ are solutions of (10.17), when $f^+(x)$ and $f^-(x)$ are any differentiable functions. In Chapter 1 we wrote these solutions as $v^+(r - Ut)$ and $v^-(r + Ut)$. Since the functions are arbitrary, there is no real difference.

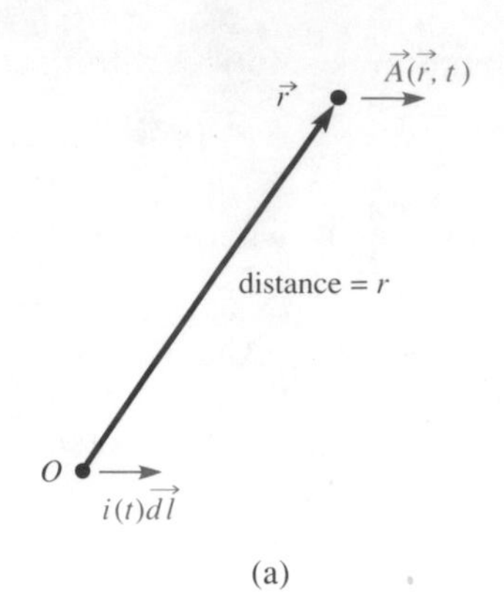

(a)

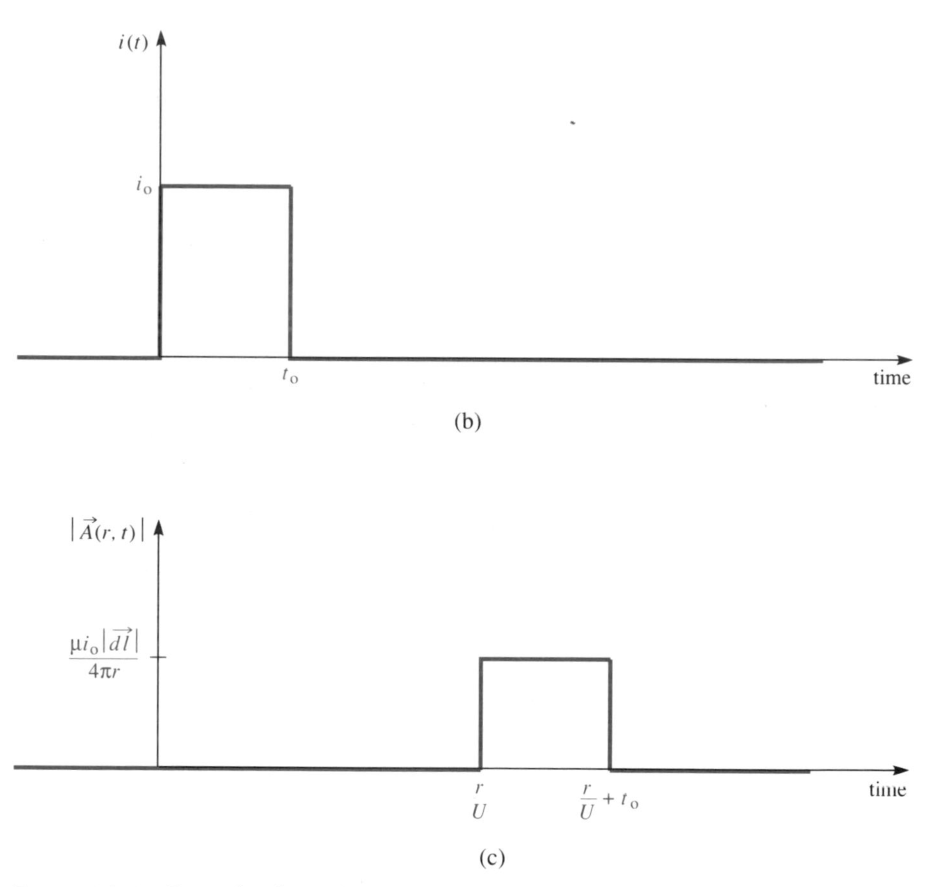

(c)

Figure 10.1 Example of retarded vector potential. If the observation point is as shown in (a), and the current at the origin is the rectangular pulse shown in (b), the retarded vector potential at the observation point is as shown in (c).

corresponding to its distance from $\vec{r}$, the observation point. Thus

$$\vec{A}(\vec{r}, t) = \frac{\mu}{4\pi} \int \frac{i(\vec{r}', t_R)\, \vec{dl}}{|\vec{r} - \vec{r}'|} \tag{10.22}$$

where

$$t_R = t - \frac{|\vec{r} - \vec{r}'|}{U} \tag{10.23}$$

By means of (10.22) we can find the electric and magnetic fields everywhere, provided the source currents are known.

EXAMPLE 10.1

A wire lies on the z axis between $z = a$ and $z = \infty$ (where $a > 0$). The current in this wire is $I_o S(t)$, where $S(t)$ is the unit step function defined by

$$S(t) = 0 \quad \text{for} \quad t < 0$$
$$S(t) = 1 \quad \text{for} \quad t > 0$$

Find the vector potential at the origin.

Solution From (10.22)

$$\vec{A}(t) = \frac{\mu \vec{e}_z}{4\pi} \int_a^\infty \frac{i(z, t_R)\, dz}{z}$$

In this case $t_R = t - (z/U)$. Hence

$$\vec{A}(t) = \frac{\mu \vec{e}_z}{4\pi} \int_a^\infty \frac{S\left(t - \dfrac{z}{U}\right) dz}{z}$$

For $t < (a/U)$ the argument of S is less than zero over the entire range of integration, and the integral vanishes. (Physically this is because for $t < (a/U)$ the signal has not yet had time to arrive.) For $t > (a/U)$, the argument of S is positive for $z < Ut$. Thus

$$\vec{A}(t) = \frac{\mu \vec{e}_z}{4\pi} \int_a^{Ut} \frac{dz}{z}$$

$$= \frac{\mu \vec{e}_z}{4\pi} \ln \frac{Ut}{a} \quad \text{for} \quad t > \frac{a}{U}$$

and zero for earlier times.

10.2 THE ELEMENTARY DIPOLE

As our first example of an antenna, let us choose the very simple current distribution known as an *elementary dipole* (or *Hertzian dipole*). This consists of an infinitesimally long wire $\vec{dl}$ carrying current $I_o \cos \omega t$. To be specific, let $\vec{dl} = h\vec{e}_z$. Then from (10.22),

$$\vec{A}(r,t) = \frac{\mu I_o h \cos \omega \left(t - \dfrac{r}{U}\right)}{4\pi r} \vec{e}_z \tag{10.24}$$

where r is the distance from the antenna to the field point. The phasor representing $\vec{A}$ is

$$\vec{\mathbf{A}}(r) = \frac{\mu I_o h e^{-jkr}}{4\pi r} (\vec{e}_r \cos\theta - \vec{e}_\theta \sin\theta) \tag{10.25}$$

where

$$k = \omega/U = k_o = \omega\sqrt{\mu\epsilon}$$

Here in order to use spherical coordinates we have expressed $\vec{e}_z = \vec{e}_r \cos\theta - \vec{e}_\theta \sin\theta$. The magnetic field is now

$$\begin{aligned} \vec{\mathbf{H}}(r,\theta) &= \frac{1}{\mu} \nabla \times \vec{A} \\ &= \frac{jkI_o h \sin\theta\, e^{-jkr}}{4\pi r} \vec{e}_\phi + \frac{I_o h \sin\theta\, e^{-jkr}}{4\pi r^2} \vec{e}_\phi \end{aligned} \tag{10.26}$$

From (10.12) the scalar potential Φ is

$$\begin{aligned} \Phi &= -\frac{1}{j\omega\mu\epsilon} \nabla \cdot \vec{A} \\ &= \frac{I_o h}{4\pi j\omega\epsilon} \cos\theta\, e^{-jkr} \left(\frac{1}{r^2} + \frac{jk}{r}\right) \end{aligned} \tag{10.27}$$

and from (10.9) the electric field is

$$\begin{aligned} \vec{E}(r,\theta) &= -\nabla\Phi - \frac{\partial A}{\partial t} \\ &= \left[\frac{I_o h}{4\pi} e^{-jkr} \left(\frac{2\eta}{r^2} + \frac{2}{j\omega\epsilon r^3}\right) \cos\theta\right] \vec{e}_r \\ &\quad + \left[\frac{I_o h}{4\pi} e^{-jkr} \left(\frac{j\omega\mu}{r} + \frac{\eta}{r^2} + \frac{1}{j\omega\epsilon r^3}\right) \sin\theta\right] \vec{e}_\theta \end{aligned} \tag{10.28}$$

where $\eta = \sqrt{\mu/\epsilon}$.

At large distances ($kr \gg 1$) from the antenna, terms in (10.26) and (10.28) containing r^{-2} or r^{-3} become negligible. These terms constitute what are known as the *induction fields* of the antenna; they are local and do not contribute to what we normally think of as radiation. The remaining terms, which die off as $1/r$, constitute the antenna's *radiation fields*. Thus the radiation fields of the elementary dipole are

$$\vec{\mathbf{H}} = \frac{jkI_o h}{4\pi r} e^{-jkr} \sin\theta\, \vec{e}_\phi$$
$$\vec{\mathbf{E}} = \frac{j\omega\mu I_o h}{4\pi r} e^{-jkr} \sin\theta\, \vec{e}_\theta \tag{10.29}$$

In the radiation zone, where induction fields are negligible, the time-averaged Poynting vector is

$$\vec{S} = \tfrac{1}{2}\,\text{Re}\,[\vec{\mathbf{E}} \times \vec{\mathbf{H}}^*]$$
$$= \frac{\eta k^2 I_o^2 h^2}{32\pi^2 r^2} \sin^2\theta\, \vec{e}_r \tag{10.30}$$

This result describes the distribution of radiated power as a function of direction. No power is radiated along the polar axis, and radiation is maximum in the equatorial plane. A polar plot of the radiation pattern, $|\vec{S}(\theta)|$ as a function of θ, is shown in Fig. 10.2.

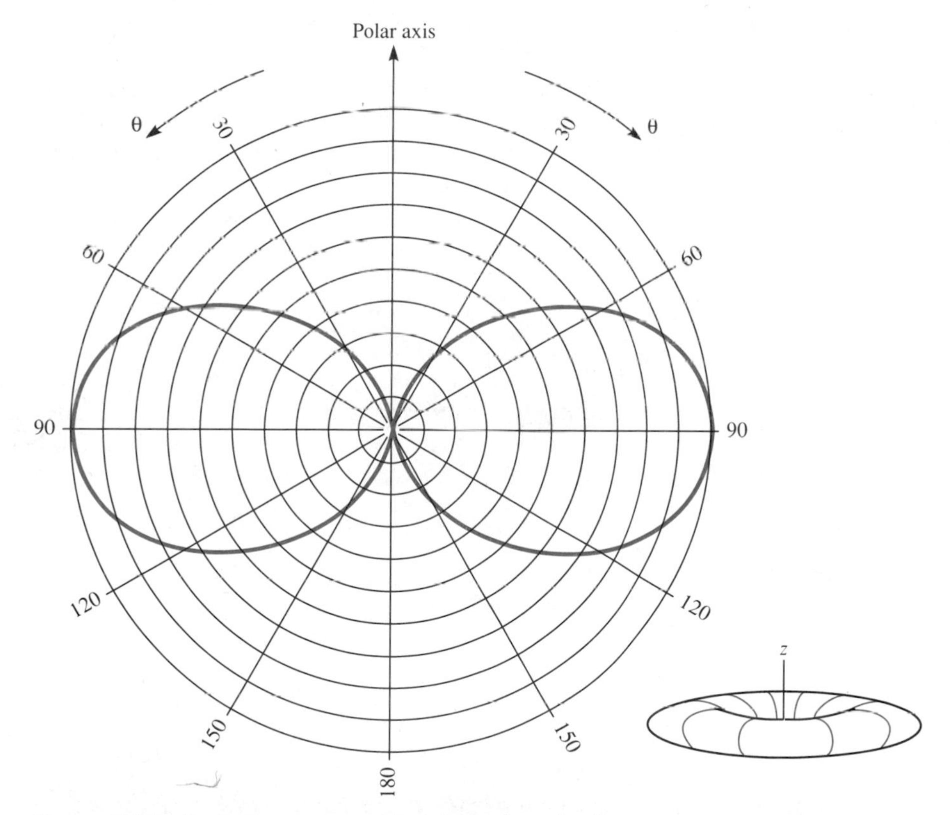

Figure 10.2 Radiation pattern of the elementary dipole.

The total power radiated by the dipole is

$$P = \int_S \vec{S} \cdot \overrightarrow{dS}$$

where the surface of integration is a sphere of radius r. Thus

$$P = \int_0^\infty |\vec{S}(\theta)| 2\pi r^2 \sin\theta \, d\theta$$

$$= \frac{\eta k^2 I_0^2 h^2}{12\pi}$$

$$= \frac{\eta \pi I_0^2}{3} \left(\frac{h}{\lambda}\right)^2 \quad \textbf{(10.31)}$$

where $\lambda = 2\pi/k$ is the free-space wavelength at the frequency in question. The factor (h/λ) must be small to justify the assumptions used in the calculation. We note that the total radiated power is independent of r, the radius of the sphere over which integration took place, which is as one might expect.

DIRECTIVITY AND GAIN

In many cases one wishes to radiate power in a particular direction. For example, in a radio communications link only the power radiated toward the receiver is useful; power radiated in other directions is wasted. This requirement leads to the figure of merit for antennas known as *directivity*. The directivity D is defined as the ratio of the radiated intensity, at some distance r, in the direction at which radiation is maximum, to the intensity that would be radiated at the same r if the antenna radiated in all directions equally. (We recall that "intensity" is power per unit area.) Thus

$$D = \frac{|\vec{S}(r, \theta_{max})|}{\left(\dfrac{P}{4\pi r^2}\right)} \quad \textbf{(10.32)}$$

where P is the total radiated power. This ratio is always greater than one,[3] and can be many thousand for highly directive antennas such as the parabolic reflector or "dish." Often the directivity is stated in decibels. For the elementary dipole we have, from (10.30) and (10.31)

$$D = \frac{\left(\dfrac{\eta k^2 I_0^2 h^2}{32\pi^2 r^2}\right)}{\left(\dfrac{\eta k^2 I_0^2 h^2}{(12\pi)(4\pi r^2)}\right)}$$

$$= \frac{3}{2} \quad \text{or } 1.76 \text{ dB} \quad \textbf{(10.33)}$$

[3]Interestingly, it can be shown that it is impossible to construct a truly isotropic radiator. Thus the directivity of any antenna must be strictly greater than one.

This antenna has a very low directivity, because of its extremely small size. In general, higher values of directivity are obtained by making antennas larger with respect to wavelength.

EXAMPLE 10.2

The distance from the earth to the moon is approximately 3.9×10^5 km. A laser transmitter is capable of projecting a spot on the moon that is 2.6 km in diameter. Find the approximate directivity of this transmitter.

Solution Let P be the total transmitted power. Assuming that the 2.6-km spot is uniformly illuminated,

$$|\vec{S}(r, \theta_{\max})| \cong \frac{P}{\pi(1.3 \times 10^3)^2}$$

and from (10.32) the directivity is

$$D \cong \frac{\dfrac{P}{\pi(1.3 \times 10^3)^2}}{\dfrac{P}{4\pi(3.9 \times 10^8)^2}}$$

$$= 3.6 \times 10^{11} \qquad \text{or } 115.6 \text{ dB}$$

Because of this very high directivity, lasers are potentially useful in space communications.

The power radiated by an antenna must be furnished by some external source, usually in the form of a voltage applied to the antenna's terminals. However, antennas are not 100% power-efficient; some power is lost in its resistance or in other ways. Another common figure of merit, the *gain*, is similar to the directivity but takes losses into account. Let the power efficiency e of the antenna be given by

$$e = \frac{P}{P_i} \tag{10.34}$$

where P is the total radiated power and P_i is the power furnished to the antenna by the external source. The antenna's gain G is then defined by

$$G = eD \tag{10.35}$$

Gain is a more realistic estimator of antenna performance than directivity. For example, in some antennas, high directivity is obtained by using very large currents at certain points. However, these large currents lead to large I^2R losses. The gain of the antenna accurately indicates the improvement in intensity at the receiver that is obtained by using the antenna, while the directivity does not.

10.3 LONG-WIRE ANTENNAS

Elementary dipoles are mainly of theoretical interest. However, practical antennas are often made using thin linear wires, the lengths of which are comparable to wavelength. The radiated fields of such an antenna can be found using (10.22), provided that the currents on the antenna are known.

Let us consider a wire antenna that extends from z_1 to z_2 along the z axis, and carries a current that is sinusoidal in time:

$$I(z, t) = I_o(z) \cos[\omega t + \psi(z)] \tag{10.36}$$

The details of the current distribution, described by $I(z)$ and $\psi(z)$, depend on how the antenna is made. The phasor corresponding to this current is

$$\mathbf{i}(z) = I_o(z)e^{j\psi(z)} \tag{10.37}$$

The current at the retarded time t_R is

$$i(z, t_R) = I_o(z) \cos\left[\omega\left(t - \frac{|\vec{r} - \vec{r}'|}{U}\right) + \psi(z)\right] \tag{10.38}$$

and the phasor representing the current at the retarded time is

$$\mathbf{i}_R(z) = I_o(z)\, e^{j\psi(z)}\, e^{-jk|\vec{r} - \vec{r}'|} \tag{10.39}$$

Using (10.39) in (10.22) we find the phasor representing the vector potential

$$\vec{\mathbf{A}}(\vec{r}) = \frac{\mu\vec{e}_z}{4\pi} \int_{z_1}^{z_2} \frac{I_o(z)\, e^{j\psi(z)}\, e^{-jk|\vec{r} - \vec{r}'|}\, dz}{|\vec{r} - \vec{r}'|} \tag{10.40}$$

We now introduce a very useful mathematical simplification known as the *dipole approximation*. This is illustrated in Fig. 10.3. We see that the

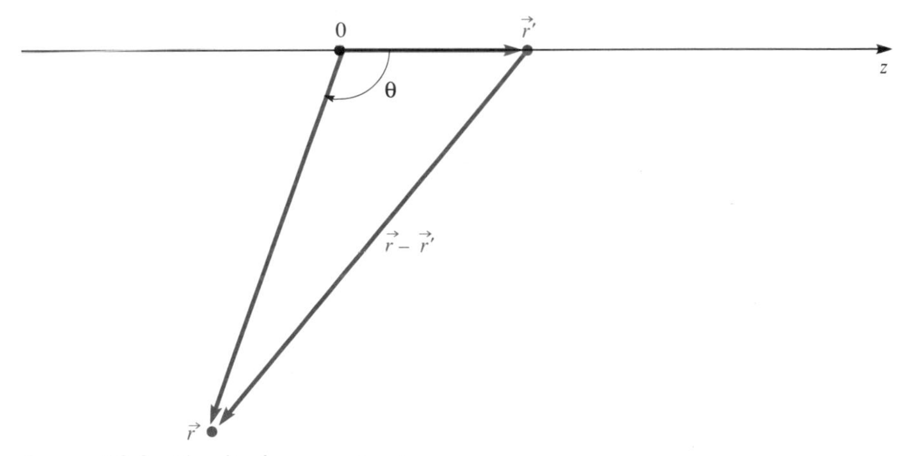

Figure 10.3 The dipole approximation.

distance $|\vec{r} - \vec{r}'|$ from the source point (located at z) to the field point is given by the law of cosines:

$$|\vec{r} - \vec{r}'| = \sqrt{r^2 + z^2 - 2rz\cos\theta} \qquad \textbf{(10.41)}$$

where $r = |\vec{r}|$. Since the distance r to the field point is usually very much larger than the length of the antenna, (10.41) can now be simplified by means of the binomial theorem:

$$|\vec{r} - \vec{r}'| = r\sqrt{1 + \left(\frac{z^2}{r^2}\right) - 2\left(\frac{z}{r}\right)\cos\theta}$$

$$\cong r\left(1 - \frac{z}{r}\cos\theta\right) \qquad \textbf{(10.42)}$$

(where the very small term (z^2/r^2) has been neglected). Now, using (10.42) in (10.40), we have

$$\vec{\mathbf{A}}(r, \theta) = \frac{\mu e^{-jkr}\vec{e}_z}{4\pi r}\int_{z_1}^{z_2} I_o(z)e^{j\psi(z)}\, e^{jkz\cos\theta}\, dz \qquad \textbf{(10.43)}$$

The reader will note that $|\vec{r} - \vec{r}'|$ in the denominator has been replaced simply by r (which is not a function of z and hence can be taken outside the integral). Why did we not use (10.42) instead? The reason is that the term containing (z/r) in (10.42) is very small and would have little effect on the integral. However, in the numerator, $|\vec{r} - \vec{r}'|$ appears in the exponent. Here both terms of (10.42) must be retained because even a small change in the value of $|\vec{r} - \vec{r}'|$ can cause a large change in the exponential function. For instance, when z varies over half a wavelength, e^{jkz} undergoes a complete change of sign.

To make use of (10.43) we must still know the amplitude and phase of the current everywhere on the antenna, and this brings us to one of the great difficulties in the study of antennas. In general one does not know the current distribution. It is influenced by the field distribution, and to be rigorous, fields and currents are all unknowns, which must be solved for simultaneously. However, in the case of linear wire antennas, it has been found that the current behaves approximately as it does on a transmission line. That is,

$$\mathbf{i}(z) \cong I^+e^{-jkz} + I^-e^{jkz} \qquad \textbf{(10.44)}$$

Long-wire antennas can be of any length, but work best when certain particular lengths are used. The example most often encountered is the *half-wave dipole antenna*. This is a long-wire antenna of length $\lambda/2$, fed at the center as shown in Fig. 10.4. At the open-circuited ends of this antenna the current must vanish; thus $\mathbf{i}(z = \lambda/4) = 0$, from which we find that $I^+ = I^-$. Thus

$$\mathbf{i}(z) \cong I\cos kz \qquad \textbf{(10.45)}$$

where I is the current at the center.

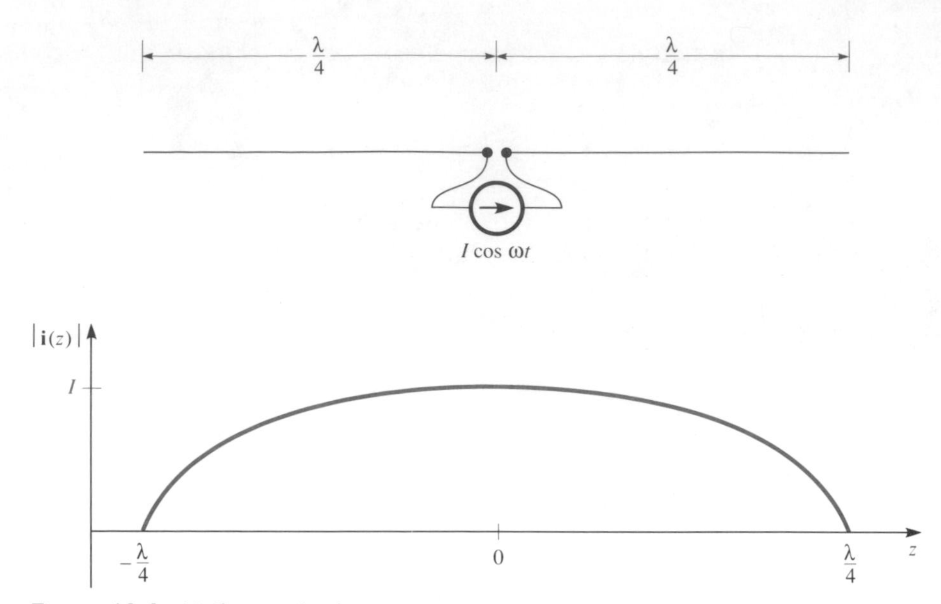

Figure 10.4 Half-wave dipole antenna.

We are now ready to calculate the fields using (10.43). Comparing (10.45) with (10.37), we have

$$I_o(z) = I \cos kz$$
$$\psi(z) = 0 \tag{10.46}$$

and (10.43) becomes

$$\vec{\mathbf{A}}(r, \theta) = \frac{\mu e^{-jkr} \vec{e}_z I}{4\pi r} \int_{-\lambda/4}^{\lambda/4} \cos kz \, e^{jkz \cos\theta} \, dz \tag{10.47}$$

The complex exponential can be expanded by means of Euler's formula $e^{jx} = \cos x + j \sin x$; the integral containing the sine term then vanishes by symmetry. The remaining integral of $\cos kz \cos(kz \cos\theta)$ is then evaluated straightforwardly, with the result

$$\vec{\mathbf{A}}(r, \theta) = \frac{\mu e^{-jkr} I}{2\pi k r} \frac{\cos\left(\dfrac{\pi}{2}\cos\theta\right)}{\sin^2\theta} \vec{e}_z \tag{10.48}$$

The electric field of the antenna can now be found using (10.9) and (10.12). However, many of the terms obtained in this way belong to the induction fields, which are at present not of interest. In general, the only terms of

$$\vec{\mathbf{E}} = -\frac{j\omega}{k^2} \nabla(\nabla \cdot \vec{\mathbf{A}}) - j\omega \vec{\mathbf{A}} \tag{10.49}$$

which decrease as $1/r$ are

$$\mathbf{E}_\theta = -j\omega\mathbf{A}_\theta$$

$$\mathbf{E}_\phi = -j\omega\mathbf{A}_\phi \tag{10.50}$$

Thus the electric radiation field of the half-wave dipole is

$$\vec{\mathbf{E}} = \frac{j\eta I}{2\pi r} e^{-jkr} \frac{\cos\left(\dfrac{\pi}{2}\cos\theta\right)}{\sin\theta} \vec{e}_\theta \tag{10.51}$$

At large distances from the antenna the expanding spherical waves gradually become plane waves. (See Fig. 8.1.) Thus, from (8.27) the time-averaged Poynting vector is

$$\vec{S} = \frac{|\mathbf{E}_\theta|^2 \vec{e}_r}{2\eta} = \frac{\eta I^2}{8\pi^2 r^2} \frac{\cos^2\left(\dfrac{\pi}{2}\cos\theta\right)}{\sin^2\theta} \vec{e}_r \tag{10.52}$$

The radiation pattern is shown in Fig. 10.5 with the pattern of the elementary dipole also shown for comparison. (The amplitudes of the two patterns are

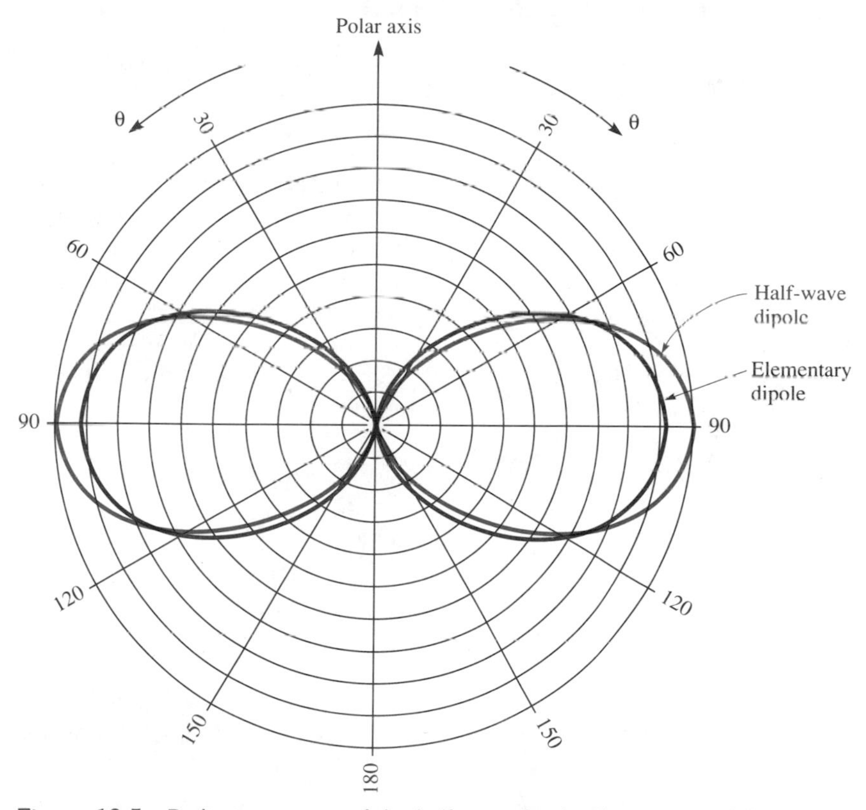

Figure 10.5 Radiation pattern of the half-wave dipole. The pattern of the elementary dipole is also shown for comparison. The amplitudes of the two patterns are chosen to make the total radiated powers equal.

scaled to make the total radiated powers equal.) Both patterns are doughnut-shaped, and in fact there is little difference between them. The half-wave dipole radiates slightly more intensely into the equatorial plane. The total radiated power is

$$P = \int_{0}^{\pi} 2\pi r^2 \sin\theta \,|\vec{S}|\, d\theta = \frac{\eta I^2}{4\pi} \int_{0}^{\pi} \frac{\cos^2\left(\frac{\pi}{2}\cos\theta\right)}{\sin\theta}\, d\theta \qquad \textbf{(10.53)}$$

By numerical integration the value of this integral is found to be 1.22; hence (assuming $\epsilon = \epsilon_o$, $\mu = \mu_o$)

$$P = 36.56 I^2 \qquad \textbf{(10.54)}$$

From (10.32), (10.52), and (10.54) we find that the directivity of the half-wave dipole is 1.64, or 2.15 dB. This small value is only slightly larger than that of the elementary dipole; evidently it is not for the sake of directivity that one would prefer this antenna. The popularity of the half-wave dipole has more to do with its convenient driving-point impedance.

RADIATION RESISTANCE

We have not yet considered the voltage that appears across the antenna terminals (that is, across the current source) in Fig. 10.4. Our guess of the current distribution, (10.45), was based on the idea that the arms of the antenna behave like transmission lines. Based on that idea, one might expect that at the center, where current is maximum, the voltage would be zero. However, this is impossible because the power produced by the current source would be zero, and there would be no way to supply the radiated power. This seeming contradiction shows that the transmission-line approximation cannot be exactly correct.

At any rate, the *real* part of the impedance seen by the current source can be found from energy conservation. Let the antenna impedance be

$$Z_A = R_A + jX_A \qquad \textbf{(10.55)}$$

Neglecting ohmic loss inside the antenna, we have, using (10.54)

$$\frac{1}{2}\,\text{Re}\,(VI^*) = \frac{R_A}{2} I^2 = 36.56 I^2$$

Thus the real part of Z_A, known as the *radiation resistance*, is

$$R_A = 73.1 \text{ ohms}$$

Unfortunately this argument tells us nothing about X_A, the reactance of the antenna, and more advanced theoretical methods must be used.[4] For

[4]For a readable summary of these methods see Chapter 12 of S. Ramo, J. R. Whinnery, and T. Van Duzer, *Fields and Waves in Communication Electronics*. New York: Wiley, 1984. For data see H. Jasik, *Antenna Engineering Handbook*. New York: McGraw-Hill, 1961.

center-fed antennas, it is found that X_A vanishes at antenna lengths near (but not exactly) integer multiples of $\lambda/2$. At such lengths the antenna is said to be "resonant." Resonant antennas are often used to facilitate impedance matching and avoid unnecessary ohmic loss. (See Problem 2.7.) Interestingly, it is found that the thickness of the antenna wire affects X_A and the resonant length. However, R_A is nearly independent of wire thickness.

Long-wire antennas, such as the half-wave dipole, are used primarily at radio frequencies (say 0.1–30 MHz), in applications where high directivity is not required. They are simple, easy to construct, and have good efficiency and a convenient driving-point impedance. A typical installation of a half-wave dipole is shown in Fig. 10.6. The antenna (which at AM radio broadcast frequencies—about 1 MHz—would be about 150 meters long) is suspended between two tall supports. The transmitter, which is below at ground level, is connected by means of a transmission line. A standing-wave ratio near unity can be obtained by using a transmission line with a characteristic impedance of 73 ohms. (Coaxial line of this impedance is commercially available.) The transmitter is usually designed to drive a pure resistance. Residual reactance is removed by a lumped-circuit matching network such as the "pi network" shown in the figure. This network can also transform the real part of the impedance to the value for which the transmitter is designed, if this is something other than 73 ohms.

The reader may inquire as to whether an elementary dipole could be used in place of a half-wave dipole, with great saving of space. Sometimes they are; for instance, automobile AM radio antennas, which are much

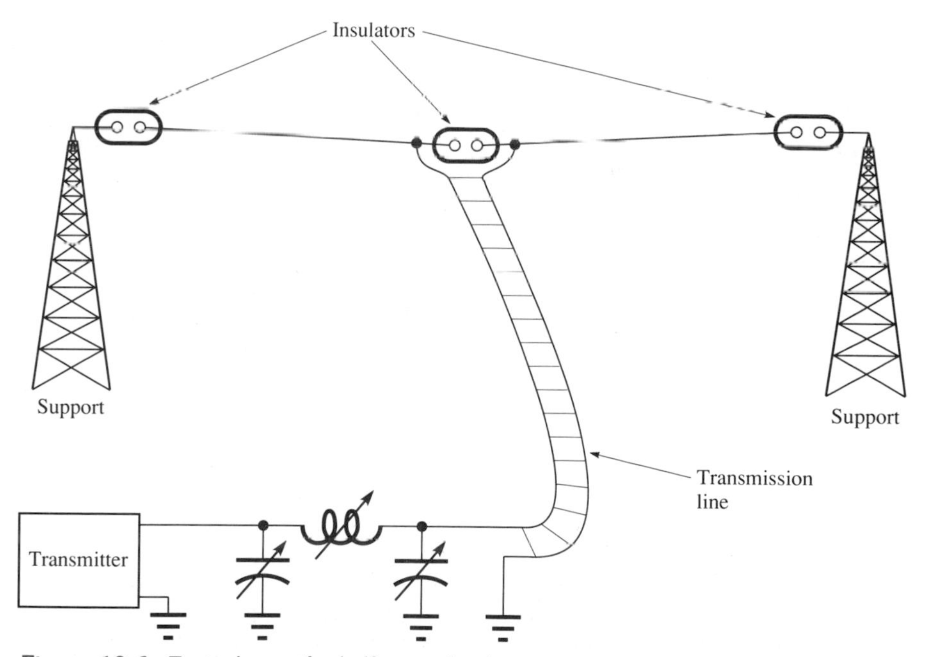

Figure 10.6 Typical use of a half-wave dipole antenna.

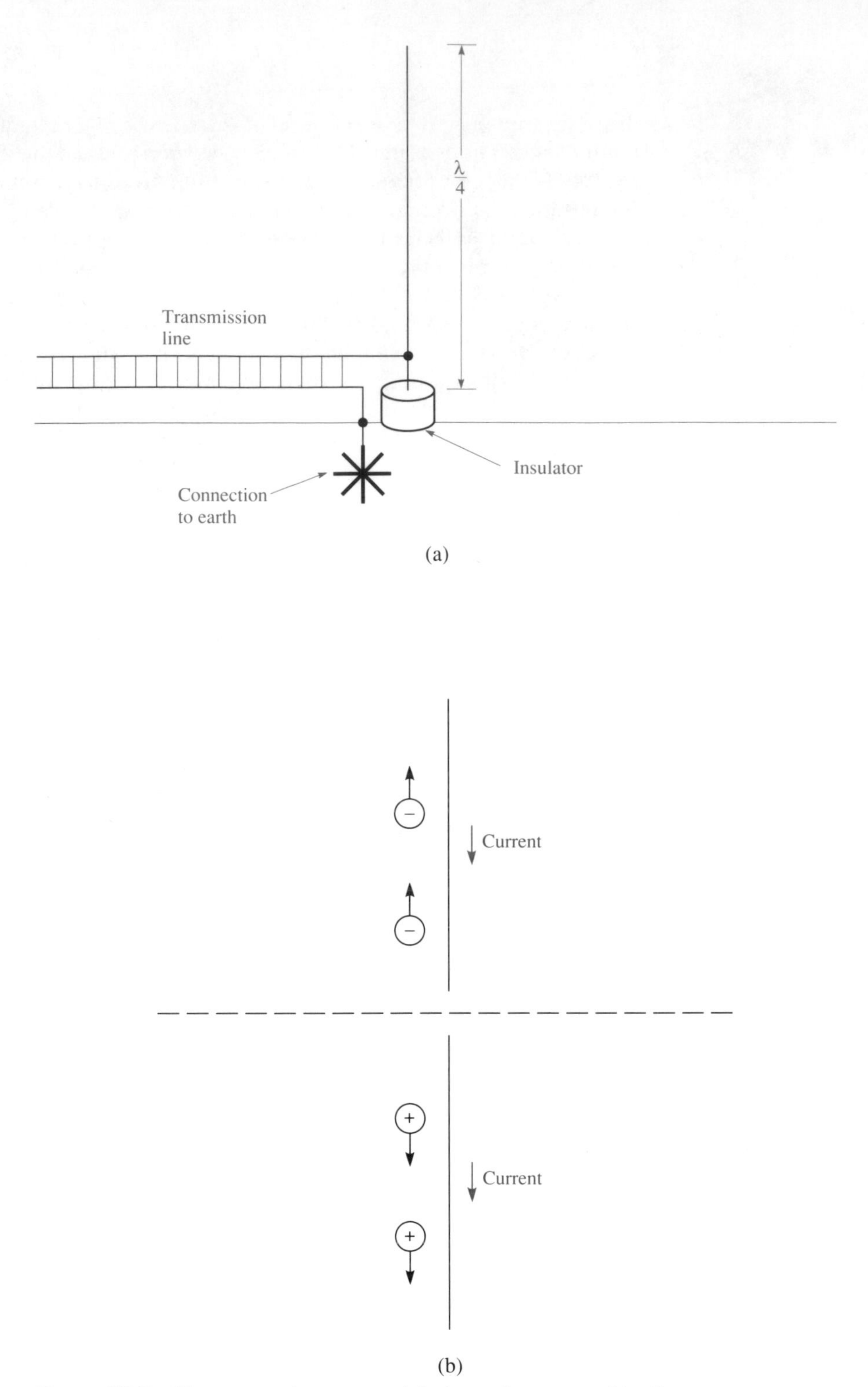

Figure 10.7 The monopole antenna: (a) physical structure; (b) effective structure including image.

smaller than wavelength, are in essence elementary dipoles. Comparing (10.31) with (10.54), we see that a much larger current is required in an elementary dipole to achieve the same radiated power, or to put it another way, the radiation resistance of an elementary dipole is very small. Moreover, the driving-point reactance will be very large. Thus the required matching circuit is difficult to design, and there may be considerable ohmic loss because of the high driving-point current that is required. (If the resistance of the matching circuit is larger than the very small radiation resistance, more power will be lost than radiated!) These difficulties are somewhat less severe when the antenna is used for receiving. The necessary matching circuits are easier to build at the low power levels typical in receiving, and it is easier to match to the high input impedances obtainable in receivers.

A common variation of the half-wave dipole is the quarter-wave vertical (or *monopole*) antenna, shown in Fig. 10.7. This antenna makes use of the fact that the earth is a reasonably good conductor at radio frequencies. Thus the vertical antenna is accompanied by an image, as shown in Fig. 10.7(b). Note that a movement of negative charges upward in the antenna is imaged by a downward movement of positive charges in the image, so the direction of current is the same in both. Thus the effective currents and radiation pattern are identical with those of the half-wave dipole. However, it is oriented differently with respect to the earth, and the polarization of the radiated field is different, with $\vec{E}$ vertical instead of horizontal.

EXERCISE 10.2

Show that the radiation resistance of the quarter-wave vertical antenna is approximately 36.5 ohms.

10.4 ANTENNA ARRAYS

To obtain better directivity, several antennas can be combined to form an *array*. The individual antennas, which are called the *elements* of the array, are positioned in such a way that their radiated fields in the desired direction add in phase, while radiation in undesired directions tends to cancel.

Let us imagine that we have two identical antennas, each with radiated field $\vec{E}_A(\vec{r})$, where $\vec{r}$ is the vector from the antenna to the field point. As seen from a large distance away, each antenna acts as a source of spherical waves. Thus $\vec{E}_A(\vec{r})$ can be expressed as

$$\vec{E}_A(\vec{r}) = f(r,\theta)e^{-jkr} \tag{10.56}$$

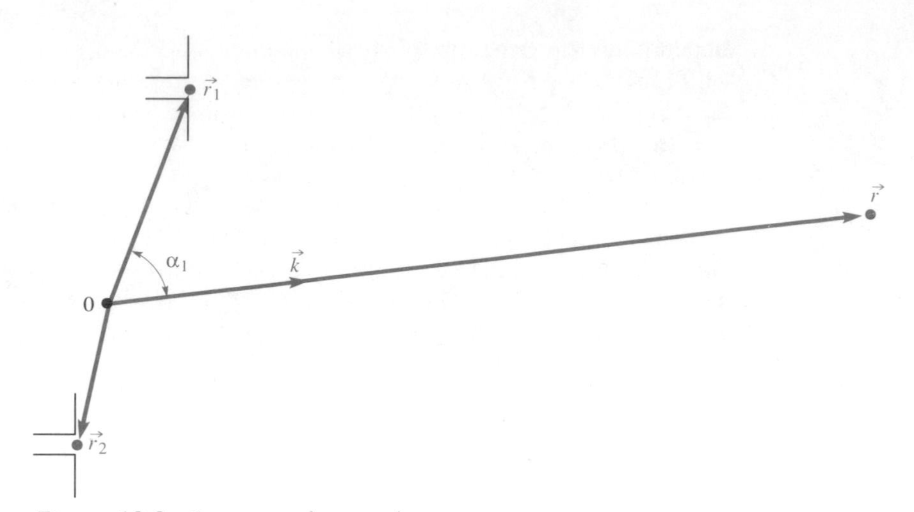

Figure 10.8 Geometry of a two-element antenna array.

where $r = |\vec{r}|$ and f is a slowly varying function.[5] Note that (10.29) and (10.51) are of this form. If the two antennas are located not at the origin but at $\vec{r}_1$ and $\vec{r}_2$ (see Fig. 10.8), the total radiated field is the sum of their two contributions:

$$\begin{aligned}\vec{E}(\vec{r}) &= f(|\vec{r} - \vec{r}_1|, \theta_1)\, e^{-jk|\vec{r}-\vec{r}_1|} + f(|\vec{r} - \vec{r}_2|, \theta_2)\, e^{-jk|\vec{r}-\vec{r}_2|} \\ &\cong f(r,\theta)(e^{-jk|\vec{r}-\vec{r}_1|} + e^{-jk|\vec{r}-\vec{r}_2|})\end{aligned} \tag{10.57}$$

Again, we simplify the exponential functions by means of the dipole approximation. Let α_1 be the angle between $\vec{r}_1$ and $\vec{r}$, and let $r_i = |\vec{r}_i|$. Then

$$\begin{aligned}k\,|\vec{r} - \vec{r}_1| &= k\,(r^2 + r_1^2 - 2rr_1 \cos\alpha_1)^{1/2} \cong kr\left(1 - \frac{r_1}{r}\cos\alpha_1\right) \\ &= kr - \vec{k} \cdot \vec{r}_1\end{aligned} \tag{10.58}$$

where $\vec{k} = k(\vec{r}/r)$ is a vector of magnitude k in the direction of $\vec{r}$. Thus

$$\vec{E}(\vec{r}) \cong f(r,\theta)\, e^{-jkr}\,(e^{j\vec{k}\cdot\vec{r}_1} + e^{j\vec{k}\cdot\vec{r}_2}). \tag{10.59}$$

Obviously this result can be generalized to arrays containing any number of elements. We note that the radiated field is the product of three factors. The first, $f(r,\theta)$, depends on the properties of the individual antennas, and is known as the *element factor*. The third factor, which in general is

$$\sum_i e^{j\vec{k}\cdot\vec{r}_i}$$

[5] "Slowly varying" means that it changes little if $\vec{r}$ moves a distance comparable with the size of the array.

depends on the positions of the elements and is known as the *array factor* (or *space factor*). The other term, e^{-jkr}, tells us that the radiated field consists of spherical waves propagating outward from the origin.

As an example, let us consider an array consisting of two parallel elementary dipoles, side by side, a distance $\lambda/4$ apart, as shown in Fig. 10.9. The dipole located at ($x = \lambda/4$, $y = 0$, $z = 0$) has a current I; the dipole at the origin has a current $Ie^{j(\pi/2)}$. Both currents are in the z direction. This particular arrangement is chosen to create a directive antenna pattern. Radiation from the dipole at the origin falls back 90° in phase as it travels in the $+x$ direction. Since the driving current in that dipole is advanced 90° with respect to the other, the fields from the two dipoles go off in the $+x$ direction in phase with each other, and thus add. On the other hand, the field from the dipole at $x = \lambda/4$ is originally 90° behind that of the other dipole, and by the time it has traveled to the origin it is 180° behind. Thus the fields transmitted in the $-x$ direction will cancel. The result should be a directive antenna pattern concentrated in the $+x$ direction.

Using (10.29) in (10.59) and taking the phases of the currents into account we have

$$\mathbf{E}_\theta = \frac{j\omega\mu I h}{4\pi r} \sin\theta\, e^{-jkr}\left(j + e^{jk(\frac{\lambda}{4})\sin\theta\cos\phi}\right) \qquad \textbf{(10.60)}$$

where the direction $\theta = \pi/2$, $\phi = 0$ is chosen to coincide with the x axis.

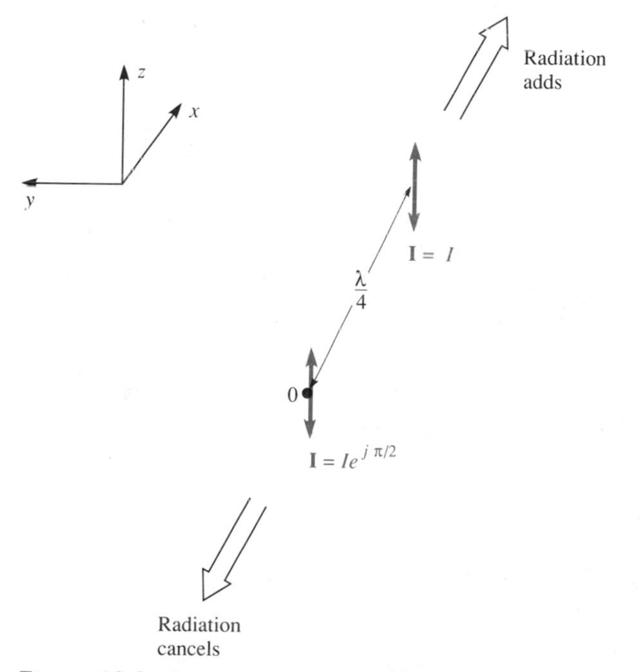

Figure 10.9 An array consisting of two parallel elementary dipoles.

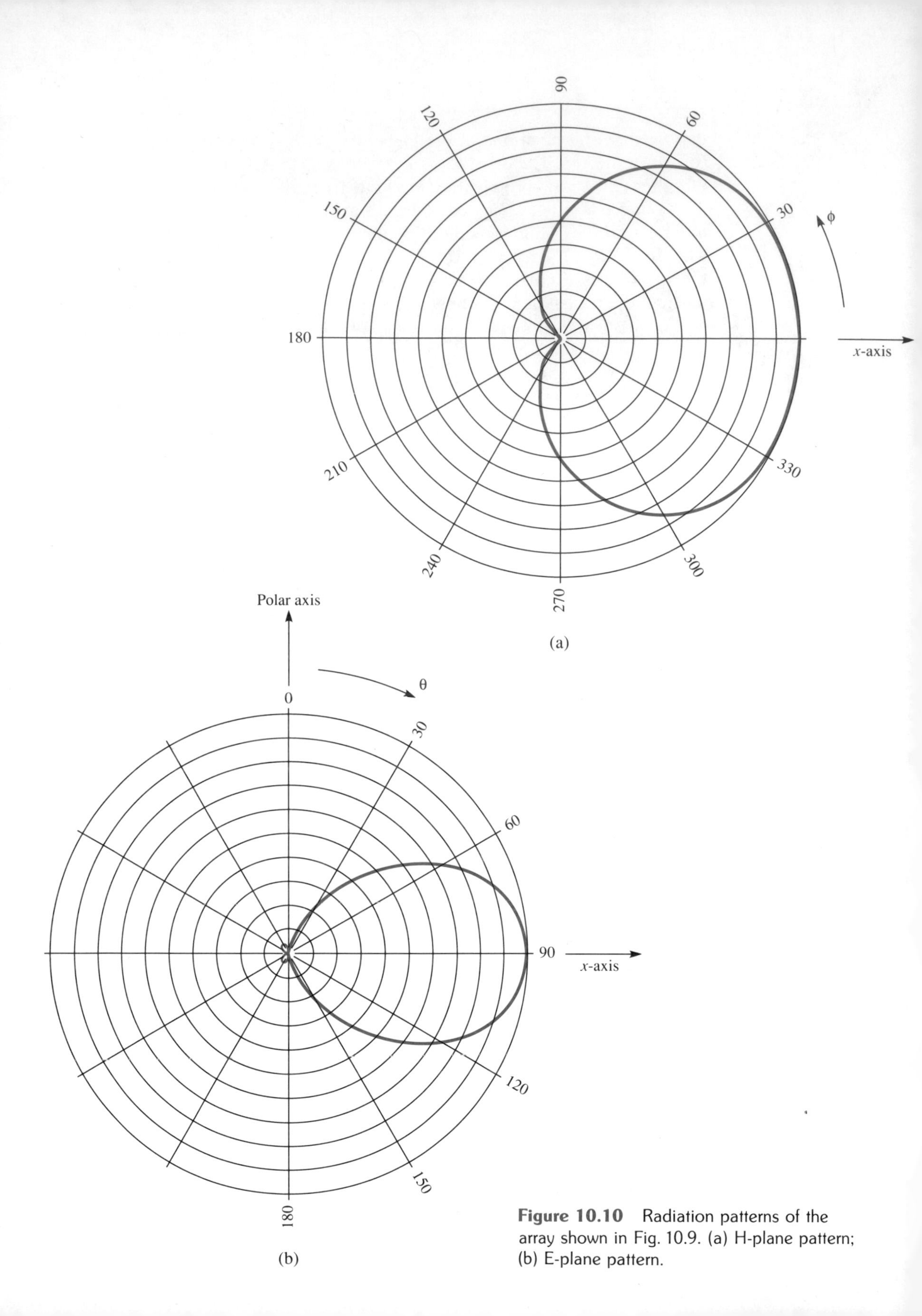

Figure 10.10 Radiation patterns of the array shown in Fig. 10.9. (a) H-plane pattern; (b) E-plane pattern.

The time-averaged intensity is

$$S(r, \theta, \phi) = \frac{|\mathbf{E}_\theta|^2}{2\eta} = \frac{(\omega\mu I_o h)^2}{32\pi^2\eta r^2} \sin^2\theta \left| -1 + e^{j\frac{\pi}{2}(\sin\theta\,\cos\phi + 1)} \right|^2$$

$$= \frac{(\omega\mu I h)^2}{16\pi^2\eta r^2} \sin^2\theta \left\{ 1 + \sin\left[\frac{\pi}{2}(\sin\theta\,\cos\phi)\right]\right\} \tag{10.61}$$

The element factor $\sin^2\theta$ contributes nulls in the $+z$ and $-z$ directions. To study the pattern in the equatorial plane we set $\theta = \pi/2$ and obtain

$$S\left(r, \frac{\pi}{2}, \phi\right) = \frac{(\omega\mu I h)^2}{16\pi^2\eta r^2}\left[1 + \sin\left(\frac{\pi}{2}\cos\phi\right)\right] \tag{10.62}$$

Figure 10.10(a) shows the power in the equatorial plane, that is, as a function of ϕ in the plane $\theta = 90°$. Maximum radiation is in the x direction, that is, in the direction $\theta = 90°$, $\phi = 0$. The magnetic field lies in the plane of Fig. 10.10(a); hence this is known as the *H-plane pattern*. Figure 10.10(b) shows the power as a function of θ, in the plane $\phi = 0$. The electric field lies in the plane of Fig. 10.10(b); hence it is known as the *E-plane pattern*. The total radiated power is

$$P = \int_{\theta=0}^{\pi}\int_{\phi=0}^{2\pi} S(r, \theta, \phi)\, r^2 \sin\theta\, d\theta\, d\phi$$

$$= \frac{(\omega\mu I h)^2}{16\pi^2\eta}\int_0^{\pi} d\theta\, \sin^3\theta \left\{2\pi + \int_0^{2\pi} \sin\left[\frac{\pi}{2}(\sin\theta\,\cos\phi)\right] d\phi\right\} \tag{10.63}$$

The integral over ϕ vanishes by symmetry, giving

$$P = \frac{(\omega\mu I h)^2}{8\pi\eta}\int_0^{\pi}\sin^3\theta\, d\theta$$

$$= \frac{(\omega\mu I h)^2}{6\pi\eta} \tag{10.64}$$

From (10.32), (10.62) and (10.64) we find that the directivity is 3, or 4.77 dB.

The method outlined in this section can be used to treat other problems. For example, a long-wire antenna of arbitrary length can be regarded as a linear array of elementary dipoles.

EXAMPLE 10.3

A long-wire antenna lies on the z axis between $z = -\frac{1}{2}$ and $z = \frac{1}{2}$. It carries a current described by the phasor $\mathbf{i}(z) = Ie^{-jkz}$. Find the electric field radiated by the antenna, by treating it as a linear array of elementary dipoles.

Solution Extending (10.59) from a sum into an integral over z, and using (10.29), we have

$$\vec{\mathbf{E}}(r, \theta) = \frac{j\omega\mu\vec{e}_\theta}{4\pi r} e^{-jkr} \sin\theta \int_{-L/2}^{L/2} (Ie^{-jkz})\, e^{jkz\cos\theta}\, dz$$

$$= \frac{jI\omega\mu\vec{e}_\theta}{4\pi kr} e^{-jkr} \sin\theta \left\{ \frac{\sin\left[kL\left(\sin^2\frac{\theta}{2}\right)\right]}{\sin^2\frac{\theta}{2}} \right\} \quad (10.65)$$

Let us consider the case of an antenna that is long compared with wavelength, so that $kL \gg 1$. In that case the function of θ in (10.65) is maximum at an angle close to zero. Using the approximation $\sin\theta \cong \theta$, valid at small angles,

$$S(\theta) \propto \frac{\sin^2\left(\frac{kL}{4}\theta^2\right)}{\theta^2}$$

which has its first (and largest) maximum when

$$\frac{kL}{4}\theta_{\max}^2 = 1.166$$

$$\theta_{\max} = 0.86\sqrt{\frac{\lambda}{L}}$$

We see that the strongest radiation from this antenna is in a cone making an angle on the order of $\sqrt{\lambda/L}$ with the z axis.

10.5 APERTURE ANTENNAS

At microwave frequencies, above 1 GHz, broadcasting is unusual; communication at these frequencies is usually point-to-point. Thus highly directive antennas are required. In this section we shall see that high directivities can be obtained if the size of the antenna is large compared with wavelength. At frequencies above 1 GHz ($\lambda = 30$ cm) this requirement becomes increasingly easy to satisfy.

The usual antenna employed for this purpose is the *parabolic reflector antenna,* or microwave "dish," shown in Fig. 10.11(a). Radiation is fed to the antenna through a hollow metal waveguide. At its end this flares outward to form a structure known as a *feed horn.* The purpose of the horn is to illuminate the surface of the parabolic reflector. The reflector then acts like a focusing lens operating in reverse, creating a beam which is nearly a plane wave.

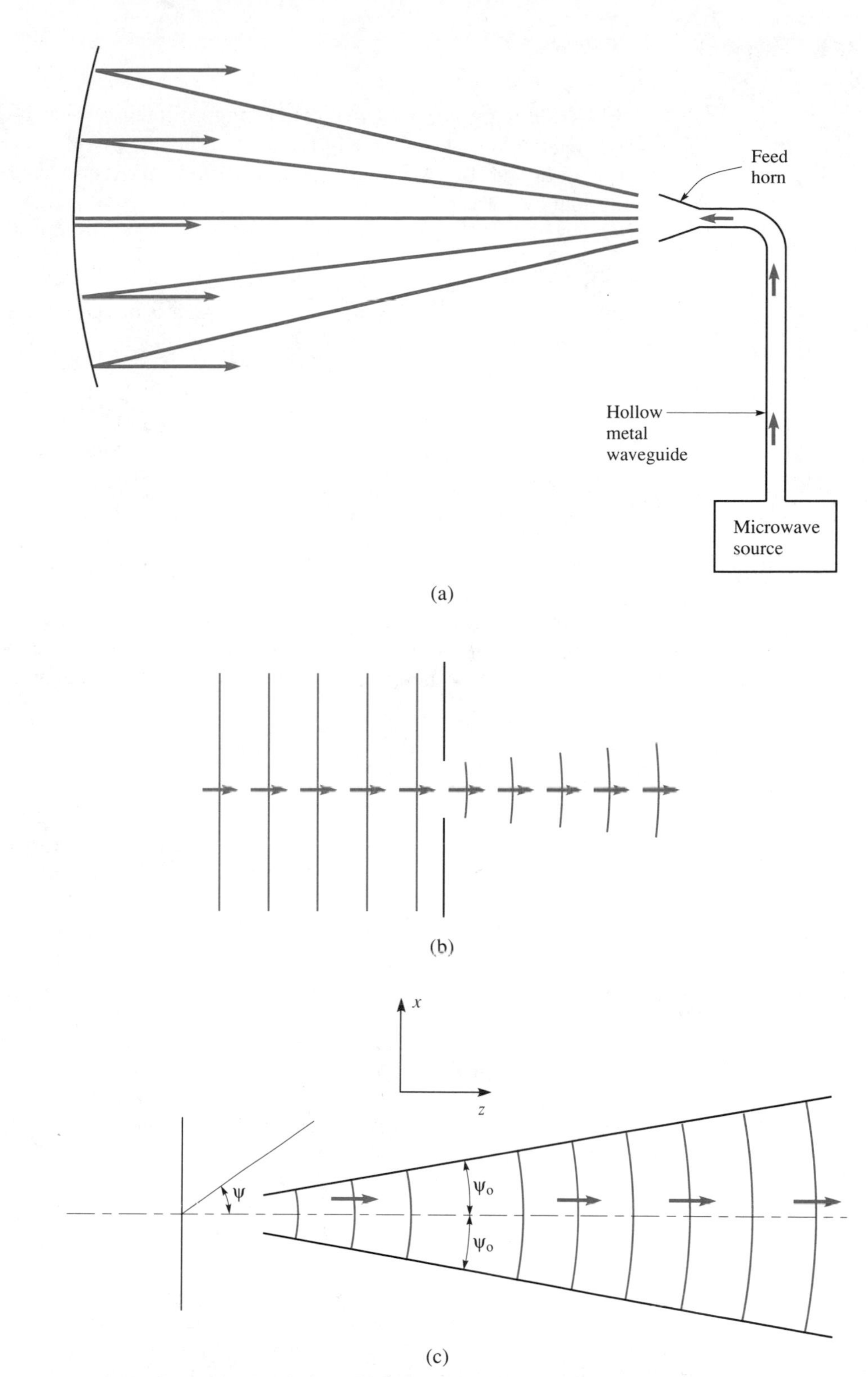

Figure 10.11 The parabolic reflector antenna, or "dish." (a) physical structure; (b) equivalent structure; (c) diffracted waves from aperture.

If the feed and reflector are well designed, the beam emerging from the reflector resembles a plane wave of limited extent. That is, it resembles what would be obtained if an infinitely broad plane wave were to pass through a circular hole in a wall, as shown in Fig. 10.11(b). In such a case, it is found that the emerging beam does not retain the diameter of the aperture; on the contrary it becomes wider as it travels. In physics this behavior is described by the theory of *diffraction*. After the beam has traveled a great distance from the aperture, it becomes a steadily diverging beam composed of spherical wavefronts that are nearly planar, as shown in Fig. 10.11(c). The angle ψ_o, which describes the spreading of the beam, determines the gain of the antenna. The main objective of this section is thus to calculate ψ_o.

In the theory of diffraction one begins with Maxwell's equations, and, after suitable approximations,[6] arrives at a formula which gives the field at a distant point in terms of an integral over the aperture. Let the wave normally incident from the left in Fig. 10.11(b) be linearly polarized. We shall assume that all electric fields are in the same direction as the incident electric field; there is then only a single scalar field component, say $\mathbf{E}_x$, to be concerned with, and we may omit the subscript and simply refer to the field as **E**. This is what is known as a scalar theory of diffraction. Scalar theory tends to be quite accurate so long as the aperture is large compared with wavelength. It is then found that the diffracted field is given by

$$\mathbf{E}(\vec{r}) = \frac{je^{-jkr}}{\lambda r} \int_{S'} \mathbf{E}(\vec{r}\,')e^{j\vec{k}\cdot\vec{r}\,'}\, dA' \tag{10.66}$$

Here $\vec{r}$ is the field point, $r = |\vec{r}|$, $k = \omega/c$, $\vec{k} = k\vec{r}/r$, and the surface integral is taken over the aperture; the position of a source point in the aperture is $\vec{r}\,'$ (see Fig. 10.12). This integral is known as the *Kirchhoff integral*.

The Kirchhoff integral can be interpreted physically in terms of *Huygens' principle*, which states that each point on a wavefront radiates as though it were an elementary dipole oriented in the direction of the field. Comparing (10.66) with (10.59) and (10.29), we see that the Kirchhoff integral has exactly the same form as the radiation from a large array of elementary dipoles, all excited in the same phase and filling the plane of the aperture.[7]

Microwave dish antennas are usually circular, and correspond to circular apertures. However, we can avoid mathematical difficulties by considering instead a square aperture of side $2a$ in the plane $z = 0$. Equation (10.66) becomes

$$\mathbf{E}(x, y, z) = \frac{je^{-jkr}E_o}{\lambda r} \int_{-a}^{a} \int_{-a}^{a} e^{\frac{jk(xx' + yy')}{r}}\, dx'\, dy' \tag{10.67}$$

[6]See, for instance, J. D. Jackson, *Classical Electrodynamics*. New York: Wiley, 1962.

[7]The factor $\sin\theta$ from (10.29) disappears because it is assumed that most of the radiation will be at values of θ near 90° (i.e., the rays are bent only slightly as they pass through the aperture). Thus we set $\sin\theta = 1$.

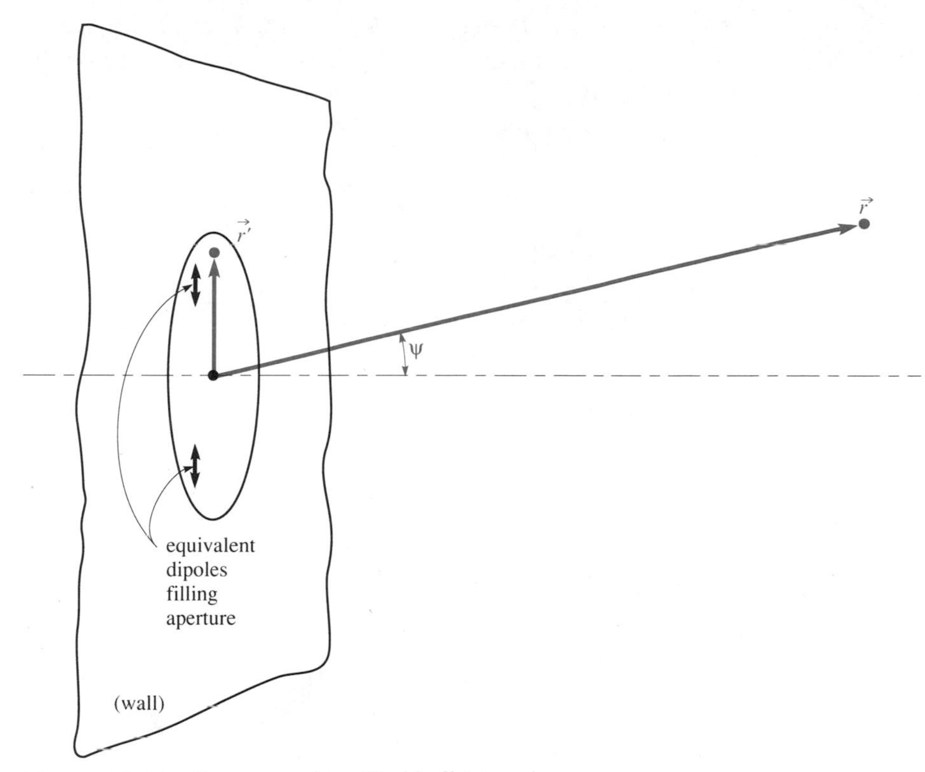

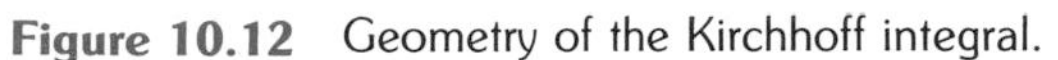

Figure 10.12 Geometry of the Kirchhoff integral.

which upon integration yields

$$\mathbf{E}(x, y, z) = \frac{4je^{-jkr}E_o a^2}{\lambda r} \frac{\sin\left(ka\dfrac{x}{r}\right)}{\left(ka\dfrac{x}{r}\right)} \frac{\sin\left(ka\dfrac{y}{r}\right)}{\left(ka\dfrac{y}{r}\right)} \qquad \textbf{(10.68)}$$

Let us consider the radiation pattern in the plane $y = 0$. We note that $x/r = \sin\psi \cong \psi$, where ψ is the angle with the normal to the wall, as shown in Fig. 10.11(c). Hence

$$\mathbf{E}(r, \psi) = \frac{4je^{-jkr}E_o a^2}{\lambda r} \frac{\sin(ka\psi)}{(ka\psi)} \qquad \textbf{(10.69)}$$

and the intensity is

$$S(r, \psi) = \frac{8E_o^2 a^4}{\lambda^2 r^2 \eta} \frac{\sin^2(ka\psi)}{(ka\psi)^2} \qquad \textbf{(10.70)}$$

This function is sketched in Fig. 10.13. The radiation pattern consists primarily of a strong central lobe between $-\psi_o$ and ψ_o, with small side lobes at larger values of ψ. The limit of the main lobe, ψ_o, according to (10.70), corresponds to $ka\psi_o = \pi$; hence

$$\psi_o = \frac{1}{2}\left(\frac{\lambda}{a}\right) \tag{10.71}$$

Thus to obtain a narrow beam, we must make the aperture large compared with wavelength. The general form of this result holds true for apertures of

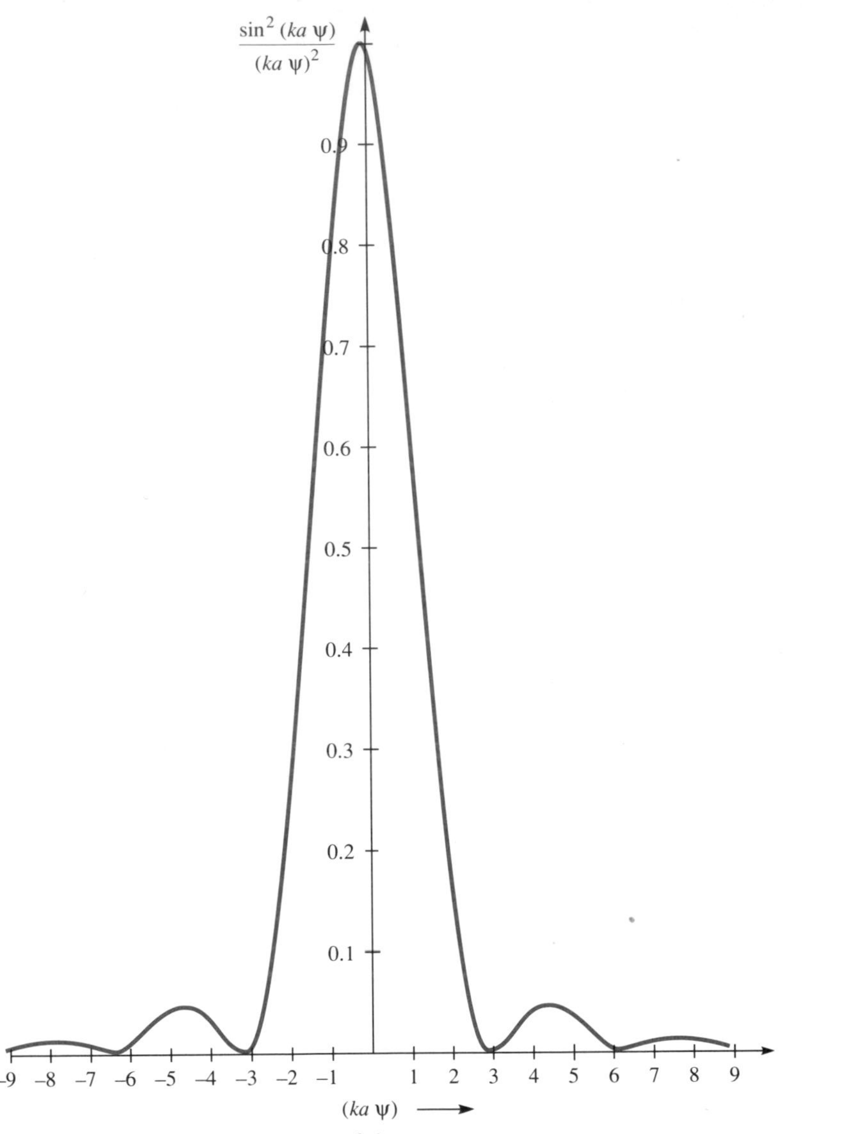

Figure 10.13 Radiation pattern of the aperture antenna.

other shapes, although the numerical factor changes slightly. For a circular aperture of radius a, the factor $\frac{1}{2}$ in (10.71) is replaced by 0.61. The general relationship between large antenna size (compared with wavelength) and high directivity is a very fundamental one. Interestingly, it can be demonstrated from Heisenberg's uncertainty principle, without any mention of electric or magnetic fields at all!

The total radiated power, found by integrating (10.68) over $-\infty < x < \infty$, $-\infty < y < \infty$, is

$$P = \frac{E_0^2}{2\eta}(2a)^2 \tag{10.72}$$

which is the same as the power incident on the aperture from the left. From (10.32), (10.70), and (10.72), the directivity is given by

$$D = \frac{4\pi(2a)^2}{\lambda^2} = \frac{4\pi}{\lambda^2} \times \text{area} \tag{10.73}$$

EXERCISE 10.3

A 100-watt microwave transmitter operating at 10 GHz has a square parabolic reflector antenna, each side of which is 30 cm. The receiver, 15 km distant, is equipped with an identical antenna. Find the power striking the receiving antenna.

Answer 4 μW.

10.6 RECEIVING ANTENNAS AND RECIPROCITY

Until now we have thought of antennas as devices for *transmitting,* that is, for converting electrical power into radiation with desired characteristics. However, antennas are also used for *receiving,* in which their function is exactly the opposite. In this section we shall investigate the behavior of antennas in receiving, and show that this behavior can be predicted from their transmitting behavior. Our main tool in this study will be a rather general theorem of electromagnetics known as the *Lorentz reciprocity theorem.*

RECIPROCITY

The general notion of "reciprocity" has to do with a system working equally well in either direction. For instance, a two-port network is said to be reciprocal if its operation, when port 1 is used as the input and port 2 as the output, is equivalent in some way to its operation when port 2 is used as the input and port 1 as the output. What is meant by "equivalent," however, is not a simple matter, and as we shall see (Example 10.4), some intuitive expectations of reciprocity turn out not to be correct. There is, however, a rigorous theorem of wide validity known as the Lorentz reciprocity theorem. Any system composed entirely of linear, isotropic materials will be reciprocal in the sense that it will have the characteristics predicted by this theorem.[8]

Mathematically, the reciprocity theorem follows from Maxwell's equations using a derivation similar to that of Poynting's theorem. Suppose a system contains two separate sources, designated 1 and 2. When source 1 is operating and 2 is not, the resulting fields $\vec{H}_1$ and $\vec{E}_1$ and current $\vec{J}_1$ are described by

$$\nabla \times \vec{\mathbf{H}}_1 = j\omega\epsilon\vec{\mathbf{E}}_1 + \vec{\mathbf{J}}_1 \tag{10.74}$$

$$\nabla \times \vec{\mathbf{E}}_1 = -j\omega\mu\vec{\mathbf{H}}_1 \tag{10.75}$$

Similarly, when only source 2 is on, we have $\vec{E}_2$, $\vec{H}_2$, $\vec{J}_2$, related by

$$\nabla \times \vec{\mathbf{H}}_2 = j\omega\epsilon\vec{\mathbf{E}}_2 + \vec{\mathbf{J}}_2 \tag{10.76}$$

$$\nabla \times \vec{\mathbf{E}}_2 = -j\omega\mu\vec{\mathbf{H}}_2 \tag{10.77}$$

Taking the dot product of $\vec{\mathbf{E}}_2$ with (10.74) and the dot product of $\vec{\mathbf{H}}_1$ with (10.77) and adding the resulting equations gives

$$\nabla \cdot (\vec{\mathbf{E}}_2 \times \vec{\mathbf{H}}_1) = -j\omega\epsilon(\vec{\mathbf{E}}_1 \cdot \vec{\mathbf{E}}_2) - j\omega\mu(\vec{\mathbf{H}}_1 \cdot \vec{\mathbf{H}}_2) - \vec{\mathbf{E}}_2 \cdot \vec{\mathbf{J}}_1 \tag{10.78}$$

where the vector identity expanding $\nabla \cdot (\vec{A} \times \vec{B})$ has been used. A result similar to (10.78) also exists, with "1" and "2" interchanged. Subtracting (10.78) from the interchanged equation gives

$$\nabla \cdot (\vec{\mathbf{E}}_1 \times \vec{\mathbf{H}}_2 - \vec{\mathbf{E}}_2 \times \vec{\mathbf{H}}_1) = \vec{\mathbf{E}}_2 \cdot \vec{\mathbf{J}}_1 - \vec{\mathbf{E}}_1 \cdot \vec{\mathbf{J}}_2 \tag{10.79}$$

Let us now apply the divergence theorem, using a surface of integration that is extremely large compared with the size of the system in question:

$$\iint_S (\vec{\mathbf{E}}_1 \times \vec{\mathbf{H}}_2 - \vec{\mathbf{E}}_2 \times \vec{\mathbf{H}}_1) \cdot \vec{dS} = \int_V (\vec{\mathbf{E}}_2 \cdot \vec{\mathbf{J}}_1 - \vec{\mathbf{E}}_1 \cdot \vec{\mathbf{J}}_2)\, dV \tag{10.80}$$

It can be shown that as the surface of integration becomes arbitrarily large, the surface integral on the left vanishes.[9] Thus we have

$$\int \vec{\mathbf{E}}_2 \cdot \vec{\mathbf{J}}_1\, dV = \int \vec{\mathbf{E}}_1 \cdot \vec{\mathbf{J}}_2\, dV \tag{10.81}$$

where the integrals are taken over all space.

---

[8]The most common exception to reciprocity occurs in devices containing externally applied magnetic fields, which remove isotropy by establishing a preferred direction. Isolators and circulators are non-reciprocal devices based on this principle.

[9]Far from all sources, the fields become plane waves with $\mathbf{E}_\theta = \eta\mathbf{H}_\phi$, $\mathbf{E}_\phi = -\eta\mathbf{H}_\theta$. The surface integral is then

$$\eta \iint (\mathbf{H}_{\theta 1}\mathbf{H}_{\theta 2} + \mathbf{H}_{\phi 1}\mathbf{H}_{\phi 2} - \mathbf{H}_{\theta 2}\mathbf{H}_{\theta 1} - \mathbf{H}_{\phi 2}\mathbf{H}_{\phi 1}) \cdot dS = 0$$

See R. F. Harrington, *Time-Harmonic Electromagnetic Fields*. New York: McGraw-Hill, 1961.

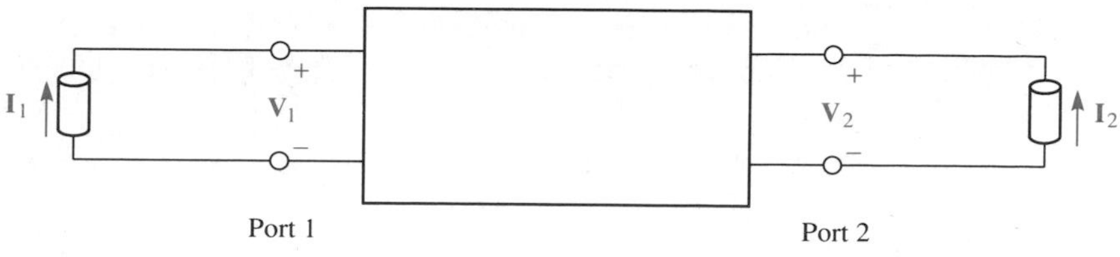

Figure 10.14

Let us now apply (10.81) to the arbitrary circuit shown in Fig. 10.14. The two ideal current sources I_1 and I_2 are imagined to be tiny cylinders, so that inside them $\vec{E}$ and $\vec{J}$ are parallel. Then for the portion of the volume integral enclosing source I_1,

$$\int_{I_1} \vec{\mathbf{E}}_2 \cdot \vec{\mathbf{J}}_1 \, dV = \int_{I_1} (|\mathbf{E}_2| \, dl)\,(|\mathbf{J}_1| \, dA) = \mathbf{V}_{12}\mathbf{I}_1 \tag{10.82}$$

Here $\mathbf{V}_{12}$ is defined as the voltage produced at port 1 by source I_2. Furthermore, for the volume enclosing source 2,

$$\int_{I_2} \vec{\mathbf{E}}_2 \cdot \vec{\mathbf{J}}_1 \, dV = 0 \tag{10.83}$$

because source I_1 produces no current through ideal current source I_2.[10] The integrals in (10.81) also include the volume of the system lying between the two ports. Portions made of perfect conductors give no contribution because $\vec{E}$ vanishes. In resistive materials the integral on the left of (10.81) becomes

$$\int_{\text{System}} \mathbf{E}_2\mathbf{E}_1\sigma_E \, dV$$

and the integral on the right is identical. Thus (10.81) becomes

$$\mathbf{V}_{12}\mathbf{I}_1 + \int_{\text{System}} \mathbf{E}_2\mathbf{E}_1\sigma_E \, dV = \int_{\text{System}} \vec{\mathbf{E}}_1\vec{\mathbf{E}}_2\sigma_E \, dV + \mathbf{V}_{21}\mathbf{I}_2$$

and finally we have the result

$$\mathbf{V}_{12}\mathbf{I}_1 = \mathbf{V}_{21}\mathbf{I}_2 \tag{10.84}$$

which is our statement of the Lorentz reciprocity theorem. In words, (10.84) states that

$$\begin{pmatrix}\text{value of}\\ \text{current source}\\ \text{at port 1}\end{pmatrix} \times \begin{pmatrix}\text{voltage produced at}\\ \text{port 1 by current}\\ \text{source at port 2}\end{pmatrix}$$

$$= \begin{pmatrix}\text{value of}\\ \text{current source}\\ \text{at port 2}\end{pmatrix} \times \begin{pmatrix}\text{voltage produced at}\\ \text{port 2 by current}\\ \text{source at port 1}\end{pmatrix}$$

[10] That is, the current through source 2 must be zero when source 2 is turned off and source 1 is turned on.

The significance of our result (10.84) is still rather obscure. Its meaning can be clarified by adopting the language of circuit theory. There it is usual to describe a two-port by means of its *impedance matrix:*

$$\mathbf{V}_1 = z_{11}\mathbf{I}_1 + z_{12}\mathbf{I}_2 \qquad (10.85a)$$

$$\mathbf{V}_2 = z_{21}\mathbf{I}_1 + z_{22}\mathbf{I}_2 \qquad (10.85b)$$

For the thought-experiment just described, in which the two sources $\mathbf{I}_1$ and $\mathbf{I}_2$ are turned on one at a time, we have from (10.85)

$$\mathbf{V}_{21} = z_{21}\mathbf{I}_1$$

$$\mathbf{V}_{12} = z_{12}\mathbf{I}_2 \qquad (10.86)$$

Substituting $\mathbf{V}_{21}$ and $\mathbf{V}_{12}$ from (10.86) into (10.84) we have

$$z_{21} = z_{12} \qquad (10.87)$$

Equation (10.87) is a statement of reciprocity, expressed in the language of circuit theory. It applies to all systems composed of linear, isotropic materials. For systems with many ports it becomes $z_{ji} = z_{ij}$.

EXAMPLE 10.4

Figure 10.15(a) shows a linear, reciprocal two-port system with an ideal voltage source connected to port 1, and a load resistance R_L connected to port 2.

(a) Find the current $\mathbf{I}_2$ that flows through R_L, in terms of the components of the impedance matrix.

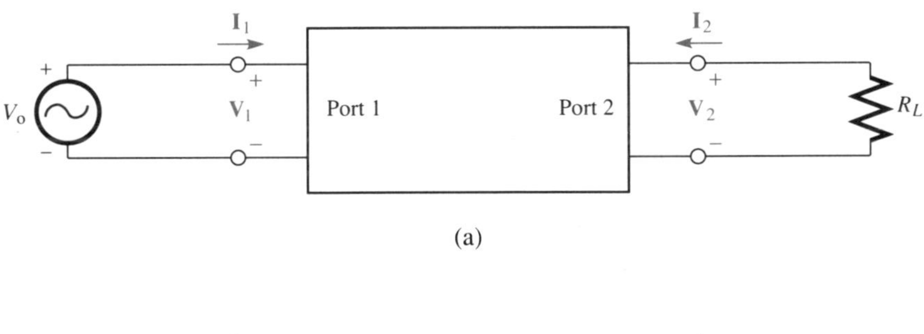

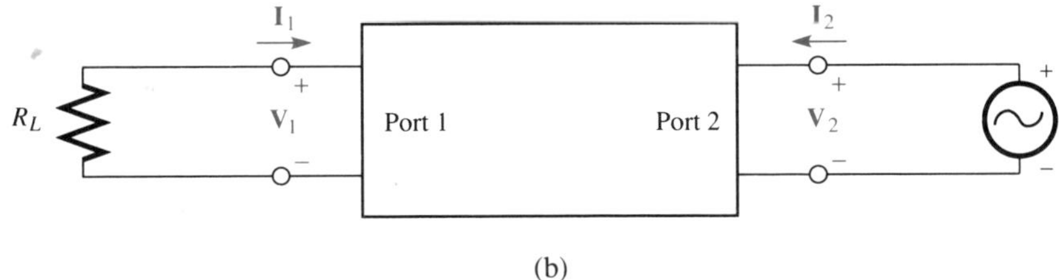

Figure 10.15

(b) The voltage source and the load are now interchanged, as shown in Fig. 10.15(b). Determine whether or not the current that flows through R_L is the same as in part (a).

Solution

(a) From (10.85a) we have

$$\mathbf{I}_2 = \frac{1}{z_{12}} (\mathbf{V}_1 - z_{11}\mathbf{I}_1) \tag{10.88}$$

and from (10.85b), using $\mathbf{V}_2 = -\mathbf{I}_2 R_L$,

$$\mathbf{I}_1 = \frac{1}{z_{21}} (\mathbf{V}_2 - z_{22}\mathbf{I}_2) = -\frac{R_L + z_{22}}{z_{21}} \mathbf{I}_2 \tag{10.89}$$

Substituting (10.89) into (10.88) and solving for $\mathbf{I}_2$, we find

$$\mathbf{I}_2 = \frac{z_{21}}{z_{12}z_{21} - z_{11}(R_L + z_{22})} V_o \tag{10.90}$$

(b) Interchanging source and load has the same effect as interchanging subscripts 1 and 2 in expression (10.90). We observe that even though $z_{21} = z_{12}$, this expression does *not* remain unchanged when the subscripts are interchanged. Thus the current through R_L is *different* from that in part (a).

The slightly disappointing result of part (b) shows that one must be careful in interpreting the word "reciprocity." If a system is reciprocal, then $z_{12} = z_{21}$; but what *that* in turn implies, is not always easy to guess.

APPLICATION OF RECIPROCITY TO ANTENNA THEORY

Let us now consider a system composed of two antennas, as sketched in Fig. 10.16(a). Each antenna has a pair of terminals; thus, we may regard the pair of antennas as a two-port system, as in Fig. 10.16(b). Since the medium between the antennas is linear and isotropic, we expect that reciprocity will apply.

One of the antennas will be used for transmitting and the other for receiving. The transmitting antenna is driven by a Thèvenin source V_S, Z_S, and the receiving antenna is provided with a load Z_L, as shown in Fig. 10.16(c). We shall assume that source and load are impedance matched in the sense that $Z_S = z_{11}^*$, and $Z_L = z_{22}^*$. By straightforward calculations similar to Example 10.4, we obtain

$$\mathbf{I}_2 = \frac{z_{21} V_S}{z_{12}z_{21} - 4R_{11}R_{22}} \tag{10.91}$$

where $R_{11} = \text{Re}\,(z_{11})$, $R_{22} = \text{Re}\,(z_{22})$. The power delivered to Z_L is thus

$$P_L = \frac{|\mathbf{I}_2|^2 R_{22}}{2} = \frac{|z_{21}|^2 R_{22} |V_S|^2}{2|z_{12}z_{21} - 4R_{11}R_{22}|^2} \tag{10.92}$$

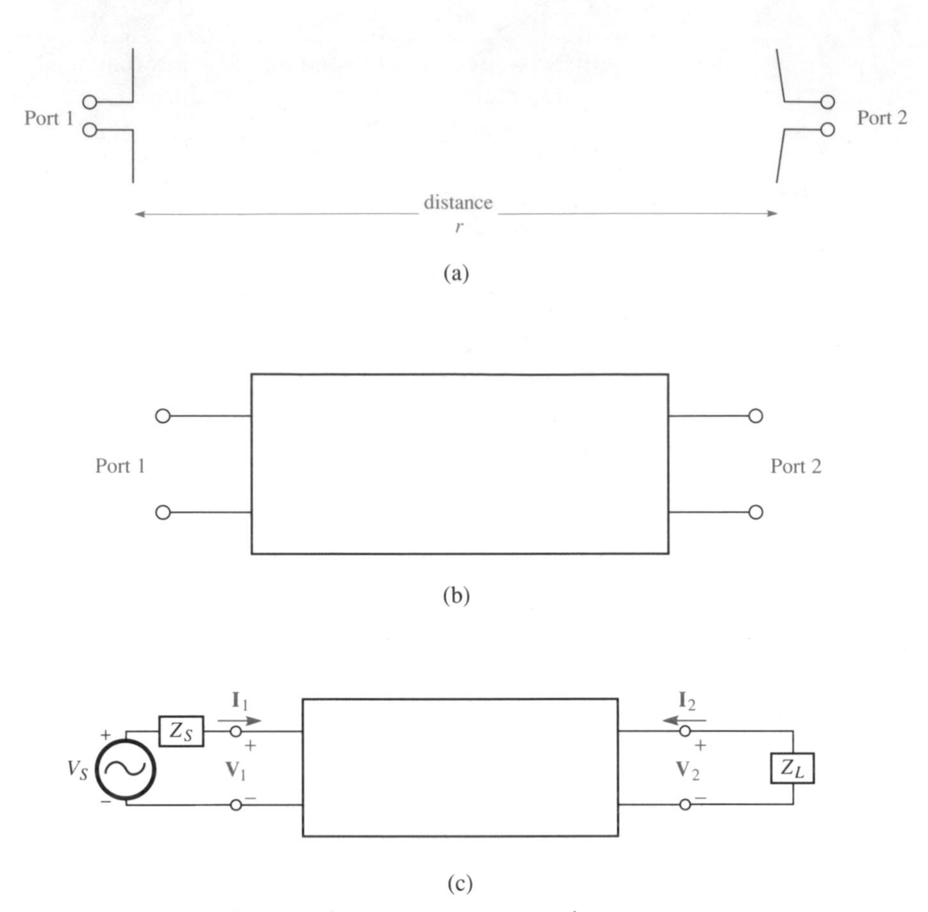

Figure 10.16 Application of reciprocity to antenna theory.

Expressions (10.91) and (10.92) can now be simplified through the use of physical intuition. We are interested in antennas that are very far apart, and therefore very weakly coupled. Thus current $\mathbf{I}_1$, for example, produces an extremely small voltage at port 2, much smaller than the voltage it produces at port 1. From this sort of reasoning we conclude that z_{12} and z_{21} are very small numbers that can be neglected in comparison with R_{11} and R_{22}. Thus expressions (10.91) and (10.92) become

$$\mathbf{I}_2 = -\frac{z_{21} V_S}{4R_{11}R_{22}} \tag{10.93}$$

$$P_L = \frac{|z_{21}|^2 |V_S|^2}{32R_{11}^2 R_{22}} \tag{10.94}$$

We next wish to compare P_L with the input power to the transmitting antenna, which is

$$P_i = \tfrac{1}{2} \operatorname{Re} (\mathbf{V}_1 \mathbf{I}_1^*) \tag{10.95}$$

Using (10.85b), $\mathbf{V}_2 = -\mathbf{I}_2 Z_L$, $Z_L = z_{22}^*$, and (10.93), we obtain

$$\mathbf{I}_1 = -\frac{2R_{22}\mathbf{I}_2}{z_{21}} = \frac{V_S}{2R_{11}} \tag{10.96}$$

$$\mathbf{V}_1 = V_S - z_{11}^*\mathbf{I}_1 = V_S\left(1 - \frac{z_{11}^*}{2R_{11}}\right) \tag{10.97}$$

Hence, from (10.95), (10.96), and (10.97) the input power is

$$P_i = \frac{|V_S|^2}{8R_{11}} \tag{10.98}$$

and

$$\frac{P_L}{P_i} = \frac{|z_{21}|^2}{4R_{11}R_{22}} \tag{10.99}$$

Here we can observe an effect of reciprocity. From (10.99) we see that P_L/P_i will have the same value if the roles of the two antennas are interchanged. That is, using 2 as a transmitter and 1 as receiver amounts to interchanging subscripts 1 and 2 in (10.99), an operation which leaves the result unchanged. (The numerator is unchanged because of reciprocity.) We shall soon make use of this result.

In order to describe the ability of a receiving antenna to capture power, it is customary to ascribe to it an *effective area* A_e. The significance of this parameter is that, by definition, the power conveyed to a matched load at the antenna terminals is

$$P_L = A_e S \tag{10.100}$$

where S is the intensity (power per unit area) of the nearly plane wave produced by the transmitting antenna. However, from (10.32) and (10.35) we have

$$S = \frac{GP_i}{4\pi r^2} \tag{10.101}$$

where G is the gain of the transmitting antenna. Thus when antenna 1 is the transmitter and antenna 2 the receiver,

$$A_{e2} = \frac{P_L}{S} = \frac{4\pi r^2}{G_1}\frac{P_L}{P_i} \tag{10.102}$$

Now at last we are ready to make use of reciprocity. Let us interchange the roles of the two antennas. In that case,

$$A_{e1} = \frac{4\pi r^2}{G_2}\frac{P_L}{P_i} \tag{10.103}$$

However, from (10.99) and our knowledge that $z_{12} = z_{21}$, we can see that P_L/P_i remains the same when antennas 1 and 2 are interchanged. Thus P_L/P_i

has the same value in (10.102) and (10.103). Dividing one of these by the other, we have

$$\frac{A_{e2}}{A_{e1}} = \frac{G_2}{G_1} \tag{10.104}$$

In other words, the *effective area of an antenna is proportional to its gain.* To find the constant of proportionality, it is only necessary to find A_e/G for any antenna. It is fairly evident that for a large, lossless aperture antenna, the effective area is equal to the geometrical area of the antenna. Thus from (10.73) the constant of proportionality is $4\pi/\lambda^2$, and the general result, valid for all antennas, is

$$A_e = \frac{\lambda^2 G}{4\pi} \tag{10.105}$$

EXAMPLE 10.5

A communications system uses two impedance-matched antennas, directed at each other, with gains G_1 and G_2, separated by a distance r. The power of the transmitter is P_o. Find the power supplied to the receiver's input terminals. Obtain a numerical result for the case of two lossless half-wave dipoles working at 15 MHz, separated by 200 miles, if $P_o = 1$ kW.

Solution From (10.32) and (10.35), the intensity S at the receiving antenna is given by

$$S = \frac{G_1 P_i}{4\pi r^2}$$

where P_i is the power furnished by the transmitter to the antenna terminals. Using (10.100) and (10.105), we find the receiver's input power:

$$P_L = A_{e2} S = \frac{\lambda^2 G_1 G_2}{16\pi^2 r^2} P_i \tag{10.106}$$

This result is known as the *Friis transmission formula*. For the numerical example, we have seen that the gain of a lossless half-wave dipole is 1.64. Thus from the Friis formula we have $P_L = 66$ nW.

This result is based on the assumption that the antennas are in free space. In actual communications at 15 GHz, wave propagation would be strongly influenced by reflections from the earth and the ionosphere.

Similar reasoning can be used to show that *the pattern of an antenna when it is used for receiving is the same as its pattern when used for transmitting*. Let there be two antennas, as shown in Fig. 10.17. Antenna 1, with

Figure 10.17 Demonstrating that the receiving and transmitting patterns of an antenna are identical.

gain G_1, is pointed toward antenna 2. However, antenna 2 is directed away from antenna 1 at some arbitrary angle θ. In this case the effective area of 2 will be a function $A_{e2}(\theta)$, dependent on θ. The power received by 2 will be given by

$$\frac{P_{L2}}{P_i} = \frac{G_1}{4\pi r^2} A_{e2}(\theta) \tag{10.107}$$

Now let us imagine that 1 is used for receiving and 2, still misoriented at angle θ, is used for transmitting. The power received by 1 will be

$$P_{L1} = S_2(\theta) \frac{\lambda^2 G_1}{4\pi} \tag{10.108}$$

where $S_2(\theta)$ is the intensity produced at 1 by antenna 2 when it is used as a transmitter. We have seen, however, that $P_{L2}/P_i = P_{L1}/P_1$. Thus from (10.107) and (10.108),

$$\frac{G_1}{4\pi r^2} A_{e2}(\theta) = \frac{\lambda^2 G_1 S_2(\theta)}{4\pi P_i}$$

from which we have

$$A_{e2}(\theta) = \frac{r^2\lambda^2}{P_i} S_2(\theta) \tag{10.109}$$

This result shows that the antenna's effective area depends on θ in the same way as its transmitted radiation; that is, that its receiving and transmitting patterns are the same.

Finally, we can show that the impedance of the antenna is the same for receiving and for transmitting. For transmission, the input voltage is

$$\mathbf{V}_1 = z_{11}\mathbf{I}_1 + z_{12}\mathbf{I}_2 \tag{10.110}$$

However, the last term of this equation represents the effect of currents in the receiving antenna on the voltage at the terminals of the transmitting antenna. Since the antennas are far apart and very weakly coupled, this term is obviously negligible, and

$$Z_{in} = \frac{\mathbf{V}_1}{\mathbf{I}_1} = z_{11} \tag{10.111}$$

When antenna 1 is the receiver, we can write

$$Z_{out} = -\frac{\mathbf{V}_1(\text{oc})}{\mathbf{I}_1(\text{sc})} \tag{10.112}$$

where $\mathbf{V}_1(\text{oc})$ is the open-circuit voltage at port 1 and $\mathbf{I}_1(\text{sc})$ is the short-circuit current. From (10.110) we have $\mathbf{V}_1(\text{oc}) = z_{12}\mathbf{I}_2(\text{oc})$. For the short-circuit condition, we have

$$\mathbf{V}_1(\text{sc}) = z_{11}\mathbf{I}_1(\text{sc}) + z_{12}\mathbf{I}_2(\text{sc}) = 0 \tag{10.113}$$

Hence

$$Z_{out} = \frac{z_{12}\mathbf{I}_2(\text{oc})}{\left(\dfrac{z_{12}\mathbf{I}_2(\text{sc})}{z_{11}}\right)}$$

However, it is again physically obvious that opening or shorting the terminals on the receiving antenna (antenna 1) can have no effect on the current in the transmitting antenna (antenna 2). Thus

$$\mathbf{I}_2(\text{oc}) = \mathbf{I}_2(\text{sc})$$

and consequently

$$Z_{out} = z_{11} \tag{10.114}$$

Comparing with (10.111), we see that the antenna's input impedance when used as a transmitter is equal to its source impedance when used to receive.

10.7 POINTS TO REMEMBER

1. The fields of a time-varying current are most conveniently found from a vector magnetic potential. This potential can be found by means of a Green's function integral similar to the one used in magnetostatics. An important difference, however, is the fact that in the time-varying case the potential is *retarded* with respect to its source. That is, if the source current changes, the potential does not change until a time τ later, where τ is the time required for light to travel from the source to the point where the potential is being found.

2. By means of the retarded potential, the fields of various current distributions can be found. The simplest example of an antenna is an

infinitely short wire carrying a sinusoidal current, which is known as a Hertzian dipole. Straight wires of finite length are known as long-wire antennas.

3. The pattern of an antenna is the relationship (usually described by means of polar graphs) of radiated intensity at a fixed distance to the angle at which intensity is measured.

4. The directivity of an antenna is the ratio of the radiated intensity in the direction at which intensity is maximum to the average radiated intensity. The product of the directivity and the power efficiency is known as the gain.

5. Several antennas can be combined together to form an antenna array. The pattern of an array contains a factor describing the pattern of the individual antennas, called the element factor, and another describing the positions of the elements.

6. Highly directive antennas for use at high frequencies can be made using parabolic reflectors. These have fields similar to those produced when a plane wave passes through an aperture, and thus are known as aperture antennas. They are analyzed by means of a formula known as the Kirchhoff integral. The half-angle at which the beam of a circular aperture antenna spreads is given by

$$\psi_o = 0.61 \left(\frac{\lambda}{a}\right)$$

where a, the radius of the reflector, is assumed much larger than the wavelength λ.

7. The Lorentz reciprocity theorem is a powerful general theorem relating electromagnetics to circuit theory. From this theorem information can be gained about the behavior of antennas when they are used for receiving. The pattern of an antenna used for receiving is the same as its pattern when used to transmit. The effective area of an antenna, A_e, is its capture cross section for energy arriving in the form of plane waves. The effective area is related to the gain through

$$A_e = \frac{\lambda^2 G}{4\pi}$$

REFERENCES

1. S. Ramo, J. R. Whinnery, and T. Van Duzer, *Fields and Waves in Communication Electronics*. New York: Wiley, 1984.
2. R. S. Elliott, *Antenna Theory and Design*. Englewood Cliffs, N.J.: Prentice-Hall, 1981.
3. J. D. Kraus, *Antennas*. New York: McGraw-Hill, 1950.
4. E. C. Jordan and K. G. Balmain, *Electromagnetic Waves and Radiating Systems*, 2nd ed. Englewood Cliffs, N.J.: Prentice-Hall, 1968.

The following, more advanced references contain derivations of the Kirchhoff integral:

5. J. A. Stratton, *Electromagnetic Theory*. New York: McGraw-Hill, 1941.
6. J. D. Jackson, *Classical Electrodynamics*. New York: Wiley, 1962.

With regard to microwave circuits and reciprocity, see reference 1 and

7. R. E. Collin, *Foundations for Microwave Engineering*. New York: McGraw-Hill, 1966.

or at a more advanced level,

8. R. F. Harrington, *Time-Harmonic Electromagnetic Fields*. New York: McGraw-Hill, 1961.

PROBLEMS

Section 10.1

10.1 A very short wire of length h at the origin carries a current $i(t) = I_o \cos \omega t$ in the x direction. Find the vector potential $\vec{A}$ at the position $x = \dfrac{100\pi c}{\omega}$, $y = z = 0$. (Assume $U = c$ = velocity of light in vacuum.)

*10.2 Two very short identical wires of length h are located at the origin and at $x = a$, $y = z = 0$. Both wires carry current $I_o \cos \omega t$ in the x direction. Find $\vec{A}$ at $x = 2a$, $y = z = 0$. Assume that $a = \dfrac{\pi U}{\omega}$.

10.3 Show that if the magnetic vector potential $\vec{A}$ is sinusoidal in time and is known, the electric field can be found using

$$\vec{\mathbf{E}} = -\frac{jU^2}{\omega} \nabla(\nabla \cdot \vec{\mathbf{A}}) - j\omega\vec{\mathbf{A}}$$

*10.4 The phasor representing the magnetic field of a sinusoidal plane wave is $\vec{\mathbf{H}} = H_o e^{-jkz} \vec{e}_y$.
(a) Find a vector potential $\vec{\mathbf{A}}$ that corresponds to $\vec{\mathbf{H}}$.
(b) Does your $\vec{\mathbf{A}}$ lead to the correct electric field for the plane wave?
(c) What is Φ?

Section 10.2

10.5 Suppose we wish to use a standard car radio antenna (length = 1 m) as a transmitting antenna at 1 MHz. The power to be transmitted is 10 W. What antenna current would be required?

*10.6 Consider a small cube of material, with side a and conductivity σ_E, oriented with its edges along the rectangular axes. A plane wave with wavelength λ ($\lambda \gg a$) passes through the material. Assume that the electric field inside the material is $E_o e^{-jkz} \vec{e}_x$.

(a) What current flows in the material?
(b) How much power does it radiate?
(c) How does the radiated power depend on frequency?
This phenomenon is known as *scattering* of electromagnetic waves. The preferential scattering of high-frequency light by air molecules is what makes the sky blue.

10.7 The current of an elementary dipole is aligned with the z axis of a rectangular coordinate system. Sketch the radiation pattern
(a) in the x-z plane
(b) in the y-z plane
(c) in the x-y plane.

10.8 The antenna of Problem 10.5 has a resistance of 2×10^{-3} ohm. What is its efficiency? Its gain?

*10.9 The resistance of the wire used to form an elementary dipole is R_L ohms/m. Find the efficiency of the antenna if its length is h and the wavelength is λ. Does efficiency increase or decrease as h increases?

Section 10.3

**10.10 Assume that $\vec{\mathbf{A}} = \dfrac{e^{-jkr}}{r} f(\theta, \phi)$, where f is any function of θ and ϕ. Verify that the only terms of $\vec{\mathbf{E}}$ that decrease as $1/r$ are those given in (10.50).

10.11 Verify that the directivity of a half-wave dipole is 2.15 dB.

*10.12 Find the power lost in the resistance of a half-wave dipole at wavelength λ, if the resistance per unit length is R_L and the current amplitude at the center is I_0.

*10.13 Estimate the efficiency of a half-wave dipole operating at 3 MHz, assuming that its resistance is determined by the skin effect. The wire is made of copper and is 2 mm in diameter.

*10.14 A quarter-wave vertical antenna uses a perfectly conductive metal groundplane. Find the current that flows in the groundplane, at distances sufficiently great that induction fields can be neglected.

Section 10.4

*10.15 Plot and discuss the pattern of a traveling-wave long-wire antenna like that in Example 10.3, if $L = \lambda/2$. (Use a programmable calculator.) At approximately what value of θ is the radiation maximum? Is the pattern different from that of the half-wave dipole? What is the physical difference between the two antennas?

10.16 Use the method of Example 10.3 to re-derive the field of a half-wave dipole.

*10.17 Two elementary dipoles have currents in the z direction that have the same phase. The dipoles are located at $(0, -\lambda/4, 0)$ and $(0, \lambda/4, 0)$. Let the polar axis coincide with the z axis and the $\phi = 0$ direction with the x axis. Find and sketch the radiation pattern. Explain the direction of maximum radiation physically.

*10.18 Sketch a polar plot of the antenna pattern for a traveling-wave long-wire antenna, as in Example 10.3. (Major lobes only.) Assume that $L = 30\,\lambda$. Verify the result for θ_{max} obtained in the example. (Numerical or graphical methods can be used.)

Section 10.5

*10.19 Usually the illumination of a microwave dish by a feed horn is nonuniform. Consider a square reflector with sides a, illuminated in such a way that the field in the aperture, instead of being a constant E_o, is the function $E_o e^{-(x^2+y^2)/b^2}$, where $b \ll a$. Find the far-field pattern $S(\psi)$ in the plane $y = 0$.

*10.20 Consider an aperture consisting of an infinitely long slit bounded by $x' = \pm a$. Let the phasor representing the illumination of the slit (that is, the field incident from behind the screen) be $\mathbf{E}_i(x')$. Show that at a distance z from the aperture, the field $\mathbf{E}(x)$ is approximately the Fourier transform of the incident field $\mathbf{E}_i$, provided that $z \gg x$. (This result is used in holography.)

Section 10.6

10.21 Find the four elements of the Z matrix for the network shown in Fig. 10.18. Verify that the network is reciprocal.

10.22 Show by means of effective area that the directivity of a large circular aperture antenna of radius a is given by

$$D = \frac{4\pi^2 a^2}{\lambda^2}$$

10.23 You are designing a communication system to operate over a path length L and are restricted to circular aperture antennas with maximum radius a and a transmitter with maximum power P_T. However, you may choose the frequency f. Will the receiver's input power P_R be

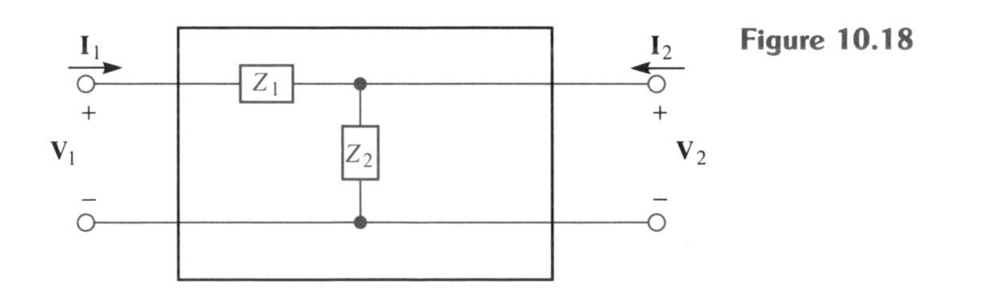

Figure 10.18

larger at high frequencies or at low frequencies? How does P_R/P_T depend on f? (Use the result of the preceding problem.)

10.24 A home television antenna has a gain of 12 dB. It is located 20 miles from a 1000-W TV transmitter, with a transmitting antenna we may imagine is isotropic. The frequency is 55 MHz (Channel 2). Estimate the input power to the TV receiver.

$$\int \frac{dx}{(a^2+x^2)^{3/2}} = \frac{x}{a^2\sqrt{a^2+x^2}}$$

$$\int \frac{a}{u\sqrt{a^2+u^2}}\,du = -\frac{1}{u}\ln\left(\frac{a+\sqrt{u^2+a^2}}{u}\right)$$

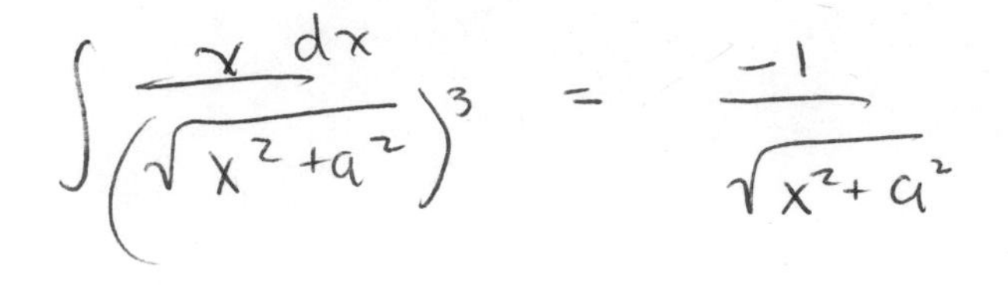

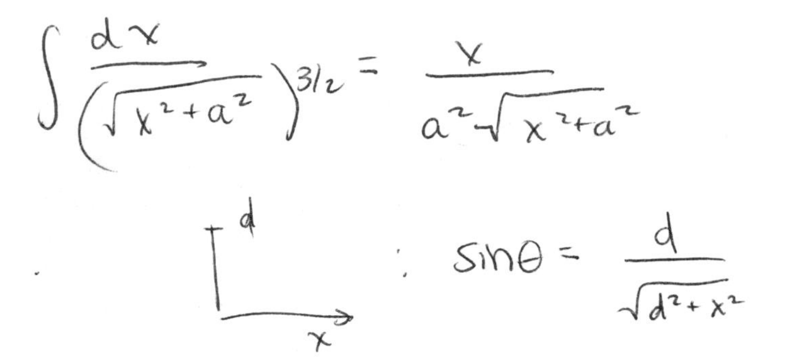

APPENDIX A

Curvilinear Coordinates

Problems in electromagnetics often possess symmetry of one sort or another. When this is so, considerable effort can be saved by using a coordinate system that has the same symmetry as the problem at hand. For example, consider a circle of radius a centered at the origin. A circle, of course, has a high degree of circular symmetry. If we are asked for the equation describing the circle, and we use rectangular coordinates, which are rather unsuitable, we obtain the fairly complicated quadratic equation $x^2 + y^2 = a^2$. On the other hand if we use polar coordinates we have the much simpler equation $r = a$.

Curvilinear coordinate systems are those in which changing the value of a coordinate moves one along a curve, rather than a straight line. There are many systems of curvilinear coordinates, some of them quite obscure.[1] Here we shall concern ourselves only with cylindrical coordinates and spherical coordinates, which are the only ones encountered in everyday work. Curvilinear coordinate systems have distinct peculiarities as compared with rectangular coordinates. For instance, the unit vectors do not always point in the same direction: the direction of a unit vector depends on where you stand.

CYLINDRICAL COORDINATES

These are among the easiest to understand, since they amount to a combination of rectangular coordinates and the planar system known as "polar" coordinates. The three coordinates are r, ϕ, and z. The latter, z, is identical

[1]For instance ellipsoidal coordinates, bispherical coordinates, toroidal coordinates. For an extensive discussion see P. M. Morse and H. Feshbach, *Methods of Theoretical Physics*. New York: McGraw-Hill, 1953.

with z in rectangular coordinates. Positions in the x-y plane are determined by the values of r and ϕ, as shown in Fig. A.1. If the rectangular coordinates of point P are x, y, z, its cylindrical coordinates are given by

$$r = x^2 + y^2$$
$$\phi = \tan^{-1}\frac{y}{x} \qquad \textbf{(A.1)}$$

The conversion in the opposite direction is obtained from

$$x = r\cos\phi$$
$$y = r\sin\phi \qquad \textbf{(A.2)}$$

(The value of z, the third coordinate, is the same in both systems.) The unit vectors are always directed in the direction one moves if the corresponding coordinate is increased, as shown in Fig. A.2. The three unit vectors are mutually perpendicular, and thus dot products between any two different ones vanish. Their cross products obey the right-hand rule and are given by

$$\vec{e}_r \times \vec{e}_\phi = \vec{e}_z$$
$$\vec{e}_\phi \times \vec{e}_z = \vec{e}_r \qquad \textbf{(A.3)}$$
$$\vec{e}_z \times \vec{e}_r = \vec{e}_\phi$$

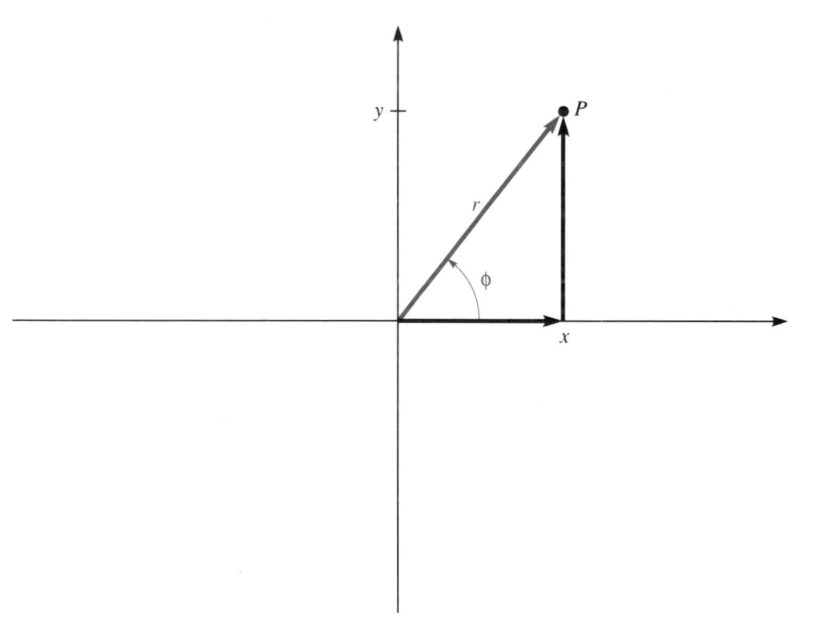

Figure A.1

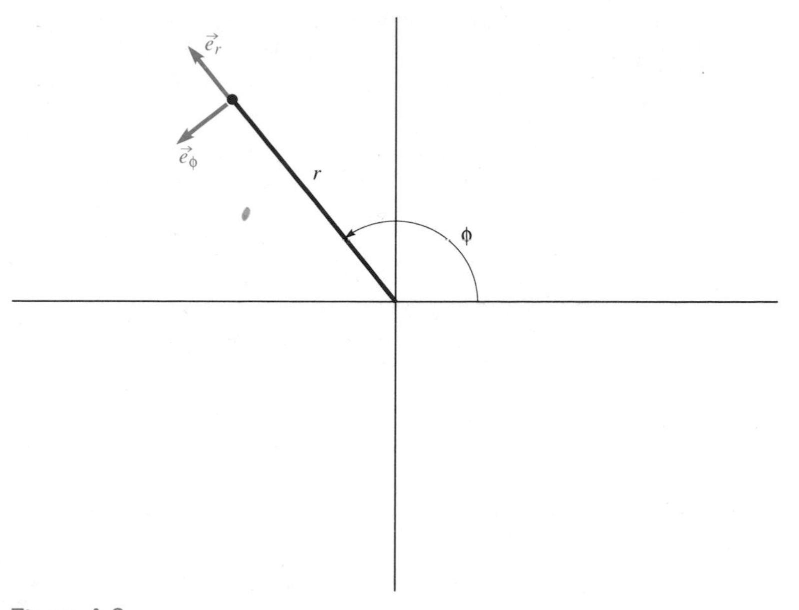

Figure A.2

The constant-coordinate surfaces are cylinders centered on the z axis (when r is held constant); planes containing the z axis (ϕ held constant); and planes perpendicular to the z axis (z held constant).

Incremental distances in the r and z directions are simply dr and dz. However, $d\phi$ cannot be a distance in the ϕ direction because ϕ is an angle and lacks the dimension of length. The incremental distance in the ϕ direction (see Fig. A.3) is the distance one moves as ϕ changes from ϕ to $(\phi + d\phi)$. As we see from the figure, this distance is $r\,d\phi$. The incremental vector $\vec{dl}$ that appears in line integrals is therefore given by

$$\vec{dl} = dr\,\vec{e}_r + dz\,\vec{e}_z + r\,d\phi\,\vec{e}_\phi \tag{A.4}$$

The increment of area in a plane of constant z is (see Fig. A.4)

$$dA = (dr)(r\,d\phi) = r\,dr\,d\phi \tag{A.5}$$

while on a cylinder of constant r it is $dA = r\,d\phi\,dz$. The differential volume, used in volume integrations, is

$$dV = r\,dr\,d\phi\,dz \tag{A.6}$$

EXERCISE A.1

Sketch and describe the curve $r = a$, $z = b\phi$, where a and b are given positive real constants.

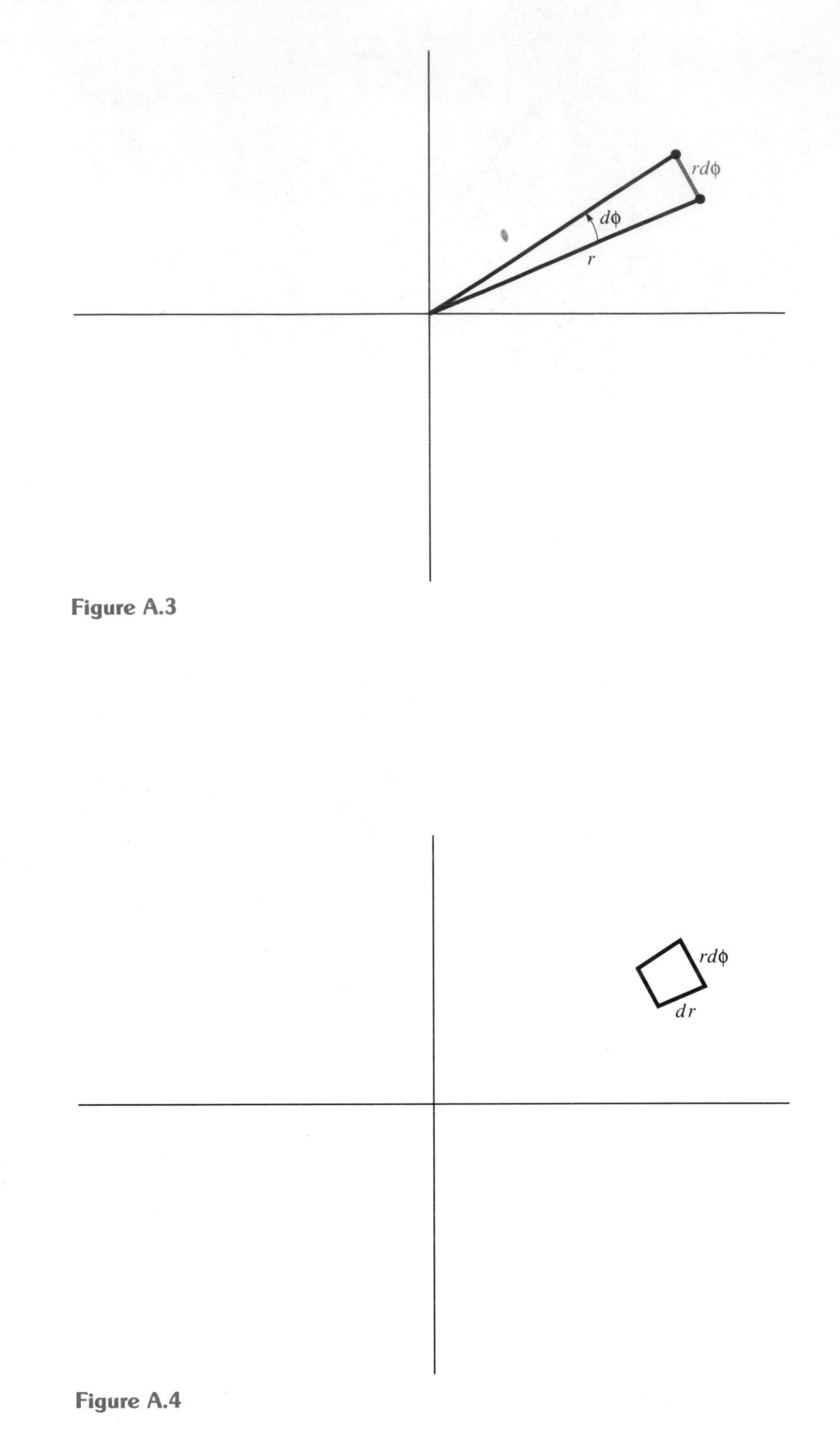

Figure A.3

Figure A.4

EXERCISE A.2

Let $f(r, \phi, z) = r\cos^2\phi$. Find $\int f dV$, where the integral is taken over the cylindrical volume $0 < r < a$, $0 < z < c$, $0 < \phi < 2\pi$.

Answer $2\pi^2 a^3 c/3$.

SPHERICAL COORDINATES

Although these are less familiar than rectangular or cylindrical coordinates, they are extremely useful. They are especially important in radiation problems. An antenna at the origin normally generates spherical waves, and spherical coordinates are the natural ones to use.

The three spherical coordinates are r, θ, ϕ. The first of these specifies the distance from the origin to the point in question. Thus surfaces of constant r are spheres centered on the origin. The other two coordinates are angles which indicate in which *direction* one must travel a distance r from the origin, in order to reach the point in question. The first of these angles, θ, is measured from a given axis called the *polar axis*, as shown in Fig. A.5(a) and θ itself is known as the *polar angle*. Surfaces of constant θ are *cones*, as shown in Fig. A.5(b). Note that the range of θ is $0 < \theta < 180°$ (not 360°).

EXERCISE A.3

Describe the surface corresponding to (a) $\theta = 90°$; (b) $\theta = 160°$.

All points having a specified value of r lie on the corresponding sphere, and all points with a specific θ lie on the cone with the corresponding vertical angle. When both r and θ are specified, the allowable points lie on the intersection of the sphere and cone, which is a circle (Fig. A.6(a)). Note that the radius of this circle is $r\sin\theta$. The function of the remaining coordinate ϕ is to select a particular point on this circle. Figure A.6(b) is a view down the polar axis (from the top in Fig. A.6(a)) toward the origin. The circle in Fig. A.6(a) is seen centered on the polar axis. The position of point P is specified by the angle ϕ, which is measured with respect to some direction chosen to correspond to $\phi = 0$. Note that the direction of the polar axis is *out* of the page in Fig. A.6(a). The surfaces of constant ϕ are half-planes containing the polar axis and passing through the origin.

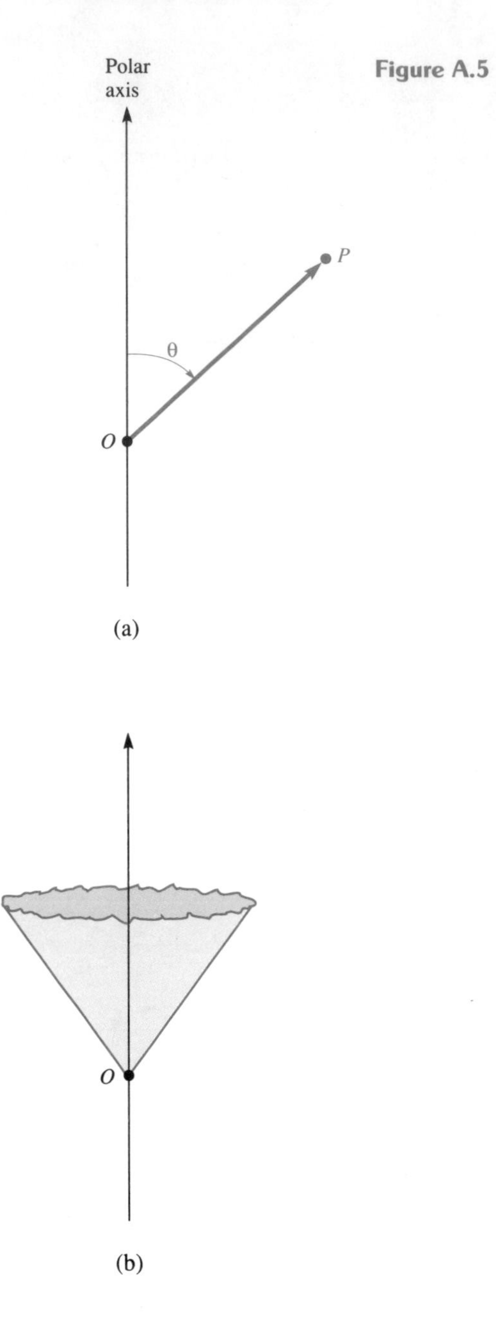

Figure A.5

EXERCISE A.4

Describe (a) the curve corresponding to $\theta = 60°$, $\phi = 90°$; (b) the curve corresponding to $r = 1$, $\phi = 30°$; (c) the curve $\theta = 30°$, $r = b\phi$ (where b is a given positive real constant).

Figure A.6

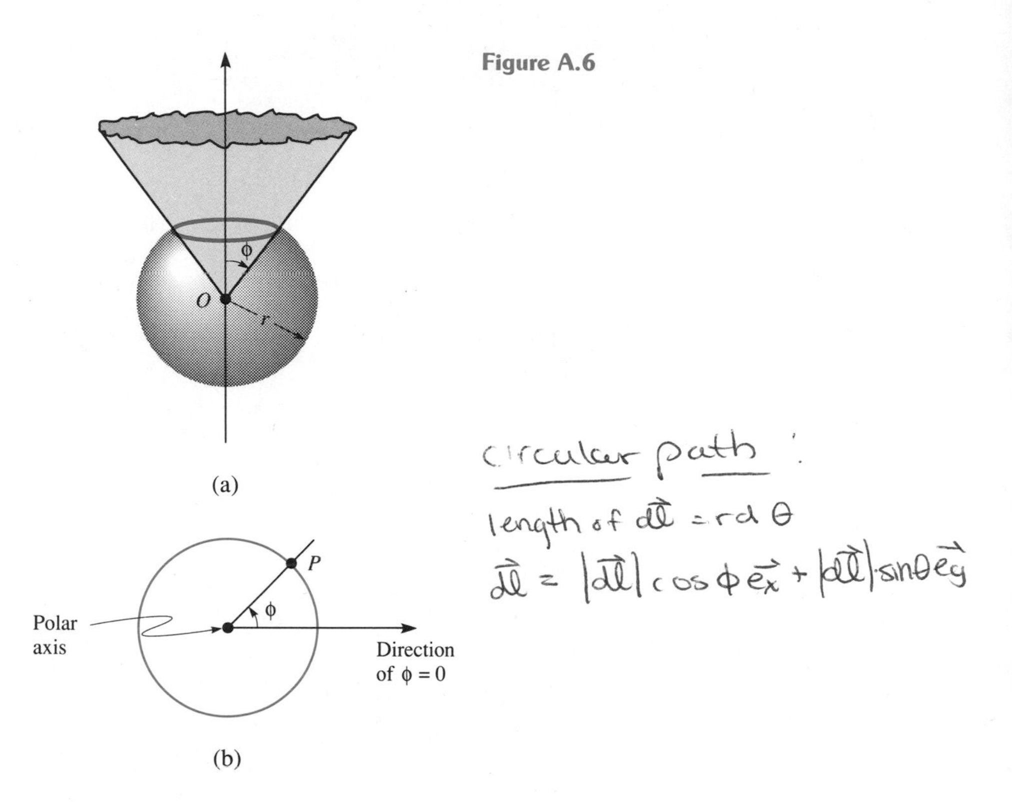

Let us choose the polar axis to coincide with the z axis of a rectangular coordinate system, and let the $\phi = 0$ direction coincide with the x axis. Then the spherical coordinates of the point (x, y, z) are given by

$$
\begin{aligned}
r &= (x^2 + y^2 + z^2)^{1/2} \\
\theta &= \cos^{-1}\frac{z}{(x^2 + y^2 + z^2)^{1/2}} \qquad \textbf{(A.7)} \\
\phi &= \cos^{-1}\frac{x}{(x^2 + y^2)}
\end{aligned}
$$

The inverse relationships are

$$
\begin{aligned}
x &= r\sin\theta\cos\phi \\
y &= r\sin\theta\sin\phi \qquad \textbf{(A.8)} \\
z &= r\cos\theta
\end{aligned}
$$

The unit vectors in spherical coordinates are mutually perpendicular and point in the direction one moves if the corresponding coordinate increases, as shown in Fig. A.7(a). Their cross products obey

$$
\begin{aligned}
\vec{e}_r \times \vec{e}_\theta &= \vec{e}_\phi \\
\vec{e}_\theta \times \vec{e}_\phi &= \vec{e}_r \qquad \textbf{(A.9)} \\
\vec{e}_\phi \times \vec{e}_r &= \vec{e}_\theta
\end{aligned}
$$

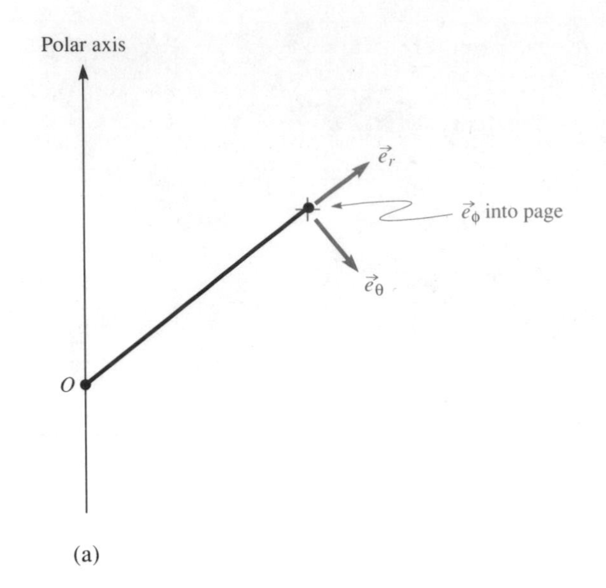

Polar axis
$\vec{e}_r$
$\vec{e}_\phi$ into page
$\vec{e}_\theta$
O

(a)

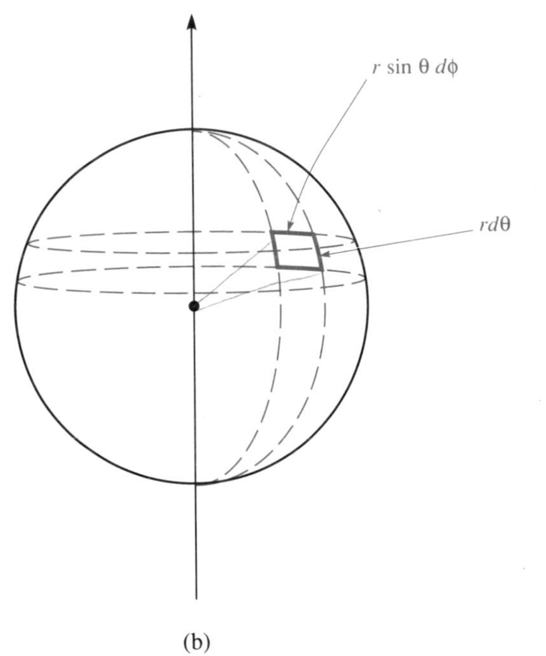

$r \sin \theta \, d\phi$
$rd\theta$

(b)

Figure A.7

The incremental distance is

$$\vec{dl} = dr\,\vec{e}_r + r\,d\theta\,\vec{e}_\theta + r\sin\theta\,d\phi\,\vec{e}_\phi \tag{A.10}$$

(which the reader may verify, using trigonometry). An increment of area on a spherical surface (Fig. A.7(b)) is

$$dA = (r\,d\theta)(r\sin\theta\,d\phi) = r^2\sin\theta\,d\theta\,d\phi \tag{A.11}$$

and the increment of volume is

$$dV = dA\,dr = r^2\sin\theta\,dr\,d\theta\,d\phi \tag{A.12}$$

EXERCISE A.5

Verify by integrating (A.11) and (A.12) that the area of a sphere is $4\pi R^2$ and its volume is $(4/3)\pi R^3$.

PROBLEM

*A.1 Bearing in mind that $d\phi$ is by definition small, verify that the incremental length shown in Fig. A.3 is $r\,d\phi$. (Use the law of cosines and take the limit as $d\phi \to 0$.)

APPENDIX B

Summary of Boundary Conditions

References to text equations are given in parentheses.

I. At an interface between two dielectrics:

(a) $E_{T1} = E_{T2}$ (6.19)

If there is no charge on the interface,

(b) $D_{N1} = D_{N2}$ (5.23)

If there is charge on the interface,

(c) $D_{N2} - D_{N1} = \sigma$ (4.14)

(*N.B.: If either of the materials is conductive, it is likely that there will be charge on the interface.*)

II. At an interface between two conductors:

For non-time-varying currents,

(a) $J_{N1} = J_{N2}$ (6.9a)

For sinusoidally time-varying currents,

(b) $\mathbf{J}_{N2} - \mathbf{J}_{N1} = -j\omega\sigma$ (6.9a)

III. At the surface of a perfectly conducting metal:

(a) $E_T = 0$ (Sec. 4.4)

(b) $D_N = \sigma$ (4.30)

For sinusoidally time-varying fields,

(c) $\mathbf{B}_N = 0$ (7.36)

(d) $\vec{\mathbf{I}}_W = \vec{n} \times \vec{\mathbf{H}}_T$ (7.34)

(*N.B.:* III(d) *also holds for imperfect conductors much thicker than the skin depth.*)

IV. At the interface of two magnetic materials:

(a) $B_{N1} = B_{N2}$ (5.22)

Provided that neither of the materials is an ideal, perfectly conducting metal,

(b) $H_{T1} = H_{T2}$ (5.18)

APPENDIX C

Electrical Energy Expressed as an Integral of the Field

In our discussion of Poynting's theorem, use was made of equation (8.14),

$$W_E = \int \frac{\epsilon}{2} |\vec{E}|^2 \, dv \tag{8.14}$$

According to this equation, W_E, the work required to assemble a set of charges, is equal to an integral involving the fields produced by those charges. We now present a derivation of this result.

The work required to assemble a distribution of charge is the same as the energy that can be recovered if all the charges are allowed to once again move away to infinity. Suppose that when fully assembled the charge distribution is $\rho(\vec{r})$ and the electrostatic potential it produces is $V(\vec{r})$. Let us now imagine that at every point a small fraction of the charge density ρ/N (where N is a large number) is detached and allowed to move away to infinity. The potential energy of a charge q at a place where the potential is V is qV. Thus the energy released when ρ/N moves to infinity is

$$\delta W_1 = \int \frac{\rho}{N} V \, dv$$

Now the remaining charge density has been reduced to $(N - 1)/N$ of its original value, and hence V is everywhere $(N - 1)/N$ of what it was originally. When the next increment of ρ/N is allowed to go off to infinity, the energy obtained is

$$\delta W_2 = \int \frac{\rho}{N} \frac{(N - 1)V}{N} \, dv$$

The total energy obtained when all N increments of charge have been removed will thus be

$$W = \int \rho V\, dv \left[\frac{1}{N} + \frac{1}{N}\frac{(N-1)}{N} + \cdots + \frac{1}{N}\frac{1}{N}\right]$$

$$= \int \rho V\, dv \frac{1}{N^2}[N + (N-1) + (N-2) + \cdots + 1]$$

$$= \int \rho V\, dv \frac{1}{N^2}(N+1)\frac{N}{2}$$

(The last step follows from adding the numbers of the sum in pairs. The first and last add to $N + 1$; so do the second and next to last. When N is even there are $N/2$ such pairs. It is easily shown the result also holds when N is odd.) Thus, since N is very large,

$$W = \tfrac{1}{2} \int \rho V\, dv \tag{C.1}$$

Now using[1] $\epsilon\nabla \cdot \vec{E} = \rho$, (C.1) becomes

$$W = \frac{\epsilon}{2}\int (\nabla \cdot \vec{E}) V\, dv$$

Making use of the vector identity $\nabla \cdot (V\vec{E}) = \vec{E} \cdot \nabla V + V\nabla \cdot \vec{E}$, we can further transform our expression into

$$W = \frac{\epsilon}{2}\left[\int \nabla \cdot (V\vec{E})\, dv - \int \vec{E} \cdot \nabla V\, dv\right] \tag{C.2}$$

The first integral on the right of (C.2) can be shown to vanish by converting it to a surface integral using the divergence theorem. Since the integral is over all space, the surface of integration is so distant that V and $\vec{E}$ are very small all over it. On the other hand, the area of the surface of integration becomes very large, so how can we be sure the surface integral converges? To show this we note that when we are at a very great distance R from all the charges, V decreases approximately as $1/R$, and $\vec{E}$ as $1/R^2$. Thus the product $V\vec{E}$ decreases as $1/R^3$. On the other hand, the area of the spherical surface of integration is $4\pi R^2$. Hence the surface integral will decrease as $(1/R^3)R^2 = 1/R$ and approach zero when the surface grows infinitely large.

Thus we are left with the final term of (C.2). Since $-\nabla V = \vec{E}$, we have

$$W = \frac{\epsilon}{2}\int (\vec{E} \cdot \vec{E})\, dv$$

which completes the proof.

---

[1]For simplicity we are assuming, as elsewhere in this book, that ϵ is a scalar, independent of position.

SOLUTIONS TO EVEN-NUMBERED PROBLEMS

Chapter 1

1.6 $A = 2.24\text{ V} \qquad \phi = 109^\circ$

1.12 $$\rho = \frac{(Z_{o2} \| Z_{o3}) - Z_{o1}}{(Z_{o2} \| Z_{o3}) + Z_{o1}}$$

$$\frac{P_R}{P_I} = |\rho|^2$$

$$\frac{P_2}{P_I} = \frac{Z_{o3}}{Z_{o2} + Z_{o3}}(1 - |\rho|^2)$$

$$\frac{P_3}{P_I} = \frac{Z_{o2}}{Z_{o2} + Z_{o3}}(1 - |\rho|^2)$$

1.14 $$\mathbf{v} = \frac{Z_o \mathbf{v}_S}{Z_o \cos kl_1 + jR_S \sin kl_1}$$

1.28 $v_C(t) = V_o(1 - e^{-t/Z_oC})$

Chapter 2

2.2 SWR $= 3.55 \qquad d = 1.23$ cm

2.8 0.86 m

2.16 $0.62e^{-j(29.5^\circ)}$

2.18 $1, 0, j, 0.44 + 0.14j, 0.23 - 0.15j$

2.20 $R'_L = 0.30$ at $0.232\,\lambda$ or 3.4 at $0.482\,\lambda$

2.22 $Y = 0.0078 + 0.14j\text{ ohm}^{-1}$

2.24 4800 ohms

2.26 $45 + 14.5j$

2.28 $b = 0.062\,\lambda \qquad d = 0.091\,\lambda$

2.30 $0.16 - 12.3j$ ohms

2.32 $b = 0.158\,\lambda \qquad a = 0.29\,\lambda$

Chapter 3

3.2 $\vec{e}_x - \vec{e}_y$

3.10 (a) $-\vec{e}_z$ (b) $\vec{e}_z$

3.14 $x = -3.42 \qquad y = 9.4 \qquad z = 6$

3.16 $\vec{V}_1 \cdot \vec{V}_2 = 1 \qquad \psi = 78.7°$

3.20 $\vec{V}_1 \cdot \vec{V}_2 = -9 \qquad \vec{V}_1 \times \vec{V}_2 = -6\vec{e}_r - 3\vec{e}_\theta$

3.28 $\dfrac{V_o}{2}$

3.32 $Ka^2\left(\dfrac{aA}{3} + B\right)$

3.34 $\dfrac{4}{3}Bab^3$

3.36 $\dfrac{\pi A a^4}{4}$

3.38 3

Chapter 4

4.2 (a) 0 (b) $\dfrac{q}{\epsilon a^2}(.056)\vec{e}_y$ (c) $\dfrac{q}{\epsilon a^2}(.014\vec{e}_x + .087\vec{e}_y)$

4.4 $\dfrac{\tau a \vec{e}_y}{2\pi\epsilon b\sqrt{b^2 + a^2}}$

4.6 $\dfrac{\sigma\vec{e}_x}{2\pi\epsilon}\ln\left(\dfrac{b}{a}\right)$

4.8 $\dfrac{\tau d\vec{e}_x}{2\epsilon a^2}$

4.10 $\dfrac{\rho\vec{e}_z}{2\epsilon}[c + \sqrt{b^2 + (d - c)^2} - \sqrt{b^2 + d^2}]$

4.12 $E_x = \dfrac{A}{4}(x^4 - x_0^4)$

4.20 $r < R$: $E_r = \dfrac{\rho r}{2\epsilon} \qquad r > R$: $E_r = \dfrac{\rho R^2}{2\epsilon r}$

4.24 $\dfrac{\pi\sigma_E B a^4}{2}$

4.28 $2\pi\sigma(\sqrt{b^2 + h^2} - h)$

4.30 (a) $V = \dfrac{\tau}{4\pi\epsilon} \ln \left[\dfrac{v(1 + \sqrt{1 + u^2})}{u(1 + \sqrt{1 + v^2})} \right]$

(b) $E_x = \dfrac{\tau}{4\pi\epsilon} \left(\dfrac{v}{1 + \sqrt{v^2 + 1}} \dfrac{1}{\sqrt{v^2 + 1}} - \dfrac{u}{1 + \sqrt{u^2 + 1}} \dfrac{1}{\sqrt{u^2 + 1}} \right]$

$E_y = \dfrac{\tau}{4\pi\epsilon} \left(\dfrac{1}{1 + \sqrt{v^2 + 1}} \dfrac{1}{\sqrt{v^2 + 1}} - \dfrac{1}{1 + \sqrt{u^2 + 1}} \dfrac{1}{\sqrt{u^2 + 1}} \right)$

where $u = \dfrac{x}{y + c}$ $\quad v = \dfrac{x}{y - c}$

4.36 0.7 pF, approximately

4.38 $C = \dfrac{\epsilon \rho_o I}{V}$

4.40 $V = \dfrac{Q}{4\pi\epsilon} \left[\dfrac{1}{\sqrt{(x - a)^2 + (y - b)^2}} + \dfrac{1}{\sqrt{(x + a)^2 + (y + b)^2}} - \dfrac{1}{\sqrt{(x - a)^2 + (y + b)^2}} - \dfrac{1}{\sqrt{(x + a)^2 + (y - b)^2}} \right]$

Chapter 5

5.2 9.00×10^{-7} A/m

5.4 $\dfrac{2\sqrt{2} I \vec{e}_z}{\pi a}$

5.6 $I \vec{e}_z$

5.8 $\mu \pi a^2 H_o$

5.10 $\dfrac{2}{3} \mu H_o a^3$

5.12 0

5.18 (a) yes; (b) yes; (c) no; (d) no; (e) yes

5.22 $\vec{A} = -\dfrac{B_o}{2} z^2 \vec{e}_y$

5.24 $-\dfrac{\mu I a}{2\pi b}$

5.26 (a) $\vec{J} = \dfrac{2H_o \vec{e}_z}{a}$

(b) $\vec{A} = -\dfrac{\mu H_o}{2a} r^2 \vec{e}_z$

5.28 $2\pi a^3 A$

5.30 $D = 5.07$ A/cm^2

5.32 49.1°

5.34 $\dfrac{\sigma_{E1}}{\sigma_{E2}} E_1\vec{e}_x$

5.36 yes; no

5.38 $\dfrac{L}{\mu\pi a(a + b)}$

Chapter 6

6.2 $\vec{E} = \dfrac{2At}{\epsilon_o}\vec{e}_z$

6.4 $\rho(x, t) = -At$

6.6 $I_W(t)$; 0

6.8 $V_{AB} = -Ka^2U$

6.10 (a) $\vec{H} = \dfrac{E_o ak}{\mu\omega r}\cos(\omega t - kz)\vec{e}_\phi$; (c) same

6.12 54.7°

6.14 continuous

6.16 $\dfrac{\mu\pi c^2 d^2}{4h^3}$

6.18 $\dfrac{\mu\pi a^2 b^2}{2(h^2 + b^2)^{3/2}}$

6.20 20.5 H

6.26 (a) $E_{T1} = E_{T2}$; (b) $\dfrac{E_{N1}}{\rho_1} = \dfrac{E_{N2}}{\rho_2}$

(c) $\sigma_{12} = D_{N1}\left(\dfrac{\rho_2}{\rho_1} - 1\right)$

Chapter 7

7.2 $\vec{E} = Ax^2e^{j30°}\vec{e}_y - Bze^{-j90°}\vec{e}_x$

7.4 $\vec{E}(x, y, z, t) = A_o\cos(\omega t + kx)\vec{e}_x + B_o\cos(\omega t - kx + \phi)\vec{e}_y$

7.6 $\vec{H}(r, \phi, z, t) = 3.606e^{-j33.69°}\left(\dfrac{r}{a}\right)(j\vec{e}_r + \vec{e}_\phi)$

7.8 (a) $k_x x + k_y y + k_z z = 0$

7.14 (a) 1.7 cm (b) 41.6 μm (c) 1.32 μm

7.16 $\vec{J} = J_o\vec{e}_z\exp[(1 + j)(0.89x + 0.45y)/\delta]$

7.18 $\dfrac{\partial^2}{\partial r^2}J_z(r) + \dfrac{1}{r}\dfrac{\partial J_z}{\partial r} - j\omega\mu\sigma_E J_z = 0$

7.20 (b) $2.4\,Z_S$

7.22 0.017 Ω/m; 2 Ω/m

7.24 (a) $\vec{H} = \dfrac{1}{2\pi a}\,\vec{e}_\phi \cos\omega t$

7.26 $\vec{\mathbf{H}} = -\dfrac{I}{2\pi a}\,\vec{e}_\phi$

7.28 $\dfrac{3}{2}\,R_S H_0^2 a^2$

7.34 $\dfrac{E_o}{\eta}\,e^{-j\omega\sqrt{\mu\epsilon}z}\,\vec{e}_y$ (positive-going wave)

7.36 $\mathbf{E}_z = E_o \exp[-j\omega\sqrt{\mu\epsilon}\,(0.65x + 0.76y)$

7.40 (a) $\dfrac{1}{j\omega hC}$ (b) $-jZ_o \cot(\omega\sqrt{LC}\,h)$

Chapter 8

8.2 $\vec{\mathbf{E}} = \dfrac{E_o}{\sqrt{2}}(\vec{e}_x - \vec{e}_y)e^{jkz}$ $\quad\vec{\mathbf{H}} = \dfrac{E_o}{\sqrt{2}\eta}(-\vec{e}_x - \vec{e}_y)e^{jkz}$

$\vec{E}(t) = \dfrac{E_o}{\sqrt{2}}(\vec{e}_x - \vec{e}_y)\cos(\omega t + kz)$

8.8 $\vec{\mathbf{H}} = \dfrac{E_o}{\eta}(\vec{e}_y - j\vec{e}_x)$ $\quad\vec{H}(t) = \dfrac{E_o}{\eta}(\cos\omega t\,\vec{e}_y + \sin\omega t\,\vec{e}_x)$

8.12 $W = \dfrac{2\pi\epsilon abV^2}{b - a}$

8.14 $F = \dfrac{\epsilon_o AV^2}{2L^2}$ (attractive)

8.22 0

8.24 (a) $\vec{\mathbf{E}} = 2E_o \cos kz\,\vec{e}_x$ $\quad\vec{\mathbf{H}} = -2j\dfrac{E_o}{\eta}\sin kz\,\vec{e}_y$

(b) $\dfrac{\mathbf{E}_x}{\mathbf{H}_y} = j\eta \cot kz$ (c) 0

(d) $E(t) = 2E_o \cos kz \cos\omega t\,\vec{e}_x$ $\quad H(t) = \dfrac{2E_o}{\eta}\sin kz \sin\omega t\,\vec{e}_y$

(e) an open-circuit load

8.26 $\dfrac{P_T}{P_I} = \dfrac{4\eta_1\eta_2}{(\eta_1 + \eta_2)^2}$ $\quad\dfrac{P_R}{P_I} = \left[\dfrac{\eta_2 - \eta_2}{\eta_2 + \eta_1}\right]^2$

8.28 $\frac{\lambda}{2n}$

8.30 $Z_{\max} = -2.24\ \text{cm}$

8.38 (a) $\vec{\mathbf{H}} = \frac{k_{o2}}{\omega\mu} e^{-jk_{o2}Ax} e^{-k_{o2}Bz} \vec{e}_y$

(b) $\frac{1}{2}\,\text{Re}(\vec{\mathbf{E}} \times \vec{\mathbf{H}}^*) = \frac{E_2 k_{o2} A}{2\omega\mu} e^{-2k_{o2}Bz} \vec{e}_x$

8.42 0.401%

8.44 (a) $\frac{E_0^2}{2\eta_1} \cos\theta_1$

(b) $P_T = P_I(1 - |\rho_1|^2)$

where $\rho_1 = \frac{Z_{L1} - Z_{o1}}{Z_{L1} + Z_{o1}}$ $\qquad Z_{o1} = \eta_1 \cos\theta_1$

$$Z_{L1} = \eta_2 \cos\theta_2 \left[\frac{\eta_3 \cos\theta_3 \cos(k_2 \cos\theta_2 L) + j\eta_2 \cos\theta_2 \sin(k_2 \cos\theta_2 L)}{\eta_2 \cos\theta_2 \sin(k_2 \cos\theta_2 L) + j\eta_3 \cos\theta_3 \cos(k_2 \cos\theta_2 L)}\right]$$

8.46 21%

Chapter 9

9.6 $\vec{\mathbf{H}} = -\frac{A}{\eta r} \vec{e}_\phi$

9.8 0.11 nH/cm

9.14 $\frac{\pi^2}{8} \frac{K^2 \omega k \epsilon ab}{(\omega^2\mu\epsilon - k^2)^2} \left[\left(\frac{m}{a}\right)^2 + \left(\frac{n}{b}\right)^2\right]$

9.18 $U_G = c\sqrt{1 - (\omega')^{-2}}$

9.20 (a) $H_z(t) = K \cos\frac{\pi x}{a} e^{-\alpha z} \cos\omega t$

$$E_y(t) = \frac{\omega\mu a K}{\pi} \sin\frac{\pi x}{a} e^{-\alpha z} \cos(\omega t - \pi/2)$$

$$H_x(t) = \frac{j\alpha K a}{\pi} \sin\frac{\pi x}{a} e^{-\alpha z} \cos(\omega t - \pi/2)$$

where $\alpha \equiv jk$

(c) $L = \alpha^{-1} = |k|^{-1}$

9.22 $f_R = \frac{U_P}{2d}$

9.26 $Z_o = 56.5\ \Omega$ $U_P = 0.34c$

9.28 $f_1 = 3.5\text{ GHz}$ $f_2 = 7.0\text{ GHz}$

9.32 (a) $U_P \cong 0.31c$ $U_G \cong 0.29c$

(b) $U_P \cong 0.33c$ $U_G \cong 0.33c$

9.36 $0.35c$

Chapter 10

10.2 $$\vec{A}(t) = \frac{h\mu I_o \vec{e}_x}{4\pi a}\left[\tfrac{1}{2}\cos(\omega t - 2\phi_o) + \cos(\omega t - \phi_o)\right]$$

where $\phi_o \equiv \dfrac{\omega a}{c}$

10.4 $\vec{\mathbf{A}} = \dfrac{j\mu H_o}{k} e^{-jkz}\,\vec{e}_x$ $\Phi = 0$

10.6 (a) $\vec{\mathbf{i}} = E_o \sigma_E a^2 \vec{e}_x$

(b) $P = \dfrac{\eta\pi E_0^2 \sigma_E^2 a^6}{3\lambda^2}$

(c) as f^2

10.8 $I_o = 47.7\text{A}$ $E = 95\%$ $G = 1.425$

10.12 $\dfrac{\lambda I_0^2 R}{8}$

10.14 $\vec{\mathbf{I}}_W = -\dfrac{jI}{2\pi r} e^{-jkr}\vec{e}_r$

10.24 2.87 nW

INDEX

Page numbers followed by the letter "N" refer to footnotes. Citations of the form "P10.38" refer to problems. The abbreviation "ff" denotes "and following pages."

$\nabla \times \vec{V}$ = vorticity of $\vec{V}$

$\vec{E} = -\nabla V$

$q\vec{E}$ = force on stationary charge

⊥ to $d\vec{l}$ and $\vec{r} - \vec{r}'$ = Direction of $\vec{H}$

$-\nabla V$ = direction of greatest change in V

$\nabla \cdot \vec{V}$ = Net flow of $\vec{V}$ out of a unit volume

$q(\vec{v} \times \vec{B})$ = force on a moving charge

$\vec{E}$ is ∥ to $\vec{r} - \vec{r}'$

$\nabla^2 V$ = divergence of gradient of V

$J = \sigma_E \vec{E}$

---

complex #'s:

$z = a + jb = re^{j\theta}$

$a = r\cos\theta$

$b = r\sin\theta$

$\frac{b}{a} = \tan\theta$ ($\theta = \arctan(\frac{b}{a}) = \angle z$

phasor: $v(z,t) = Re[Ae^{-jkz}e^{j\omega t}]$

$v^+(z) = Ae^{-jkz}$ (Right moving Phasor rotate CW)

$v^-(z) = Ae^{jkz}$ (left, neg. moving, Phasor goes CCW)

Euler's Formula:

$\cos x = \frac{1}{2}(e^{jx} + e^{-jx})$

$\sin x = \frac{1}{2j}(e^{jx} - e^{-jx})$

$\vec{A} \cdot \vec{B} = |\vec{A}||\vec{B}|\cos\phi$

$|\vec{A} \times \vec{B}| = |\vec{A}||\vec{B}|\sin\phi$

$\vec{v} \cdot d\vec{l} = |\vec{v}||d\vec{l}|\cos\phi$

$d\vec{l}$ length $= r\,d\theta$

$e^{j\omega t} = \cos\omega t + j\sin\omega t$

TRIG

$$\sin(A \pm B) = \sin A \cos B \pm \cos A \sin B$$

$$\cos(A \pm B) = \cos A \cos B \mp \sin A \sin B$$

$$1 + \tan^2 A = \sec^2 A$$

$$\cos A = \frac{1}{\sec A}$$

$$1 + \operatorname{ctn}^2 A = \csc^2 A$$

$$\tan(A \pm B) = \frac{\tan A \pm \tan B}{1 \mp \tan A \tan B}$$

$$\sin 2A = 2 \sin A \cos A$$

$$\sin 3A = 3 \sin A - 4 \sin^3 A$$

$$\sin nA = 2 \sin (n-1) A \cos A - \sin (n-2) A$$

$$\cos 2A = 2\cos^2 A - 1 = 1 - 2\sin^2 A$$

$$\cos nA = 2 \cos (n-1) A \cos A - \cos (n-2) A$$

$$\sin^2 A = \tfrac{1}{2}(1 - \cos 2A)$$

$$\cos^2 A = \tfrac{1}{2}(1 + \cos 2A)$$

$$\frac{d}{dx} \tan u = \sec^2 u \frac{du}{dx}$$

$$\frac{d}{dx} \sin u = \cos u \frac{du}{dx}$$

$$\frac{d}{dx} \sec u = \sec u \tan u \frac{du}{dx}$$

$$\int \tan x \, dx = -\log \cos x + C$$

$$\int \sin^2 x \, dx = \tfrac{1}{2} x - \tfrac{1}{2} \sin x \cos x + C$$

$$\int \cos^2 x \, dx = \tfrac{1}{2} x + \tfrac{1}{2} \sin x \cos x + C$$

VECTOR IDENTITIES

$$\nabla(\phi + \psi) = \nabla\phi + \nabla\psi$$

$$\nabla \cdot (\vec{A} + \vec{B}) = \nabla \cdot \vec{A} + \nabla \cdot \vec{B}$$

$$\nabla \times (\vec{A} + \vec{B}) = \nabla \times \vec{A} + \nabla \times \vec{B}$$

$$\nabla(\phi\psi) = \phi\nabla\psi + \psi\nabla\phi$$

$$\nabla \cdot (\psi\vec{A}) = \vec{A} \cdot \nabla\psi + \psi\nabla \cdot \vec{A}$$

$$\nabla \cdot (\vec{A} \times \vec{B}) = \vec{B} \cdot \nabla \times \vec{A} - \vec{A} \cdot \nabla \times \vec{B}$$

$$\nabla \times (\phi\vec{A}) = \nabla\phi \times \vec{A} + \phi\nabla \times \vec{A}$$

$$\nabla \times (\vec{A} \times \vec{B}) = \vec{A}\,\nabla \cdot \vec{B} - \vec{B}\,\nabla \cdot \vec{A} + (\vec{B} \cdot \nabla)\,\vec{A} - (\vec{A} \cdot \nabla)\,\vec{B}$$

$$\nabla \cdot \nabla\phi = \nabla^2\phi$$

$$\nabla \cdot \nabla \times \vec{A} = 0$$

$$\nabla \times \nabla\phi = 0$$

$$\nabla \times \nabla \times \vec{A} = \nabla(\nabla \cdot \vec{A}) - \nabla^2\vec{A}$$

$$\nabla(\vec{A} \cdot \vec{B}) = (\vec{A} \cdot \nabla)\vec{B} + (\vec{B} \cdot \nabla)\vec{A} + \vec{A} \times (\nabla \times \vec{B}) + \vec{B} \times (\nabla \times \vec{A})$$

$$\vec{A} \cdot \vec{B} \times \vec{C} = \vec{B} \cdot \vec{C} \times \vec{A} = \vec{C} \cdot \vec{A} \times \vec{B}$$

$$\vec{A} \times (\vec{B} \times \vec{C}) = \vec{B}\,(\vec{A} \cdot \vec{C}) - \vec{C}\,(\vec{A} \cdot \vec{B})$$

Stokes' theorem $\int_S \nabla \times \vec{A} \cdot \vec{dS} = \oint_l \vec{A} \cdot \vec{dl}$

Divergence theorem $\int_V \nabla \cdot \vec{A}\, dV = \oint_S \vec{A} \cdot \vec{dS}$

Cylindar

$V = \pi r^2 L$

Surface area - lateral $= 2\pi r L$

End area $= \pi r^2$

Sphere

$V = \frac{4}{3}\pi r^3$

$A = 4\pi r^2$

Sphrere

$\int_S dA = 4\pi x^2$

$\int_V dV = \frac{4}{3}\pi x^3$

Cylinder

Surface Area : $2\pi r L$

end area = πr^2

$V = \pi r^2 L$